Science Networks · Historical Studies
Founded by Erwin Hiebert and Hans Wußing
Volume 26

Springer Basel AG

Giorgio Israel
Ana Millán Gasca

The Biology of Numbers

The Correspondence of Vito Volterra on Mathematical Biology

Springer Basel AG

Authors' addresses:

Giorgio Israel
Dipartimento di Matematica
Università di Roma "La Sapienza"
Piazzale Aldo Moro, 2
I-00185 Roma
email: giorgio.israel@uniroma1.it

Ana Millán Gasca
Viale dei Promontori 278
I-00122 Roma
e-mail: amillan@tiscalinet.it

This publication has been made possible by the financial support of the grant Confinanziamento MURST 1999.

A CIP catalogue record for this book is available from the Library of Congress, Washington D.C., USA

Deutsche Bibliothek Cataloging-in-Publication Data
Israel, Giorgio:
The biology of numbers : the correspondence of Vito Volterra on mathematical biology / Giorgio Israel ; Ana Millán Gasca. - Basel ; Boston ; Berlin : Birkhäuser, 2002
ISBN 978-3-0348-9447-0 ISBN 978-3-0348-8123-4 (eBook)
DOI 10.1007/978-3-0348-8123-4

Originally published by Birkhäuser Verlag, Basel · Boston · Berlin 2002
Softcover reprint of the hardcover 1st edition 2002
Member of the BertelsmannSpringer Publishing Group
Cover design: Micha Lotrovsky, CH-4106 Therwil, Switzerland
Cover illustration: Vito Volterra
Layout: *wiskom*, D-88046 Friedrichshafen
Printed on acid-free paper produced from chlorine-free pulp. TCF ∞

ISBN 978-3-0348-9447-0

9 8 7 6 5 4 3 2 1

To Alberto

I myself, who unfortunately am unable to judge your magisterial mathematical work, am astonished by the brilliant ideas you have introduced in the biology of numbers.

(Jean Régnier to Vito Volterra)

Foreword

The modern developments in mathematical biology took place roughly between 1920 and 1940, a period now referred to as the "Golden Age of Theoretical Biology". The eminent Italian mathematician Vito Volterra played a decisive and widely acknowledged role in these developments. Volterra's interest in the application of mathematics to the non physical sciences, and to biology and economics in particular, dates back to the turn of the century and was expressed in his inaugural address at the University of Rome for the academic year 1900/01 (VOLTERRA 1901). Nevertheless, it was only in the mid-twenties that Volterra entered the field in person, at the instigation of his son in law, Umberto D'Ancona, who had confronted him with the problem of competition among animal species, asking him whether a mathematical treatment was possible. From that time on, until his death in 1940, Volterra produced a huge output of publications on the subject.

Volterra's specific project was to transfer the model and the concepts of classical mechanics to biology, constructing a sort of "rational mechanics" and an "analytic mechanics" of biological associations. The new subject was thus to be equipped with a solid experimental or at least empirical basis, also in this case following the tried and tested example of mathematical physics. Although very few specific features of this reductionist programme have actually survived, Volterra's contribution was decisive, as is now universally acknowledged, in encouraging fresh studies in the field of mathematical biology. Even today, the primary reference in the literature of the field of population dynamics consists of Volterra's work and the descriptive schemata (the "models", in modern parlance) he proposed.

Many historical studies have been written on Volterra's contribution to the field of mathematical biology, although much work remains to be done both as regards an overall synthesis and the specific analysis of the ways in which his contribution was developed. Of particular interest is Volterra's concern in the biological issues forming the empirical basis of his research. The present book aims to fill this historiographic gap by providing an exhaustive collection of the correspondence between Volterra and numerous other scientists on the topic of mathematical biology. The correspondence presented is drawn from the Vito Volterra Archive conserved at the Accademia Nazionale dei Lincei. It involves many important researchers who were active in the biological field at the time, such as R. N. Chapman, U. D'Ancona, C. S. Elton, G. F. Gause, V. A. Kostitzin, A. J. Lotka, R. Pearl and D. W. Thompson. In some cases (such as the correspondences with D'Ancona and M. Brelot) the letters allow an accurate reconstruction to be made of the genesis of Volterra's work. In other cases, it sheds an extremely interesting light on the issue of the relationship between mathematical formulation and empirical problems, which is certainly one of the most difficult and controversial and over which Volterra toiled at length.

The book begins with an introductory essay by Ana Millán Gasca, which consists of a revised and extended version of an article published in the journal *Historical Studies in the Physical and Biological Sciences* (MILLÁN GASCA 1996) and which aims at giving a picture of the research field of biomathematics in the "Golden Age", and shows the importance of the correspondence in this context. This is followed by a transcript of the correspondence ordered by the correspondent's name. Each item is preceded by a biographical profile of the correspondent and accompanied by notes containing information and references to facilitate understanding. Excluded from the transcripts of the correspondence are the parts referring to purely personal matters devoid of any general interest. The book is concluded by a catalogue of the correspondence containing the references to the Volterra Archive classification.

We wish to thank the Presidency of the Accademia Nazionale dei Lincei for permission to publish the correspondence and Professor John L. Heilbron for authorizing the reproduction of Ana Millán Gasca's article. We wish also to thank for their help in finding biographical documentation the Bentley Historical Library of the University of Michigan (Ann Arbor), Cristiana Baldazzi, Paul Banfield (Queen's University Archives), Angela Hartwig (Universität Rostock), Paolo Israel, Eberhard Knobloch (Technische Universität Berlin), Diana Ritchmond (University of Minnesota, Department of Entomology), Roberto Scoth, John Witter (University of Michigan, Ann Arbor). The biographies and the notes were translated in English by Ian McGilvray. The publication of the book has been made possible by the financial support of the grant Cofinanziamento MURST 1999 ("La storia della matematica in Italia").

It is our hope that the book will be found useful not only by science historians but also by all those – in particular, biomathematicians and biologists – with an interest in the origins of and events in a branch of learning that has undergone an astonishing development. Many of the problems discussed – in particular that of empirical verification – appear extremely topical even today and in some cases could even fuel reflection on topics still open to research.

Contents

1 Mathematical Theories versus Biological Facts: A Debate on Mathematical Population Dynamics in the 30s

The idea of applying mathematics to biological or biomedical problems goes back to the Enlightenment period, as part of the general project of using mathematics as an understanding tool in every exploration of natural or human facts.

However, little progress in this direction was made until the second half of the 19th century. The transformation of classical natural history produced by the arrival of the theory of evolution, with its dynamical conception of life phenomena, opened the way to "numbers" in biological research. Starting from the work of Francis Galton (1822–1911) and Karl Pearson, founder of the journal Biometrika, statistical analysis and elementary quantitative techniques were applied to the handling of biological information, in connection with the study of evolution. By the turn of the century, the way to mathematization had been cleared and important contributions followed one another. Raymond Pearl furthered biometric research in genetics and with his work at Johns Hopkins University on the demography of *Drosophila*, in which he introduced into theoretical biology the extended used of the logistic curve. Ronald Ross (1857–1932) applied quantitative methods in epidemiology in his studies on malaria in the second decade of the 20th century, followed by William O. Kermack and A. G. McKendrick. William R. Thompson and Royal N. Chapman are examples of American entomologists interested in quantitative and mathematical methods. The development of animal ecology enhanced interest in the study of populations. Researches related to economical aspects and aimed at the development of biological regulation and control procedures, as in economical entomology and in fishery management. Economical entomology studies had a great development in the United States at the beginning of the 20th century[1]; in Europe, the *Conseil international pour l'exploration de la mer* was founded in 1902 in Copenhagen.

Moreover, in the period prior to the Second World War quite a lot of seminal work towards a systematic and organic development of mathematical research in biology was produced, as well as reflections on a suitable mathematical methodology. Galton's biometrics foundational assumption of the statistical quantification of the laws of evolution was developed by Pearl and, above all, by the English

[1] On the development of mathematical studies on population ecology in the United States, see Kingsland 1985.

statistician Ronald A. Fisher (1890–1962), who introduced a whole set of mathematical tools to deal with problems in population genetics. The work of Fisher, as well as that of John B. S. Haldane (1892–1964) and Sewall Wright (1889–1988) offered a statistical model of evolution that combined Mendelian heredity with natural selection, a contribution that would play an important role in the emergence of the so-called "evolutionary synthesis"[2]. Most trends in mathematical oriented research in biology during the second half of the 20th century derived from these and other, lesser known, contributions made during this period. The Italian astronomer Giovanni Schiaparelli (1835–1910), in collaboration with the biologist Tito Vignoli (1828–1914), published in 1898 a memoir on geometrical forms and natural organic forms; the same geometrical approach to the living world informed *On Growth and Form* by D'Arcy Wentworth Thompson[3]. Another approach, which lies at the foundation of biomathematics, emerged from the work of a diverse group of scientists. The Dutch engineer Balthazar Van der Pol's (1909–1959) first mathematical model of the heartbeat; Alfred J. Lotka's *Elements of Physical Biology*[4]; and the set of general principles of a mathematical biology developed by Vito Volterra, and his son-in-law, the ichthyologist Umberto D'Ancona.

The approaches of these studies met strong opposition. This may bee seen in the antagonisms and conflicts that accompanied the emergence of biometrics at the Royal Society Committee for Conducting Statistical Inquiries into the Measurable Characteristics of Plants and Animals at the turn of the century, in the controversies around Pearl's logistics in the 1920s and in Lotka's academic and institutional difficulties and the indifference to his book[5]. An interesting debate on the application of mathematics to biology took place in the 1920s and 1930s, in different places in the world and in different specific scientific contexts. We will analyse here the reception of Volterra's biomathematics in the framework of this broader debate.

Some preliminary remarks are necessary. The merely "instrumental" applications of mathematics to biology, that is, the use of elementary computations, must be distinguished from efforts towards a conceptual application of mathematics to biology, that is, the building of a genuine methodology of biomathematical research, as in the works of D'Arcy Thompson, Lotka, Fisher and Volterra. As part of his general scientific reductionist program Volterra aimed his biomathematical research to transfer the conceptual apparatus of mechanics to biology. In his early studies he applied ordinary differential equations and integro-differential equations to build a "rational mechanics of biological associations"; in the 1930s he introduced the variational approach of mathematical physics to get an "analy-

---

[2] See PROVINE 1971 and MAYR, PROVINE (eds.) 1980.

[3] VIGNOLI, SCHIAPARELLI 1898; THOMPSON D'A. W. 1917. See also FREGUGLIA 1998.

[4] VAN DER POL, VAN DER MARK 1928; LOTKA 1925. See also ISRAEL 1996, 1998.

[5] During the 1928 International Congress of Entomology the distinguished biologist Stephen Forbes, congratulating R. N. Chapman on his work, spoke about the breakthrough that his ideas represented: "Forbes reminded the audience that ecologists had often not known when to pause from the gathering of facts to draw a conclusion from the fragments they studied, confident that it would have general applicability" (KINGSLAND 1985, p. 129).

tical mechanics of biological associations"[6]. Many biologists questioned the legitimacy of applying mathematical concepts and methods in biology. They felt that biology, as a natural science concerned with living beings, could not follow the blind mathematical laws of physics or submit to the simplification requirements of a mathematical formulation. The threat from mathematics grew as it was applied to foundational questions, in particular the theory of evolution. As a matter of fact, all the above mentioned authors had in mind evolution as the central problem, thus following in some sense Galton's original motivations.

Volterra held a determinist point of view expressed in differential equations, whereas Fisher, Haldane and Wright used extensively probabilistic techniques. The contrast between their approaches went back to the first attempts to apply mathematics to non-physical sciences[7]. The latter dominated with the emergence of biometrics. In fact, the foundational principle of biometrics as it has been built up by Galton and Pearson was based on the statistical concept of correlation, considered as a generalisation of causation. For Volterra, statistical and probabilistic methods could be very useful, but less reliable than infinitesimal analysis: he saw mathematical biology as a more advanced stage than biometrics. Pearson could not agree. He wrote in 1934[8]:

> [...] that there was a category broader than causation, namely correlation, of which causation was only the limit, and that this new conception of correlation brought psychology, anthropology, medicine and sociology in large parts into the field of mathematical treatment. It was Galton who first freed me from the prejudice that sound mathematics could only be applied to natural phenomena under the category of causation. Here for the first time was a possibility – I will not say a certainty of reaching knowledge – as valid as physical knowledge was then thought to be – in the field of living forms and above all in the field of human conduct.

Lotka, like Volterra, had physics as a reference, but did not consider *only* mechanics: he took also into account the energetic approach of thermodynamics, and in fact, considered evolution as the expression of the second law of thermodynamics[9].

An important point in all these discussions was the fit of biomathematical equations and laws to experience. Practical problems in agriculture (like plagues) as well as in fishery motivated an indistinct interest in quantitative methods in biology. Volterra's himself began with a problem laid out by D'Ancona regarding data from fishery in the Adriatic. However, many biologists did not trust the possible effective results of mathematical research. On the other hand, from a purely theoretical point of view, this matter was related to the far-reaching transformation that science was undergoing in the first decades of the 20th century. Classical science aimed to provide a unified objective description of natural phenomena and, in this sense, formulated likely mathematical laws that were subjected to comparison with empirical evidence, that is, with observation

---

[6] See ISRAEL 1990, 1991b.

[7] See Israel 1993a.

[8] Speeches Delivered at a Dinner Held in University College, London, in Honour of Professor Karl Pearson, 23 April 1934 (Cambridge, Cambridge University Press 1934), pp. 22–23.

[9] For a general appraisal, see ISRAEL 1982b, 1988, 1993b.

or experimentation data. Volterra - and Lotka also - wanted to extend this schema to biology. Moreover, Volterra wanted to use this research to strengthen the validity of such approach in science in general[10]. Yet the reductionist paradigm of classical science, represented by mechanics and mathematical physics, was in crisis in the first decades of the century. The new results of relativity and quantum mechanics questioned basic assumptions regarding space, time and movement as well as the representation of phenomena as continuous processes or the principle of objectivity. At the same time, the foundational crisis of mathematics gave rise to a reformulation of mathematics itself and its relationship to the other natural sciences. Pure mathematics became a completely axiomatized body of knowledge which formulated many of its leading theoretical problems independently of the other sciences. Mathematics was no more the language of nature and did no more aim to provide a global conception which could explain natural phenomena. But it could provide mathematical models, that is, formal structures which could throw light on partial aspects of phenomena. The power of mathematical modelling lies in the ability to identify the mathematical analogies among phenomena of different nature – physical or chemical, or biological, economical, psychological, sociological. Mathematical analogy replaced the ancient mechanical analogy that aimed at giving an exact description of the real world. Thus, the model must prove to be useful to understand or to gather meaningful information from the real word, but it need not prove to be "true" in the classical sense of empirical verification[11].

These profound changes in mathematical practice, particularly encouraged by the French mathematicians of the so-called "Bourbaki group", emerged during the first third of the 20th century. Volterra's biomathematics can be seen in this sense as an anomaly in a general trend. He was really interested in empirical evidence of his theory, and an intense debate took actually place among biologists and mathematicians, particularly on his mathematical laws of biological associations. In the next section we consider the core of this theoretical debate, that is, Volterra's deduction of fluctuations in populations arising from biological interactions alone. In sections 2 and 3 we describe Volterra's early contacts with the biological community, particularly pioneers of ecology and population dynamics. The first experimental data regarding Volterra's equations gathered in the 30s are the subject of section 4. The scientific collaboration among Volterra, D'Ancona, the French biologist Jean Régnier and the Russian geophysicist Vladimir A. Kostitzin is considered in sections 5 and 7. Sections 6 and 7 deal with further theoretical discussions regarding at large the mathematical *démarche* in biology.

[10] In fact, Volterra's reductionism was more open than that of many other physico-mathematicians of his time. His introduction of physical equations for "hereditary" phenomena was cause of disagreement with illustrious colleagues. See ISRAEL 1984b, 1991a.

[11] See ISRAEL 1984a, 1991c, 1996. On the case of economics, see INGRAO, ISRAEL 1990.

1.1 Volterra's Biomathematics, or the Daring of a Conservative

Volterra's published his first results on variations and fluctuations of the number of individuals in animal species living together in 1926 (VOLTERRA 1926a). He tried to explain the increase in population of Selachians and other predators in the Adriatic Sea during World War I as compared with the preceding and following periods, as well as the decrease in their prey, which feed on vegetables or invertebrates. This fact had been noticed by his son-in-law Umberto D'Ancona in his study of very detailed fishery statistics gathered at Venice, Trieste, and Fiume between 1905 and 1923. These statistics represented almost the entire fishery in the area in that period and allowed him to evaluate the impact of the almost complete fishing standstill during the War. The study related to the ongoing debate about the need for fishery regulation. The main point was that after fishing resumed at the end of the War, no global increase occurred in the fish market, but only a relative increase of some species and decrease of others. According to D'Ancona, the suspension of fishery displaced the biological equilibrium in the Adriatic Sea in favour of the voracious species. Volterra tried to give a theoretical explanation of these variations in the relative numbers of a biological association, beginning with the following simplified hypothetical case of a two-species association[12]. The first species alone, finding unlimited food resources in the environment, would multiply indefinitely following a Malthusian law, that is, the rate of increase (in mathematical terms, the derivative with respect to time t of the population N_1, $\mathrm{d}N_1/\mathrm{d}t$) would be proportional to the number N_1 of individuals of the species (being ε_1 the coefficient of proportionality). The second species feeds upon the first, but would die out if left alone, that is, the rate of increase of the number of individuals $\mathrm{d}N_2/\mathrm{d}t$ would be negative and proportional to the number N_2 of individuals of the species (with ε_2 coefficient of proportionality). Using what Volterra called "principle of encounters", the interaction between both species could be measured by the number of possible encounters among individuals, that is, the product N_1N_2 which would affect negatively with a coefficient γ_1 the rate of increase of the first species (preys) and positively with a coefficient of proportionality γ_2 the rate of increase of the second species (predators). This leads to the following system of two non-linear differential equations of the first order (where ε_1, ε_2, γ_1 and γ_2 are "life" constants):

$$\begin{cases} \dfrac{\mathrm{d}N_1}{\mathrm{d}t} = \varepsilon_1 N_1 - \gamma_1 N_1 N_2 \\ \dfrac{\mathrm{d}N_2}{\mathrm{d}t} = \varepsilon_2 N_2 + \gamma_2 N_1 N_2 \end{cases} \tag{1}$$

Volterra summed up the analysis of this system of ordinary differential equations in the following three laws: (I) both curves are periodic functions of

[12] Volterra pointed out that he had begun his work on D'Ancona problem in late 1925 (see. VOLTERRA 1931a, préface). D'Ancona's statistical studies were published in the monograph D'ANCONA 1926.

time, with equal periods depending only on the coefficients ε_1 and ε_2, and of course of the initial numbers of both species; (II) the average number of individuals of both species tends to constant values; and (III) if individuals of both species are destroyed uniformly and proportionately to their number, the average number of individuals of the eaten species increases, and the average number of those of the eating species decreases.

Now this third law seemed to be in perfect agreement with D'Ancona observations regarding the effect of fishery in the Adriatic Sea. This kind of agreement was in Volterra's mind the clue of the whole study, the point where theoretical research met biological facts, where a careful thinking should concentrate in order to set research next steps and, at the same time, by means of which theoretical research would be able to show practical usefulness. His comments on the third law show the kind of biological-mathematical discussion he was thinking about (VOLTERRA 1926b):

> Law III is, however, true only up to a certain limit. It is evident that if the destruction of both species continue, their exhaustion will ensue. It is therefore necessary to ascertain up to just what point it is profitable to destroy both species in order to obtain the greatest augmentation in the average number of the eaten species. We arrive in this manner at a curious example of a mathematical *upper limit* without the existence of a *maximum*. There is in fact a limit of destruction beyond which both species are exhausted. If we remain below it, the average number of the eaten species grows as this limit is approached; but once the limit is reached, the eating species tends to exhaustion and the fluctuation ceases, while the number of individuals of the eaten species tends asymptotically towards a value which is less than the average formerly reached.

This analysis represented to Volterra the application to biology of the methods of 19th century mathematical physics. According to it, applied sciences – as opposed to exact sciences – could only reach the category of "true science" if mathematized; but, at the same time, mathematical developments were only legitimated by empirical verification and, without that, they came down to empty abstractions – thus following Joseph Fourier's classical point of view. He schematized the method as follows: (1) identification of elementary, characteristic properties of the natural phenomenon under study; (2) setting up of the general description thus obtained in mathematical terms, to obtain a certain number of differential equations; (3) analytical developments, using eventually qualitative methods to deal with nonlinear equations; (4) comparison of results with empirical facts.

The first three steps were called by Volterra "rational phase", the fourth, "applied phase". The schema was circular: after step 4, the analysis returned to step (1) and improved the mathematical formulation of step (2). Hence, to follow this methodology Volterra needed the collaboration of biologists, to give plausibility to steps (1) and (2), and to provide facts for step (4), whereas only for step (3) the mathematician could work alone. Volterra's background in the biological field was meager, save for a firm belief in Darwinian natural selection, and he relied strongly for biological matters on D'Ancona. Reciprocally, neither D'Ancona, nor most of the other biologists with whom Volterra cooperated could discuss the mathematics involved. The only exception was Vladimir Kostitzin, a "naturalist mathematician" as he described himself, who had previously worked in integro-differential equations - a field in which Volterra had made great contributions - which were used by both of them in biomathematical research. Kostitzin shared Volterra's interest in the problem of Darwinian evolution, and also questioned biometric research in the manner of Fisher, Wright and Haldane. He played an important role in the further development of Volterra's views on the import of the theory of biological associations, or, as they liked to say, the struggle for life, on the natural selection. From the very beginning of his biomathematical research, Volterra thought that he was giving mathematical form to Darwin's intuition of phenomena of vital competition, in proof of which he used to quote a passage from *The origin of species,* dealing with hunting[13]:

> The amount of food for each species of course gives the extreme limit to which each can increase; but very frequently it is not the obtaining food, but the serving as prey to other animals, which determines the average number of a species. Thus there seems to be little doubt that the stock of partridges, grouses, and hares on any very large estate depends chiefly on the destruction of vermin. If not one head of game were shot during the next twenty years in England, and at the same time if no vermin were destroyed, there would in all probability be less game than at present, although hundreds of thousands of game animals are now annually shot.

Furthermore, Volterra couched his conclusions in the terminology of "the preservation of favoured in struggle for life" (VOLTERRA 1926b, pp. 559–60):

> Besides the case dealt with above, a study of variations in the number of individuals of two associated species can also be made in all cases in which the species interact either favourably or injuriously, in all possible degrees or combinations. All such cases can be classified in distinct types, and in each of these it is possible to follow the numerical variations of the two species by the help of formulae, or of diagrams to correspond. It is easy to see from these diagrams *which species is winning in the struggle for existence, and which of them is in process of extinction.*

1.2 Early Reactions

Soon after the publication of his first paper on this subject, Volterra started an intense activity of international diffusion of his biomathematical results. In June 1926, in a letter to the mathematical physicist Joseph Larmor, he described his theory of associations and its confirmation by the dynamics of the Adriatic fishery. He asked Larmor to put him in touch with the editor of *Nature,* because he wanted to publish an English short version of his Italian paper. In September 1926 he also wrote to Pearson about the possibility of preparing such a paper for *Biometrika*[14]. A paper was actually published only in *Nature* (VOLTERRA 1926b). In 1927 and 1928, Volterra published three notes in the *Rendiconti* of the Accademia dei Lincei and several abridged versions of his work addressed to the general scientific public, particularly to biologists, in *Scientia* and the journals of

[13] Volterra quoted from DARWIN 1882, pp. 53–54. See VOLTERRA 1926b, p. 559.

[14] Cat. no. 380 and no. 396.

the Italian Comitato Talassografico, the Società Italiana per il Progresso delle Scienze, the Societé d'Océanographie de France and the Conseil international pour l'exploration de la mer. Besides the French and English expositions, a Russian translation was published in Moscow[15].

At the time Volterra published his first biomathematical results, field and laboratory data on variations in numbers of individuals were already available that took into account not only the relation of the individual organism with the environment, but also the associations of species in the natural world and their interactions, such as competition and equilibrium situations. Moreover, zoologists were acquainted with fluctuations, that is, periodical variations in numbers of animals related to cyclical environmental factors, even if the nature of this relationship was not always clear. Now Volterra's main point was that such fluctuations could arise from biological interactions alone. It was this mathematical result that many biologists questioned.

The friction of Volterra's entry into the biological scene was compounded by his neglecting to cite biomathematical studies previous to his own and related to it, such as Ross' contributions and, more significantly, Lotka's. This omission caused a priority dispute with Lotka, who had used the same differential equations (1) to analyze the problem of the biological association of a parasite population and its host, developing W. R. Thompson's ideas[16]. In 1927 Pearson's son, Egon Sharpe Pearson (1895–1980), published an account of the application of differential equations to biological associations in the pages of *Biometrika,* but he criticized D'Ancona's analysis of fishery statistics, and questioned whether these could actually provide a confirmation of Volterra's laws. Friedrich S. Bodenheimer (1897–1959) argued that only environmental factors caused fluctuations in animal populations, not internal factors of competition and struggle for life[17]. Appreciations of Volterra's work also came very soon, mainly from English and American biologists working in quantitative biology of populations. After reading the note for *Nature,* Larmor advised Volterra to contact D'Arcy Thompson, who might help in putting the results in a more readable form for biologists. Thompson showed interest, even if stressing the distance between his approach and Volterra's mathematics[18]. In fact Thompson was concerned with morphology

[15] VOLTERRA 1927a, b, c, d, e, f, g, 1928a, b. Volterra's results were also discussed at the Association Française pour l'Avancement des Sciences 1927 meeting and the French mathematician Joseph Pérès published a summary of them (PÉRÈS 1927).

[16] Volterra to Richard Gregory (editor of *Nature*), draft, enclosed in Cat. no. 416. He continued to call attention to the fact that he did not know Ross and Lotka's work as late as 1936 (see VOLTERRA 1936f). D'Ancona didn't help writing to him in 1930: "After so many time I think that there is no need to insist on that point, I would better say: independently from these works, or following a different approach" (Cat. no. 86). The prey-predator equations (1) are known as Volterra-Lotka equations. On Lotka-Volterra priority controversy, see ISRAEL 1982b, 1988.

[17] PEARSON 1927, BODENHEIMER 1928.

[18] D. W. Thompson wrote: "I am only a Biologist, but I have some little interest in Mathematics of an elementary kind. One of my little papers I should like to offer you" (Cat. no. 338). And after a few months: "I have now read it with very great interest, – all except some of the more difficult parts; and it has interested me very greatly" (Cat. no. 339).

and morphogenesis, and ad-hoc topological and geometrical methods. But he liked the idea of developing a physico-mathematical method that avoided finalism[19]. In France, Edouard Monod-Herzen, who had recently published on this same subject the book *Science et esthétique,* wrote in December 1927[20]:

> I was highly impressed to find in your Note CXXXI the presence, among other things, of strong oscillations or long successive alternations, which in different ways, seem constantly to be encountered when one is dealing with living matter. (I shall not even mention that you have succeeded in subjecting this phenomenon to calculus!). In a totally different field I have some knowledge of recent work by a Russian physiologist, Manoukhine, which can be considered as an extraordinary completion of Metchnikoff's work on phagocytosis. Manoukhine demonstrated that the white corpuscles, or leukocytes, are *alternatively* produced in large quantities (leukocytosis), and *then* destroyed en masse (leukocytolisis). These phenomena of appearance and subsequent destruction are closely linked to the opposite actions of the spleen and the liver. They become visible above all in the case of disease and Manoukhine derived from this a number of remarkable applications by (*appropriately*) acting on the spleen with X rays. Probably by means of a powerful mathematical analysis developed by a scientist like yourself, these alternating phenomena can also be subjected to calculation and your research has raised much hope!

William R. Thompson sent in July 1927 his mathematical papers on parasitism to Volterra, emphasizing again the contrast between elementary and higher mathematics (Cat. no. 350). Thompson had begun his attempts to apply mathematics to biological control problems working at the European Parasite Laboratory of the U.S. Bureau of Entomology at Hyères in France. He had published his first results few years before in France, because he thought American journals would not accept the mathematics.

Another person in Thompson's line of work, Royal Norton Chapman, Director of the Division of Entomology and Economical Zoology of the University of Minnesota Department of Agriculture, wrote to Volterra about his research in the biology of populations, specifically concerning the flour beetle, *Tribolium confusum.* Chapman had travelled to Europe during 1926–27, and he had met Fisher in England and Thompson in France[21]:

> I have been very much interested in your recent publications on the fluctuations of organisms in natural systems. It happens that I have been doing experimental work in very much the same line. I am sending you under separate cover, a reprint of one of my recent papers. You will notice that I have been able to develop a technique whereby I can produce under experimental conditions, the same phenomena which you have been calculating.

---

[19] "Morphological differentiation, that is, the indefinite variety of forms, arouses no feeling of ecstatic astonishment in D'Arcy Thompson. It is rather the possibility of relating morphological plurality to a procedure governed by a small number of formalizable laws which stimulates his solitary research work" (MAZZOCUT-MIS 1995, p. 24).

[20] Cat. no. 271 (the Mémoire CXXXI is VOLTERRA 1927d). See MONOD-HERZEN 1927. In the letter he considers physiological problems, which had given rise to the earliest and most direct applications of physico-mathematical methods. See also VOLTERRA 1927f.

[21] Cat. no. 67. See CHAPMAN 1928, where Volterra's, Lotka's, W. R. Thompson's and Pérès' papers were cited.

Chapman studied the demography of the flour beetle from an ecological point of view, that is, he was interested in quantifying the environmental impact on the population. He had developed an analogue of Ohm's law in electricity to deal with his data, introducing the idea of environmental resistance (analogue to electrical resistance of a conductor) and of biotic potential (as the potential difference in the conductor). His American colleague Raymond Pearl was also interested in demographic studies, but always from the perspective of genetic aspects and the problem of evolution. Pearl made use of the statistical techniques which have been developed during the 19th century for human demographic studies in connection with actuarial problems. But he had also reintroduced a little-known tool, the logistic curve first considered by the Belgian François Verhulst (1804–1849), a follower of Adolphe Quetelet (1796–1874). Pearl actually used the equation of this curve, that Verhulst had obtained starting from a differential equation.

Pearl and Chapman were dealing with different aspects of the same problem. Their point of view differed, however, and they questioned each other's methodology. Chapman thought that *Drosophila's* genetic variability hindered Pearl's experiments, whereas for Pearl this variability was an important advantage. Pearl found misleading Chapman's experimental design, which aimed at calculating the values of the parameters characterizing the different species whose behaviour obeyed the same abstract mathematical equation. Chapman reciprocated by doubting the validity of expressing Pearl's experiments[22].

Chapman found in Volterra's work the echo of his own concepts and methodology. Volterra offered an alternative formulation of the influence of internal (biotic) or external (environmental) factors in the development of populations, extending the analysis to biological associations of any number of species. Moreover, Chapman entered into the spirit of Volterra's research, aiming to provide a methodology which could give to biology the same rigor of mathematical physics, as it has been done previously for other disciplines like chemistry[23], and he shared his confidence in the actual applicability of a deterministic point of view (Cat. no. 69):

> I think that you will be able to understand the possibilities of such experimental work as we are doing here at the University of Minnesota. One very interesting possibility in this experimental work is that we are able to duplicate experiments and get essentially the same results with very little variations as you can see by looking at the curves which I am sending to you. The present work represents only a very small beginning, but we have plans for extending it.

---

[22] See KINGSLAND 1985, pp. 95–96 and Cat. no. 69. Kingsland shows that Pearl was interested in a qualitative appraisal of the influence of inherited and environmental factors in the evolution of populations.

[23] "I am greatly interested in the work which you have done. I believe that it is of great importance to the field of biology. In fact, I think it is comparable to what Dr. Willard Gibbs did for the field of physical chemistry and I hope that unlike his work, yours will come immediately to the attention of the biologists. In the next few days I hope to send on to you some of the results which I have recently gotten in experimental work which substantiate your calculations and I want then to take up with you further the possibility of cooperating with you in this study." (Cat. no. 68).

Chapman's contribution to the diffusion of Volterra's studies was very important. He included references to the mathematical theory of the struggle for life in his own publications, and his Minnesota laboratory served as a meeting point for scientists interested in population research: Karl Friederichs and Filippo Silvestri, both defenders of Volterra's studies, were his guest between 1928 and 1930[24]. Moreover, Chapman provided exactly what Volterra wanted from comparisons of theoretical analysis with empirical facts. He proposed to design experiments to verify different theoretical cases by means of particular chosen species. This was the only strategy for answering most of the criticisms of Volterra-D'Ancona thesis on the origin of fluctuations, which originated mainly from possible alternative interpretations of field observations in wild animals or insects. Volterra asked Chapman for more details, in view of his invited lectures on biomathematics at the newly founded Institut Henri Poincaré in Paris. In his answer Chapman mentioned the research on fluctuations carried out by the British zoologist Charles S. Elton, and Pearl's experiments on the *Drosophila*. On these later he observed (Cat. no. 69):

> These experiments, however, I think are open to considerable amount of doubt and I question myself whether it is worth while spending much time in making any calculations on the basis of experiments in which there is question, as to the dependability of the data.

Chapman's experiments concerned the evolution of *Tribolium confusum*. Nonetheless he told Volterra that it could be viewed as a variant of his prey-predator case (Cat. no. 69):

> One possibility which I think would be interesting to you is to start a culture similar to the one shown in my Graph 1 and at the time that the pupae begin to appear to remove them from the culture. We would then have an adult population which would act as the origin of the egg and larval population. The eggs and larvae then should fluctuate as in the case which you have calculated for two species one of which feeds upon the other. The egg life would end when a larva hatched from it and the larval life would end when it emerged as a pupae. We would then take the pupae out of the culture. The egg population should then come up again, then the larval population would come up and these fluctuations should then continue as you have calculated. [...]
>
> P.S.: I am interested in finding a method of calculating the chance of a larva or adult beetle finding an egg when it is known how many eggs are present per gram of flour and also how many adult and larval beetles are present. We must, of course, first obtain some exact experimental data on the rates of eating eggs at various concentrations.

He mentioned also that his collaborator John Stanley was working on an experiment on the competition between two species for the same food, using *Tribolium confusum* and *Tribolium ferruginium,* and he assured Volterra that he had "the rather ambitious hope that we may be able in time to set up experiments to substantiate practically all the cases that you have calculated".

---

[24] See KINGSLAND 1985, p. 234.

Chapman informed Samuel A. Graham, a professor at the Ann Arbor School of Forestry and Conservation of the University of Michigan, of Volterra's work (Cat. no. 184). Graham expressed to Volterra his firm confidence in the biomathematical approach, particularly regarding practical problems of biological control (Cat. no. 183):

> This is an extremely interesting method of attack on this problem and is, I believe, a method that has a wide possible application. We have only recently ventured into the field of quantitative biology and all of us, I believe, are feeling about for satisfactory means of expression.

In January 1929 he reported Volterra on the state of American entomological studies on vital competition, regarding mostly parasitism (Cat. no. 185):

> Your letter of the second of January has been received and I shall be very glad indeed to give you what information I have that may be of help to you in your work on fluctuations. I am afraid, however, that the material is too fragmentary to be of much use to you. I know of no quantitative studies that have been made in America concerning the results of competition between two species contending for the same food although we all realize that this kind of competition does occur all about us. For instance we know that such barkbeetles as *Dendroctonus ponderosae* may, in some instances, be reduced in numbers as a result of competition with certain Cerambycidae. The Dendroctonus beetles by killing trees provide an abundant supply of food for those cerambycids which feed on the inner bark and cambium of injured and recently killed trees but cannot live successfully in healthy trees. These cerambycids come into direct competition with the barkbeetles since they are both contending for the same food. Under the favourable food conditions brought about by a Dendroctonus outbreak, the cerambycids will increase in numbers until they may consume such a large proportion of the cambium and inner bark that the barkbeetle broods will find insufficient food for development, with the result that many of the larvae will perish. Dr. M. W. Blackman of Syracuse University, Syracuse, New York, has studied this interrelationship on the Kaibab Forest, Utah. He has not as yet published his materials but I am sure that you can obtain information concerning his results by writing to him.

He mentioned research carried out at the U. S. Bureau of Entomology in Hawaii regarding two parasites *Diachasma tryoni* and *Opius humilis* of the Mediterranean fruit fly *Ceratisis capitata* . Although the second parasite had proved much more effective in destroying the fly, its effectiveness was so reduced by competition with the first, that it was of little interest in fruit fly control. Regarding prey-predator fluctuations, Graham mentioned quantitative studies by J. D. Tothill in Canada on the effect of parasite and predator agents in the emergence and end of a webworm outbreak, *Hyphantria cunea.* He added that the effect of the interaction of parasitic and predatory organisms upon a host could be studied quantitatively using the great amount of data available at the Melrose Highlands Gypsy Moth Laboratory:

> The number of egg masses per acre as determined in their scouting work could be used as an index of abundance of the moths from year to year and the percentage of eggs, larvae, and pupae parasitized by the various parasitic species together with a record of the number of individuals examined each year would serve as an index of parasite fluctuation.

Finally, Graham described his own research using Chapman's terminology to distinguish external and internal factors:

> Most of my work in the past has been concerned with the influence of physical environmental factors upon the relative abundance of certain species and only in the last few years have I been collecting data bearing upon the interrelation of biotic factors.
>
> One series of experiments, started under my direction, I am watching with great interest. These should in the course of a few more seasons produce valuable data regarding the influence of parasites upon a host. These experiments deal with the interrelations of the pine tipmoth *Rhyacionia frustrana* and its parasites in an artificially established forest in the naturally treeless sand-hill region of Nebraska. Each year since 1924 we have recorded the number of tips killed by this moth on a series of sample plots. This will serve as an index of moth abundance. We have also records of the number of parasites of a large series of species in samples taken from these plots. During the second year of the work a parasite that appears to be much more effective than any naturally present was introduced. Since that time this species has increased tremendously, raising the percentage of parasitism on tipmoth larvae in western yellow pine, *Pinus ponderosa,* from between 10 and 20 per cent to about 60 per cent. The condition of equilibrium has not as yet been reached and so we do not know what the end result will be, but whatever it is the information collected during the course of the study is bound to be interesting.
>
> I have also been studying the larch sawfly quantitatively but the data concerning its interrelations with other organisms are still in the formative stage and are not adequate to serve as a basis for reliable conclusions.

The biologists concerned themselves with interactions of all sorts of species. The British zoologist Charles S. Elton, a pioneer of animal ecology, was favourable to Volterra's attempt, but emphasized the need to build a methodology of interaction between mathematicians and biologists (Cat. no. 168)[25]:

> It appears to me that the study of marine animals is indeed more likely to provide the first materials for the application of your theories, since they are probably less subject to interference from external factors. In this connection I enclose an Abstract of a paper which may have escaped your notice, by being published in a small English Journal. I am sending by separate post some numbers of these "Ecology Abstracts".
>
> I hope to publish shortly some work upon the fluctuations in numbers of Canadian wild animals (rabbits, lynx etc.) upon which I have managed to get more light by examining the old records of the Hudson's Bay Company. I do not know, however, whether these figures and facts will be sufficiently accurate or relevant, to provide the material you would wish.
>
> I agree completely with your main thesis i. e. that external factors are not in themselves necessary to cause fluctuations in the populations of wild animals. In fact I have been led to postulate both the existence of an external climatic cycle, and also a process similar to the ones you have described from mathematical data, in explaining the rabbits-lynx 10-year cycle in Canada. It will give me great pleasure to send you this report when it appears.

---

[25] In the second edition of his book *Animal Ecology* (Elton 1927), Elton mentioned biomathematical studies of populations dynamics, but pointed out the lack of effective confirmation by experimental data.

> I shall probably be in a position to obtain a good deal of material concerning these mammalian cycles, during the next few years, and shall attempt to procure figures which may be of use to you in carrying further the analysis of these problems.
>
> My only regret is that I am unable to follow the purely mathematical arguments, and must confine myself to the working out of an adequate technique for obtaining biological data which may be of use to the mathematician.

In 1932 Elton asked Volterra to publish his results on his newly founded *Journal of Animal Ecology;* and one year later he published one of the first papers in biomathematics by the Australian biologist Alexander John Nicholson (1895–1969, NICHOLSON 1933). As with the fisheries studied by D'Ancona, the coexistence of hares and lynxes in Canada could be investigated because of commercial accounting registers of a private company. These data, with D'Ancona's, provide almost the only examples of free competition in nature in which the oscillatory trend (even if not strictly periodical) foreseen by Volterra's law I has been confirmed[26].

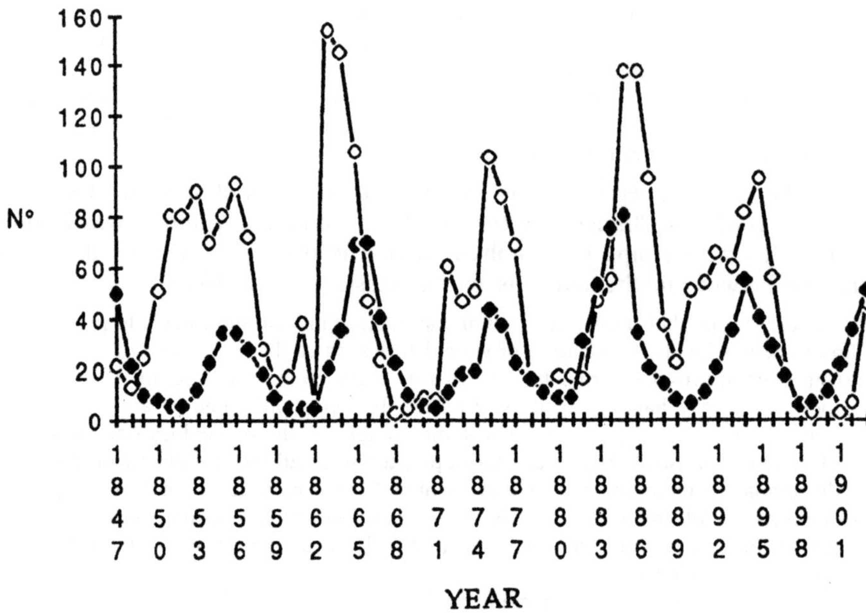

Figure 1 Hare-lynx 10-years fluctuations in Canada. White dots curve for hares, black dots curve for lynxes (Source: ISRAEL 1990, p. 108)

26 See LEIGH 1968. Fig. 1 is obtained from data taken from MACLULICH 1937, which are analyzed in Leigh's paper. Leigh rightly observes that this "is not a prey-predator oscillation" because in fact it is not periodic. However, as the Figure 1 shows, the oscillatory trend represents one of the best empirical fits of Volterra's law.

1.3 Developing Likely Mathematics

Volterra and D'Ancona published their theories in 1931 in a book, *Leçons sur la théorie mathématique de la lutte pour la vie,* based on Volterra's lectures at the Institut Henri Poincaré and edited by the French mathematician Marcel Brelot (VOLTERRA 1931a). This book extended the 1927 monograph published by the Comitato Talassografico Italiano, and presented in an organic exposition Volterra's research. In the Introduction, the mathematical theory of struggle for life was placed in the framework of mathematical and quantitative studies in biology as well as in that of studies on biological associations. Ross, Lotka and W. R. Thompson studies were mentioned, thus filling the gap in Volterra's first publications, and emphasizing that he was not aware of these works at the time of his 1926 paper. Friederichs' remarks on the application of biomathematics in agronomic zoology in his recent book were stressed[27]. It is important to see how the main point on fluctuations in the numbers of individuals of species in biological associations was stated, as it was related to the method of inquiry, that is, the old statistical or biometric method versus the new "biomathematical" one (VOLTERRA 1931a, p. 3):

> The question appears to be very complex. Of course, there are periodical environmental circumstances like those, for example, which depend on the succession of the seasons, and which set up *forced* or *external* oscillations in the number of individuals of different species. In addition to these periodical external actions that have been studied in greater detail using statistical methods, are there not others of an *internal* nature with their *own* periods, which are independent of the external causes and are superimposed on them? Observation would seem to substantiate this. In this way *certain infectious diseases* like malaria undergo fluctuations during their existence which cannot be accounted for solely in terms of variations in the environmental conditions. We shall see how mathematical investigation explains and confirms this observation. The phenomena are schematized by using hypotheses that, although perhaps unsophisticated and simple, allow very elementary mathematical development to be performed and one generally disregards the variations in the external actions in order to study what may be termed the purely *internal* phenomena.

The chapter of conclusions was completed with a detailed historical and bibliographical appendix on this main issue which was prepared by D'Ancona (VOLTERRA 1931, pp. 202–210). He started from Darwin his review of studies on the causes of fluctuations in animal populations, on equilibrium states in biological associations and on modifications of equilibrium caused by the introduction of new species. On biological equilibria regarding the problem of harmful insects in agriculture he cited among others Paul Marchal (1862–1942), Maurice Caullery (1868–1958), Antonio Berlese (1853–1927), A. Chigi, W. R. Thompson, Bodenheimer and Friederichs; as to human and animal epidemiology he mentioned works by Ross, Lotka, Elton, E. Martini and M. Sella; on hydrobiology, his research field, he made reference to his own work and that of August Thienemann (1882–1960), Gustavo Brunelli (1881–1960), Aage Johannes

[27] FRIEDERICHS 1930b. Friederichs had already taken part in the debate with an answer to Bodenheimer criticism: see FRIEDERICHS 1929.

Christian Jensen, C. G. J. Petersen, R. Monti, and Carlo Marchi[28]. Then he considered studies which had identified periodical or irregular fluctuations, citing among others Elton, Angelo Celli, Jacob Peter Jacobsen and A. C. Johansen, and Fernando de Buen. Bodenheimer's point of view was opposed to Friederichs' favourable opinion. He concluded (VOLTERRA 1931, p. 210):

> As you see, biologists chose to seek the cause of these fluctuations in the variation of physical environmental factors. However, it emerges from the work of Mr Volterra that the mere fact that several species coexist can produce the fluctuations. Of course, it cannot be ruled out that, in nature, both these kinds of causes can contribute to determining fluctuations and it is to be hoped that this can be demonstrated by observation and experience. In any case, D'Ancona's observation concerning the relative increase of the more voracious species corresponding to the suspension of fishing activities can be explained only by acknowledging that fluctuations occur between the devouring and devoured species even in an unchanging environment.

Therefore, the validity of D'Ancona analysis was emphasized as the only empirical evidence of the internal fluctuations, but the need for new observational or experimental proofs was also clearly stated.

During the planning of the publication of the book, D'Ancona told Volterra that a good interdisciplinary book should present examples of practical applications; but the final version of the book did not meet this requirement: it was basically a mathematical book addressed to mathematicians. However, a specific reference was made to Chapman's studies, as regards to the influence of the number of individuals of a species on its coefficient of growth

$$\frac{1}{N}\frac{\mathrm{d}N}{\mathrm{d}t} .$$

Volterra's first generalization of the prey-predator model (1) gave rise to unstable situations and boundless population increase in finite intervals of time. The equations had to be modified, considering the (negative) influence of the number of individuals. So, for the case of a single species, Volterra put forward the following equation:

$$\frac{1}{N}\frac{\mathrm{d}N}{\mathrm{d}t} = \varepsilon - \lambda N \tag{2}$$

where ε, λ are constants characteristic of the chosen species. This differential equation has the solution:

$$N = \frac{\varepsilon N_0 e^{\varepsilon t}}{\varepsilon + N_0 \lambda (\varepsilon e^{\varepsilon t} - 1)} \tag{3}$$

where N_0 is the initial number of individuals of the species.

The maximum number of individuals obtainable occurs when $t \rightarrow \infty$ and is given by $N = \varepsilon / \lambda$. These considerations were already included in Volterra's former papers. In the book he added the remark that unbounded growth was

[28] Marchi's studies, which followed D'Ancona's approach, using information gathered in Sardinia, were also mentioned in the Introduction. But D'Ancona's private opinion on them was rather skeptical (See Cat. no. 85).

"unacceptable for the biologist", and explained equation (2) in biological terms: "the development of a species in a limited environment is more difficult when the species is very numerous, because its density of distribution grows up to the point to modify the conditions of life" (VOLTERRA 1931a, pp. 198–99). This idea of density was that used by Pearl to justify the introduction of the logistic curve[29]. In fact Volterra equation (2) corresponded exactly to the increase or decrease of a single species considered logistic and not exponential (as it was considered in the basic prey-predator model (1))[30]. But Volterra did not cite Pearl[31]. Instead, he emphasized the correspondence between Chapman's law ("l'existence d'un état limite, causé par la résistance du milieu, qui est assez rapidement atteint" (VOLTERRA 1931a, pp. 201–202)) and the limit $N = \varepsilon/\lambda$. Brelot and Volterra had discussed Chapman's papers, and Brelot had noticed the problems raised by Chapman's formulation[32]:

> [...] I wish to recount the experiments carried out by Chapman before writing this last page. Firstly, we are still dealing with *bounded* environments, the size of which affects coefficients other than ε and I have emphasized this point thourough the work except perhaps in the few pages in your possession. To say that species is numerous from the point of view of its influence on the others or on itself means that it is *dense*, because it is implicit that the environment is bounded either naturally or artificially by conception. Furthermore, if this were not done the number of individuals would tell us nothing.
>
> What therefore appears of interest is the *density* of the flour in the environment or, given a definite environment, its total quantity. However, this assumes that it is mixed with something else. Otherwise, whether large or small, this total quantity would not be enough to affect the coefficient of growth ε,
>
> $$\frac{1}{N}\frac{\mathrm{d}N}{\mathrm{d}t}$$
>
> in the case of a relatively *non* numerous species. Moreover, if we take the equation
>
> $$\frac{1}{N}\frac{\mathrm{d}N}{\mathrm{d}t} = \varepsilon - \lambda N\,,$$

---

[29] Pearl developed a deep analysis of the biological meaning of the logistic curve (urged also by the many criticisms): he first stated the influence of the density of the population on the reproduction; then, in 1932, he explained that encounters among individuals altered the food intake, energy output in muscular activity, and oviposition (PEARL 1932; see also KINGSLAND 1985, pp. 77–83).

[30] VOLTERRA 1931a, p. 32. This same reasoning was introduced for the case of whatever number of species (p. 65).

[31] Discussion on Pearl's work occurred in Volterra-D'Ancona correspondence only in late 1934 (see infra § 6). Cat. no. 386 shows that they were already in touch before the publication of Volterra's lectures in 1931. The correspondence resumed only around 1937.

[32] Cat. no. 23. In seems that on this issue also Volterra asked D'Ancona's opinion; but D'Ancona pointed only out the importance of the problem of fluctuations larvae-eggs (Cat. no. 85). In fact a short sentence on the particular situation regarding *Tribolium confusum* problem of eggs, larvae and adults appears in the book, and this problem would have been considered in the new book published in 1935 (see infra § 1.5). A 1929 manuscript by Chapman entitled "The Trend of Insect Population from the Viewpoint of Biotic Potential and Environmental Resistance" was cited, as well as CHAPMAN 1928.

λ once again refers to an environment of a *precisely defined* size and depends on this quantity since the corrective term depends solely on the *species density*. I thus wish to consult you on the subject of this flour which is not said to be mixed with anything else or suitably *dispersed* in a sufficiently large space.

Presented in this way the law of Chapman may seem a commonplace, as in given quantity of *pure* flour, the limited number of individuals is obviously proportional to this *total* quantity, provided that the flour is always arranged in the same way or piled up. [...] Do you think that the quantity of flour per unity of volume influences ε in a *precisely* proportional way, as is acknowledged. We get:

$$\frac{\lim N}{\varepsilon} = \frac{1}{\lambda}$$

and Graham's law holds only thanks to this proportionality. This is a somewhat dubious hypothesis.

Of course, I also take it for granted that the quantity of flour is *kept constant*; otherwise the insects, after having eaten everything, would disappear.

No mention is made in the note of whether the quantity of flour is kept constant or whether the impurities due to the insects are removed.

In the general abstract case of an association of n species, the extension of this idea of logistic development of the single species, or, as Volterra named it using the mechanical analogy, a growth under "friction" among individuals of every single species, led to one of his fundamental theoretical classifications. This was the distinction between *conservative* (where the number of individuals does not influence the increase of population) and *dissipative* associations. A second basic assumption in the generalization of the prey-predator case to the association of n species was introduced, concerned the effect of encounters on both populations: the proportion between the number of destroyed preys and of new predators was supposed constant, or, the effect of encounters taken to be proportional to the number of encounters.

W. R. Thompson wrote to Volterra, when the book was forthcoming, that in the case parasite-host populations the result of the "encounters" was not so simple (Cat. no. 354):

The likelihood of encounter increases as the number of parasitic individuals (predators) and their hosts (prey) increases: however, at the same time, there is an increase in the probability that two or more parasites will visit the same host individual. Let us assume that each encounter is followed by the laying of eggs. It follows that several eggs from the parasite will accumulate in the same host. Generally speaking, if x eggs are distributed among N hosts, the numer of parasitized hosts will be $N(1 - \exp(-x/N))$. Or, *only one*, or, at most, a maximum of m of these eggs can develop in the same host: the others will die. The fraction of encounters favourable to N_2 is therefore not constant: it will decrease as the numbers of the two species increase. On the other hand, the number of N_2 needed to destroy one N_1 individual also increases.

The case is very frequent among insects and I wonder whether it is not actually the general case. For example, in the case of fish it is conceivable that 2 or 3 predators may appear simultaneously and share the same prey so that none of them takes fully advantage of it.

However, this hypothesis of constant proportion, the "equivalents hypothesis" (because equivalent values were assigned to the different species) was not explicitly justified in the book. In a further book published by Volterra and D'Ancona in 1935 (VOLTERRA, D'ANCONA 1935) a long introductory note was devoted to this important point, emphasizing the fact that the consideration of the most general case was a dead-end approach, because no meaningful deduction could be drawn. The general underlying problem was that of the relation between verisimilitude and empirical validation. Volterra had stated clearly the right methodology of biomathematical research: the distinction between rational phase and applied phase was the key to avoid contradiction between mathematical theory and biological evidence. This was the only way to handle in this context a mathematical tool, infinitesimal calculus, based on the abstract concept of limit and on the use of continuous magnitudes, which allowed the deduction of general results but introduced many unrealistic elements. Mathematical concepts regarding populations numbers, such as limit, oscillations, or boundary, could have no realistic biological translation (VOLTERRA 1927f, p. 100):

> [...] we should like to warn readers about the objections that could be raised and put in a false light the results of the preceding research, thus making it appear incorrect and meaningless. [...] For example, in the case of the two species one of which eats the other, we observe that a periodical cycle is set up in which the two species oscillate around median values. It might be objected that it is easy to conceive of the devouring species as being so numerous and voracious as to rapidly destroy all the individuals in the other species one by one and therefore make such oscillations impossible.
>
> However, it will be recalled that the law of the cycle derives from the proposition that a species which lacks food can only disappear in an infinitely long period and this may seem even further removed from actual reality than the law itself. The reason for this is that among the hypotheses on which the whole treatment is based there is one according to which the number of individuals is a continuously varying positive number, while it can in fact only be a whole number and can never be less than unity. We must therefore acknowledge that, if the number of individuals of a species is reduced to a sufficiently small size it must be assumed to be zero and the fact of extending its value is merely a theoretical construct lacking of any real meaning. [...] if the coefficient of voracity λ_1 is very large as well as the initial value of N_2, N_1 can rapidly become less than 1, which in practice amounts to its reduction to zero so that the cycle, which could theoretically continue, cannot be closed and would have to cease at this point.

A problem of this kind, regarding an example of a three-species association, caused a disagreement between Brelot and Volterra during the preparation of the book. The example, which Volterra had considered in his paper in *Nature* (VOLTERRA 1926b), proposed an island where a carnivorous population N_1 feeds upon an herbivorous species N_2, and this feeds on a certain plant species N_3. Brelot disaagreed with Volterra's mathematical treatment; moreover, he thought that a pure mathematical study was perhaps impossible, and certain postulates had to be added to the mathematical reasoning to guarantee its biological plausibility. The biomathematical equations of the conservative association, that is, neglecting interaction between individuals of the same species, were:

$$\begin{cases} \beta_1 \dfrac{dN_1}{dt} = (-\alpha_1\beta_1 + aN_2)N_1 \\ \beta_2 \dfrac{dN_2}{dt} = (-\alpha_2\beta_2 - aN_1 + bN_3)N_2 \qquad \alpha_1, \alpha_2, \alpha_3, a, b > 0 \\ \beta_3 \dfrac{dN_3}{dt} = (\alpha_3\beta_3 - bN_2)N_3 \end{cases} \tag{4}$$

and the general integral was:

$$N_1^{\beta_1 b} N_3^{\beta_3 a} = Ce^{(\alpha_3\beta_3 a - \alpha_1\beta_1 b)t} \tag{5}$$

The case $(\alpha_3\beta_3 \mathrm{a} - \alpha_1\beta_1 \mathrm{b}) > 0$ corresponded to a boundless increase of the three species, unacceptable from the biological point of view. To stop the increase, Volterra introduced a "friction" for N_3 and reformulated the equations. As for the remaining case, this was Brelot's explanation (Cat. no. 16):

> In the case $(\alpha_3\beta_3 \mathrm{a} - \alpha_1\beta_1 \mathrm{b}) < 0$, all we know is that $N_1^{\beta_1 b} N_3^{\beta_3 a} \to 0$ and that N_3 cannot go on being small after a certain point has been reached.
>
> This is not enough to conclude that $N_1 \to 0$. But if N_1 did not tend toward 0, it would undergo oscillations like N_3 and would occasionally take on very small values tending toward 0. This would mean that the species, if they did not disappear, would undergo an infinite number of *temporary* disappearances. If *we acknowledge* that a species cannot survive a relatively large number of temporary disappearances, N_1 will certainly disappear definitively (since N_3 cannot).

According to Brelot, the biological postulate stated in the last sentence made it possible to speak about the extinction of a species which did not take place from a strictly mathematical point of view. He put forward two other postulates for the conservative case. Volterra wrote at least three drafts of his answer to Brelot, progressively polishing his opinion; he confirmed a methodology of research centered on the clear separation between mathematical development and biological facts (Cat. no. 359):

> As far as my lectures are concerned, I think it necessary to change the part referring to what you call the *biological postulates. At least* it would be necessary to retain a single postulate but keep it quite separate from the mathematical part and its consequences. Furthermore, mathematics can take us much further and *do away with the need for postulates.*

Biological evidence should only guide the introduction of likely mathematical premises (the *mise en équations)* and control the final results, but should not interfere with the theoretical developments. As to the rational phase – and the book was part of it – problems should be solved "abandoning oneself, as it were, to the analysis procedures and considering the hypotheses as if they were absolutely confirmed. Having obtained the solution, it would be advisable to discuss it at a later stage and, if certain limits seem to have been exceeded in the solution, so that the hypotheses may appear too far removed from reality, it will be necessary to abandon the solution or else modify it" (VOLTERRA 1927f, p.101).

"Postulates are for the most part suggested by Satan to render you lazy", wrote Volterra to Brelot in the first draft of the letter. As to the main problem, the exhaustion of a population, he proposed this solution, which came after the complete mathematical development (Cat. no. 359):

> All this is pure mathematics. At this stage we shall leave the mathematical part and go on to the biological issue. It may be said from this point of view that:
>
> 1°) A species becomes extinct if the number of individuals can become as small as one likes (Definition)
>
> 2°) If a species becomes extinct it may be neglected after a certain point. (This is an intuitive principle and could even be called a *biological postulate*). However this principle can lead to consequences that clash with the mathematical laws. This is why it may be applied to biological issues but not to mathematical issues. The two issues must therefore be considered as different and dealt with separately in order to avoid giving rise to clashes and confusion.
>
> This brings us back to the case 1°)[33] Since N_1 can approach zero as closely as we wish, by *definition* this species could be said to be extinct and if we apply the principle that it may be neglected, we find ourselves in the well known case of the two species, one of which feeds on the other. The well known fluctuations will therefore occur.

The published version followed roughly this approach: Volterra claimed that a population can be considered as exhausted not only when the function N (number of individuals) tended mathematically to zero, but also when it took arbitrary small values as time went on. Still, the expression "irregular disappearance" – linked to a formal definition of a particular kind of limit, called *smallest limit* – instead of "temporary disappearance" appeared in the book, perhaps because of the objections raised by D'Ancona (Cat. no. 86):

> What I do not fully understand is the "temporary disappearance" of a species, see sheet 4. Perhaps some explanation would be appropriate. If I have understood correctly it should be interpreted as follows. A species can decrease until it has almost completely disappeared. However, a few individuals, at least a couple, remain, which can multiply as soon as the conditions become favourable again. Is that it? It would therefore be appropriate to assert that what one means is an almost complete disappearance with the cessation of reproduction. However, since a few individuals survive, they could reproduce again. Of course, in the latter case the duration of the apparent disappearance will be of importance. If it does not last very long the surviving individuals can again reproduce; if it lasts a long time this will be prevented by aging or death.
>
> Again for the same reason I marked sheet 11. Instead of "two individuals" it would be preferable to refer to "a couple". In the case of lower species, the same purpose may be served by a simple suspension of reproduction, as actually does happen in many cases, or else by encystment or egg preservation, in other words "latent life". In some cases these periods of latent life (some are known because of the unfavourable physical conditions of the environment) can persist for quite some time.
>
> The biological postulate contained in the same sheet is however unconvincing, namely that "a species cannot withstand a relatively large number of temporary disappearances". Biologically speaking, I consider the duration of the disappearances and their frequency to be more important. If they are very frequent they may be more effective as in the intervening periods the species is unable to reproduce adequately. If

[33] This is the case in which $(\alpha_3\beta_3 a - \alpha_1\beta_1 b) < 0$.

they occur at longer intervals they may even be quite numerous without the species actually becoming definitively extinct. There are species that disappear once or twice every year because of the unfavourable conditions.

Thus, the problem of interactions between individuals of a single species in case of great population density, and the opposite problem of a radical decrease in the number of individuals were the most significant issues regarding likelihood dealt with in the book. A specific remark was introduced in the Introduction, following Volterra's guidelines, as stated in the journal *Scientia* in 1927[34]:

> Having expressed this in the form of an equation, the next step is to make a purely mathematical study using differential or integrodifferential equations to obtain the properties of the integral *N* functions.
>
> It then remains to draw the biological conclusions from these mathematical properties. It must be stressed that the equations have no value or biological meaning unless the *N* functions lie within certain limits, which are essential for validating the hypotheses leading to the form of the equations.
>
> Conclusions that are valid for the biologist can be drawn from the integral properties only if the latter remain within these limits. The equations will cease to have any meaning as soon as one of the *N* functions becomes too small; they may cease to correspond to reality if certain *N* values become too large, bringing into the limited environment investigated a density of individuals that is so large that it profoundly modifies the actual conditions of existence.
>
> It should also be observed if the fact that an *N* function exceeds these limits it can have biological consequences. For example, if the *N* for a species remains positive but tends towards zero when the time is indefinitely long, while the other functions will stay close to appropriately large numbers, it may be inferred that the species will disappear and become definitively extinct. This is because, as soon as the species is too small and the equations can no longer be applied to it, they will continue to mean that the species will always in future be in unfavourable circumstances which tend to destroy the surviving individuals rapidly.

The weakest point of Volterra's theory was the dependence of the results concerning n species associations on the even or odd number n of species. This difference was caused by the mathematical structure, because the involved differential equations were not linear. The frail explanation given was that in the case of an even number of species it was possible to couple the species two by two, while in the odd case there at least one species would act very differently. Kostitzin traced this problem to Volterra's interpretation of "vital coefficients", as he called the constants in the biomathematical equations; he suggested to free those coefficients from particular restrictions which limited possible biological interpretations of the equations (V. A. Kostitzin 1934, pp. 35–36).

The Italian entomologist Filippo Silvestri (1873–1949) submitted to D'Ancona and Volterra a case of double parasitism, an example of biological association of three species of insects (Cat. no. 99):

---

[34] Volterra 1931a, pp. 7–8. This comments were not included in the 1935 book: for example, for the case $(\alpha_3\beta_3 a - \alpha_1\beta_1 b) < 0$, it was claimed that the first species tends asymptotically to 0 and that the meaning of this condition was that the food resources offered by the vegetal species to the carnivorous by means of the herbivorous is not enough to balance the natural trend to exhaustion of the carnivorous.

> Silvestri has asked to be remembered to you. He would like to present to you a case of biological struggle to see if it can possibly be subjected to mathematical treatment. There is a discussion among entomologists as to whether it is useful or harmful to introduce into a country more than one species of insects that destroys insects that are harmful to a plant. Let us assume that these insects are harmful to species A. The latter is parasitized by two independent species B and C having different habits and that, although not actually enemies, nevertheless compete for the same food. These species lay their eggs in the larvae of species A. From the egg hatches a larva of species B or C which eats the larva A and, when it has completely devoured it, becomes an adult individual. If only a single egg of the species C or B is laid on a larva A, the larva B or C that emerges will succeed in developing as it has enought food. If, on the other hand, two or more eggs of B or C are laid on larva A, either it or those of the two species will not find enough food and will ultimately die. The two species B and C have different habits and therefore have different coefficients or growth and differ in their effectiveness in the struggle against A. Some entomologists, for instance, say that as B and C having different habits and destroying the species A in different ways and under different conditions, help each other and complete the natural struggle. Others maintain that the two species, by competing for the same food, are actually fighting each other and end up by diminishing the effectiveness of the struggle by the individual species.
>
> Silvestri agrees with the former; however, he would like to know whether, by applying your calculations, it is possible to reach some conclusions.

Although many biologist were ready to offer ecological insights or examples to be related with biomathematical results, it is evident that the complexity of any simple case often demanded a laborious reelaboration task. As for Silvestri's example, Volterra apparently thought that his three species model could be applied to this case[35].

Volterra hoped that an English translation of his book could be published, as he told D'Arcy Thompson, Elton, Chapman and Stanley, but all his efforts were in vain. Chapman reproduced Volterra's paper in the *Journal du Conseil international pour l'exploration de la mer* in appendix to his book *Animal Ecology, With Special Reference to Insects* (Chapman 1931). But the collaboration project that was advanced in their correspondence never took effect. In 1931 Chapman moved to Hawaii. In this same year Volterra's troubles in fascist Italy began: he was dismissed of his position as professor at the University of Rome because of his refusal to take the fascist oath and removed from the Presidency of the Accademia dei Lincei. His scientific contacts changed. He dropped Elton's proposal to prepare a paper for his *Journal of Animal Ecology*. He looked abroad to overcome his isolation in Italy, but his main point of reference was Paris, where he had an important network of relationships: he was always welcomed by the active Paris mathematical community, and his biomathematical theories have enjoyed a good reception.

In any case, at the beginning of the 1930s American biologists were giving biomathematics the cold shoulder[36]. The most significant case is that of W. R. Thompson, who in those years reversed his former position and engaged in an

---

35 See Falcone, Israel 1985. In this paper an analytical and numerical study is developed around the example put forward by Silvestri.

36 See Kingsland 1985.

acute criticism of mathematical population dynamics and population genetics. It is likely that the eminently applied character of pioneer ecological studies in the United States moved biologists away from biomathematics: statistical biometric techniques were ready available for direct application, whereas applications of biomathematics looked complex and doubtful. Only after the World War II, interest in the mathematical biology of populations would revive in the United States. In the meantime, most of the experimental and theoretical research was developed in Europe.

1.4 In Search of Laboratory Verification

Until 1930, most of the available quantitative experimental data on populations dynamics concerned the development of a single species. Chapman tried to extend the experiments to the case of two species during the 1930s (CHAPMAN 1933). But the first systematic, mathematically oriented experimental studies on populations dynamics in biological associations were undertaken in Moscow by a young biologist of the Moscow University Zoological Institute, Georgii F. Gause. He considered microbiologic species from an ecological point of view.

Many convergent circumstances explain why this project was undertaken in Moscow. The University of Moscow had an important tradition of zoological studies emphasizing ecological aspects: Gause's first studies dealt with animal abundance in relation to habitat. A general approach to ecology was developed in the 1920s by the geochemist Vladimir I. Vernadsky (1863–1945), author of important contributions on the notion of biosphere[37]. Gause was in touch with Vernadsky during that period, and with Vladimir Kostitzin, director of the Geophysical Institute of Moscow. Also biometrical studies were being developed in Moscow, and the clarification of the Darwinian concept of natural selection concerned many among biologists[38]. Gause's teacher, Vladimir V. Alpatov, worked with Pearl, and Gause knew Pearl and Chapman studies[39]. Finally, Volterra's work in mathematics was wellknown in the Soviet Union. A number of mathematicians, like Kostitzin, followed Volterra's mathematical physics, particularly the application of integral and integrodifferential equations. Volterra's fame explains the quick diffusion of his biomathematical studies. Volterra had made only an elementary use of the qualitative analysis of nonlinear ordinary differential equations, a way of attacking the kind of problems which was initiated by the French mathematician Henri Poincaré (1854–1912); while elsewhere this kind of analysis had been dropped, it was being developed in the Soviet Union, often in connection with applied issues, and was also most suitable for the biomathematical applications. The leader in this field was Andrei N. Kolmogorov (1903–1987), who wrote on Volterra's biomathematics[40].

---

[37] See WEINER 1988. On Vernadsky, see BALADINE 1982 and BAILES 1990.

[38] See ADAMS 1980a, b.

[39] An analysis of Gause's work and his influence on the modern American biomathematical school – starting from George E. Hutchinson – can be found in KINGSLAND 1985.

[40] See KOLMOGOROV 1936. Kolmogorov described Volterra's equations (1) as the "first approximation of actual state of things". Instead of using those equations or whatever others using any

In November 1932 Gause wrote to Volterra about his experiments (Cat. no. 177):

> I am continuing now the experimental analysis of the struggle for existence and several papers are now ready for publication. One among them deals with the experimental study of carnivorism, and the integro-differential equations obtained are based on the principles discovered by you. The experimental research showed, however, that this case is a complicated one, and the quantitative analysis may be given at present for the simplest conditions only.
>
> I would wish to point out that the simultaneous logistic differential equations which have been obtained by me in the experiments with yeasts and are in agreement with the observed data, have been solved only with the aid of the approximate numerical method. At the same time there is now independent "chemical" confirmations of their correctness, and it is very important, therefore, to analyse them more closely.
>
> I am working in the field of experimental biology, and I am not qualified enough to analyse more closely the mathematical part of the problem, and it would be very interesting if it will be possible for you to investigate these equations in the future, and to find their solution in the general form.

Gause shared Chapman's methodological point of view: the struggle for life had to be studied in the laboratory, under suitable artificial conditions for evaluating the reciprocal influence of environmental resistance and biotic potential. He was able to put forward a comparison between theoretical previsions and calculated numbers which remains still today a basic point of reference[41]. He developed the idea of a qualitative experimentation, aimed at establishing several fundamental types of reciprocal action among coexisting species: this corresponded to the qualitative mathematical analysis suitable for the involved quadratic differential equations. Gause remained in touch with Volterra, but worked alone in the Soviet Union with some assistance of Russian mathematicians, such as Aleksandr Adolfovich de Witt (d. 1937).

Volterra followed these developments with great interest. Experimental data captured his attention. In early 1932 he had contacted – through Émile Roux (1853–1933), director of the Institut Pasteur – Jean Régnier, head pharmacist of the Ambroise Paré Hospital (Boulogne Billancourt) and professor at the Paris Faculty of Pharmacy. Régnier had carried out research on microbial multiplication and was developing laboratory techniques of numerical quantification (Cat. no. 283)[42]:

---

particular dependence among derivatives and numerical values of populations N_1, N_2 - the choice of which, he said, was "wholly arbitrary" –, he introduced a general pair of equations

$$\begin{cases} \dfrac{dN_1}{dt} = K_1(N_1,N_2)N_1 \\ \dfrac{dN_2}{dt} = K_2(N_1,N_2)N_2 \end{cases}$$

and presented a purely qualitative analysis of these general equations.

41 See PAVÉ 1994, p. 98.

42 Concerning Régnier see the introduction to his correspondence with Volterra and LAMBIN 1969. On his research work see also RÉGNIER, DAVID, KAPLAN 1932. Of prime importance in our analysis of Régnier's work were the documents conserved by his collaborator Suzanne Lambin

As I told you the other day, it is possible for me to attempt to verify the laws you have found, in the field of bacterial competition. As soon as I can I shall therefore begin quantitative research on both "two species competing for the same food", and "two species of which one is harmful to the other". As I have already explained, we shall be hindered by the fact that bacteriological media are rapidly rendered unsuitable for bacterial multiplication without there necessarily being any shortage of food. Therefore if the phenomena we are seeking does not occur rapidly, that is, in the first ten or twelve hours after seeding, we will be compelled to turn to more complicated techniques with which I shall have to become familiar (continuously renewed media).

Volterra and D'Ancona exchanged many letters on biological literature regarding fluctuations, and on reactions to biomathematics. In October 1933 D'Ancona wrote to Volterra that the only objections to his theory were Bodenheimer's, and that he would perhaps respond to them (Cat. no. 105). A few days later he wrote (Cat. no. 106) concerning a paper of A. J. C. Jensen (JENSEN 1933):

I too have received Jensen's note and have skimmed through it. He observes fluctuations in the abundance of several fish species in Denmark and discusses the causes of such fluctuations. He considers them not to be biological fluctuations as he found that the predator species (Gadidae) have a life cycle that is quite different from that of the species investigated; he would therefore like to attribute the fluctuations to physical causes. Of course the fact that different cycles are found in several coexisting species is not a good reason for allowing physical causes and excluding biological ones; when there are only two species, a devouring one and a devoured one, we will have cycles of equal amplitude but out of phase, but when the association is formed by many species, the relationships among which are not clearly defined, there is no way of knowing how the individual cycles will behave.

In the case of Jensen's fluctuations I would say that it is difficult to establish whether they are fluctuations due to physical causes or to biological causes. Of course, both exist in nature, but in most of the cases observed in nature it is difficult or even impossible to establish how much is due to one and how much to the other.

However, Jensen cites your research and acknowledges the existence of fluctuations due to biological factors of coexistence.

The majority of French biologist defended a point of view explicitly anti-darwinist, but there was in Paris a small group of biologist-mathematicians who was going to play an important role in developing and encouraging mathematical studies in biology[43]. The leader of this group was Georges Teissier; others included his younger colleague Philippe L'Héritier (1906–1994) and Vladimir Kostizin, who in 1926 had moved to Paris with his wife, the parasitologist Julie Kostitzin. The three of them had a mathematical background (Teissier and L'Héritier were former students at the École Normale) and had subsequently turned to biological problems. Teissier was subdirector of the Station Biologique de Roscoff; L'Héritier completed his studies with a Rockefeller Fellowship in the United States in 1931–32. After his return, he and Teissier began experimental research in genetics, creating a new experimental technique, the population cage.

subsequently the first woman to be appointed as a professor at the Paris Faculty of Pharmacy, and an interview with her at her Paris home, on February 1992.

[43] See MAYR 1980; BUICAN 1984.

As for Kostitzin, he was a mathematician and geophysicist; he had been very active in the revolutionary movements in Russia from the beginning of the century, and after the October Revolution he became an important member of the scientific establishment. He was very attracted by Vernadsky's global approach to natural geological and biological processes, and his emphasis on the importance of quantitative studies of life and its relationships with biogeochemical cycles of transformation of matter on Earth. By the way, Vernadsky had many scientific ties with Paris: his wellknown lectures on geochemistry at the Sorbonne in 1922–23, invited by Vice-chancellor Paul Appell (1855–1930), where published in 1924 in the series "Nouvelle collection scientifique" directed by the mathematician Émile Borel (1871–1956)[44]. After leaving Moscow, Julie Kostitzin began to collaborate with the Laboratory of Anatomy and Comparative Histology at the Sorbonne, while her husband, who was in touch with several Paris mathematicians, sought an academic position. It seems that Kostitzin organized this kind of voluntary exile when the first symptoms of the politization of the scientific milieu began to appear[45]. The damage produced by the atmosphere of suspicion, by censure and purges, to individual scientists as well as the whole scientific community of the Soviet Union thenceforth is wellknown, particularly in the case of biological sciences. But not even the apparently aseptic mathematical sciences escaped from this plague: as an example, in the late 1920s and the beginning of the 1930s a society of materialist mathematicians was organized at Leningrad, in opposition to the old Moscow mathematical society, considered as hostile to proletarian interests[46]. Eventually, Gause himself had to give up his biomathematical research because he was suspected of antipatriotic links with foreign researchers.

In Paris Kostitzin continued his research on mathematical physics, and became interested in biomathematics. In 1931 he published his first biological paper, in collaboration with his wife, on the statistical analysis of parasitism (J. KOSTITZIN, V. A. KOSTITZIN 1931), and in 1932 he published a note on the application of differential equations to a geological problem (V. A. KOSTITZIN 1932). Both notes are representative of the subjects developed by Kostitzin in years 1930–1935, and are the core of two biomathematical monographs published in that period. In 1933 the series "Actualités scientifiques et industrielles", published in Paris by Hermann, began a new subseries, "Exposés de biométrie et de statistique biologique", directed by Georges Teissier. The first issue, Teissier's *Dysharmonies et discontinuités dans la croissance* (TEISSIER 1984), offered some elementary examples of the utility of quantitative biological studies; next issues were, in 1934, Kostizin's *Symbiose, parasitisme et évolution (étude mathématique)* (V. A. KOSTITZIN 1934a), L'Héritier's *Génétique et évolution, analyse de quelques études mathématiques sur la selection naturelle* (L'HÉRITIER 1934) and Lotka's *Théorie analytique des associations biologiques* (LOTKA 1934). In 1934 a new

---

44 VERNADSKY 1924. See POLUNIN, GRINEVALD 1988.

45 See ISRAEL, MILLÁN GASCA 1993 and SCUDO, ZIEGLER 1976.

46 Among the most important events of this period were the arrest and deportation of D. Egorov, president of the Moscow mathematical society, in 1930, and the lawsuit against the renowned mathematician Nikolai N. Luzin (1883–1950), in 1936.

subseries on ecology directed by Marcel Prenant (1893–1983) started with the monograph *Adaptation, écologie et biocœnotique,* written by Prenant himself (PRENANT 1934), who mentioned recent biomathematical studies on the interactions between living beings and emphasized the importance of this approach together with experimental studies. Indeed, Paris enjoyed in the 1930s a lively and dynamic scientific atmosphere, open to the exploration of new ideas. Many single biologists, and specially many mathematicians, showed a real interest in the mathematical study of biological, geological or social phenomena, and in the link between this study and the problem of evolution[47].

Volterra was soon asked to prepare a monograph for Teissier's series. D'Ancona did the writing (VOLTERRA, D'ANCONA 1935). At the beginning of 1934 the first draft of *Les associations biologiques au point de vue mathématique* was ready, an abridged version of the 1931 book with almost no mathematical detail. D'Ancona had explained his conception of such a book, which would be useful to mathematicians and to biologists too, already in 1929. He had pointed out the central issue of describing the fit of statistical data with theoretical results (Cat. no. 84):

> It would perhaps be appropriate to choose from the literature some concrete examples containing statistical data and try to apply your calculations to them to see whether, by applying them, it can be established, for example, in nature, whether they are fluctuations of an interspecific nature or else due to environmental factors. Because it can never be excluded a priori for examples in nature that fluctuations may be due to periodical variations in the environmental conditions. How could such an investigation be made if a statistical data set referring to a given number of years was available? I consider this to be the first practical problem to solve.

Brelot had a much more rigid idea of separation of mathematical research from qualitative studies of biological associations. He applied this point of view to the writing of the 1931 book, particularly to the introduction and conclusion. He even defined D'Ancona participation in the book an "apparent collaboration of a professor of natural sciences" (Cat. no. 28)[48] and reacted to D'Ancona's insistence on the need for an organic exposition of mathematical results and biological facts, showing surprise that "the greatest interest and the greatest part of the chapter *'conclusion'* of a *mathematical* book consists of a historical exposition and a bibliography of *biological* works" (Cat. no. 42)[49]. A mediation was finally found, and biological comments by D'Ancona were added to the conclusion, but in such a way that the difference between mathematical conclusions and biological considerations was clear. Brelot, who as a mathematician was close to the group Bourbaki, adopted this new approach to the classical problem of the relation between mathematical theory and applications. But Volterra did not share Brelot's

---

[47] Jacques Hadamard (1865–1963), Borel and many other French mathematicians were well aware of the development of biomathematics; as an amusing example, Kostitzin told Volterra about a problem of "mathematical sociology" suggested by Henri Lebesgue (1875–1941) in an informal meeting (see Cat. no. 209).

[48] Brelot included several personal additions to the final version of the *Leçons*, and published independently some of his own results: BRELOT 1931, 1932.

[49] D'Ancona's point of view is also explained in Cat. no. 92.

point of view and had not liked the final version of his Paris lectures, because it did not meet his original aim of offering a mathematical answer to biologist's problems. The problem with this book did not concern the mathematical contents but the scarce attention to the fitting of equations to biological data, beyond the bibliographical and historical survey. In March 1930 D'Ancona wrote to Volterra (Cat. no. 88):

> I am sorry that you are not satisfied with the edition of the French book. I thought that the mathematical notes had been put in by you for the benefit of mathematicians. To tell the truth I do not believe that a naturalist will find this French edition clearer than the preceding Italian notes. The results presented in the latter were comprehensible also for naturalists. Of course, we do not possess sufficient mathematical training to understand exactly how the operations were carried out. In order for naturalists to understand how the operations were carried out, it would be necessary to ensure they were preceded by the grounds for calculations and this would turn the note into a treatise. Or else everything would have to be presented in a more elementary form, which I do not think is possible.
>
> Therefore perhaps the way you did it in Italian was the best of all. If anything, some footnotes could be provided to explain more clearly a few points or a mathematical concept that was beyond us.
>
> But the way Mr Brelot has gone about it seems to me not to be the best way to make the note accessible to naturalists. However, we naturalists generally have only a very elementary mathematical training and therefore I do not think that many of us can apply or follow this kind of calculation.
>
> The majority of us can make use of mathematics by simply applying ready-made formulae as is done, if I am not mistaken, also in statistics. Therefore, in order to encourage the dissemination of the paper among naturalists, it would be advisable in this case to pinpoint some equations to be applied to individual cases and to give examples of such applications. It might be a good idea to do this later. For example, apply your calculations to Chapman's numerical data and see whether the theoretical results can be made to fit the empirical data of population increase.

The point of view summarized in this letter was the basis of the new book. However, D'Ancona emphasized the practical instrumental aspect, whereas Volterra was increasingly interested in theoretical aspects regarding experimental data. Moreover, as we shall see, D'Ancona and Volterra's views regarding the fit of theoretical results with experience would gradually diverge. During year 1934 and the first half of 1935 many revisions and additions were introduced in D'Ancona's first draft, mostly in connection with experimental results, obtained by Chapman, Gause, and Régnier.

1.5 Régnier's Experiments

Régnier experiments introduced new aspects in the study of vital competition, which were typical of the microbiological species studied. In June 1933 he announced his first experiments (Cat. no. 284):

> [...] the research in which you are directly interested has been under way for two months now. We had however to spend a few more weeks on acquiring sufficient skills in counting two microbial species developing simultaneously.

Finally, for the time being, experiments continue as far as possible on the the first case envisaged, i.e. species competing for the same food. As a start I chose B. Coli and Staphylococcus, which are morphologically dissimilar. However, these experiments, the results of which I shall send you as soon as possible, will certainly be insufficient. Here, indeed, the question is more complex than elsewhere, as in a bounded environment, multiplication comes to a halt by itself, and the products released somehow or other by one of the microbes could hinder the development of the other. I thus have to solve these two problems without delay, which will certainly require a prolonged effort.

This was the starting point of the study of microbial antagonism, as Régnier named it. He deduced the existence of this antagonism by comparison between the mixed culture of *Bacillus coli* and *Staphylococcus aureus* and the separate cultures. Many technical problems had to be faced: for example, the difficulties involved in the individuals reckoning, due to the lack of appropriate instruments; and the problem of providing constant renewal of the culture medium. In fact, the first numerical data gathered by Régnier and his collaborator Suzanne Lambin had to do with development in a medium not continuously renewed (Cat. no. 285):

I am enclosing the preliminary results of the tests carried out at your suggestion. The first two tests clearly show the inhibiting power of B. Coli on Staphylococcus multiplication. As I said, these are obviously only preliminary tests; nevertheless, I hope that these few figures can give you some interesting indications.

I have classified this (hitherto comparatively little studied) antagonism, Coli-Staphylococcus, among the group of antagonisms between species competing for the same food. I still have to prove this. To do so, I shall pursue my research in a different direction, particularly in continuously renewed media.

Régnier and Lambin's experimental data did not fit any of the theoretical cases, and therefore, demand a reformulation of the equations. D'Ancona was rather pessimistic (Cat. no. 104):

I have examined Mr Régnier's tables. From the bacteriological point of view they are interesting in themselves and I think they are worth publishing as a preliminary note. However, as proof of your laws, the elements contributed by Régnier's experiments are too few and such as to provide scant evidence. Rather than two species competing for the same food, they seem more like mutually harmful species. Perhaps also here some fluctuations will appear, but it would be necessary to develop the media further. However, I realize that this is not easy and I fear that a proper demonstration will be difficult to achieve because of the experimental difficulties. In any case, however, any results obtained by Régnier will always be interesting and he should go on.

After some months Régnier and Lambin were able to produce accurate experimental curves corresponding to the development of each single species in separate or mixed cultures, in order to establish comparisons and deduce the effect of coexistence. They prepared a paper on the *Bacillus coli*'s inhibitory action on the *Staphylococcus*'s development and asked for a mathematical analysis (a theoretical justification of their bacteriological study) to be presented to the *Comptes Rendus de l'Académie des Sciences* (Cat. no. 286). Volterra and D'Ancona discussed these results, which disagreed with the standard logistic model and with Volterra's competition in the case of separate growth analysis for mixed culture (Cat. no. 115):

> I see that Régnier's research points to action harmful to Staphylococcus and not to Bacillus. Régnier claims that it cannot be a case of one species feeding on the other and this seems correct to me. However, I do not know whether it is simply a case of two species competing for the same food. In this case coexistence could be harmful to both species. As Régnier points out it is rather a harmful action by Bacillus against Staphylococcus. Conceivably the former produces catabolic or toxic substances that damage the latter.
>
> The trend in the development curves of the individual species is similar to that of the curves already produced by Régnier, one of which has been reproduced. I believe that in this note Régnier describes a case that had already been dealt with in the manuscript tables he sent us previously.
>
> The S type growth of a population has been verified experimentally by Pearl; nevertheless, all the cases considered referred to a population living in a uniform environment that was however gradually being saturated with individuals that had a constantly decreasing quantity of food available. An S curve occurs when the space and food decrease with an increasing number of individuals; but the quality of the medium and the food remain unchanged.
>
> On the other hand, during bacterial culture in which the medium is not renewed there is a modification of the terrain which acts both as an environment surrounding the bacterium, at least in part, and as food for it. What modifications occur is difficult to ascertain; certainly the catabolic products of the bacteria, which are also toxic, accumulate in the culture medium. It is also possible that changes occur in the colloidal nature and chemical composition of the medium. All these conditions are difficult to ascertain but are definitely revealed by the modifications in culture growth. This is a known fact in bacteriology and indeed, in order to cultivate bacteria, it is necessary to transplant them periodically, that is, the medium has to be renewed.
>
> The two cases should therefore be kept separate: 1) that in which available food and space decrease, but the quality of the medium and food remain unchanged, 2) that in which, as well as the reduction in space and food, there is also a change in the medium and the food.
>
> To the first case we apply the considerations you made previously about the S curve, and to the second the considerations you are now making and the curve corresponding to Régnier's empirical one.
>
> It will be interesting to see the results obtained by Régnier for non renewed media. It will probably be in the form of an S curve.

Thus Volterra prepared a note on the effect of the accumulation of toxic material in the culture in the numerical growth of a single species. He added to the differential form of the logistic equation a new integral term to describe this hindrance to population growth. He obtained the following integro-differential equation for the population $N(t)$[50]:

$$\frac{\mathrm{d}N(t)}{\mathrm{d}t} = \left(\varepsilon - hN(t) - \int_0^t N(t) f(t-\tau)\,\mathrm{d}\tau\right) N(t) \tag{6}$$

ε, h, constants.

50 VOLTERRA 1934. Volterra's note was published after RÉGNIER, LAMBIN 1934.

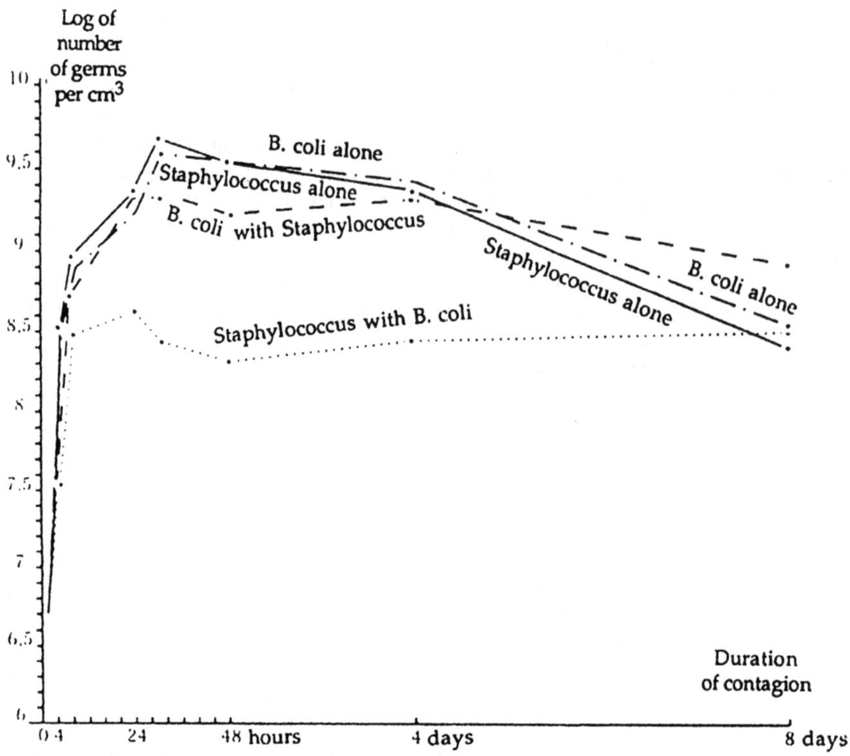

Figure 2 Growth of *Bacillus coli* and *Staphylococcus aureus* alone or in the presence of each other after Régnier and Lambin (Source: VOLTERRA, D'ANCONA 1995, p. 104)

Kostitzin called f the "function of residual action" (V. A. KOSTITZIN 1937, p. 67). Volterra called the term $hN(t)$ the "Pearl effect" or the "Verhulst-Pearl effect" (that is, the modification of the growth coefficient due to the increase of the number of individuals as in Verhulst and Pearl's logistic curve) and the integral term the "effect of intoxication". In his note to the *Comptes Rendus* he solved the equation of pure intoxication (without Pearl effect), taking f as a constant. He extended his analysis to the case of two species living together, that is, Régnier's microbial antagonism. He claimed that it was possible to calculate approximately both coefficients of toxic effect of a single species towards the other using data from the growth curves but he did not do it. Subsequently this example of curve was introduced in the new book, in chapters regarding the development of a single species and two species[51]. Volterra did develop his analysis of the case of a

[51] See VOLTERRA, D'ANCONA 1935, pp. 21–23, 32–33. Volterra had already used integro-differential equations in another biomathematical problem. In fact he had introduced a modification of the basic equations where the predator population growth was not an immediate effect of the available food (encounters with preys) but depended on the food available in the past (Ibid., pp. 51–53). He named it the "hereditary case", by analogy with the hereditary mechanics, and

single species, because the theoretical symmetric curve corresponding to the effect of intoxication alone (see fig. 3, curve B) did not fit Régnier and Lambin experimental curve for the separate cultures of *Bacillus* or *Staphylococcus*, which declined gently from the maximum (see figure 2). He succeeded in presenting an approximate solution to the general case, the superposition of the Pearl effect and intoxication, see figure 3, curve C)[52].

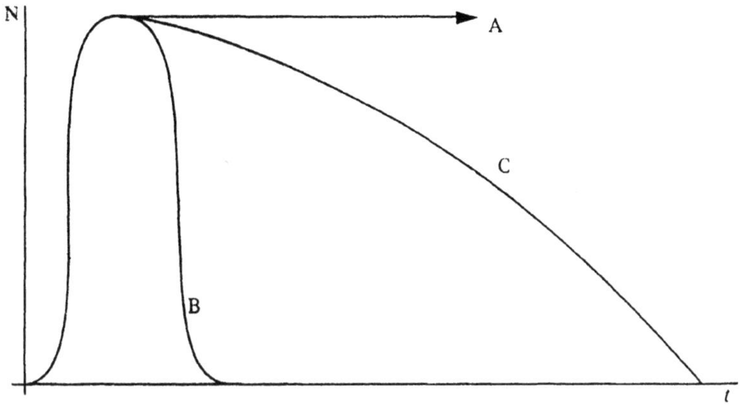

Figure 3 Population growth theoretical curves: A. Logistic curve; B Pure intoxication curve; C Mixed Pearl's and constant intoxication curve (Source: VOLTERRA, D'ANCONA 1995, p. 24)

Although Régnier appreciated Volterra's work, he was not convinced that D'Ancona's had made the right biological interpretation of his data. He thought of many other possibilities, besides the one suggested by D'Ancona, that could modify the growth rate of the culture (Cat. no. 288):

> Although I cannot fully understand the mathematical reasoning behind what you say, it seems to be based on the idea that the law governing the growth in the number of bacteria is not dependent on the depletion of food but on the release of substances harmful to microbial growth which is an integral part of microbial life itself.
>
> That, I think, is the idea that tends to emerge. However, this idea still does not satisfy me entirely. Indeed, in some experiments that still have to be verified, in which I showed from the outset the maximum allowable number of microbes per cm^3, I found no perceptible growth. In any case the number of bacteria is far from having doubled, as would theoretically have been expected, it was merely a question of exhaustion of food supplies or the release of harmful *metabolic* by-products.
>
> If these experiments are confirmed, we would have to abandon the idea of viewing the multiplication of microscopic beings as being as simple as the multiplication of the other living beings studied so far. It would probably be necessary to take into account

later on he spoke about "historical actions" to avoid misunderstandings with the biological concept of inheritance. See VOLTERRA 1939a.

52 Curves in figure 2 were drawn by D'Ancona on the base of Régnier and Lambin data and in figure 3 on the base of qualitative considerations concerning the theoretical curves, under Volterra's direction. Discussions and drafts of these illustrations can be found in the correspondence.

physical forces that play a very important role at their level: surface tension and above all electrical charges. It is highly likely that a bacterium needs a certain amount of free space around it and that, starting from a given concentration, which varies according to the bacterium and the medium, multiplication ceases.

Several years ago, in my laboratory, Madame Kaplan showed that bacteria seem very sensitive to the number of neighbours they have and that, in the same volume of nutritive medium, they govern their multiplication much more slowly than if they were more numerous. Furthermore, this is true from the beginning of their growth, namely, for the time in which the effect of food (its depletion or the release of waste) has not had time to take effect.

There may perhaps be other kinds of physical influences, e.g. radiation?

Perhaps there is a further contribution by the bacterium of *spreadable* substances that are toxic for itself (toxins...)?

However this may be, it certainly seems that it will be necessary to solve this first problem before we can really understand what goes on in the case of microbial antagonism. In the first place, it would seem necessary to find out why Staphylococcus was harmed by Bacillus coli, since the latter is not affected by Staphylococcus. In this particular case, nutritive substances may play a role. One of my pupils, Mr Léger, has installed in our laboratory some apparatus that allows a constant delivery of nutritive matter and the continuous removal of the waste, without affecting the number of microbes. It will be extremely interesting to see what happens to their development in these new conditions: rate, maximum number, decrease in number, etc.

Later, Volterra introduced Régnier to Jean Perrin's laboratory, where he began experimental research in collaboration with Nine Choucroun on physical factors in bacteria development, such as the action of metallic ions and electrical charges (Cat. no. 291).

1.6 Does Mathematics Need a Biological Justification?

In this way, Volterra became interested in the problem of the demographic development of a single species, and specially in Pearl's contributions on this case. As we have noticed, he had never mentioned Pearl's work in his publications until then, and also the first comment in D'Ancona-Volterra correspondence came in late 1934, after the publication of Gause's book on the "struggle for existence"[53]. Gause had written to Volterra to announce the publication of his book, and after a few months he published also a volume in Teissier's series (Gause 1935). He was able to illuminate Lotka's and Volterra's mathematical results of vital competition thanks to a more careful ecological treatment, including the key concept of niche, and a deep mathematical analysis of the nonlinear equations involved. Volterra was enthusiastic about Gause's work, judging from the review that he wrote for *Scientia* at the request of Gause himself; but he seemed even more interested in Pearl's preface to Gause first book, in which the biomathematical studies on the struggle for life were placed in the context of evolutionary studies[54].

---

[53] Gause 1934. In Cat. no. 115, Pearl's studies on the development of a single species, and a book of his, probably R. Pearl 1925, are mentioned.

[54] Volterra 1936f. See Cat. no. 181.

Gause's experiments suggested to Volterra an inquiry into immigration currents in a prey-predator association. This was published in Volterra's and D'Ancona's book of 1935, as well as an analysis of the development of a single species, taking into account the age of individuals, which was suggested by Chapman's experiments with *Tribolium confusum*. Volterra attached great importance to the research carried out by the biologists which were in touch with him: Régnier, Gause, and Chapman. He even suggested to devote a whole chapter to Régnier's and Gause's results. D'Ancona was not so enthusiastic. In January 1935 he wrote (Cat. no. 118):

> In the light of what you have told me of Régnier's and Gause's work I am not sure that it would be of any use to add a full chapter devoted to these two authors. I consider it somewhat exaggerated. Their new work has not added much to their already cited works. They are basically re-elaborations of things we have already cited. It will certainly be appropriate to update the chapter citing also the new works and expanding a few figures where necessary.
>
> However I consider that your work is important in itself as a theoretical approach to the problem of coexistence, if necessary even independently of any experimental verification of its laws. Of course, the value of the latter will increase as soon as they have been empirically verified. However, it does not seem to me that either Régnier or Gause have demonstrated the essential part of your treatment, namely that fluctuations are caused by coexistence. Everything they have found can be accounted for by your theory, although it cannot be said that it is demonstrated by the work of Régnier and Gause. Therefore I think that if you attribute too much worth to these works you will end up by diminishing your own. I deem it inappropriate that you should link your theory to anyone else's experimental work. As soon as a definite experimental elaboration is available it should be exploited. For the time being it would seem that only a number of attempts, mostly in agreement, exist and can appropriately be cited but that it is better not to overestimate.
>
> In any case if you are thinking of making a more extensive elaboration of the work of these two authors I am quite happy to do it, but it seems exaggerated to devote an entire chapter to them.

One week after writing this letter, D'Ancona found new ground for disagreeing with Régnier, who had begun his experiments in renewed media (Cat. no. 119)[55]:

> I have seen the figure of Régnier's new experiment. It is indeed a similar curve to Pearl's. I do not understand why Régnier does not accept this likeness. Pearl's curve does not depend on this condition or on some others. If you remember correctly, I predicted that in the case of non renewed media a Pearl curve would be the result. This is in agreement with everything you have written so far. Therefore, in the chapter on the growth of a single species these new studies by Régnier can be taken into account.
>
> In correcting the proofs I will take Régnier's and Gause's works into consideration and add what is necessary.
>
> I see from you previous letter that you consider me to be looking jealously at the work of these two authors and that I am trying to place my own research in the limelight. Far from it. I would be delighted if precise experimental demonstrations existed for your mathematical theories. In my opinion Régnier's research does not prove your laws; it is mainly concerned with the growth of a single species and this was something

[55] See also Cat. no. 287 and note 17.

that was already known. You can account for this growth through your theory but the latter is not demonstrated by the above experiments because the study of a single species was only the starting point of your treatment. The fact of finding it agrees with Régnier's experiments is no more conclusive that finding that it agrees with the preceding ones of Pearl.

With regard to the coexistence of two species Régnier appears not to have come up with very conclusive results.

Gause's experiments seem more interesting as they refer to two coexisting species. Nevertheless, they too (at least from what I have seen so far) do not demonstrate that fluctuation is an effect of coexistence.

Of course, my observations on fishery in the Upper Adriatic should provide stronger support for your theory as they ought to prove an essential point in it, the 3rd law. Unfortunately, however, my statistical observations may be interpreted in the way I did, but also in a different way, as Pearson, Bodenheimer and Gause think. Therefore, I too acknowledge that there is some truth in the criticism. My observations can be interpreted in the sense of your theory but this is not an undeniable fact, only an interpretation.

This is what I meant. You should in no way think that I want to diminish the experimental research on which your theory rests. However, I think one must be very cautious about accepting this experimental research as proof. If it is accepted too readily it could be refuted by someone else.

In all these issues your theory is in no way affected. It is a theory that follows a logical and plausible approach, and is supported by a large number of known and plausible data. Therefore, it can be considered a working hypothesis that can be used as a source of new investigations and that holds even though it is not supported by empirical data. Of course, that latter could confer greater authority upon it. However, one must be cautious about accepting this evidence, and make sure that it is reliable and conclusive. Otherwise it is better not to link your theory to an experimental basis, which is certainly less solid than the theory itself.

I apologize for writing to you at such length, but I was very interested in getting your opinion on what I am saying. I appreciate what Régnier and Gause have done, as they certainly worked seriously. However, it would mean weakening your theory to seek too much support from them, and even from me. What emerges from all our work is only partial confirmation.

D'Ancona had evolved towards a more modern "modellistic" point of view, resembling Brelot's, in which mathematical theory was considered an independent development to be subjected only to a rather soft likelihood control. The utility of mathematics was not questioned, and neither subordinate to a direct empirical verification. Gause's research program seemed meaningless or useless from this point of view. After some weeks, D'Ancona wrote on Chapman's research a more favourable opinion (Cat. no. 133)[56]:

I am returning to you also Chapman's work. They seem to have been done very well. They confirm his previous findings, but also make fresh contributions. He finds that two fluctuations build up during coexistence in a homogeneous medium containing larvae and adults, in which the latter eat the former. It is true that the latter are not always clearly apparent in his diagrams, although I think he draws the correct conclusion. Furthermore, he seems to thereby demonstrate that in a homogeneous

[56] See Cat. no. 372, where Volterra mentioned CHAPMAN 1933 and stated: "Mr. Chapman verified the laws of fluctuation that I gave".

medium fluctuations due to a relationship of coexistence take place, thus proving his second law.

Stanley had published several mathematical papers on *Tribolium confusum*, considering three phases of individual growth (adults, larvae and eggs) and the fact that adults could feed on younger individuals; and he couched his analysis in differential equations (STANLEY 1931, 1932b). His results inspired a study, included in *Les associations biologiques*, of an ideal population made up of two different types, both containing adults and young, in two different cases: first, when the groups competed for the same food, and second, when one was a predator and the other was his prey. The first case was represented by plaice *(Pleuronectes platessa)* in the North Sea, during the Great War (VOLTERRA, D'ANCONA 1935, p. 73):

> The research carried out on this fish species by Danish (Petersen, Johansen and Smith), German (Heincke, Mielk, Bückmann, Henking) and British (Borley, Thursby-Pelham) scientists has revealed that, during the war, the percentage of larger and older individuals increased. On the other hand, there was a slackening in the rate of growth of individuals. These facts were interpreted as follows: before the war the intensification of fishing activities led to a reduction in population; the number of older individuals was much reduced and this had made it easier for younger individuals to find food and thus grow more quickly. During the war the suspension of fishing activities allowed the older individuals to survive and thus made it more difficult for the younger ones to find food and grow.

After all this reelaborations, Volterra and D'Ancona's monograph had a much more biological flavour than the *Leçons* of 1931: many concrete examples of the struggle for life were drawn up, and the methodological problem of the interaction between theory and biological facts was adumbrated. The approach was much more Volterra's than D'Ancona's. In the first, introductory chapter, the utility of mathematical methods in biological research was explained, together with the need for comparisons of mathematical results with biological evidence. Moreover, deductive biomathematical studies were distinguished from quantitative experimental research. Biomathematics aimed at obtaining biological laws by deduction, but should be aware of the danger of branching off from biological facts (VOLTERRA, D'ANCONA 1935, pp. 8–9):

> In order to avoid this danger it is necessary to start off from premises grounded as firmly as possible on experience. Furthermore, as the theory develops, the results, as they become available, must be compared with indications of experience in order to verify their conformity or at least their probability.
>
> On the other hand, one must not be too concerned if ideal elements are conceived and if one places oneself in ideal conditions which do not correspond entirely to natural elements or conditions. This is a necessity and it is sufficient to recall the applications of mathematics to mechanics and to physics which led to such important and even practically useful results. In rational mechanics and mathematical physics one in fact assumes the existence of frictionless surfaces, absolutely flexible and inextensible wires, perfect gases, etc.[...]
>
> Indeed as far as biological associations are concerned, mathematical calculus, developed in this way, allows us to formulate laws and perceive relations that can be verified. In the following pages we shall see to what extent the laws formulated in this

way agree with the experimental data available to us. They could be considered as a good start on experimental demonstration.

This is why the solution to the problems of the coexistence between different species, although necessarily having to be inferred mainly from observation and experience, can find firm support in mathematical calculation and prediction, thanks to which the experimental research can be directed and the results of experiment and observation can be more clearly interpreted.

Among the simplifications Volterra employed was ignoring age or other characteristic of individuals (which were supposed invariable in time), and periodicity of births and deaths. References to biological evidence and examples of the different theoretical cases remained at the level of pure qualitative description (including curves profiles): no effective calculation or explicit comparisons with empirical data for single theoretical results – D'Ancona's target – were included. Nevertheless, the book ends stressing the empirical validity and the relevance of biomathematical results. The tone may be seen in their discussion of other's further interpretation of D'Ancona data (extended until 1932) and work. Egon Pearson and Bodenheimer's criticism was presented (VOLTERRA, D'ANCONA 1935, p. 81). Here is what they say about criticisms put forward by the Soviet researcher Aleksej N. Sewertzoff (1866–1936) (VOLTERRA, D'ANCONA 1935, pp. 83–84):

[Sewertzoff], who studied the increase of mammals and birds in the wild state, believes that Volterra's considerations do not agree with all the cases found in nature. He points out that: 1) in biophage and phytophage species fluctuations do not occur even though they are wild carnivorous or herbivorous species; at best, this hypothesis is justified in the case of foot and mouth disease in which the sick or infection spreading animals play a destructive role; 2) the fluctuation curves are not symmetrical in the herbivores, the ascending branch being much longer than the descending one; 3) to a certain point there is a degree of synchronization between the increase or decrease of predatory animals, on the one hand, and the variation in the number of the species on which they prey, on the other.

To tell the truth we are unable to ascertain whether these objections are well grounded. It is a known fact that, in nature, fluctuations occur among both carnivorous and herbivorous animals. That these fluctuations can be caused also by phenomena related to their common existences is obvious after what has been said. The fact that the fluctuation curves are not symmetrical in no way contradicts Volterra's claims; quite the contrary, the asymmetry of the curves emerges clearly from the equations. As far as the synchronization between the curves referring to the predators and those referring to the prey is concerned, has it actually been proved, or was it not simply an isolated case where such a synchronism was purely accidental?

Sewertzoff nevertheless accepts Volterra's third law, which he considers to be exact, although continuing to believe that Volterra's premises are not justified.

The following fact was emphasized: biologists had tried to identify a periodicity in populations fluctuations, but fluctuations often were actually irregular, as Volterra's laws for *n* species associations foresaw. Biomathematical studies by other authors, as Brelot, Kostitzin, and the Italians Elena Freda (1890–1978) and Beppo Levi (1875–1961), were mentioned. Mathematization of population dynamics and regulation as practiced at the University of Sidney by the physicist Victor A. Bailey (1895–1964) and the biologist Alexander J. Nicholson

(1895–1969) came in the following comment (VOLTERRA, D'ANCONA 1935, pp. 82–83)[57]:

> Bailey also subjects to mathematical treatment the relations between two species, of which one is the host and the other the parasite. He discusses Volterra's general equations which he considers to contain a lack of exactitude because (as illustrated earlier), for the sake of simplicity, no account is taken of the animals' age. Bailey believes that a state of equilibrium is not possible when the number of parasite species is lower than that of the host species. When the association consists solely of two species the state of equilibrium at a first approximation is unstable.
>
> Bailey's criticism of Volterra's equations can be rebutted by acknowledging the difference between the case treated by Bailey and that imagined by Volterra. In Bailey's case the parasite lays its eggs inside the body of the individuals of the other species. Consequently, if the latter disappears, the individuals of the other species cannot hatch. But this is not what is implied in Volterra's hypothesis, in which coefficient ε_1 is not the coefficient of birth but the difference between the coefficient of birth and that of death: as a result, ε_1 can even be zero and birth could take place just the same: it would be sufficient for the deaths to make up for the births.

Shortly after Volterra and D'Ancona's monograph, Gause's *Vérifications expérimentales de la théorie mathématique de la lutte pour la vie* was published in Teissier's series. Gause considered the whole matter from a very general point of view. He mentioned the analogy observed by the Dutch engineer Balthazar Van der Pol (1899–1959) between the equations of electrical oscillations he had obtained in 1925 and the equations of the struggle for life. Van der Pol had pointed out a mathematical analogy among these phenomena – and several others listed by him – and had in this way anticipated the modern point of view of mathematical modelling. Gause, as Volterra or Lotka, was very prudent as regards to these formal analogies[58].

Gause introduced more elements from the biological problem in the mathematical formulation in order to improve its verisimilitude. In the case of two species competing for the same food, he distinguished species which share the ecological "niche" (food, habits, and way of life) from those that do not. He obtained Volterra-Lotka equations for particular conditions of the coefficients in the case of species sharing the same ecological niche. These new considerations allowed him to design more careful and precise experiments corresponding to the different theoretical cases: for example, for the Volterra-Lotka model, a mixed culture of two very close species of protozoa, *Paramecium caudatum* and *Paramecium aurelia* was used.

As to the prey-predator classical case, Gause observed again that Lotka-Volterra's description could only fit a very primitive association, since only chance governed the reciprocal actions between species. Thus, statistical molecular physics worked but no specific adaptations of predators, which could intensify theirs effectiveness, could be taken into account. In this way, he questioned the principle of encounters on which the whole Volterra-Lotka theory was based. Gause mentioned George W. Salt's experiments, which showed that parasites did

---

[57] See BAILEY 1931, NICHOLSON 1933, and NICHOLSON, BAILEY 1935.

[58] Gause 1935, pp. 9–10. See Israel 1993b, 1998.

not follow the "blind" behaviour of lifeless particles, but could distinguish healthy from infected hosts and avoid latter. He also observed (GAUSE 1935, pp. 42–43):

> Although Lotka and Volterra introduced this principle in the equations of their mathematical theory, they were fully aware that it could not be justified in practice. Lotka thus wrote in 1925 that many, and perhaps all, organisms possess some power of selection and are to some extent independent of pure chance. This introduces an extremely peculiar complication into the mechanisms governing living beings, a complication which does not exist in the statistical dynamics of molecular physics. Not only is the living organism capable of accomplishing at the macroscopic level achievements similar to those that in the world of molecules are allowed to figments of the imagination, such as Maxwell's demon, but this power "to beat chance", as it were, is possessed to varying degrees by the different living organisms, and the mechanisms underlying the systems containing living matter must necessarily take into account, in addition to this capacity, also the degree to which this capacity is possessed, as it plays an important part in the determination of the place occupied by the various biological species in the scale of evolution.
>
> Taking this idea even further, Lotka points out that the senses represent a means for defeating chance. Their function is to replace chance with choice, to replace random attack with deliberate collision.

Gause presented an example of a primitive Lotka-Volterra prey-predator system, where classical fluctuations were observable: a population of *Paramecium bursaria,* which fed on the small cells of yeast *Saccharomyces exigus*. However, in other cases a variety of adaptations occurred as the association evolved, so that the effectiveness of the predators increased to the point that it could cause the extinction of the victims. He noticed the appearance of what Van der Pol called "relaxation oscillations".

1.7 Kostitzin on Intoxication and the Logistic Curve

In the last and most troubled years of his life, from 1935 until his death on October 11, 1940, Volterra devoted much energy to biomathematics. In 1936 he published a new group of theoretical developments, independent of experimental data, regarding what he called the analytical mechanics of biological associations. A systematic exposition of his new results appeared under the title "Principes de biologie mathématique" in *Acta Biotheoretica*. Closer to experimental observations, particularly Régnier's, were two papers on the logistic curve and population growth published in Pearl's journal *Human Biology* (VOLTERRA 1938b, VOLTERRA 1939b). The starting point of the first of them was a remark made by D'Ancona on the possibility of explaining the demographic evolution of a single population applying the method of encounters to mating[59].

Volterra continued to attend to Régnier and Lambin's experimental work, now in collaboration with Vladimir Kostitzin, who in this period became his

[59] See Cat. no. 160, 367, 161. Volterra met Pearl in London in November 1937, and decided to send the paper to his journal. Kostitzin pointed out (Cat. no. 215) that the basic idea of this paper has already been advanced by Léon Brillouin in BRILLOUIN 1935, and by himself in V. A. KOSTITZIN 1937b.

main interlocutor on biomathematics. Already in his first book, which concentrated on symbiosis and parasitism (V. A. KOSTITZIN 1934), Kostitzin had noticed that the study of variations of vital coefficients and the possible stable states for biological associations could throw light on the problem of evolution. In his second monograph for Teissier's series, he had drawn the general geophysical and geochemical background of the problem of life in the manner of Vernadsky, for which mathematical analysis was also suitable (V. A. KOSTITZIN 1935a). In 1935 he began to work out a general exposition of his ideas, starting from the mathematical analysis of the cycles of organic matter, oxygen and nitrogen, and explaining then recent research on population growth. He took into account the influence of environment and interactions among species, as well as the development of the organism, all intended as part of the general process of evolution (V. A. KOSTIZIN 1937b)[60]. He shared with Volterra the idea of translating into biology the classical approach of mathematical physics, although he also favoured Lotka's energetic approach, which came from chemical physics and included the link between entropy and evolution.

Kostitzin read with great attention Volterra's analysis of the integrodifferential equation of a species intoxicated with products of individuals chemical activity. In August 1935 he wrote to Volterra communicating the exact solution of equation (6); he also gave the formula of the maximum value reached by a population before initiating the decrease toward extinction. Following Volterra's advice, he presented a note on his result to the *Académie des Sciences*[61]. With this success in hand, Kostitzin looked for further biological applications of integral and integrodifferential equations (Cat. no. 192):

> I am currently writing a book on mathematical biology for the Colin series and, as I write it, I am jotting down remarks and interesting observations for further development. In this way I have noticed that the term integral (hereditary in the mathematical sense) must be introduced in equations referring to the growth of organisms.
>
> This seems perfectly appropriate as far as ageing is concerned, and I have succeeded in demonstrating it in the case of embryonic growth. Perhaps in a few weeks time I shall be asking you to present one or two further notes to the Academy of Sciences.

In late 1935 Régnier contacted Kostitizin, following Volterra's advice (Cat. no. 294, 194). In his *Biologie mathématique* Kostitzin compared theoretical calculations and experimental data for *Bacillus coli* and *Staphylococcus*. He deduced from this comparison the values of vital coefficients in equation (6), using his exact solution: for example, for the *Bacillus*, $\varepsilon = 1.14$, $c = 0.023$ and $h \le 0.003$, where $c = f(t)$ in equation (6). As to the agreement of theoretical and experimental values of N in Régnier's experiment, which lasted 192 hours, he noticed that it was very good at the beginning and the end, and bad between 8

[60] An English translation appeared two years later: V.A. KOSTITZIN 1939a. The book has been more recently translated into Japanese by the Institute of Population Problems: V. A. KOSTITZIN 1981.

[61] Cat. no. 191. The mathematical content of the letter is reproduced without modifications in V. A. KOSTITZIN 1935b.

and 24 hours. He explained this disagreement by ignored factors, like lack of food and space[62].

Régnier decided to prepare with Suzanne Lambin a long memoir presenting their research on microbial antagonism in the context of microbiological studies as well as in that of quantitative studies on the life of cells, organisms, and populations. But Volterra's differential equations of competition with the effect of intoxication remained unsolved (Cat. no. 297)[63]:

> [...] I have just finished our article, giving a longer treatment to the work of Verhulst... Deschamps... which are related to the growth of populations and organisms. I have also extended your views on the association of two species competing for the same food. Lastly I have considered *Gause's* recent work (single niche, different niche...) and demonstrated that the case of antagonism among microbes must be included in the third category examined by Gause, that is in exactly the same conditions as those defined by you. As Staphylococcus is far from being completely expelled by Bacillus coli, the obvious conclusion, which in any case could be seen *a priori*, is that antagonism among microbes is dependent on a larger number of factors than antagonism among other species, for instance, animals. You yourself have, to boot, taken these several factors perfectly into account in your equations (Verhulst-Pearl effect, auto-intoxication of one of the species, intoxication of one species by the other...), and it is regrettable only that your equations should necessarily have become so complicated that they are difficult to solve.
>
> I have continued to emphasize the fact that the Volterra-Kostitzin curve is superimposable on the experimental curve in the case of a pure species and have discussed more fully the implications of this superimposition, as well as illustrating the services that this superimposition could provide in the future. There is no doubt, in fact, that by bringing the two curves as close together as possible quantitatively (the qualitative similarity has already been established) we will obtain for the equations values of the action of the different factors. It will be enough, subsequently, find the experimental value (all the other factors having been eliminated if possible) of an action in some hypothesis or another so that, after comparison, one or other of the hypotheses can be eliminated. As soon as I can, I shall therefore resume my study of microbial multiplication in special conditions: continuously renewed medium, etc. Thus continuing hitherto unpublished experiments.
>
> However, all these additions and discussions have lengthened our article enormously. It will be further increased by the discussion of recent work: *Ludwig, Richards....* that Mademoiselle Lambin is currently translating. So much so that I am not at all sure that Monsieur Mesnil will be able to accept it for Annales de l'Institut Pasteur. If not, I shall submit it to other journals that are not afraid of long articles, or perhaps I shall even offer it to Georges Teissier, whom I have met, for the biological statistics series in which you published your work together with Mr d'Ancona.

The work on this paper continued in 1937, and its publication was also delayed because Régnier expected that Kostitzin and Volterra would be able to enlighten mathematically the central point, that is, competition in the particular conditions of a microbiological culture. These continuous delays carried as a consequence

[62] V. A. Kostitzin 1937b, pp. 66–72; see Israel, Millán Gasca 1993.

[63] His first reference to this long paper is in Cat. no. 292. W. Ludwig and Oscar W. Richards, together with Gause, have carried out pioneering studies on the microorganisms population growth, in particular for protozoa and yeasts: Ludwig 1928, Richards 1928.

that, with the beginning of the war – and the destruction of Régnier's laboratory – the monograph was never published[64].

Moreover, despite all the talent mobilized to fit for the case of a single species, major discrepancies persisted. In 1937 Kostitzin introduced a further generalization of equation (6), in order to preclude the theoretical result of intoxicated populations exhaustion in an infinite lapse of time[65]. In cases of very intense multiplication or chemical activity, Kostitzin claimed, the average intrinsical growth could be considered as proportional to a power p^k of the number p of individuals of a population, and factors p^l, p^m, should be introduced in the equation (6). Thus, the following equation resulted:

$$\frac{\mathrm{d}p(t)}{\mathrm{d}t} = \varepsilon p^k - hp^{k+l} - cp^k \int_0^t p^m(\tau)\,\mathrm{d}\tau \ , \quad e, h, c = \text{constants} \qquad (7)$$

Here again Kostitzin gave a formula for the maximum values reached by the population.

In July 1938 Lambin sent to Volterra new data on the influence of the quantity of nutritive substance on *Bacillus coli* multiplication (Cat. no. 252). The French biologists have drawn an S-shaped curve of maxima reached by populations growing in media with different quantities of nutritious substances. Volterra was in very low spirits: he was 78 years old, had many health problems and faced the dramatic political situation that led to the enacting of the racial laws in that same year. But he discussed the new data with D'Ancona and put forward an analysis based on Kostitzin generalization (7) of his theoretical curve for intoxication. If Kostitzin's equations fitted Régnier's experimental phenomenon, then the dependence of the maxima on the concentration of nutritious substance should be a logistic curve, asymptotic to a parallel of the axis of abscissa, when the abscissa – representing the concentration of nutritious substance – tended to infinity. But Volterra found that the maxima tended to infinity (Cat. no. 402):

> The theoretical maxima curve in which only intoxication is taken into account cannot be used to verify your curve. It must therefore be concluded that the trend of the maxima plot cannot be explained in terms of intoxication. There must be other causes which produce your nice curve and they must be found. Is it space, as Mr D'Ancona suspects? Forgive me for having so little to say. It is already a great deal for the state of mind that I am in.

[64] In Cat. no. 300, he asked Volterra to prepare a preface for the monograph, as he had done for Kostitzin's book; after two weeks, he insisted upon the need to complete the research (Cat. no. 301). S. Lambin conserved a first draft of the first page and the introduction of the planned work. The title was: *Évolution des cultures microbiennes. Vie des microbes et phénomènes généraux caractérisant la vie des populations;* and the subtitle: *Dénombrement des microbes, Multiplication et variation des germes. Antagonisme entre bactéries. Rapprochement entre les données numériques caractérisant la vie des microbes et celles qui caractérisent la vie des cellules, des organismes et des populations.* The contents of the work were presented in the introduction, where the interest of mathematical treatment and of experimental information provided by microorganisms was emphasized.

[65] V. A. Kostitzin 1937a. See Cat. no. 205.

After Kostitzin examined Régnier's data (Cat. no. 310), he was convinced that the curve of intoxication could be used and suggested to consider carefully the dependence of the coefficients on the concentration q of peptone (Cat. no. 218, 219). He developed some numerical calculations, that allowed him to improve his idea of the dependence on the nutritive substance (Cat. no. 220):

> I have made a few preliminary calculations which show right from the start that one must not be presumptuous: in fact, in my previous letter I said that the coefficients h and c vary only slowly as a function of the concentration q, and the following table shows that, while this is true for c, it is quite incorrect for h. This coefficient decreases rapidly, but (and this is something important) it flattens out at about 0.003 for high concentrations of peptone. It is clear a priori that competition weakens when food becomes abundant although I find the rapidity of this decrease astonishing. The coefficient of multiplication ε increase from simple to quadruple, while the coefficient of toxic action c decreases from double to simple, which is quite normal, as food makes [the] organism more resistant. All the coefficients have a distinctly asymptotic trend and tend towards finite limits. It is odd that the coefficients c and ε computed according to Mr Régnier's early experiments (abundant food regime) were 0.023 and 1.14, respectively –which shows a very interesting agreement (see my book p. 70–71). After all, my calculations fully confirm the point of view set out in my previous letter and based on the existence of an upper limit of food assimilation.

So Kostitzin had found an agreement between the vital coefficients in the first case of abundant food. He prepared a note, including a numerical table, which was published under the names of Kostitzin and Volterra (V. A. KOSTITZIN, VOLTERRA 1938), after a note by Régnier and Lambin (RÉGNIER, LAMBIN 1938) – even if Volterra wanted to publish it under the name of Kostitzin alone. Kostitzin tried to improve the theoretical description of functions $\varepsilon(q)$, $h(q)$, $c(q)$, which replaced the former constant values when considering the concentration of peptone q. And he told Volterra about new results which rested on new experiments by Régnier (Cat. no. 219)[66].

But all these projects were dramatically aborted. The final group of letters of 1938–1940 shows an interesting collaborative effort of three scientists of different but complementary backgrounds, which was darkened by dramatic events in Europe. Kostitzin and Régnier sent often to Volterra messages of support and sympathy. A few months later Volterra died at his house of Ariccia, near Rome, in complete isolation. In June 1941 Kostitzin was interned in the Compiègne civil prisoners camp; after being released in 1943, he took part in the Resistance and, wanted by the Gestapo, together with his wife, he had to hide for the last nine months of the war. Régnier saw the destruction of his laboratory and died in 1946.

[66] In later letters, Kostitzin considered the general problem of the integrodifferential equation (6) (Cat. no. 222, 224, 225). See V. A. KOSTITZIN 1939b.

1.8 Mathematics, Biology and Evolution

Volterra evolved from an interest in the biological sciences, which pivoted on D'Ancona, to a much deeper concern about the biological aspects of his research: not only he was more and more attentive to biological literature regarding the fitting of his equations to empirical evidence, but he paid also attention to different approaches and contributions to quantitative and mathematical biology and, above all, to their relevance to the theory of evolution. In his 1936 review of Gause's work in *Scientia,* Volterra underlined Pearl's foreword to Gause's *The Struggle for Life,* namely Pearl's views on current changing attitudes in the scientific world as to the theory of evolution, related specially to new trends in genetics, to the development of statistical population studies and to the conviction that the struggle for life and natural selection fell within demographic dynamics (VOLTERRA 1936f). In 1901, in his first text on the application of mathematics to non-physical sciences, Volterra had spoken about biometrics as a "first step" of a suitable new field of mathematical biology, intended as the theoretical counterpart of experimental biology. He considered biometrics as preliminary because it was based on probability, not on a solid and reliable branch of mathematics as infinitesimal analysis: this point of view rested on Volterra's mechanist philosophy of science (VOLTERRA 1901). By the late 1930s, however, he was emphasizing biometrics as a rational study of the mechanisms of evolution. Here is an example, from his lecture at the international meeting of mathematicians held at Paris in July 1937 (VOLTERRA 1938a):

> It is after 1900 that the statistical and mathematical research applied to biology was intensified. The work was in statistics and it was precisely in 1900 that a distinguished British mathematician, Karl Pearson, founded the journal "Biometrica" which rendered sterling service to science. It was Pearson who recognized that the problems raised by the theories of evolution, transformism and natural selection, should be considered from the demographic point of view. However, he was not immediately understood by the scientists. It took some time to convince them of this.
>
> According to Pearl, one of the greatest living biostatisticians, the theories in question had seemed to be on their deathbed several years earlier.
>
> However, the ideas of evolution, the struggle for existence, had initially aroused great interest. They were the object of innumerable papers which exceeded everything else that had been written on topical issues.
>
> However, despite the contributions by distinguished authors, the papers and experiments made on this subject over more than half a century had not led to any significant results. This is the opinion expressed by Pearl who adds that this is the new path followed by genetics and that it is the new studies on populations and demographic dynamics that have revived transformism and the struggle for life, showing that these theories retain a surprising vitality.
>
> At the present time, these studies are making constant progress both experimentally and in practice, as well as from both the mathematical and the theoretical points of view. Ross, and in particular Lotka, Elton, Gause, Thompson and numerous others must be mentioned for their calculations, their experiments, their material struggle against insects harmful to agriculture.
>
> Among the more recent publications I would like to mention the excellent book by Mr. Kostitzin in the Colin collection.

D'Ancona did not share Volterra's conviction that mathematical population dynamics was important for evolutionary studies, or, indeed, that the then current form of the theory of evolution was sustainable. Thus, in a letter of February 1935, when discussing the opinion of the French biologist Maurice Caullery (CAULLERY 1931, 1935), he wrote (Cat. no. 122):

> With regard to evolution, I do not believe that any zoologist today can objectively claim to be Darwinist. This phase is now behind us. You can be an evolutionist but no longer a Darwinist.
>
> Evolution itself has now entered the sphere of experimentation and only this can be a basis for discussion. In actual fact, the results obtained so far are not very promising, and so the skepticism of the majority of zoologists is quite justified.
>
> Nevertheless evolution may still be acknowledged as the only plausible hypothesis available to us and until we have something better it will have to be admitted as a working hypothesis.
>
> But nothing more. Of course progress will continue to be made and new hypotheses will be formulated that can provide a better account of the facts. Evolution will probably come back in another form. But I don't think it will be a Darwinian form as this would be a retrograde step.
>
> In any case one cannot claim to be a Darwinist or its opposite: this expression has become non-scientific. It is only at the experimental level that the discussion can be carried on and this is quite different from what it was at the time of Darwin.

And in a further letter, in connection with Pearl's views, D'Ancona made clear where he thought the problem lay (Cat. no. 125):

> You ask me whether I agree with Pearl's foreword? Not completely. Not that I wish to detract from Darwin's extraordinary merit. His theory is extremely ingenious and has made the greatest contribution ever done to zoological research. There is no doubt that much of his research, in addition to the stimulus it gave, has remained, among other things also in the specific case of the theory of the struggle for survival as a function of the economy of nature. But that the struggle for survival should have been important for evolution is something no one believes in any more. Not even Darwin would believe in it if he returned to the world of the living.

Caullery's views on the limits of Darwinian natural selection were widespread in France[67]. Teissier, L'Héritier and Kostitzin were among the few scientists working in France which took part in the intellectual trend that would lead to the so-called evolutionary synthesis. It can be interesting to consider here the explanation given by L'Héritier to Ernst Mayr when asked on the reason of his, and Teissiers's, acceptance of Darwinism (MAYR 1980, p. 322):

> L'Héritier told me that he and Teissier had no difficulty in accepting the Darwinian interpretation, particularly the noninheritance of acquired characters and the principle of natural selection, because as members of a mathematics department they had not been exposed to the prevailing French ideas and natural selection was an eminently reasonable process for a mathematician.

It is easy to imagine that also for Volterra, as a mathematician, natural selection was an eminently reasonable, or rational process. After a long parenthesis in D'Ancona-Volterra correspondence between late 1935 and the first months of

67 See BOESIGER 1980 and LIMOGES 1980.

1936, Volterra resumed the discussion, and received from his son-in-law a reaffirmation of scepticism (Cat. no. 139):

> I read with interest what you wrote in your last letter. However, I think that Caullery's opinion, which in any case is not his alone but reflects that of the vast majority of zoologists throughout the world, is perfectly objective. Of course many are unwilling to give up such a fascinating and apparently perfectly logical and justified theory. Furthermore, in England, also partly for reasons of national sentiment, they still take Darwin's words as proven fact.
>
> What should instead be acknowledged is the fact that, from 1859 to the present day, above all thanks to Darwin, zoological knowledge has made great strides. What at the time appeared to be securely documented no longer seems so certain today. Therefore we must seek new evidence, new facts. Darwin's theory cannot be resuscitated without support from new facts. The old facts have been exhaustively examined and discussed. No new ones are yet forthcoming.
>
> At the present time the only form of evolution that has benefited from a certain experimental demonstration is that of mutation. But this does not prove Darwinian evolution.
>
> I am pleased to hear you are reading Jennings' book. It is not the best in the world. In particular, it is not a popularizing work as it assumes some knowledge of the subject. However, I would not know what else to recommend.

Volterra and D'Ancona did not collaborate again when Ludwig von Bertalanffy (1901–1972) invited them in early 1937 to prepare a new book for his series (Cat. no. 150). Strangely enough, on March 18, 1937, D'Ancona wrote that he was happy to know that Volterra agreed to von Bertalanffy's proposal; but after three days, he had to accept Volterra's drastic refusal to appear as a co-author (Cat. no. 152, 153). In spite of von Bertalanffy's insistence, the book was finally written by D'Ancona alone (D'ANCONA 1939) and Volterra did not even contribute a preface (Cat. no. 155, 157 and 166). However this change of opinion could not be traced back to the scientific disagreements between Volterra and D'Ancona. A careful reading of the correspondence suggests that it was related to the fact that the book was to be published in Germany (Berlin), and that was clearly inacceptable for Volterra for political and "racial" reasons.

Soon another opportunity for disagreement between Volterra and D'Ancona arose around the foundations of Darwinism. It concerned a scientific documentary on the struggle for life sponsored by the mathematical section of the new Palais de la Découverte. Kostitzin worked on the screenplay; he submitted it to Volterra, who sent it to D'Ancona, who did not spare any criticism (Cat. no. 151):

> I am returning to you the film script on the struggle for life. I have read it and am horrified. If the intent was to produce a parody it would be acceptable, but certainly not something serious.
>
> From fluctuations you jump to Mendel's laws and mutations and then on to the evolution of horses. Enough to make your hair stand on end. It is one of those examples of popularization that have discredited the theory of evolution. However, fortunately these days this type of popularization is no longer fashionable and everyone is convinced that if you want to talk of evolution it is necessary to be more cautious and serious. Mr Kostitzin has obviously not progressed beyond the immediate post-Darwinian period.

For my part I believe that anyone associating with such a work runs the risk of appearing ridiculous. I beg you therefore not to include my name and I advise you not even to allow your own to be used.

Your research on biological associations are not a suitable topic for popularization. To assert that a big fish eats the smaller one is something obvious that everyone has been familiar with for a long time. The merit of having put interspecific relations into concrete form is yours, but these relations cannot be translated into elementary expressions, and it is even more difficult to express them in the banal form of a cartoon. You end up being grotesque, as happened to Mr Kostitzin.

What I do believe is that this research, which is something serious, must be kept within the bounds of serious investigation and not delivered up to superficial persons.

I apologize for writing in such a categorical style but I would like to convince you of the true significance of this mess caused by Mr Kostitzin.

As it can be deduced from D'Ancona's following letter, Volterra defended the idea of the film and pointed out Kostitzin's scientific qualities, which gave D'Ancona another opportunity to repeat his criticism (Cat. no. 152):

With regard to Kostitzin I am sorry to say I do not agree with you. I believe that the topic does not lend itself to cinematographic popularization without becoming trite and stating the obvious. In my opinion this is what Kostitzin has done; by mixing together Mendel, mutations and the evolution of horses, he made a ridiculous mess. I thus do not consider it proper to associate serious scientific research with such a grotesque parody.

There is no denying that Kostitzin is a good mathematician, but it would seem that he does not go beyond popularization as far as biology is concerned, of which, in the present case, he does not seem to have assimilated the subject matter.

I am not familiar with him and so I do not consider it proper to make a precise critique of his script. For my part I would say that this kind of popularization is inappropriate.

D'Ancona's disagreement hit what for Kostitzin were the central issues. On the one hand, Kostitzin presented vital competition, as well as the counter-phenomena of mutual assistance derived from social and moral instincts, in the framework of a general process, the relentless struggle for survival, that took place in the natural world. On the other hand, he presented this struggle, along with genetic processes, as the basic mechanisms of selection or elimination that governed evolution. To consider together the genetics of populations and population dynamics was an advanced idea, though only roughly formulated. Kostitzin found it difficult to display these complex views by means of the cinematographical medium (Cat. no. 204):

The script is certainly defective for it is extremely difficult to illustrate a set of complicated problems in the short space of 10–12 minutes; it is even more difficult to regulate this mixing so that everything is well balanced. It nevertheless seems that the Mendelian issue, which in any case takes only 40 seconds, is properly treated since, from the transformist point of view, the destiny of the mutated groups appearing in a mother population is at least as important as the destiny of the associations of non parentally related groups. Therefore, if it is possible to show that also here there exists an elimination mechanism, all is well. As far as the mutual assistance among species is concerned, this problem was much debated by Darwin himself, who, as a highly conformist Englishman, made every effort to defend his theories against any criticism of immorality. One of the British Darwinists, who was more consequential than Darwin,

> – Sutherland – wrote a very interesting paper on moral instinct and social instinct (in this connection, the fact of reading this work when I was at high school in Smolensk cost me four Sundays' punishment). It does not appear possible to complete the script with even a summary illustration of these extremely complicated problems although one could add a few images to show mutual assistance among species and include a few phrases to explain the biological importance of this phenomenon. Tomorrow I shall see Mr Dufour – a science film director – to whom I shall mention this.

Before reaching its definite form, the film "Images mathématiques de la lutte pour la vie" underwent many revisions[68]. Jean Painlevé (1902–1989), a pioneer of scientific cinema and director of the French Institut de Cinematographie Scientifique, engaged actively in it. The final version presented a survey of biomathematical studies; summarized the most important examples of population growth, for a single population, or for biological associations of two or three species; animated mathematical graphics illustrated theoretical ideas, while laboratory or field images explained the biological examples related to the different cases. The film examined the probability that different "results" of struggle for life might be realized in nature. It concluded that, from the mathematical viewpoint, probability of stable or cyclical equilibria was low and decreased when the number of species increased, that is, in modern parlance, when the complexity of the system became higher.

The mathematical analysis of natural selection was the goal of Kostitzin's research program in mathematical biology. Natural selection combined with Mendelism, as in Fisher, Haldane and Wright's work, became his main research subject in the years before the war. He did not agree with most practitioners of mathematical genetics about their choice of statistical techniques. After the publication of his 1937 book *Biologie mathématique,* he wrote to Volterra (Cat. no. 205):

> My book brings me both joy and sorrow.
>
> *Joy:* I have received a proposal to write an article on mathematical biology for *Tabulae Biologicae* – a sort of German biological encyclopaedia published in the Netherlands.
>
> *Sorrow:* there are biologists (just as there are physicists) who consider the analytical method much less important than the statistical method. Without intending to disparage probabilism, I think, for instance, that the statistical study of Paris floods, accompanied by the demonstration of their Gaussian distribution, is worth much less than the study of their causes accompanied by the forecasting, even if inaccurate, of disasters.

He took the occasion of the presentation of Fisher's research in L'Héritier's book *Génétique et évolution,* to shape his objections (Cat. no. 216):

> The research of Mr. R. A. Fisher pursues the same object as yours and mine, although it differs entirely from it in that it uses a purely statistical method. He performs an enormous amount of work to obtain a tiny proportion of what the analytical method can give you in a few lines. He belongs to this species of statisticians who are willing to use the exponential function but avoid using derivatives and integrals even though the error committed in both cases is exactly the same. It appears to me that the overuse of the statistical method is a disease of our times or rather a fashion. Some time back I

[68] On the film, see Millán Gasca 1993.

received the visit of a young mathematician who wanted to learn more about population theory. Here is his method: let $f(p, q)$ be the probability that a population equal to p at time t becomes equal to q at time $(t + 1)$; this function does not depend directly on t; it is necessary to find the most probable law of population growth. You see where he got this odd idea of likening an organic process to a set of tests similar to shuffling a deck of cards. Indeed these statistical analogies make the present mathematical generation incapable of going from the molecular level to the molar level. This young man went away extremely put out by my critical remarks and he exhibits his card shuffling talents in one of the best scientific establishments in Paris. It is a pity because he is talented.

Kostitzin shared Volterra's methodological views on the mathematization of the non-physical sciences. He had an even greater interest in the problem of evolution than Volterra and remained closer to the biological facts (Cat. no. 226). Also he thought more highly of Lotka's chemical physical approach to the problem of the evolution of life[69]:

How do you expect him not to be somewhat annoyed when he sees his works, although very recent, neglected and unjustly passed over? You do not come into it, but it is true nonetheless. Lotka is perfectly well aware that he deserves more than this, and he is right. His book "Elements of physical biology" and his notes of the same period contain many interesting thoughts, remarks and considerations, as well as the germs of further research. Even if he was not the first to introduce equations of this form, he has made a nice study of them for the case of two species which contains insights into more general problems. His considerations on the absence of contradiction between the existence of living matter and the law of entropy are still not sufficiently well appreciated by biologists and natural philosophers.

And this was his reaction after reading Volterra's last biomathematical studies, in which he built an analytical mechanics of biological associations (Cat. no. 203):

I have read with great pleasure and astonishment your note on the "Principles of Mathematical Biology". My naturalist's mind does not agree with your interpretation of vital coefficients, although my mathematician's mind grasps the philosophic part of your work quite readily: the region of life phenomena that you delimit is too narrow, but within these limits you construct a rational biology similar to holonomous, canonical, rational mechanics, in which economic conditions and extremal principles are verified. These are giant's steps.

1.9 Final Remarks

In 1946, almost a year after the end of the war and nearly six years after Volterra's death, Kostitzin wrote to his widow, Virginia Almagià[70]:

You are no doubt aware that in recent years mathematical biology had fallen out of favour with biologists and statistical methods (R. A. Fisher, Haldane) were in vogue.

[69] Cat. no. 195. D'Ancona had told Volterra that Kostizin gave too much importance to Lotka and had not considered enough his contributions in his monograph V. A. Kostitzin 1934 (Cat. no. 107).

[70] Cat. no. 235. The only paper on the topics that he had studied together with Volterra and Régnier that Kostitzin published after the war was V. A. Kostitzin 1956.

> Fortunately, this situation will not last forever. Ecological studies are again bringing analytical methods to the fore. For instance, the workers at the Marseille Biological Station (main object, the Camargue region) attentively study your husband's publications and mine and are very interested in the work of Mr D'Ancona, which is unfortunately unavailable in France.

In fact, after a brief eclipse, biomathematical studies were gradually resumed in the 1950s, especially in the United States. They made use not only of the classical biometrical methods and differential equations, but also of new mathematical tools such as game theory, systems theory and information theory. The recent revival of interest in nonlinear equations has also had a great impact in the field of biomathematics. Nevertheless, with the emergence of the modellistic approach in applied mathematics, the debate on the empirical verification of mathematical theories has been muted. In ecology, for example, a bifurcation can be noticed between an ecological "engineering", which makes wide use of mathematical tools and computer simulation to solve practical problems, and theoretical ecology, which develops abstract "possible" models without revising them continuously in the light of experiments or observations. Thus, the problem persists and is far from having been solved. Robert K. May's discussion of the problem of complexity and stability in model ecosystems is an example. In 1972, May pointed out that theoretical previsions show that the stability of a system decreases when its complexity, that is, the number of involved variables or species increases; whereas in nature very complex systems, such as the Amazonian forest, are very stable, and very simple systems, such as a field of fruit trees, show great instability[71].

This profound change in the philosophy of applied mathematical research that consolidated after the World War II has also affected the attitude of biologists toward biomathematics. Theoretical proposals of mathematicians inspired a real interest in many biologists – entomologists, zoologists, microbiologists – in the 1920s and 1930s. Interpretations differed of course. Chapman or Régnier accepted classical reductionism as a ideal of scientificity and as a viable model for the application of mathematics to the life world; many others were willing to discuss the fit of mathematical description to biological facts, and consider mathematics as a possible source of methodological renovation. Today the situation is completely different. A good example is Ernst Mayr's insistence that mathematics has a little relevance to the biology of evolution[72]. Mayr's views are based on the affirmation of the autonomous epistemological status of biology, in which categories as finalism have a prominent role. A useful comparison can be made between the developments of biology and economics. Mathematical studies began in both about the same time but were much more intensively pursued in economics. In this field the classical historical method of research was gradually superseded by a mathematical approach that was thought to raise the level of scientificity and rigour of the discipline. So in the 1920s and 1930s, while Volterra was trying to extend the classical mechanist approach to biology, the modellistic entered economics and has since prospered greatly, although it has

---

[71] MAY 1972, 1974.

[72] MAYR 1982, chapter 1 (see the paragraph "Why biology is different").

paid little attention to the problem of the fit of theoretical results and models with economical actual facts[73].

While mathematics has played a most important role in economics in the second half of the 20th century, biologists and biomathematicians have largerly ignored one another. Some biomathematicians are optimistic that their discipline may soon re-establish the lost connection. Simon Levin, editor of the series "Lecture Notes in Biomathematics" has written, in his prologue to volume 100, which has ended the series (LEVIN (ed.) 1994, p. ix):

> Mathematical biology has changed a great deal in the last decade, mainly for the better. Whereas twenty years ago, it was peripheral to many subdisciplines of biology, its success is marked today by ever increasing integration into areas of application. Mathematical biologists, who once found in each other the most eager listeners and constructive critics, now see their work guiding experiments, and shaping the conceptual foundations of almost all areas of biology. Indeed, in some areas, the integration is so complete that there are no clear distinctions between theoretician and experimentalist: every respectable researcher must be a little of each. And in areas where that is not yet true, it will be so within a decade more.

Other biomathematicians consider instead the situation with disappointment. Compare last quotation with the following "bitter" reflection, written only six years before, by the authors of a book on a classic topic in biomathematics, biological oscillators (GLASS, MACKEY 1988, p. 182):

> Somehow, a myth has arisen (which we believe is accepted by the great majority of practising biologists) that detailed mathematical and theoretical analyses are not appropriate in biology. Certainly the mathematical training of most biologists and physicians is minimal. Yet, if the complex dynamic phenomena that occur in the human body were to arise in some inanimate physical system – let us say in a laser, or liquid helium, or a semiconductor– they would be subjected to the most sophisticated experimental and theoretical study.

The isolation of biomathematics from the biological audience to which they are addressed has led to a reconsideration of the place of finalism and freedom in mathematical models. Although Volterra's mechanist strong program of mathematization in ecology was given up in the 1940s, most biomathematical models retained Volterra's and Lotka's original view of biological interactions as the particle kinetics of statistical mechanics. Subsequently other trends coming from the same pioneer period have been revived. The French mathematician René Thom has explicitly acknowledged the influence of D'Arcy Thompson's thought on his research in the theory of catastrophes and his ideas about "qualitative" mathematical biology. Lotka's eclectic views, particularly his ideas on thermodynamic aspects of biological systems, have had an important influence upon scientists such as Nicholas Rashevsky, Ludwig von Bertalanffy and Ilya Prigogine.

The thermodynamic point of view has introduced non-deterministic aspects into the mathematical study of life phenomena. Recent studies on chaos had been also interpreted in this sense, that is, as a breaking with determinism. But Thom has pointed out that it is misleading to associate the idea of randomness with that of freedom, and thus probability as an alternative to determinism. This idea

---

[73] See INGRAO, ISRAEL 1990.

echoes Gause's comparison of the purely casual interactions on which biomathematical equations were based and the existence of a margin of choice in the behaviour of even very simple organisms. John Maynard Smith has brought up this issue in connection with the problem of stability and complexity in ecology mathematical models: considering the problem put forward by May, he has written that if there are in nature stable complex ecosystems, this can only be due to the strongly non arbitrary character of the interactions that take place there (SMITH 1974).

Smith has also taken into account the second great problem central to the interwar debate, that is, the oversimplification introduced by mathematical models, which exclude many aspects of a phenomenon. Moreover, the mathematician has to exclude many important aspects, if he wants to build a model, whereas the biologist likes better to consider the phenomenon in its entirety. This explains, he claims, why biologists took so long to appreciate the work of Volterra and Lotka and why today ecologists still question the need for mathematical theories in their discipline. An explanation of the drifting away of biologists from their early positive reception of Volterra's and Lotka's biomathematics may be that the problem of empirical verification of applied mathematical theories has only now come to the fore and that the intricate methodological and theoretical aspects were not discussed after the war. The formal mathematical structures need then to be "refilled of meaning" (OSTER, GUCKENHEIMER 1976, pp. 327–328):

> Biological systems tend to be considerably more complex than those studied in physics or chemistry. In analyzing models, one is frequently presented two alternatives: either resorting to brute force computer simulation or to reducing the model further via such drastic approximations as to render it biologically uninteresting. Neither alternative is attractive. Indeed, the former alternative is hardly viable for most situations in ecology since sufficient data is rarely available to quantitatively validate a model. This contrasts starkly with the physical sciences where small differences can often discriminate between competing theories. The situation is such that many ecologist seriously question whether mathematics can play any useful role in biology. Some claim that there has not yet been a single fundamental advance in biology due to mathematical theory. Where complex systems are concerned, they assert that the appropriate language is English, not mathematics. A typical attitude among biologist is that models are useful only insofar as the explain the unknown or suggest new experiments. Such models are hard to come by.
>
> In the face of such cynism, perhaps mathematicians who would dabble in biology should set themselves more modest goals. Rather than presenting the biological community with an exhaustive analysis of an interesting model, it might be better to produce a "softer" analysis of a meaningful model. From this view point, the role of mathematics is not to generate proofs, but to act as a guide to one's intuition in perceiving what nature is up to. This is no excuse for avoiding hard analysis where it can be done, but as models mimic nature more closely it becomes harder to prove theorems.

Should mathematics be assigned a pure hermeneutic role, as Oster and Guckenheimer have suggested, or a descriptive-predictive role? Mathematical modelling has given up the ambition of obtaining a universal code to substantiate mathematical descriptions. However, in the meantime a singular "perversion" of the classical views on the relationship theory-verification has emerged. Mario

Bunge has pointed out the confusion between the ontological and the epistemological levels which leads to project over reality the theoretical conclusions of mathematical modelling, obviating the need of verification (BUNGE 1979). Thus, observations or experiments are lacking, but anyway the properties of the mathematical schema and of reality are confounded. This problem, as well as the one regarding the meaning of determinism, are characteristic of the present debate and did not occur in the interwar period. A *regard en arrière,* back to the early decades of our century, recalls the need and value of a philosophical inquiry into the use of mathematics and the progress of science as knowledge of reality.

CORRESPONDENCE

2 Vito Volterra

Vito Volterra was born in Ancona on 3 May 1860, the only son of the merchant Abramo Volterra and Angelica Almagià. Volterra was only two when his father died, leaving him and his mother in dire financial straits, as a result of which they had to go and live at the house of his uncle, Alfonso Almagià, a Bank of Italy official. They moved first to Turin and then to Florence, where Volterra spent much of his youth and where he attended the Scuola Tecnica "Dante Alighieri" and later the Istituto Tecnico "Galileo Galilei". Here he had a brilliant teacher, the physicist Antonio Roiti (1843–1921), who was to have a decisive influence on his education. Through Roiti he got a job as assistant in the Physics Laboratory of the University of Florence, thus avoiding the fate of having to become a bank clerk (to make both ends meet) and thus renounce his scientific studies for which he had such an extraordinary aptitude. He had already shown signs of this aptitude at the age of 13 when, after reading the Jules Verne novel *From the Earth to the Moon*, he had tackled the calculation of the trajectory of a projectile in the gravitational fields of the Earth and the Moon, that is, a restricted version of the extremely difficult three-body problem.

After completing his studies in 1878, Volterra entered the Faculty of Mathematical, Physical and Natural Sciences of the University of Florence and two years later passed his entrance examination to the Scuola Normale Superiore of Pisa. Here his teachers were the great mathematicians Ulisse Dini (1845–1918) and Enrico Betti (1823–1892). Dini gave him a solid grounding in the field of mathematical analysis, although Dini's "rigour" did not express the view of mathematics most congenial to Volterra's temperament. He was indeed more fascinated by Betti's teaching, which influenced his scientific outlook, and in particular by one of the latter's most distinctive traits, that is his tendency to maintain close links between mathematical analysis and mathematical applications, his interest in the problems of mathematical physics and the primary role he always assigned to the latter in the determination of mathematical problems. Years after, when speaking of Betti, he said: "Those who knew Betti, not only through his work, but also through his conversation, will recall that although he used to talk Mathematics, he often thought Physics. Like one of those flashes of lightning in the night that illuminate the path ahead, a word that escaped him would suddenly reveal this natural disposition of his mind."

In 1882, Volterra graduated in physics with a thesis on hydrodynamics and immediately became Betti's assistant. In 1883, at the age of only 23, he won the competitive examination for the chair of professor extraordinary of rational mechanics at the University of Pisa and, after Betti's death, he succeeded him in the chair of mathematical physics. This was the beginning of a rapid academic career which led him first, in 1892, to the University of Turin as professor of rational mechanics and then, in 1900, to the University of Rome where he took over the chair of mathematical physics previously held by Eugenio Beltrami (1835–1899). The same year he married Virginia Almagià with whom he lived in Rome until the end of his days.

The period between 1887 and 1908 was one of great scientific production for him. He wrote more than sixty scientific papers ranging from topics of terrestrial mechanics to matters of rational mechanics, the theory of elasticity, electrodynamics, the theory of linear differential equations. But the most significant work was the introduction of the concept of "line function", which opened up the way to the study of what would later be called "functional analysis", with a study linked to the theory of integral and integro-differential equations.

In Rome, Volterra gradually became also an important public and political figure. In 1905 he was nominated Senator of the Kingdom and took part in many parliamentary debates, above all on issues related to the organization of the University and scientific research. At the outbreak of World War I he adopted a position favourable to intervention on the side of the Anglo-Franco Allies and, when Italy entered the conflict he enlisted in the engineers corps, although now aged 55. He was actively involved in calculus problems related to the firing of guns mounted on dirigibles and others related to aerial warfare. These activities were certainly at the origin of his idea to set up an Inventions and Research Bureau (initially dedicated to wartime applications) presided over by himself and which was later transformed into the Research Committee (Comitato di Ricerche), the first nucleus of the National Research Council (Consiglio Nazionale delle Ricerche), founded in 1923.

These initiatives were only one aspect of an extremely intense public activity. As early as 1897 he founded the Italian Physics Society (SIF, Società Italiana di Fisica). In 1906 he founded the Italian Society for the Progress of Science (SIPS, Società Italiana per il Progresso delle Scienze), which, following the example of similar foreign societies, set out to gather together all those (university staff, secondary school teachers, engineers, economists, technicians, etc.) who were interested in promoting forms of exchange and scientific collaboration aimed above all at establishing and disseminating national scientific culture. In 1907 he founded the Regio Comitato Talassografico Italiano, took an active part in the reorganization of the schools of engineering and promoted legislative action for the control of radioactivity and the creation of a national telegraph and telephone network. He was also appointed President of the Bureau International des Poids et Mesures. In 1887 he was elected corresponding member of the Accademia Nazionale dei Lincei and in 1899 became an ordinary member. From 1920–23 he was Vice-President of the Academy and in 1923–26 its President.

In October 1922, fascism came to power in Italy. Volterra's liberal and democratic ideas, so close to the models of the French and English-speaking societies, and which fostered an enlightened view of the progressive function of science in society, inevitably clashed with the nationalistic and autarkic authoritarianism of the regime. Not only did he oppose all legislative bills introduced by the fascist government before the Senate but, in 1925, he signed the intellectuals' manifesto against fascism drawn up by Benedetto Croce (1866–1952) and published on 1 May that year in *Il Mondo*. Moreover, in 1923, in his capacity of President of the Accademia Nazionale dei Lincei, he promoted an initiative aimed at opposing the Giovanni Gentile (1875–1944) reform of secondary school education, which he saw as the expression of a tendency to suffocate science and subordinate it to humanistic culture, engendering a dangerous conflict between the two cultures.

In 1931, when state functionaries were obliged to take an oath of loyalty to the fascist regime, Volterra was one of the thirteen university professors who refused to accept this imposition. As a consequence he was expelled from the University of Rome and, in 1932, from all Italian cultural and scientific institutions. In obvious disapproval of such persecutory measures, in 1936, the Pope Pius XI appointed him member of the Pontificia Accademia delle Scienze.

The period marked by the conflict with the fascist regime (from 1922 until his death) represented for him another important phase of his scientific production. It was starting in 1925 that Volterra embarked upon his mathematical biology research, based on the biological work of his son in law, Umberto D'Ancona, with whom he developed a relationship of scientific collaboration. These were the circumstances in which he began a new and extremely fertile scientific undertaking based on the idea of introducing into the field of population dynamics those methods that had proved their worth in mechanics and mathematical physics. However, he was now practically cut off completely from his country's scientific activity. Volterra was now one of the internationally better known scientists and was frequently invited abroad. He gave lectures and courses in Paris (one of his favourite scientific centres), in Spain, in Romania, in Belgium, in Czechoslovakia and in Switzerland. After 1931, it may be said that he spent most of his time outside Italy, except for the periods he lived in his house in Rome and above all in his villa in Ariccia (a township in the Roman castles area), where he had put together an impressive library containing thousands of ancient and modern books, first editions and a collection of his manuscripts and correspondence. These materials are now conserved partly at the "Vito Volterra" Centre of the Brandeis University (Mass., U.S.A), and partly at the Accademia Nazionale dei Lincei, as well as at the Rome National Library.

In 1938, Volterra was directly affected by the promulgation of the racial laws, as he was Jewish. This was a final, extremely harsh blow. The same year he began to suffer from phlebitis, which limited him in his movements and activities, but only partially in his scientific studies. He died in his Ariccia villa on 11 October 1940, in a state of complete isolation, amid the silence of the Italian scientific community, which was now intimidated by or at the service of the regime. The correspondence of those years (including that of his wife Virginia, in the years immediately following his death) testify to this isolation but also to the great respect and affection felt for him abroad. One of the rare obituaries announcing his death in Italy was published in the news bulletin of the Jewish community of Ancona[1].

[1] See also E. Volterra 1976 and the biographical note in V. Volterra, U. D'Ancona 1995.

3 Presentation of the Correspondence

The following transcripts contain the most significant correspondence between Volterra and a number of correspondents that is related to his research work on biomathematics. As already mentioned, this correspondence is conserved in the Volterra Archive at the Accademia Nazionale dei Lincei in Rome (cf. ISRAEL 1982a, PAOLONI 1990).

The transcription of the letters was carried out with the greatest fidelity to the text, without intervening in any way, even in the case of spelling and accent errors (more frequent in the letters written in French). Such errors are thus to be considered as authentic in the originals and not as errors of transcription or misprints in the present edition. Only in those few cases in which the interpretation of a word in the original was impossible has the illegible word been omitted and replaced by the indication [illeg.].

The full content of the letters has been transposed as the correspondence is mainly scientific in nature, except for a few short, strictly private passages which have been omitted. Such omissions actually refer solely to the correspondence between Volterra and Umberto D'Ancona.

The correspondence has been arranged in alphabetical order of correspondent, is preceded by a short relevant biographical note and contains notes providing bibliographic references to several issues and texts mentioned by the authors, as well as essential information for understanding several passages contained in the letters. We did not deem it necessary to go beyond this information, and have expressed no value judgments concerning the content of the epistolary exchange. In any case, such judgments are sometimes expressed in the introductory essay.

Some indication needs to be given concerning the numbering of the correspondence. At the time that the collection was re-ordered and archived, before being deposited at the Accademia Nazionale dei Lincei, each letter was numbered in chronological order. Reference to this numbering is essential and must not be overlooked, as it is useful in materially tracing each single letter contained in the correspondence. However, this original numbering contains several errors. In certain cases the date of the letter was not correctly identified. In other cases, although the letter is undated, and has thus been classified as such, the interpretation of the content or of other contiguous letters allows the date to be determined, albeit only approximately (year or even month). In general, the original numbering of the letters does not correspond exactly to the actual chronological order of the letters. In the present edition of the individual correspondences we have arranged them in correct or at least plausible chronological order, except in those rare cases of undated letters that were impossible to date with certainty and that have consequently been placed at the end of the correspondence. However, to facilitate the rapid retrieval of every letter contained in the original Archive we have marked the transcription with the number of the catalogue that the reader will find at the end of the book (indicated as Cat. no.). This catalogue contains all the information necessary to identify the letter's position in the Volterra Archive (see also the presentation of the catalogue for information on abbreviations used). The Accademia Nazionale dei Lincei also possesses a computerized catalogue of the Volterra collection that was completed in the late ‘eighties and represents an essential working tool.

4 Letters between Marcel Brelot and Vito Volterra

Marcel Émile Brelot was born on 29 December 1903 at Châteauneuf-sur-Loire (France). He was sickly as a child and had delicate health for the rest of his life. His parents moved to Cépoy (Montargis) where he received his secondary schooling. He won admission to the École Polytechnique and the École Normale Supérieure, and entered the latter in 1924. His fellow course members and future friends were distinguished mathematicians such as André Weil, Jean Dieudonné, Henri Cartan, Claude Chevalley. These were all to become the main constituents of the "Nicolas Bourbaki group" who proposed a radical axiomatic approach. Brelot was influenced by these trends which profoundingly innovated the French mathematical scene and his passion for the axiomatic view remained with him for the rest of his life.

Towards the end of his studies at the École Normale Supérieure he won a Rockefeller scholarship which he used in Rome in 1929–1930, under the direction of Volterra, and in Berlin, in 1930–31, under the direction of Erhard Schmidt (1876–1959). This is the period to which most of the correspondence with Volterra contained herein refers and during which he undertook the editing of Volterra's first book on biomathematics (VOLTERRA 1931a). In 1931 he presented his thesis on a topic in mathematical analysis proposed by Émile Picard (1856–1941) and he worked at the Centre National de la Recherche Scientifique and then at the Institut Français of Berlin. During the period 1933–38 he had the job of organizing courses and was then "maître de conférences" at Algeri. In 1938 he was appointed professor at Bordeaux, where he remained until 1942, then moving to Grenoble. In 1953 he was appointed "maître de conférences" and then professor in Paris where he settled for the remainder of his life, occupying the university chair until 1974.

Brelot's numerous posts include that of secretary of the French national committee of mathematicians and the chairmanship of the Société Mathématique de France which he held in 1960. He was awarded four prizes by the Académie des Sciences of Paris, to which he was elected in 1974. Brelot's more important initiatives include the transformation of the journal *Annales de Grenoble* into *Annales de l'Institut Fourier* which rose to become one of the mostly widely cited mathematical journals at the international level.

He died in Paris on 3 August 1987.

Brelot's scientific activities were concentrated in the field of mathematical analysis mathematics and are contained in over 150 publications as well as in university coursebooks and seminars. He also contributed to the mathematical theory of the struggle for life, following in the wake of Volterra's research.

However, the central argument of Brelot's research was the theory of potential, of which, at the height of his career, he was the greatest specialist. After his arrival in Paris he oriented the research of many workers in this direction, founding a "Potential Theory Seminar" which he directed until his death and the proceedings of which were systematically published. As Gustave Choquet has pointed out, he made Paris the world centre of research on the theory of potential. The topics to which he made the most important contributions were the problem of Dirichlet, the theory of capacity, studies of subharmonic functions on the boundary of an open set and, in particular, on ramified boundary. He introduced the theory of harmonic spaces named after him and provided proof for a well-known theorem of convergence for subharmonic functions.

Although he never became part of the "Bourbaki" group he was said to be more Bourbakist than the Bourbakists. In all his research he followed the axiomatic method, the application of which he attempted to extend also to fields such as rational mechanics and fluid dynamics[1].

* * *

4.1 Brelot to Volterra, Paris, 13.5.1929 *Cat. no. 13*

Paris, lundi 13 mai 1929

Monsieur et Cher Maître,

Monsieur Vessiot m'a proposé récemment de demander une bourse Rockefeller pour l'an prochain en Italie. Je serais heureux si vous acceptiez de diriger mes recherches l'année prochaine et je vous écris dès maintenant sans attendre votre retour à Paris parce il est assez urgent de préparer le dossier. Monsieur Vessiot va vous écrire pour vous demander de lui envoyer une lettre à joindre à ce dossier où vous diriez que vous acceptez.

Je vous remercie infiniment à l'avance et je profite de cette lettre pour vous entretenir de la rédaction de votre dernier cours sur les systèmes biologiques[2]. J'ai remanié et complété le premier chapitre que je vous avais montré. J'ai dans un second traité le cas de n espèces avec l'hypothèse des équivalents et ajouté une note sur les déterminants et systèmes d'équations linéaires nécessaire pour l'intelligence du texte à des gens n'ayant qu'une faible culture mathématique. Dans un troisième chapitre, j'ai introduit, dans l'étude des n espèces, l'hypothèse que les

$$\frac{1}{N_i}\frac{\mathrm{d}N_i}{\mathrm{d}t}$$

[1] See also CHOQUET 1987.

[2] This is the future volume VOLTERRA 1931a. Much of the following correspondence refers to the editing of this book. Therefore many of the topics treated can be understood more easily by referring to it directly.

dépendent des N_i du même indice, puisque ce sont des fonctions linéaires quelconques des N_r. Puis j'ai parlé des notions d'associations conservatives et dissipatives. J'ai ajouté une note mathématique sur les formes quadratiques.

Il me reste dans un ou deux chapitres supplémentaires à rédiger ce que vous avez dit avec l'hypothèse d'hérédité. J'ai travaillé beaucoup plus que je ne prévoyais et qu'on s'en apercevra sans doute. J'ai introduit diverses modifications de détail et quelques développements. Récemment j'ai longuement cherché les conditions nécessaires ou suffisantes pour qu'un système soit dissipatif et j'ai reconnu la grande complexité de la question. J'ai abouti surtout à des résultats négatifs, mais importants à signaler. Ainsi les conditions

$$\begin{cases} p_{rr} > 0 \\ p_{rs}\, p_{sr} \leq 0 \end{cases}$$

qui correspondent à la réalité ne sont pas suffisantes, comme on le voit dans l'étude complète déjà fort compliquée du cas de 3 espèces.

Quand vous viendrez, je vous expliquerai en détail les modifications d'ordre et de démonstration que j'ai cru utiles et intéressantes à faire. Je serai sans doute alors près de terminer si je ne rencontre pas trop de recherches à faire dans la partie héréditaire et si vous n'arrivez pas avant trois semaines.

Je tiens avant tout à ne rien laisser qui donne prise à quelque objection ou qui pose des questions importantes sans réponse. Je serai heureux si vous appréciez les quelques petits perfectionnements que je m'efforce d'apporter, en profitant de mon expérience de la rédaction si difficile du cours de M. Picard.

Qu'il est difficile d'arriver à une forme qui défie la critique! Je fais de mon mieux vers cet idéal inaccessible et je vous prie de croire, Monsieur et cher Maître, à mes remerciements les plus vifs pour cette possibilité de vous consulter souvent l'an prochain sur des questions d'équations intégrales, et à l'expression de tout mon respect.

M. Brelot

Marcel Brelot, Ecole N.ale S.re, 45 Rue d'Ulm, Paris V

4.2 Brelot to Volterra, Paris, 5.6.1929 *Cat. no. 14*

Parigi, 5 di giugno

Vi prego di mandare quella lettera a Prof. Vito Volterra dove esso.
Urge fare questo. – Grazie -

Università di Paris

Prière de m'excuser car j'ignore l'Italien. M. Vito Volterra est je crois en voyage et je désirerais que cette lettre lui parvienne d'urgence.

Paris, mercredi 5 Juin 1929

Monsieur,

Je vous ai écrit, ainsi que M. Vessiot, il y a quelques semaines, pour vous prier de m'envoyer une lettre où vous diriez que vous acceptez que j'aille l'an prochain, à Rome, travailler sous votre direction. Je transmettrai cette lettre indispensable à la Fondation Rokefeller, à laquelle M. Vessiot m'a engagé à postuler pour une bourse d'une année scolaire en Italie.

Je n'ai point reçu de réponse mais j'apprends que vous êtes actuellement en voyage. Comme la lettre que je vous demande est urgente, j'écris à l'Université de Rome pourqu'on fasse ce mot. Et je vous prie donc d'envoyer directement à l'adresse suivante:

M. Tisdale 20 Rue de la Baume
(Fondation Rokefeller) Paris 8

une lettre où vous expliquez que vous me connaissez et acceptez de diriger mes recherches. Ce papier est indispensable. Vous voudrez bien également indiquer à quelle date commence l'enseignement dans les Facultés (Octobre ou Novembre) en Italie.

J'achève en ce moment la rédaction du cours de l'Institut H. Poincaré. Bien qu'absorbé par ailleurs, dans d'autres travaux d'analyse, pendant un certain temps, j'ai rédigé complètement plus des 2/3 et j'étudie la fin que j'écrirai bientôt.

J'espérais vous voir fin mai à Paris. Peut-être allez vous venir bientôt? J'aurai grand plaisir à vous montrer mon travail de rédaction et parler des recherches que je pense entreprendre.

Avec toute ma reconnaissance, veuillez croire, Monsieur et cher Maître, à mes sentiments de profond respect

M. Brelot

Marcel Brelot, Ecole N.^{ale} S.^{re}, 45 Rue d'Ulm, Paris V

4.3 Volterra to W. E. Tisdale, n.p., 1929 *Cat. no. 424*[3]

Monsieur W. E. Tisdale
Rockfeller Foundation
20 Rue de la Baume
Paris (18e)

J'ai l'honneur de vous recommander M. Marcel Brelot ancien élève de l'École Normale Supérieure qui demande la concession d'une bourse de la Fondation Rockefeller.

M. Brelot montre les meilleures dispositions pour l'étude des mathématiques, une clarté d'exposition remarquable et des vues originales, ce que j'ai pu constater pendant le temps où j'ai tenu en hiver passé des conférences à l'Institut Henri Poincaré.

[3] Draft.

M. Brelot a des projets de travaux qui pourront le conduire à des résultats intéressants dans le domaine de l'analyse mathématique, c'est pourquoi s'il obtiendra la bourse Rockefeller je serai heureux de pouvoir l'accueillir à Rome, m'intéresser à ses études et les diriger.

J'ai l'espoir que si l'on facilitera les études de M. Brelot en lui assignant une bourse Rockefeller, sa future carrière scientifique correspondra au but élevé qui se propose la fondation Rockefeller.

Veuillez accepter, Monsieur, l'expression de toute mon estime et de tout mon respect.

4.4 Volterra to Brelot, Ariccia, 17.7.1929 *Cat. no. 357*[4]

VILLINO VOLTERRA
ARICCIA (ROMA)

17 Juillet 1929

Cher Monsieur

Pour démontrer l'existence des intégrales des équations (K) et (L) page 119 je propose le méthode suivante que je prends la liberté de vous transmettre très rapidement en vous priant de vouloir bien la développer dans les détails et de manière à vous assurer qu'il n'y ait pas d'inexactitudes.

Les équations (K), (L) peuvent être remplacées par

$$N_1(t) = N_1^0 + \int_0^t N_1(\xi)\left(\varepsilon_1 - \gamma_1 N_2(\xi) - \int_0^{T_0} F_1(\tau) N_2(\xi - \tau)\,\mathrm{d}\tau\right)\mathrm{d}\xi$$

$$N_2(t) = N_2^0 + \int_0^t N_2(\xi)\left(-\varepsilon_2 + \gamma_2 N_1(\xi) + \int_0^{T_0} F_2(\tau) N_1(\xi - \tau)\,\mathrm{d}\tau\right)\mathrm{d}\xi$$

où $N_1^0 = N_1(0)$, $N_2^0 = N_2(0)$ sont des quantités positives.

Il ne suffit pas de connaître N_1^0, N_2^0 pour pouvoir déterminer $N_1(t)$, $N_2(t)$ pour $t > 0$. Il faut que $N_1(t)$ et $N_2(t)$ soient connus pour $0 > t > -T_0$ (Cfr. Mémoire Journal de Math page 260) et soient des fonctions $f_1(t)$ et $f_2(t)$ finies et continues dans cette intervalle. $f_1(0) = N_1^0$, $f_2(0) = N_2^0$. Supposons $t < \theta$ et employons la méthode des approximations successives pour calculer $N_1(t)$ et $N_2(t)$ pour $t > 0$. En posant d'abord dans les seconds membres

$$N_1(t) = f_1(t)\,, \quad N_2(t) = f_2(t)\,, \quad 0 > t > -T_0$$
$$N_1(t) = N_1^0\,, \quad N_2(t) = N_2^0\,, \quad \theta > t > 0$$

on obtiendra ainsi les fonctions

[4] Draft.

$$N_1'(t) \text{ et } N_2'(t) \quad \theta > t > 0$$

et qu'on prolongera pour $0 > t > -T_0$.

Dans les fonctions $f_1(t)$ et $f_2(t)$ remplaçons N_1 et N_2 par N_1' et N_2' dans les seconds membres des ( ) et l'on obtiendra

$$N_1'' \text{ et } N_2''$$

et ainsi de suite indéfiniment.

Remplaçons maintenant l'équation (1) par

$$N_1(t) = \overline{N}_1^0 + \int_0^t N_1(\xi)\varepsilon_1 \, d\xi$$

où $\overline{N}_1^0 = M_i e^{\varepsilon_1 T_0}$, M_1 étant le maximum des valeurs de $f_1(t)$.

La solution sera

$$N_1(t) = \overline{N}_1^0 e^{\varepsilon_1 t} \text{ pour } -T_0 < t < \theta$$

Je pense que l'on peut dire que

$$\overline{N}_1^0 e^{\varepsilon_1 t} > N_1^{(h)}(t) \quad h = 0,1,2,3,\ldots$$

De même remplaçons l'équation (2) d'abord par

$$N_2(t) = \overline{N}_2^0 + \int_0^t N_2(\xi)\left(\gamma_1 N_1^0 e^{\varepsilon_1 \xi} + \int_0^{T_0} F_1(\tau)\overline{N}_1^0 e^{\varepsilon_1(\xi-\tau)} \, d\tau\right) d\xi$$

où $\overline{N}_2^0 = M_2$ maximum des valeurs de $f_2(t)$, et après par

$$N_2(t) = \overline{N}_2^0 + \int_0^t N_2(\xi)\left(\gamma_1 N_1^0 e^{\varepsilon_1 \xi} + \int_0^{T_0} F_1(\tau)\overline{N}_1^0 e^{\varepsilon_1 \xi} \, d\tau\right) d\xi$$

dont la solution sera

$$N_2(t) = \overline{N}_2^0 e^{\frac{\gamma_2 + \Gamma_2}{\varepsilon_1} \overline{N}_1^0 e^{\varepsilon_1 t}}$$

Je pense que l'on peut dire

$$\overline{N}_2^0 e^{\frac{\gamma_2 + \Gamma_2}{\varepsilon_1} \overline{N}_1^0 e^{\varepsilon_1 t}} > N_2^{(h)}(t) \quad h = 1,2,\ldots$$

et par suite

$$N_1^{(h)}(t) < \overline{N}_1^0 e^{\varepsilon_1 \theta} = A \quad (h = 1,2,3,\ldots)$$

$$\overline{N}_2^{(h)}(t) < \overline{N}_2^0 \, e^{\frac{\gamma_2+\Gamma_2}{\varepsilon_1}\overline{N}_1^0 e^{\varepsilon_1 A}} = B$$

Supposons maintenant

$$\left| N_1^{(h+1)}(t) - N_1^{(h)}(t) \right| < a\frac{t^h}{h!}$$

$$\left| N_2^{(h+1)}(t) - N_2^{(h)}(t) \right| < b\frac{t^h}{h!}$$

a, b étant des nombres positifs.

On aura

$$\left| N_1^{(h+2)}(t) - N_1^{(h+1)}(t) \right| < \int_0^t \left(a\frac{\xi^h}{h!}\alpha + \beta\frac{\xi^h}{h!} \right) d\xi = (a\alpha + bA)\frac{\xi^{h+1}}{(h+1)!}$$

$$\alpha = \varepsilon_1 + (\gamma_1 + \Gamma_1)B$$

On tire de là la convergence de

$$N_1', N_1'', \ldots$$

De même on pourra démontrer la convergence de

$$N_2', N_2'', \ldots$$

Ce que semble prouver l'existence des solution des équations ( ) et ( ) dans toute intervalle 0 θ quelque soit θ.

Si cette démonstration est exacte on pourrait l'intercaler dans le §1 de le 4ème partie en simplifiant ce qu'on y trouve déjà.

4.5 Volterra to Brelot, Ariccia, 19.7.1929 *Cat. no. 358*[5]

VILLINO VOLTERRA *19 Juillet 1929*
ARICCIA (ROMA)

Cher Monsieur

Dans ma lettre du 18 ct. je me suis trompé en vous disant que les équations sur lesquelles il faut employer le procédé des approximations successives sont les équations (1) et (2). Les équations sur lesquelles il faut employer ce procédé sont les équations équivalentes

$$\text{(A)} \quad N_1(t) = N_1^0 \, e^{\int_0^t \left(\varepsilon_1 - \gamma_1 N_2(\xi) - \int_0^{T_0} F_1(\tau) N_2(\xi-\tau)\, d\tau \right) d\xi}$$

[5] Draft.

$$\text{(B)}\quad N_2(t) = N_2^0\, e^{\int_0^t \left(-\varepsilon_2 - \gamma_2 N_1(\xi) - \int_0^{T_0} F_2(\tau) N_1(\xi-\tau)\, d\tau\right) d\xi}$$

Alors les approximations successives

$$N_1^0,\, N_1',\, N_1'',\ldots,N_1^{(h)}$$

résultent toutes positives, ainsi que

$$N_2^0,\, N_2',\, N_2'',\ldots,N_2^{(h)}$$

Les équations par lesquelles on les majore sont les mêmes que dans ma dernière lettre, c'est à dire

$$N_1(t) = \overline{N}_1^0\, e^{\varepsilon_1 t}\,, \quad N_2(t) = \overline{N}_2^0\, e^{\frac{\gamma_2+\Gamma_2}{\varepsilon_1} \overline{N}_1^0\, \varepsilon^{\varepsilon_1 t}}$$

et alors il me semble qu'il n'y ait pas de difficulté à reconnaître que

$$N_1^{(h)}(t) < \overline{N}_1^0\, \varepsilon^{\varepsilon_1 t} < \overline{N}_1^0\, e^{\varepsilon_1 \theta} = A < C$$

$$N_2^{(h)}(t) < \overline{N}_2^0\, \varepsilon^{\frac{\gamma_2+\Gamma_2}{\varepsilon_1} \overline{N}_1^0\, e^{\varepsilon_1 t}} < \overline{N}_2^0\, \varepsilon^{\frac{\gamma_2+\Gamma_2}{\varepsilon_1} \overline{N}_1^0\, e^{\varepsilon_1 \theta}} = B < C$$

C étant une quantité plus grande que A et B.

4.6 Brelot to Volterra, Paris, 22.7.1929 *Cat. no. 15*

Paris, lundi 22 Juillet 1929

Cher Monsieur,

J'ai reçu votre lettre contenant la démonstration, le petit mémoire, et la lettre rectificative. Je n'ai point lu encore votre démonstration. Je m'y mettrai sous peu.

J'ai dû, pour rendre grand service à M. Julia, travailler pour lui sans avoir pu achever la rédaction de votre cours. J'avais entrepris une mise au point d'un cours sur la représentation conforme en collaboration avec un professeur de Louvain: mais celui ci n'étant que fort peu analyste n'a à peu près rien apporté comme travail personnel et j'ai dû passer un temps assez long à rectifier tout seul une foule de détails parfois pénibles. D'autre part, M. Julia m'avait demandé de rédiger son cours de cette année; j'ai pris un collaborateur, un de mes amis, d'ailleurs plus versé que moi dans ces questions de fonctions analytiques, de théorème de Schwarz et quantité d'inégalités; cependant je n'ai pas tout à fait fini ma part de travail car j'ai été dérangé par des examens que j'ai contribué à faire passer, et par les fêtes du 14 juillet. M. Julia est fort pressé parce qu'il veut être sûr que le livre sera paru au début de l'an prochain afin que dans son cours du 2° semestre il puisse supposer connus les résultats établis cette année, en grande partie élémentaires. Je n'ai pu lui refuser cela et c'est pourquoi j'ai délaissé la rédaction presque achevée de votre cours. Mais, d'ici une semaine, je m'y remettrai et j'achèverai en moins de 15 jours car elle est presque finie. J'étais même ennuyé de ne pouvoir liquider ce cours fin Juin et être obligé de m'interrompre un mois.

La fameuse démonstration en litige, je tâcherai de la mettre sur pied et je l'écrirai dans une note mathématique de la fin de dernier chapitre et non dans le texte même qu'elle alourdirait. D'ici quinze jours je pense donc vous envoyer la fin des mémoires et quelques jours après les figures.

J'ai obtenu la bourse Rockefeller et je vous remercie beaucoup. Peut être ces vacances ferai je une note sur des recherches à propos de cours de M. Picard dont je corrige actuellement les épreuves – cause de retard également.

Si donc je n'ai pas d'ennui avec l'autorité militaire j'irai sûrement en Italie fin Octobre. Je l'espère bien.

Dans une quinzaine, donc.

Croyez à mes sentiments les plus respectueux

M. Brelot, 45 Rue d'Ulm, Paris V

4.7 Brelot to Volterra, Boisseaux, 12.8.1929 *Cat. no. 16*

Boisseaux (Loret), lundi 12 août 1929

Cher Monsieur

Depuis quelques jours seulement, je suis débarrassé du travail que je faisais pour M. Julia: collaboration à la rédaction de deux de ses cours, l'un sur la représentation conforme, l'autre tout récent sur "quelques principes géométriques de l'Analyse" et qui contient surtout des généralisations du lemme de Schwarz sur les fonctions holomorphes dans le cercle unité et nulles à l'origine.

M. Julia désirant que ce soit publié très vite afin que dans son prochain cours, il puisse renvoyer a cet ouvrage, sans le reprendre en grande partie, j'ai dû, depuis la fin de Juin, abandonner complètement la rédaction de votre cours presque achevée pourtant.

J'en suis désolé, mais je le serais moins si vous n'étiez pas trop mécontent. Car ce ne sera qu'un retard d'un mois ou deux.

D'ailleurs, d'après ce que m'a dit M. Julia sur les habitudes de Gauthier-Villars, cette maison ne prend, ou du moins ne met en chantier aucun manuscrit de Juillet à Octobre. Peut-être donc que mon retard n'aura pas été cause d'un retard pour la publication.

Depuis quelques jours je me suis remis à l'étude de votre cours et j'ai d'abord apporté quelques modifications suivant vos indications. J'ai ajouté en note votre démonstration p. 84 (mémoire italien) ou du moins quelque chose de voisin; et c'est un peu plus simple que ce que j'avais fait sans avoir lu ce passage.

Je suis revenu sur le problème pour trois espèces en utilisant la remarque que vous m'avez faite fin Juin. Il y avait dans ce cas de 3 espèces plusieurs difficultés; mathématiquement on ne peut passer outre et arriver aux conclusions que vous donniez toutefois avec raison et que j'indiquais comme probables, en écartant certaines allures des [illeg.] comme contraires à celles que les études expérimentales fournissent d'ordinaire.

J'ai tenu à donner ces résultats - sans en donner une démonstration purement mathématique, puisqu'on ne peut probablement pas - parce qu'ils sont importants, pour le biologiste et épuisent la question.

Seulement j'ai tout de même été rigoureux en introduisant des postulats biologiques très vraisemblables et très précis. En les adjoignant au raisonnement mathématique, on peut conclure en toute rigueur sous la réserve bien spécifiée de ces postulats.

Je pense qu'ainsi tout est pour le mieux.

D' abord dans la première étude, sans coefficient d'amortissement,

$$\begin{cases} \beta_1 \dfrac{dN_1}{dt} = (-\alpha_1\beta_1 + aN_2)N_1 \\ \beta_2 \dfrac{dN_2}{dt} = (-\alpha_2\beta_2 - aN_1 + bN_3)N_2 \qquad \alpha_1, \alpha_2, \alpha_3, a, b > 0 \\ \beta_3 \dfrac{dN_3}{dt} = (\alpha_3\beta_3 - bN_2)N_3 \end{cases}$$

Dans le cas $(\alpha_3\beta_3 a - \alpha_1\beta_1 b) < 0$, on sait seulement que $N_1^{\beta_1 b} N_3^{\beta_3 \alpha} \to 0$ et que N_3 ne peut rester petit à partir d'un certain moment.

Cela ne suffit pas pour conclure que $N_1 \to 0$. Mais si N_1 ne tendait pas vers 0, il aurait des oscillations ainsi que N_3 et prendrait de temps en temps des valeurs très petits et tendant vers 0.

Cela entraînerait que les espèces auraient si elles ne disparaissent pas une infinité de disparitions passagères.

(I) Si on admet qu'une espèce ne peut résister à un nombre aussi grand de disparitions passagères, N_1 doit sûrement disparaître définitivement (puisque N_3 ne le peut pas).

Après cette petite chose, j'arrive, pour la reprise de la question avec un coefficient d'amortissement, à des choses plus intéressantes, et qui sont vraiment celles que je veux vous écrire.

$$\begin{cases} \beta_1 \dfrac{dN_1}{dt} = (-\alpha_1\beta_1 + aN_2)N_1 \\ \beta_2 \dfrac{dN_2}{dt} = (-\alpha_2\beta_2 - aN_1 + bN_3)N_2 \\ \beta_3 \dfrac{dN_3}{dt} = (\alpha_3\beta_3 - \lambda N_3 - bN_2)N_3 \end{cases}$$

Si $b(a\alpha_3\beta_3 - b\alpha_1\beta_1) - a\lambda\alpha_2\beta_2 > 0$ les trois racines de l'état stationnaire sont positives; on en déduit seulement qu'il y aura des variations bornées pour les 3 espèces et que $N_3 \to q_3$.

On peut voir que dans tout intervalle de longueur donnée y, on peut trouver séparément pour chacune des fonctions N_1, N_2, $\dfrac{dN_1}{dt}$, $\dfrac{dN_2}{dt}$, $\dfrac{dN_3}{dt}$ au moins un point où les différences avec le nombre correspondant q_1, q_2 ou 0 sont en module inférieure à ε donné à l'avance, dès que l'intervalle est postérieur à un instant T qu'on peut fixer dès qu'on connaît y et ε.

(II) Si on admet que, lorsque dans un système une espèce a un état limite fini non nul, son coefficient d'amortissement vrai $\frac{1}{N}\frac{dN}{dt} \to 0$, on peut conclure que N_1 et N_2 tendent vers q_1, q_2 (Car $\frac{dN_3}{dt} \to 0$ et cela entraîne facilement le reste).

Si $b(a\alpha_3\beta_3 - b\alpha_1\beta_1) - a\lambda\alpha_2\beta_2 < 0$, $q_2, q_3 > 0$, $q_1 < 0$ on ne peut conclure à l'épuisement de la première espèce. On peut seulement voir que N_1 ne peut tendre vers q_1.

(III) Admettons alors que dans un système biologique tel que ceux étudiés, l'écart absolu $|N - A|$ entre l'état variable d'une espèce et un état déterminé, ne peut tendre vers 0 en moyenne sans tendre vers 0 au sens ordinaire c.a.d. que si

$$\frac{1}{t-t_0}\int_{t_0}^{t} |N - A|\, dt \to 0, \quad |N - A| \quad \text{tendra verso 0.}$$

Cela entraîne que la 1ère espèce s'épuise. En effet

$$\int_{t_0}^{t} (N_3 - q_3)^2\, dt \to \infty.$$

Sinon, borne L. Et alors

$$\left[\int_{t_0}^{t} |N_3 - q_3|\, dt\right]^2 \le \int_{t_0}^{t} (N_3 - q_3)^2\, dt (t - t_0) \le L(t - t_0)$$

d'où

$$\left(\frac{\int_{t_0}^{t} |N_3 - q_3|\, dt}{t - t_0}\right)^2 \le \frac{L}{t - t_0} \quad \text{et} \quad \frac{1}{t-t_0}\int_{t_0}^{t} |N_3 - q_3|\, dt$$

tendrait vers 0, donc aussi N_3.

Aussi $\int_{t_0}^{t} (N_3 - q_3)^2\, dt \to +\infty$, donc $\int_{t_0}^{t} \lambda\beta_3 q_3^2\left(1 - \frac{N_3}{q_3}\right)^2 dt$ et cela entraîne grâce à la combinaison déduite des équations que N_1 doit tendre vers 0.

Comment trouvez vous cela?

Enfin j'ai lu votre mémoire, celui que vous m'avez envoyé en juillet; je renvoie à la 1ère partie dans ma rédaction au sujet de

$$Q(t) = q(t) + \int_{t_0}^{t} q(\tau)Q(t-\tau)\,dt$$

sans le reproduire parce que c'est un peu en marge et je songe à utiliser plus largement la seconde partie quand je rédigerai l'énergétique.

Puis j'ai étudié votre démonstration d'existence. Elle me parait inattaquable. Je vais la rédiger. D'ailleurs ce que vous avez écrit est presque assez développé (2ème lettre, naturellement).

Je vais faire une note mathématique où elle figure à coté de celle du n° 5 p. 262 (Journal de Mathématiques) que je vais donner par le cas de l'hérédité complète.

Il y a intérêt à ne considérer l'hérédité limitée qu'après le cas général en faisant $Q(t) = 0$ pour $t \geq T_0$. La démonstration du cas général est en effet plus simple et ceci s'applique à d'autres propositions analogues.

Je pense vous envoyer cette note mathématique ces jours ci encore que ce ne soit peut être pas très utile de la séparer de la fin de la rédaction.

Malheureusement il me faut partir prendre quelques vacances dans quelques jours et je ne serai de retour que vers le 5 Septembre. Je ne pourrai donc pas achever avant le 10 ou 15 Septembre. Je suis désolé que cela traîne tant mais je ne pouvais guère faire autrement. J'aurais dû me presser davantage avant Pâques de façon à avoir fini en Juin.

Je pense bien maintenant vous envoyer toute la fin et les figures, vers la mi-septembre. Vous aurez le temps de regarder cela fin septembre j'espère et le 1er octobre je pourrai remettre le manuscrit à G. Villars; à ce moment j'espère qu'il pourra le travailler presque tout de suite et que le livre paraîtra en mai prochain, au plus tard, peut être en mars, avril.

Je m'excuse encore. Je serais heureux d'avoir un mot de vous. Je serai du 22 Août au 30 ou 31 Août à

l'Hôtel des Bains à Trégastel (Côte du Nord)

où je m'arrêterai au cours de ma promenade.

En dehors de cette période vous pouvez écrire chez moi: à Boisseaux (Loiret).

J'irai sûrement en Italie vers le 15 ou 20 Octobre pour la prochaine année Scolaire et en attendant de vous voir à Rome, je vous adresse maintenant, avec ma reconnaissance pour votre appui à la Fondation Rokefeller, mes sentiments de respect.

Marcel Brelot

4.8 Brelot to Volterra, Boisseaux, 14.8.1929 *Cat. no. 17*

Boisseaux (Loiret), 14 Août 1929

Cher Monsieur

je viens de vous envoyer un petit paquet contenant un certain nombre de feuilles de la rédaction: tout ce qui concerne le problème des 3 espèces, en deux parties; l'édition de votre démonstration en note, et enfin la note mathématique en fin du dernier chapitre.

J'ai rédigé la démonstration d'existence que vous m'avez envoyée et ajouté en quelques lignes celle, toute pareille, de l'unicité.

Je pars en voyage d'agrément après demain pour 15 jours environ.

Recevez, cher Monsieur, l'expression de tout mon respect.

Marcel Brelot à Boisseaux (Loiret)

Je vous ai donné dans ma lettre récente mon adresse passagère à "Trégastel (Côte du Nord". Hotel des Bains du 23 au 31.

4.9 Volterra to Brelot, Ariccia, 20.8.1929 *Cat. no. 359*[6]

20 Août 1929

Cher Monsieur.

J'ai l'honneur de vous accuser réception de vos lettres du 12 et 14 Août et du manuscrit que vous m'avez envoyé à part.

J'ai appris avec joie que vous avez obtenu la bourse Rockefeller. D'après les conversations que j'avais eu avec M. Tisdale je n'en doutais pas, mais il vaut toujours mieux d'en être complètement assuré. Je serai heureux de vous voir travailler chez-nous et de savoir que vous arriverez en Octobre. Nous pourrons nous occuper avec tranquillité de la correction des épreuves de Gauthier-Villars. Je pense que je serai à Paris dans la seconde moitié de Septembre. Si vous serez rentré à cette époque je compte de vous voir. Maintenant vous faites très bien à vous reposer et à soigner votre santé. Vous aurez ainsi plus d'entrain pour vos travaux futurs. Parmi les sujets que vous pensiez traiter il y avait celui d'étudier les noyaux des équations intégrales et les noyaux résolvants en vue de leurs propriétés comme des fonctions analytiques [générales] d'un ou de plusieurs paramètres. Si vous aurez sous les yeux les Comptes Rendus du 27 Juillet vous y trouverez une note de M. B[illeg.] qui pourra vous intéresser.

Pour ce qui se rapporte à mes leçons je trouve qu'il faudra changer la part qui se rapporte à ce que vous appelez des <u>postulats biologiques</u>. <u>Tout au plus</u> on pourrait conserver un seul postulat mais le tenir bien séparé de la partie mathématique et de ses conséquences. D'autre part on peut aller bien loin par les mathématiques et <u>se passer</u> <u>des postulats</u>.

[Pour ce qui se rapporte à mes leçons je trouve qu'il faut réduire au minimum ce que vous appelez des postulats biologiques et tenir bien séparées les conséquences mathématiques de ce que l'on peut tirer des postulats.]

[Je pense même qu'il ne faut conserver qu'un seul postulat: celui que l'on peut regarder qu'une espèce s'épuise lorsque le nombre de ses individus peut devenir plus petit que toute quantité donnée, même si on ne peut pas démontrer qu'il tend vers zéro et en outre que lorsque une espèce s'épuise on peut la négliger à partir d'un certain instant.]

---

[6] There are 3 drafts. The first two are dated 18 August, but are practically identical to the third, dated 20 August, which is the one reproduced here. We merely add (in [] brackets) a few passages contained in the first draft (and later deleted) that appear to be particularly interesting.

[Ces suppositions me semblent assez intuitives et justifiables par le bon sens pour pouvoir les retenir, mais pas les autres.]

D'autre part on peut aller assez loin, bien loin par le procédé mathématique de manière à pouvoir se passer des autres postulats.]

[La plupart du temps les postulats c'est la suggestion de Satan pour vous rendre paresseux].

En effet reprenons le cas des trois espèces (Je me rapporte aux notations de mon mémoire italien)[7]:

1°) Si $\beta_3 ka - \beta_1 lb < 0$ (Appendice §5, N.3) on peut démontrer que à partir d'un certain instant N_3 ne peut pas se conserver au dessous d'un nombre quelconque plus petit que $\frac{\beta_2 m}{b}$, donc N_1 doit s'approcher de zéro autant que l'on veut. (Eq. 61)

Il n'existe pas donc d'état limite de l'association biologique.

2°) Si $\beta_3 ka - \beta_1 lb > 0$ et l'on ajoute le terme d'amortissement et l'on a

$ab\beta_3 k - b^2\beta_1 l - a\lambda\beta_2 m > 0$

les trois racines sont positives. Par la théorie générale N_3 doit tendre vers q_3 et N_1 et N_2 s'approcheront des racines q_1 et q_2 autant que l'on veut sans que l'on puisse démontrer qu'elles ont ces limites. Donc s'il existe un état limite celui-ci est l'état stationnaire.

3°) Si $ab\beta_3 k - b^2\beta_1 l - a\lambda\beta_2 m < 0$ alors N_3 ne peut pas tendre vers q_3. Mais on a par la théorie générale des racines négatives que j'ai donnée dans ma 5ème leçon de cet hiver (je me rapporte aux notations des pages manuscrites que je vous ai passées)

$$\left(\nu_1 e^{\nu_1}\right)\left(\frac{e^{n_2}}{n_2}\right)^{\alpha_2 q_2}\left(\frac{e^{n_3}}{n_3}\right)^{\alpha_3 q_3} = C' e^{-\int_0^t F(\xi_r)\,\mathrm{d}t}$$

Donc ν_1, n_2, n_3 et par suite les N_r et les $\mathrm{d}N_r/\mathrm{d}t$ sont des quantités bornées. Or dans notre cas

$$-\int_0^t F(\xi_r)\,\mathrm{d}t$$

se réduit à un seul terme du type

$$-\int_0^t K(N_3 - q_3)^2\,\mathrm{d}t\,,$$

K positif. Si N_3 ne tend pas vers q_3 et puisque $\mathrm{d}N_3/\mathrm{d}t$ est borné l'expression précédente tend vers $-\infty$ pour $t = \infty$ (voir démonstration pages 81–82). Donc ν_1

[7] V. Volterra 1926a.

doit tendre vers zéro. De cette manière on démontre que N_1 tend vers zéro sans besoin d'aucun postulat.

Tout cela est une partie mathématique pure. À ce moment abandonnons le problème mathématique et passons à la question biologique. On peut dire à ce point de vue

1°) Une espèce s'épuise si le nombre des individus peut devenir aussi petit que l'on veut (Définition)

2°) Si une espèce s'épuise on peut la négliger à partir d'un certain instant. (Celui -ci est un principe intuitif et on pourrait même l'appeler un postulat biologique). Mais ce principe peut conduire à des conséquences qui sont en contradiction avec les conséquences des lois mathématiques. C'est pourquoi on peut l'appliquer pour traiter la question biologique et on ne peut pas l'appliquer pour traiter la question mathématique. Il faut donc considérer les deux questions comme des questions différentes et les traiter séparément pour ne pas engendrer des contradictions et de la confusion.

En effet revenons au cas 1°) Puisque N_1 peut s'approcher de zéro autant que l'on veut par la *définition* on pourra dire que cette espèce s'épuise et si nous appliquons le principe par lequel on peut la négliger nous nous trouverons dans le cas bien connu de deux espèces dont l'une se nourrit de l'autre. On aura donc les fluctuations bien connues.

Revenons maintenant au cas 3°) dans lequel nous avons démontré sans l'aide d'aucun postulat que N_1 tend vers zéro. Abandonnons la question mathématique et passons au problème biologique. La première espèce s'épuise. Negligeons-la. Il faudra alors considérer les équations

$$\beta_2 \frac{dN_2}{dt} = (-\beta m + bN_3)N_2 \ , \quad \beta_3 \frac{dN_3}{dt} = (\beta_3 k - \lambda N_3 - bN_2)N_3$$

Les racines seront

$$q_2' = \beta_3 k - \frac{\lambda \beta_2 m}{b} \ , \quad q_3' = \frac{\beta_2 m}{b}$$

Si $\beta_3 kb - \lambda\beta_2 m > 0$ les deux racines sont positives, c'est pourquoi N_3 tend vers q_3' et N_2 s'approche de q_2' autant que l'on veut.

Si $\beta_3 kb - \lambda\beta_2 m < 0$ la racine q_2' est négative et alors on démontre, comme précédemment que N_3 ne peut pas tendre vers q_3', autrement N_2 pourrait s'approcher de la racine négative autant que l'on voudrait ce qui serait absurde. Donc on voit par une démonstration analogue à celle qu'on a fait tout à l'heure que N_2 doit tendre vers zéro et par suite que la 2° espèce s'épuise.

La nécessité de séparer la question mathématique et la question biologique a été envisagée à la fin mon article de Scientia dont vous devez avoir la traduction française[8].

Je suis d'avis de modifier la rédaction que je viens de recevoir en tenant compte des démonstrations et des remarques que j'ai l'honneur de vous communiquer par cette lettre. Naturellement il faut conserver les solutions approchées en négligeant les termes du 2° ordre comme dans le mémoire italien.

[8] VOLTERRA 1927g.

Dans la 3ème partie de mon mémoire Italien il y a au §1er des théorèmes où l'on regarde les N_r comme des fonctions analytiques de t. Par ces théorèmes on voit facilement que tout l'axe réel du temps est interne aux étoiles de Mittag Leffler relatives aux fonctions analytiques N_r. Je pense que l'on pourrait ajouter cette remarque et placer cela au commencement pour prouver que l'on peut étendre N_r pour des valeurs de t entre $-\infty$ et $+\infty$.

Veuillez accepter, cher Monsieur, l'expression de toute mon estime et de mon affection la plus sincère.

Vito Volterra

Ci-joint vous trouverez le manuscrit que vous m'avez envoyé.

4.10 Brelot to Volterra, Boisseaux, 9.9.1929 *Cat. no. 18*

Boisseaux, 9 Septembre 1929

Cher Monsieur,

Rentré depuis quelques jours, j'ai trouvé votre lettre et votre manuscrit.

Je n'ai pu encore m'occuper de cela parce que mon père s'est trouvé brusquement atteint d'une crise d'appendicite; on l'a opéré d'urgence; il y avait déjà de la péritonite et l'appendice etait gangrené. Il va aussi bien que possible mais n'est pas encore hors de danger. Ma mère étant malade, j'ai eu fort à faire. J'ai très bon espoir, d'après le chirurgien: et ces jours ci je reprendrai les questions dont vous me parlez. Je pense vous voir à Paris fin Septembre et à ce moment avoir tout terminé et discuter avec vous des derniers points.

Dans quelques jours, je vous écrirai à nouveau au sujet du cas si délicat des 3 espèces. Je pense que tout le reste vous convenait: vous ne m'en avez point parlé.

A bientôt donc. Croyez, cher Monsieur, à mes sentiments les plus respectueux.

Marcel Brelot

4.11 Brelot to Volterra, Boisseaux, 18.9.1929 *Cat. no. 19*

Boisseaux, mercredi 18 Septembre 1929

Cher Monsieur

Débarrassé des soucis pour mon père qui entre en convalescence, et un peu reposé de mes voyages et mes émotions, j'ai lu très attentivement votre lettre du 27 août.

Ce qui m'a frappé tout d'abord c'est que vous y démontrez, dans le cas

$$\begin{cases} \beta_3 ka - \beta_1 lb > 0 \\ ab\beta_3 k - b^2\beta_1 l - a\lambda\beta_2 m < 0 \end{cases}$$

que $N_1 \to 0$ sans mon postulat biologique bizarre sur la moyenne d'un écart $|N-A|$.

J'avais eu quelque peine à en venir à ce postulat afin de pouvoir conclure et j'étais à la fois heureux et confus de voir que tout s'arrangeait mathématiquement.

Tout repose sur le fait que dN_3/dt est borné. Or en regardant de plus près votre lettre j'y vois que c'est de

$$\left(\nu_1 e^{\nu_1}\right)^{\alpha_1 \chi_1}\left(\frac{e^{n_2}}{n_2}\right)^{\alpha_2 q_2}\left(\frac{e^{n_3}}{n_3}\right)^{\alpha_3 q_3} = C' e^{-\int_0^t F(\xi_r)\,dt} \qquad (1)$$

que vous déduisez que ν_1, n_2, n_3 sont bornés.

De là, bien sûr, résulte que les dérivées, d'après les équations fondamentales, sont aussi bornées et on peut conclure, comme vous faites, et comme d'ailleurs, je l'avais fait plus haut, suivant, je crois, les indications de votre lettre d'août pour le cas général avec racines toutes positives et coefficient d'amortissement pour toutes les espèces (càd ce qui il y a à votre référence au mémoire italien). Or je ne vois pas du tout comment de (1) vous pouvez déduire que les N_r sont bornés. Il me semble juste possible de voir que N_1 l'est. Je crois bien me souvenir que c'est cela qui m'avait arrêté, en voulant procéder par analogie avec le cas général précité.

Tout est là. Ne pouvant en sortir, j'avais donc introduit un postulat biologique, avec assez de peine. Regardez bien ce point, je vous prie. C'est le plus important, de beaucoup, que j'ai à vous signaler aujourd'hui.

Je continue: le seul postulat que vous désiriez conserver est celui que vous me soulignez en l'énonçant: une espèce disparaît si N peut devenir arbitrairement petit. Il diffère bien peu de ce que j'ai admis: une espèce ne peut résister à un nombre assez grand de disparitions passagères – énoncé de forme que je voulais biologique. Sans cela en effet on est plusieurs fois arrêté et on ne saurait conclure.

En ce qui concerne mon troisième postulat (que $\frac{1}{N}\frac{dN}{dt}$ tend vers 0 si une espèce a un état limite $\neq 0$), vous êtes d'avis de le supprimer. Mais alors on ne peut conclure que, lorsqu'un état stationnaire est possible théoriquement, le système tend vers cet état. J'avoue que cela me gêne. Aucun biologiste, je crois, ne s'arrêtera a cette difficulté; il considérera sûrement que, cependant, il y a bien limite, de part son intuition qui écarte les cas mathématiques d'allure spéciale. Tout ce que l'on peut dire mathématiquement, je crois, l'avoir signalé. J'ai ajouté explicitement que moyennant des hypothèses non mathématiques, on pourrait réellement conclure. Vous craignez une confusion de la partie mathématique et de la partie biologique. Je pourrais être plus net.

En ce qui concerne le fait de regarder les N_r comme des fonctions analytiques, je trouve votre remarque intéressante. Je la ferai mais elle n'est pas indispensable puisque, dans le manuscrit j'avais bien indiqué l'existence d'une solution continue dans $(t_0, +\infty)$ de par la méthode des approximations successives.

Enfin j'avoue avoir quelque répugnance pour les procédés d'approximations que vous employez dans le mémoire italien. J'etais heureux de voir que dans votre cours vous les évitez le plus possible. D'ailleurs nous avons déjà parlé du cas précis des petits mouvements et des recherches que je ferais peut-être de ce coté. Je ne veux point sans vous revoir refaire la rédaction de ce délicat et si épineux cas de 3 espèces avec un seul coefficient d'amortissement (Je crois que si l'on en mettait 3, ce serait plus long mais on pourrait conclure à l'existence de la limite dans le cas de 3 [illeg.] positives ce qui éviterait une difficulté. C'est ce fait, en pensant à prendre très petits les deux autres coefficients, qui m'encline à vouloir d'une façon ou de l'autre conclure à la limite vraie. Est on moins rigoureux en disant "Il y aurait limite si ces coefficients sont petits; il y aura donc vraiment évolution vers cet état limite stationnaire lorsqu'ils sont nuls" qu'en admettant des procédés où l'on néglige le second ordre sans justification?

Je pense vous voir à Paris d'ici peu et nous pouvons reparler de cette question. Pour le moment je vais achever le manuscrit. Je ferai les modifications quand je vous aurai vu.

Croyez que cela me gêne, les postulats biologiques. Mais comment conclure en restant rigoureux càd en précisant les hypothèses?

J'ai discuté avec vous comme avec un camarade, un égal et je pense que vous ne m'en voudrez pas de cette attitude scientifique. Je serait tout à fait heureux de vous voir pour mettre tout cela au point. C'est si difficile de discuter par correspondance.

Est-ce qu'à Rome je trouverai les ouvrages français que je voudrai: traités d'Analyse et Collection Borel? Quand vous serez à Paris, fixez moi un rendez-vous quand vous voudrez.

A bientôt j'espère.

Croyez à mes meilleurs sentiments et tout mon respect

M. Brelot à Boisseaux (Loiret)

4.12 Brelot to Volterra, Boisseaux, 26.9.1929 *Cat. no. 20*

Boisseaux, 26 sept. 29

Cher Monsieur

J'ai reçu votre réponse. Je puis aller vous voir à Paris la semaine prochaine quand vous voudrez p ex un matin à 11^h (sauf mardi)

Faites en sorte surtout que cela ne vous gêne nullement

A bientôt cher Monsieur. Croyez à mes sentiments les plus respectueux

M. Brelot, Boisseaux (Loiret)

4.13 Brelot to Volterra, Paris, n.d. *Cat. no. 9*[9]

Paris, vendredi matin

Cher Monsieur

Je pense qu' il peut vous être utile d'avoir vos notes sous les yeux pour comparer ma rédaction à votre cours. Aussi je vous rends toute la partie que j'ai

[9] N.d. but probably written at the end of September 1929. Brelot refers to a meeting with Volterra in Paris: see letter no. 4.12 (Cat. no. 20).

déjà rédigée, y compris la 7e leçon, première leçon sur les phénomènes héréditaires.

Nous parlions hier du cas de trois espèces et du postulat que j'admettais sur la "bonne allure" des courbes biologiques. Si je l'ai employé, c'est que je n'ai rien trouvé de mathématique et que je voulais indiquer les résultats auxquels vous parvenez en les utilisant. Regardez par exemple: 5e leçon page 15 bis, de vos notes où vous admettez que si N_3 tend vers q_3, nécessairement $\mathrm{d}N_3/\mathrm{d}t$ tend vers zéro.

Je pense qu'en introduisant trois paramètres λ pour les 3 espèces on pourrait se dispenser de pareille chose; mais cela compliquerait un peu.

J'ajoute que 4e leçon p. 11, vous faites bien usage de postulats analogues en déduisant $N_1 \to 0$ ou $N_1^{\beta_1 b} N_3^{\beta_2} \to 0$ et du fait que N_3 ne peut tendre vers zéro. Dans ma rédaction j'ai insisté là-dessus en disant pourquoi il convient de l'admettre du point de vue biologique.

Je tiens à vous dire tout cela, comme hier partiellement, pour vous montrer que, il m'a beaucoup répugné d'en venir à des arguments aussi peu mathématiques, mais que je l'ai fait parce que vous aviez été contraint de le faire pour avoir des conclusions plus étendues.

J'aurais beaucoup aimé pouvoir vous parler de ce sujet au moment où je l'ai rédigé; mais il est toujours temps de modifier et d'ajouter des explications plus explicites encore sur ces hypothèses non mathématiques. Un lecteur bien averti peut être fâché de quitter le sûr domaine du raisonnement mathématique, mais il ne peut s'en prendre qu'à la mauvaise nature de la question. Et d'autre part, pour les biologistes, il serait ennuyeux de supprimer des résultats intéressants sous prétexte qu'ils ont exigé des postulats que tous les physiciens admettent en matière expérimentale.

Croyez à ma bien grande reconnaissance et tout mon dévouement.

Très respectueusement

M. Brelot

4.14 Brelot to Volterra, Boisseaux, 19.10.1929 *Cat. no. 21*

Boisseaux, 19 octobre 1929

Cher Monsieur.

Je vous enverrai la fin de la rédaction d'ici 2 ou jours. J'ai, autant que possible tout traité à la fin avec l'hypothèse de l'hérédité limitée ou non. Le cas des petites fluctuations et mouvement spontané se fait très bien aussi; j'ai changé la démonstration établissant que les moyennes asymptotiques de q_1 et q_2 sont nulles, mais en suivant l'autre méthode d'aussi près que possible. J'ai été ramené à la fin à démontrer que

$$\frac{\int_0^\theta u\Phi(u)\,\mathrm{d}u}{\theta} \to 0 \quad \text{quand} \quad \theta \to +\infty$$

où $\Phi(u) \geq 0$ est décroissante quand elle n'est pas nulle et telle que $\int_0^{+\infty} \Phi(u)\, du$ a un sens, enfin pourvue de dérivée continue.

Ce n'est pas bien difficile, encore que j'ai mis assez de temps à trouver quelque chose de simple et court.

D'ailleurs je reviendrai sur les points à vous signaler avec mon envoi.

Je ne sais plus ce qu'il me faut ajouter au sujet de M. Evans, et je n'ai pas bien vu l'autre jour comment l'Etoile de Mittag-Leffer dont vous me parliez contenait l'axe réel. Mais je réfléchirai à ce point.

Je ne peux, hélas! partir avant le 6 Novembre, car, à cause d'un accident d'automobile qui m'est survenu il y a un mois, et qui n'est ni grave ni plein de responsabilité pour moi qui était en règle, je suis appelé au tribunal correctionnel le 5 nov. Cela m'ennuie beaucoup mais je suis bien forcé d'attendre.

A bientôt. Croyez à mes sentiments très respectueux

M. Brelot.

4.15 Brelot to Volterra, Boisseaux, 22.10.1929 *Cat. no. 22*

Boisseaux, mardi 22 Octobre 1929

Cher Monsieur

Voici donc la fin du manuscrit - excepté l'introduction et la conclusion que j'écrirai demain et vous enverrai aussitôt.

J'ai ajouté dans cet envoi au dernier et 4° chapitre, les quelques pages (chap III) sur le cas de 3 espèces lorsqu'on l'étudie pour la seconde fois avec un coefficient pour restreindre l'accroissement quand la 3 espèce est trop nombreuse. Comme j'avais donné, dans la 1ère étude de ce cas, le postulat fondamental sur la disparition d'une espèce quand son N tend vers 0 pour une suite particulière d'instants, avec l'énoncé suivant: une espèce ne saurait résister à un nombre assez grand de disparitions passagères, j'ai simplement renvoyé a ce texte. Vous avez accepté ma dénomination de disparitions passagères et pas critiqué je crois la 1ère étude du cas de 3 espèces après admission du postulat. Aussi n'ai je plus rien changé.

D'ailleurs, comme je pense que maintenant il n'y aura que de légères modifications à faire ou additions à apporter, il serait toujours temps, sur les épreuves de tenir compte de vos observations.

Je vous signale les points suivants:

Outre ce que je vous ai dit sur l'extension à la durée illimitée d'hérédité (qui n'a guère qu'un intérêt théorique, mais qui est assez facile et satisfait l'esprit) - et mes questions supplémentaires, j'ajoute que j'ai un peu modifié p 100 le moyen d'arriver à l'équation énergétique (58). Je crois mon calcul assez naturel, après ce qu'on a déjà fait. Le vôtre, dans le petit mémoire est fort synthétique.

Je ne vois pas bien pourquoi la méthode des rencontres ne convient pas au cas de *n* espèces. Il me semble qu'un raisonnement analogue à celui de 2 espèces dévorante et dévorée conduit aux équations que j'indique à la fin (p 101). Je vous prie de lire plus spécialement la fin pour voir si rien de simple, si aucune idée générale ne serait à ajouter pour finir.

J'ai lié aussi étroitement que possible l'enérgetique en mécanique et en biologie, faisant les calculs avant de spécifier les cas.

J'ai un peu généralisé l'énoncé de la loi de conservation des moyennes (p 90) en disant qu'on peut prendre comme instants finaux, ceux d'extrema inférieurs à un nombre fini (non nécessairement K_1 ou K_2). J'ai également un peu étendu les inégalités du n° 7.

Je n' ai pu dès le milieu du n° 6 conserver l'hérédité illimitée et je ne sais pas si c'est possible.

Je n'ai pas mis la fin du chapitre IV que vous avez vue (Note Mathématique).

J'espère bien que les premières pages ne se perdront pas. Je vous enverrai demain ou après demain l'introduction et la conclusion et dès que vous m'aurez tout retourné, je porterai le manuscrit à Gauthier-Villars.

En attendant de vous revoir à Rome vers le 7 ou 8 Novembre, je vous prié de croire à mes sentiments les plus respectueux.

M. Brelot, Boisseaux (Loiret)

4.16 Brelot to Volterra, Boisseaux, 25.10.1929 *Cat. no. 23*

Boisseaux, vendredi 25 Octobre 1929

Cher Monsieur

Voici la conclusion et l'introduction et quelques remarques.

Peut-être dans l'introduction avez vous à ajouter quelques références, comme par exemple sur les équations de Ross[10].

Dans la conclusion j'ai résumé plus brièvement que vous dans votre enseignement oral. J'ai pensé qu'il s'agissait, dans un ouvrage où l'on a sous les yeux la table des matières, de faire des remarques générales, seulement possibles après l'exposé, des considérations d'ensemble, ou des justifications a posteriori. Je n'ai pas terminé, parce que je veux vous parler des expériences de Chapman avant de rédiger cette dernière page. D'abord il s'agit toujours de milieux délimités, dont le grandeur intéresse les coefficients autres que les ε et j'ai insisté sur ce point au cours de l'ouvrage sauf peut être dans les quelques pages que vous avez. Dire qu'une espèce est nombreuse au point de vue de son influence sur les autres ou sur elle même, signifie qu'elle est dense, parce qu'il s'agit en sous entendu d'un milieu de grandeur délimitée soit naturellement, soit artificiellement par la pensée. Sans cela d'ailleurs, le nombre des individus n'apprendrait rien.

Il me semble alors que ce qui est intéressant c'est la densité de la farine dans le milieu, ou dans le milieu de grandeur donnée sa quantité totale. Mais cela suppose qu'elle est mélangée à autre chose. Sinon, que cette quantité totale soit grande ou petite, cela ne saurait influer sur $\underline{\varepsilon}$ coefficient d'accroissement $\frac{1}{N}\frac{dN}{dt}$ quand l'espèce est peu nombreuse. De plus si on prend l'équation

[10] A reference to Ronald Ross and William R. Thompson was actually included in the book.

$$\frac{1}{N}\frac{\mathrm{d}N}{\mathrm{d}t} = \varepsilon - \lambda N \ ,$$

le λ est relatif à un milieu encore une fois de grandeur bien déterminée et il dépend de cette grandeur, puisque le terme correctif ne dépend que de la densité de l'espèce. Je vous consulte donc au sujet de cette farine dont on ne dit pas qu'elle est mélangée à autre chose ou convenablement dispersée dans une espace assez grand.

Telle qu'elle est présentée la loi de Chapman peut semble banale, car dans une quantité donnée de farine pure, le nombre limité d'individus est évidemment proportionnel à cette quantité totale, si la farine est disposée toujours de la même façon ou rassemblée en tas.

Je voudrais donc voir le mémoire de Chapman[11]. Vous l'avez sans doute; alors vous pouvez me répondre.

Dès que vous m'aurez tout renvoyé, je modifierai les quelques points que vous avez peut être à me signaler puis j'irai aussitôt chez l'éditeur.

Croyez à mes sentiments les plus respectueux.

M. Brelot

Pensez vous que la quantité de farine par unité de volume influe sur ε d'une façon exactement proportionnelle, comme il est admis. On trouve:

$$\frac{\lim N}{\varepsilon} = \frac{1}{\lambda}$$

et la loi de Graham ne s'obtient que grâce à cette proportionnalité. Il y a là une hypothèse un peu douteuse.

Naturellement, je considère aussi comme sous entendu que la quantité de farine est maintenue constante; sinon évidemment les insectes ayant tout dévoré disparaîtraient.

Dans le mémoire on ne dit pas si l'on maintient constant la quantité de farine et si l'on ôte les impuretés dûes aux insectes.

4.17 Brelot to Volterra, Boisseaux, 31.10.1929 *Cat. no. 24*

Boisseaux, 31 Octobre 1929

Cher Monsieur

Je compte partir à Rome jeudi ou vendredi prochain (7 ou 8 Nov). Je serai à Paris le 6 et 7. Je vous prie de m'envoyer le mémoire pour que je le remette à Gauthier-Villars en entier. Si vous ne me l'envoyez que lundi prochain, expédiez le Rue d'Ulm 45 à l'Ecole, puisque je quitterai Boisseaux mercredi, pour Paris.

---

[11] The reference is to CHAPMAN 1928.

Mais peut être, j'espère, votre réponse est elle en chemin. Je l'attendrai pour partir à Rome.

A bientôt cher Monsieur. Croyez à mes sentiments les plus respectueux.

M. Brelot à Boisseaux (Loiret)

Je pense que vous avez bien reçu mes deux lettres contenant la $1^{\text{ère}}$, la 4e chapitre, la seconde, l'introduction et la conclusion.

4.18 Brelot to Volterra, Paris, 6.11.1929 *Cat. no. 25*

Paris, mercredi 6 Novembre 1929

Cher Monsieur

Je reçois votre mot. Je pensais bien qu'il y aurait des modifications à faire et qu'il serait préférable sinon indispensable que je vous visse. Je vais donc donner à l'éditeur tout le manuscrit et les figures et je partirai à Rome demain ou très probablement après demain 8 par le train de $\underline{15^{\text{h}}}$ Dès mon arrivée je vous rendrai visite. Je pense que vous serez à Rome même. A bientôt... Arrivederla...

Avec mes sentiments les plus respectueux.

M. Brelot

4.19 Brelot to Volterra, Roma, 23.2.1930 *Cat. no. 26*

Rome, dimanche 23 février 1930

Cher Monsieur

Je viens de demander par téléphone quand vous devez rentrer à Rome; on me répond qu'il n'est pas sûr que vous soyez revenu dimanche prochain.

Je vous consulte donc, car je voudrais présenter une Note aux Lincei à la prochaine séance, le 2 mars je pense, dimanche prochain.

Vous m'avez dit qu'il n'était pas nécessaire que vous fussiez là. S'il suffit d'une lettre de votre part à l'Académie, je vous serais donc très reconnaissant d'envoyer un mot.

Cette Note, le dernière que je puis présenter aux Lincei cette année, puisqu'elle épuise les 16 pages – et peut être serait elle trop longue encore que je ne fasse qu'indiquer les démonstrations – traite encore des sources ponctuelles de chaleur.

"Sopra lo studio delle sorgenti puntuali del calore in un piano invariante in equilibrio termico".

Au point de vue de la forme nécessaire des intégrales, de l'unicité, l'étude est achevée. Je donne des cas d'existence et de non existence et un procédé de passage à la limite donnant l'intégrale quand elle existe.

Tout cela avec une singularité du coefficient $c(x,y)$ de

$$\Delta u = c \cdot u \begin{cases} c \text{ continue } \geq 0 \\ \text{aucune hypothèse en 0 lui même} \end{cases}$$

Bien que le démonstrations soient simples j'ai eu assez de peine; et j'ai épuisé peut être les résultats intuitifs et simples sur cette notion de source en un point singulier du coefficient de rayonnement

Je serais heureux que cette Note puisse être présentée à la prochaine séance afin de pouvoir plus tard l'utiliser comme référence – sans attendre trop longtemps.

Je pense que ma note pour le C. R. a été acceptée et imprimée.

J'ai demandé à la Fondation Rockefeller l'autorisation d'aller à Paris une semaine. J'ai donné des raisons scientifiques vraies car je serais heureux de parler à quelques professeurs. Mais je désirerais aussi voyager un peu, car je suis un peu fatigué, et un peu triste dans l'ambiance de Rome qui n'a pas la vitalité joyeuse de celle de Paris. Je travaille de moins en moins depuis un mois et un peu de changement me ferait un grand bien. La Fondation R. a accepté en suspendant malheureusement ma bourse pendant cette semaine là. Mais je préfère encore m'en tenir a mon idée. Je pense que vous n'y voyez pas d'inconvénient.

En attendant de vous voir avant dimanche, ou de recevoir votre réponse, veuillez croire à mes sentiments les plus respectueux.

M. Brelot, 183 Via Veneto, Roma

4.20 Brelot to Volterra, Roma, 16.3.1930 *Cat. no. 27*

Boisseaux, dimanche 16 Mars 1930

Cher Monsieur

J'ai fait un excellent voyage et me voici dans ma famille aujourd'hui dimanche. J'ai en plus le grand plaisir à retrouver mon cher Paris et l'ambiance de l'Ecole Normale avec quelques camarades

Je suis allé chez Gauthier-Villars et j'ai appris que la fin des épreuves a été expédiée à Rome le 10 février – et je n'avais rien reçu à mon départ. La première partie avait mis près de 3 semaines à parvenir. Il faut donc attendre avec confiance. Mais c'est un peu étrange – qui donc arrête si longtemps ces épreuves?

J'ai pris tous les placards et je pense trouver le temps ces jours ci de regarder et commencer à corriger la seconde partie – bien que je n'ai pas le manuscrit. A mon arrivée à Rome je pourrai aller plus vite.

Je suis passé aussi chez Blanchard.

Il n'y a pas la gravure dans l'ouvrage de Fermat. Or attendait pour vous envoyer tout ce que vous avez demandé certaines choses difficiles à se procurer. De toutes façons je pense apporter ce qui sera prêt: ce sera plus rapide, plus sûr et évitera des frais inutiles.

J'ai vu également hier très vite M. Picard et M. Vessiot que je retournerai voir cette semaine une peu plus longuement.

Demain matin j'irai au cours de M. Julia. Le livre de M. Julia, que j'ai rédigé en collaboration avec un camarade va paraître dans quelques jours; c'est le fascicule précédant le vôtre. Celui qui suivra – à la rédaction duquel j'ai collaboré aussi – est fort avancé, d'ailleurs.

Dans une semaine au plus je devrai repartir, hélas. Et sans regretter nullement d'aller à Rome je serais pourtant très heureux de rester un peu plus longtemps en France...

Je vous prie de présenter mes respects à Madame Volterra, et à vos fils, et de croire à mes sentiments les plus respectueux

M. Brelot

4.21 Brelot to Volterra, Paris, n.d. *Cat. no. 6*[12]

Vendredi

Cher Monsieur. Je viens de chez G. Villars où l'on m'a donné le manuscrit et les épreuves car le paquet est revenu.

L'adresse était incomplète: pas de n° via Veneto......

Chez Blanchard, j'ai pris et réglé (30^+) le livre de Lagrange, le seul qu'on ait pu vous trouver.

Je pense être à Rome mardi ou mercredi soir.

Veuillez croire à mes sentiments les plus respectueux.

M. Brelot

4.22 Brelot to Volterra, Roma, n.d. *Cat. no. 5*[13]

Rome, samedi

Cher Monsieur

Voici une Note pour les Lincei - la dernière.

J'ai reçu hier la réponse de M. Bouligand me tranquillisant tout à fait au sujet du Laplacien généralisé et je puis donc dire deux mots à ce sujet - sous entendus auparavant.

Dans cette Note, après quelques lignes pour mettre au point la dernière, je donne un théorème nouveau, puis j'indique pour l'espace un développement analogue à celui que j'ai fait dans le plan. L'application du dernier théorème fournit en particulier une réponse au problème de Dirichlet extérieur pour $\Delta u = cu$ qui est fort diffèrent du même problème plan.

J'ai rédigé très succinctement, sans démonstrations à la fin - pour avoir une Note assez courte.

Je pense que vous pourrez la présenter demain matin.

Je rédige actuellement mon petit travail sur le problème biologique et je vous le soumettrai sous quelques jours. Vous me direz ce qu'il convient d'en faire et s'il faut au moins en parler dans le livre. Ce sera un peu long mais peut être plus intéressant que je ne pensais vu la généralité des hypothèses - encore que je ne

---

[12] Postcard, n.d. sent on 20, probably March 1930 (postmark is not clear).

[13] N.d. but probably written in April or May 1930.

puisse établir qu'il y a une limite dans le cas le plus général où existe un état stationnaire.

Je me demande aussi s'il ne serait pas temps de prendre quelques dispositions pour l'an prochain si je devrais par exemple écrire à M. Schmidt. Nous en parlerons lundi.

Je n'ai toujours pas de réponse de M. Montel qui attend peut être d'être fixé pour répondre.

Très respectueusement avec tous mes remerciements.

M. Brelot

4.23 Brelot to Volterra, Roma, n.d. *Cat. no.* 3[14]

Rome, mercredi soir

Cher Monsieur

Je vous présente d'abord tous mes vœux pour une prompte guérison de cette rechute. Je ne doute pas que le beau temps ne vous rende vite la santé.

J'ai reçu les seconds placards et je me suis assuré que les corrections ont été faites. Je vous les apporte. Quand vous le voudrez faites les déposer chez votre concierge; j'irai les prendre et je les reéxpedirai.

J'ai regardé la Note de M. D'Ancona. Je pense qu'il serait peut être mieux de faire une Note à part intitulée:

Notice biologique
Historique et bibliographique
par M. D'Ancona
prof. de sciences naturelles etc.

En effet le livre, introduction et conclusion se préoccupe de recherches rationnelles et des calculs et théories mathématiques; les indications données sur d'autres auteurs n'ont rapport qu'à des ouvrages où il y a une recherche de lois mathématiques.

Dans la Note de M. D'Ancona, il se préoccupe aussi de tous ceux qui ont pensé à l'influence réciproque et ont donné des idées qualitatives plus ou moins vagues, et déjà peut être formulé l' hypothèse de variations de caractère interne. Cette Note est une historique de l'idée même qui trouve son développement dans l'ouvrage.

Et puis peut être n'est il pas mauvais qu'il y ait collaboration apparente d'un professeur de sciences naturelles, et précisément le plus qualifié pour le sujet! La présentation de l'ouvrage y gagnerait, je pense. Il me semble d'ailleurs difficile de faire autrement vu la longueur de la Note de M. D'Ancona.

Si vous êtes de mon avis, j'achèverai la conclusion en renvoyant à la Note de M. D'Ancona et parlant seulement des principales expériences qui semblent vérifier vos lois. Il y en a d'ailleurs peu en dehors des statistiques de M. D'Ancona.

---

[14] N.d. but probably written in April or May 1930.

En outre, M. D'Ancona devrait alors ajouter à sa Note des indications sur ses travaux de façon que cela fasse un tout - tout au moins terminer en en parlant et renvoyant aux indications de l'introduction. Mais ce serait le lieu de développer davantage.

Ensuite, comme cela ennuierait peut être M. D'Ancona d'écrire sa Note en française, je la traduirai.

Qu'en pensez vous?

D'autre part faut il faire imprimer ma Note?

Très respectueusement

M. Brelot

P.S. J'ai écrit à M. Schmidt à Berlin

4.24 Brelot to Volterra, Roma, n.d. Cat. no. 4[15]

Roma, giovedì sera

All'illustrissimo Prof. Vito Volterra

Avrei qualche correzione da fare sulle bozze che Le ho mandate - correzione piccola ma almeno utile. Infatti, spesso, un cambiamento porta con sé un altro al quale non si pensava.

Quindi Le domando non mandare le bozze a Parigi, se, per caso, o per una ragione qualunque, Lei volesse farlo.

Con tanti auguri per Sua salute.

Rispettosamente

M. Brelot.

4.25 Brelot to Volterra, Roma, n.d. *Cat. no. 11*[16]

Rome, dimanche soir

Cher Monsieur

Voici les placards corrigés. Les remarques de Melle Freda sont en général excellentes. Parfois Melle Freda a cru à une erreur alors que simplement ce n'était pas clair ou bien parce qu'une curieuse faute d'impression changeait complètement le sens. Dans tous les cas les remarques sont fort utiles.

Je ne croyais pas qu'il y aurait tant d'imperfections. A vrai dire je n'avais pas relu le manuscrit avec un certain recul comme pour celui de cours de M. Picard auquel j'avais consacré plus de temps, en plusieurs phases.

Je m'en excuse.

Très respectueusement.

M. Brelot

---

[15] N.d. but probably written in April or May 1930.

[16] N.d. although the most probable collocation in the correspondence is here. On Elena Freda, see the letter no. 6.1 (note 6).

P.S. – un petit "miracle"; j'ai trouvé deux fautes d'impression que n'a pas vues Melle Freda. J'ai dû aussi ajouter quelque part une ligne pour éviter une petite objection que n'a faite Melle Freda.

4.26 Brelot to Volterra, Roma, 7.5.1930 *Cat. no. 28*

Rome, mercredi 7 mai 1930

Cher Monsieur.

Voici mon travail sur l'étude du problème biologique de 2 espèces quand on introduit des termes $-\lambda N$.

<u>En résumé:</u>

Je suppose que pour la 1ere espèce le λ_1 est > 0 tandis que pour la 2e espèce le λ_2 est seulement ≥ 0. Je ne limite pas l'hérédité et j'établis (si $\varepsilon_1 > 0$, $\varepsilon_2 > 0$)

1) La 1ere espèce a toujours ses fluctuations bornées ($0 < a < N_1 < b$)
2) La 2e espèce est toujours limitée supérieurement
3) Les équations de l' état stationnaire donnent
   $K_1 > 0$, K_2 que peut être ≥ 0 ou < 0.

<u>Si $K_1 < 0$, càd</u> $\lambda_1 > \frac{\varepsilon_1}{\varepsilon_2}(\gamma_2 + \Gamma_2)$ (condition indépendante de λ_2), $N_2 \to 0$ et $N_1 \to \varepsilon_1/\lambda_1$.

<u>Si $K_2 > 0$</u> la 2e espèce est limitée inférieurement (fluctuations bornées) et on a alors la loi des moyennes asymptotiques sans restriction.

Il y a aussi impossibilité que $|N_1 - K_1|$, $|N_2 - K_2|$ restent supérieures à un nombre > 0 plus longtemps qu' une certaine <u>durée</u> (indépendante de son moment initial).

J'ajoute quelques mots sur la loi de perturbation des moyennes et je démontre enfin que si $\lambda_1\lambda_2 > (\gamma_1 + \Gamma_1)(\gamma_2 + \Gamma_2)$ <u>il y a un état limite</u>, qui est l'état stationnaire.

Peut être trouverez vous cette note assez intéressante pour figurer à la fin de l'ouvrage? Je m'en remets à vous pour juger ce qu'il convient d'en faire[17]. Dans le format de l'ouvrage cette Note aurait une douzaine de pages.

Je me permets de vous porter le manuscrit aujourd'hui car on m'a dit hier que vous alliez mieux et je vous envoie mes meilleurs vœux pour votre complet rétablissement. Je vous prie de présenter mes respects à Madame Volterra et mes sympathies à vos fils; et je vous prie de croire à mes sentiments très respectueux.

M. Brelot

Je pense vous voir vendredi matin à votre cours, toujours si intéressant par ses qualités de clarté d'exposition et de langage.

---

[17] The work was published separately (letter no. 4.27). See BRELOT 1931.

4.27 Brelot to Volterra, Roma, n.d. *Cat. no. 12*[18]

Rome, lundi matin

Cher Monsieur

J'ai reçu une réponse de M. Schmidt qui accepte que j'aille auprès de lui à Berlin. D'autre part j'ai parlé de cette prolongation de bourse à M. Vessiot qui me répond que c'est maintenant difficile à obtenir, mais peut être possible tout de même si vous appuyiez fortement ma demande. Comme vous serez sans doute à Paris à la fin du mois, tout dépend donc de vous, si vous jugez convenable que cette bourse me soit prolongée - au moins pour le semestre d'hiver.

J'ai donné à M. Maggi une Note d'une dizaine de pages. Je rédige une seconde Note que je vais lui envoyer. En effet en rédigeant, des difficultés ont surgi qui m'obligent à être plus long, et j'ai fait deux notes assez longues au lieu d'une petite. Cela m'a même pris beaucoup de temps et comme les vacances commencent bientôt à l'Ist. Lombardo, je me suis pressé de rédiger. Aussi n'ai je pas ces jours dernières traduit la Note de M. D'Ancona. Je pense le faire aujourd'hui même et demain, en faisant, suivant votre désir une traduction très large, un peu développée. Je n'ai rien reçu de Gauthier Villars ces temps derniers mais j'espère que la mise en pages de la 1ère moitie ou les 2emes épreuves de la seconde arriveront bientôt.

D'autre part le professeur français M. Bouligand, de l'Université de Poitiers, avec qui j'ai correspondu ces temps derniers au sujet de quelques points très récents de la théorie des fonctions harmoniques, me demande s'il ne lui serait pas possible de présenter sous peu une Note aux Lincei, sur certaines équations aux dérivées partielles du 1er ordre. Je lui ai dit que je vous consulterais à ce sujet.

Je pense que vous vous portez maintenant tout à fait bien et je vous prie de croire, pour vous et votre famille, à mes sentiments les plus respectueux.

M. Brelot, 6 Corso d'Italia, tel 31398

Du Circolo di Palermo, on m'a écrit que l'on m'enverra les épreuves de mon petit mémoire, à la fin du mois.

M. Pincherle a inscrit ma Note du problème biologique sur la liste des mémoires à paraître dans les Annali di Mat.

Vraiment je ne pourrai jamais vous remercier assez de toutes les facilités que vous me donnez si aimablement pour publier.

4.28 Brelot to Volterra, Roma, 15.6.1930 *Cat. no. 29*

Rome, dimanche 15 Juin 1930

Cher Monsieur

J'ai reçu une réponse de M. Tisdale de la Fondation Rockfeller, à qui j'avais écrit mon désir de voir prolonger ma bourse à Berlin, en abandonnant deux mois cet été de celle que j'ai actuellement.

---

[18] N.d. but probably written in May 1930.

Il me demande en particulier une lettre de vous – comme de M. Vessiot. Vous pourrez je pense lui répondre de vive voix. Je pense qu'il n'est pas utile que d'autres professeurs de Rome lui écrivent comme M. Tisdale semble le demander. Car j'ai eu peu affaire aux autres, sinon à M. Levi Civita que j'ai consulté parfois un moment.

Je demande donc une prolongation de bourse, qui si elle ne peut m'être accordée pour l'année suivante, pourrait être réduite au semestre d'hiver – ce qui dans ce cas reviendrait à 3 mois, vu les deux mois que j'abandonnais-: septembre et octobre, pour aller prendre mes vacances dans ma famille.

Si je ne vous revois pas avant votre départ à Paris, ce qui est probable, je vous envoie mes meilleurs vœux de bon voyage et je vous prie de croire, avec ma grande reconnaissance, à mes sentiments de sympathie, de gratitude et de profond respect.

M. Brelot

Je vous rappelle mes publications de cette année qui ne sont pas toutes parues encore.

Parues et	C. R. Ac. Sc.	30 déc. 1929
envoyées		13 Janvier 1930
à la Fondation		3 février
en triple		17 février
exemplaire	R. Dei Lincei	2 février
		16 février
		2 mars
Non encore		4 mai

parues	R. del Circolo Mat. di Palermo (janvier) mémoire d'environ	25 pages
	Annali di matematica (mai) --	15 pages
	R. Istituto Lombardo (juin) --	20 pages
	(pages présumées sur la revue imprimée)	

4.29 Brelot to Volterra, Roma, 21.6.1930 *Cat. no. 30*

Rome, 21/6/30

Cher Monsieur

Je reçois une lettre de M. Tisdale me disant qu'il y a de ma part une difficulté à demander une prolongation de bourse, vu que je ne connais pas l'allemand.

Je lui ai répondu d'abord que, dans ma famille j'avais autrefois commencé un peu cette étude abandonnée à cause d'examens qui en étaient indépendants, mais que surtout je comptais, dans les mois dont je dispose, arriver à une certaine connaissance de la langue.

J'ai justement ces jours derniers commencé à prendre des leçons d'allemand et je compte d'autre part, pendant les deux mois de vacances (sept-oct) que je prendrai, faire une étude intensive de cette langue, en particulier dans ma famille (avec un oncle professeur d'allemand au lycée). Somme toute je prendrais l'engagement tacite de pouvoir satisfaire à peu près en Novembre aux conditions requises. Je demande qu'on m'accorde confiance et qu'on ne retienne pas cette difficulté afin que je prendrai toutes dispositions pour qu'elle n'en soit pas une pour moi.

J'ajoute que une difficulté analogue existait pour l'italien; et que je me suis mis rapidement à même de ne pas être gêné ici, à Rome. Comme l'Allemand est plus difficile pour un français - encore que j'ai un assez grand vocabulaire anglais - je pense compenser cela par 4 mois d'études sérieuses. Je suis le 1er intéressé d'ailleurs à connaître suffisamment la langue Allemande avant de partir à Berlin.

Je voudrais espérer qu'on me fera confiance càd qu'on continuera à m'accorder une confiance que je me suis efforcé cette année de mériter tant au point de vue de la langue, qu'au point de vue scientifique - et qu'on acceptera comme suffisant l'engagement tacite de ma part d'une étude suffisante et que je crois vraiment possible d'ailleurs, avant Novembre.

J'ajoute que, ces mois dernièrs, j'ai fait de grands progrès en anglais, sans que mon travail eut à en souffrir. J'abandonne maintenant cette étude - que je faisais sérieusement avec un professeur - pour celle de l'allemand.

J'espère que vous ne voyez pas d'obstacle insurmontable dans cette question de langue et je vous prie de croire, avec toute ma reconnaissance, à mes sentiments les plus respectueux.

Avec tous mes souhaits d'heureux séjour en France et mes respects à Mme Volterra,

M. Brelot, 6 Corso d'Italia.

4.30 Brelot to Volterra, Roma, 5.7.1930 *Cat. no. 31*

Rome, 5 Juillet 1930

Cher Monsieur

J'ai appris par vos fils que vous étiez encore à Paris où je vous écris.

Voici: il fait ici une chaleur torride qui me déprime beaucoup; et je vais être obligé d' aller m' installer dans les environs si je veux travailler. Et je ne sais si je serai mieux.

D'autre part la bibliothèque fermera bientôt de sort que mon séjour ici - ou il n'y a plus de professeurs - n'a plus de raison d'exister.

Je songe que Zurich est à peu près sur le chemin de retour et que j'y serais infiniment mieux pour travailler. Très probablement je pourrais disposer d'une bibliothèque. Peut être même y a t il des professeurs. En tous cas quelle que soit, en été, cette Université, j'aurais plus d'avantages qu'à Rome où il n'y a plus rien.

Je demande donc à la Fondation Rockefeller de me permettre de finir mon séjour de bourse à Zurich - où j'aurais de plus des facilités pour faire de l'allemand.

Peut être même y serez vous en août?

Je viens d'écrire à M. Tisdale pour obtenir cette autorisation. Je vous remercierais beaucoup de lui envoyer un mot d' avis favorable.

Je ne demande aucune subvention spéciale naturellement. J'abandonnerais même volontiers la subvention du voyage de retour. De toute façon je voudrais pouvoir quitter Rome dont le climat actuel m'est très pénible. Je ne travaille presque plus - sans prendre la bourse qui m'a été accordée. Et j'espère d'ailleurs de Zurich divers autres avantages que celui du climat. Si vous n'y allez pas, je vous serais infiniment reconnaissant de m'indiquer à Zurich des professeurs que pourraient y être et dont je serais heureux de faire la connaissance.

Je pense que ma sollicitation est très légitime et qu'on ne saurait faire aucune objection. Il n'y en a pas, il me semble.

Je suis assez content de la tournure que prennent mes recherches mais vraiment je suis très déprimé par le climat.

Je désirerais fort obtenir cette autorisation d'aller à Zurich et je vous remercie infiniment à l'avance pour votre démarche.

Je vous dois déjà tant que je ne sais plus vous remercier davantage et je vous prie de croire à mon très profond respect - et de présenter mes hommages à Mme Volterra.

M. Brelot, 6 Corso d'Italia, Roma

- Je n' ai rien reçu de G. V.[19] encore -

4.31 Brelot to Volterra, Roma, 6.7.1930 *Cat. no. 32*

Rome, 6 juillet 1930

Cher Monsieur

J'ai oublié de vous dire certaines choses dans ma lettre d'hier. Tout d'abord au point de vue d'une fin de bourse possible à Zurich, voudrez vous m'indiquer quels professeurs seraient susceptibles de m'intéresser, s'ils étaient encore là. J'espère d'ailleurs que vous irez à Zurich comme votre fils Enrico me l'a dit - auquel cas la Fondation Rockefeller me ferait sans doute pas de difficulté.

Quant à mon travail, je suis assez content. J'ai lu - avec l'aide d'un allemand, parfois - le mémoire de Liechtenstein des Rend. di Palermo (1912) sur les équations du type elliptique. J'y trouve des choses très intéressantes et très précieuses - et aussi quelques propriétés que j'ai en particulier publiées récemment avec une démonstration différente, sans le savoir. D'autre part je vais pouvoir mettre sur pied un mémoire que fera suite à celui de l'Ist. Lombardo pour l'étude de la solution au voisinage de la frontière, étude naturellement basée sur celle du cas harmonique.

Enfin je tiens à vous remercier pour m'avoir indiqué dans le courant de l'année le livre d'Evans (Cambridge Colloquium 1916) qui m'est précieux et auquel je n'aurais jamais pensé.

Bien d'autres de vos indications de ce genre m'ont été utiles, quelquefois longtemps après. Je vous remercie bien vivement.

Je songe maintenant à votre livre dont les épreuves n'arrivent pas. Il faut que je vous dise deux mots au sujet de la préface. Vous m'aviez dit qu'il me fallait, dans le texte, aux endroits où j'avais apportée une petite modification ou

---

[19] Gauthier-Villars.

addition, mettre une Note avec mes initiales; et que vous parleriez de cela dans la préface. Mais il n'y a vraiment que deux ou trois endroits au plus dans le livre, où il y a peut être effectivement une petite contribution de ma part. Il serait parfaitement ridicule de me citer pour des remarques ou modification de pure rédaction; et s'il y a de telles choses avec indication de mon nom, je le supprimerai. Quant aux deux ou trois endroits où mon nom peut figurer, il n'y a vraiment pas lieu de faire la convention des initiales qui laisserait supposer de ma part une contribution nombreuse. De sort que dans la préface, je vous prierais de ne point parler de cette question des initiales.

J'ai écrit à M. W R Thompson. Il m'a répondu que, très occupé par un congrès à Londres en ce moment, il m'écrirait dans quelques temps. Je lui avais résumé mon petit mémoire des "Annali di Matematica"[20].

M. Bouligand m'a dit récemment que vous lui avez écrit une lettre infiniment aimable et vous chargiez d'envoyer sa Note aux Lincei. Il se pourrait que j'ai dans quelque temps une petite Note qui irait bien pour le "Bollettino". Je vous en reparlerai s'il y a lieu.

Je m'excuse de vous déranger par ces longues lettres et tous les services que vous me rendez constamment. Je vous souhaite un séjour agréable à Paris, ainsi qu'à Mme Volterra. Jamais Rome ne m'a paru aussi attachante, car on oublie la nuit, le climat si déprimant, encore qu'il ne fasse guère frais le soir. Je crois que je ne pourrai plus faire autrement dans l'avenir que revenir ici quelquefois, le plus souvent possible. Qui connaît la Villa Borghese et le Pincio le soir ne saurait plus goûter les jardins des autres villes; les boulevards parisiens m'attirent bien peu quand je songe à la Via Veneto. Les jardins de Versailles n'ont pas le charme spécial de la Villa D'Este et ce ne sera pas sans une peine secrète que je quitterai votre ville - une peine dont vous étes "responsable" ... puisque c'est grâce à vous que je suis venu ici!!

Ajouterai-je qu'une des choses qui me sera la plus pénible sera de ne plus entendre parler italien; et c'est un bien grand sacrifice que l'abandonner pour l'allemand.

Si je n'ai plus l'occasion de vous revoir, ces mois ici, je veux du moins maintenant vous exprimer toute ma reconnaissance très sincère et très profonde, que je ne pourrai jamais séparer des souvenir si tenace et sûrement nostalgique d'une ville beaucoup plus attachante qu' on ne saurait croire.

Je vous serais reconnaissant de présenter mon respectueux souvenir à M. Picard à qui j'enverrai peut être ces vacances une Note pour les C.R., a laquelle je pense depuis longtemps

Veuillez croire à mes très respectueuse sympathie.

M. Brelot, 6 Corso d'Italia

---

20 Brelot 1931.

4.32 Brelot to Volterra, Roma, 7.7.1930 *Cat. no.* 33

Rome, lundi 7 juillet 1930

Cher Monsieur

Je reçois votre lettre dont je vous remercie infiniment. Je suis très heureux de voir qu'une prolongation de bourse me sera probablement accordée. M. Tisdale m'avait écrit au sujet de l'allemand, après ma réponse et semblait être satisfait de ce que je lui avais dit. Je pense qu'on ne retiendra pas cette difficulté. J'apprends l'allemand en ce moment et je m'essaie à lire, comme je vous le disais hier, les mémoires de Lichtenstein.

Je m'aperçois que mes études primitives de $\Delta u = cu$ au voisinage d'un point singulier de c peuvent s'étendre beaucoup je crois - d'une part en prenant au lieu d'un point isolé un ensemble plus général de "capacité nulle" au sens de Bouligand-Wiener-Vasilesco, - d'autre part en prenant l'équation elliptique générale. Je pense que je pourrai traiter ce problème plus étendu en suivant les mêmes idées et grâce aux travaux récents sur les fonctions harmoniques. Il y a je pense quelque intérêt a étudier les intégrales de l'équation elliptique au voisinage d'un ensemble de points singuliers des coefficients - ensemble qui dans l'espace à 3 dimensions p ex est bien plus général qu'un arc de courbe - et cela sans hypothèse sur l'allure des coefficients de l'équation dans ce voisinage.

D'autre part la lecture du mémoire de Lichtenstein me fait avancer beaucoup dans l'extension du principe de Picard (pour les fonctions harmoniques: existence unique d'une fonction harmonique nulle sur un contour, ≥ 0 à l'intérieur et singulière en un point 0, à un facteur près) quand au point 0 il y a singularité pour les coefficients de l'équation elliptique. Je sais dès maintenant démontrer l'existence.

Tout cela n'est malheureusement guère original. Je m'aperçois même en étudiant les questions connues que les procédés que j'ai employés, l'avaient déjà été sous des formes voisines, bien souvent, dans des questions analogues. Ce qui diminue la valeur de ce que j'ai fait. Je manquais évidemment au début de culture sur le sujet. Seulement je n'aurais sans doute jamais pu l'acquérir sans m'essayer à quelques recherches.

Je me demande toujours s'il me sera possible comme je l'ai demandé, d'aller qq. temps à Zurich. Ce serait vraiment une bonne chose à tout point de vue, car je vais ici bien souffrir de la chaleur; et si l'on ne peut rentrer en France, dans ma famille à la campagne - ou je n'aurai pas de bibliothèque ce qui me gênera beaucoup et m'empêchera en particulier du publier quoi que ce soit, la bibliographie manquant.

Je vais peut être ces jours ci tâcher de trouver hors de Rome quelque coin où je puisse retrouver un peu de forces car je suis assez déprimé. On dit que il en est toujours ainsi le premier été qu'on passé ici. Et quand on aime l'activité, il est vraiment désagréable de somnoler les trois quarts du temps.

Avec toute ma reconnaissance, veuillez croire à mes sentiments les plus respectueux.

M. Brelot

4.33 Brelot to Volterra, Capri, 29.7.1930 *Cat. no. 34*

Capri, 29 luglio 1930

Carissimo professore.

Non sapevo che fare per fuggire il calore di Roma e finalmente sono venuto a Capri dove, ora, studio molto il tedesco con un professore tedesco della pensione. Penso tornare tra poco in Francia ma prima a Roma. Lei sa forse che, infatti, ho ottenuto sei mesi di borsa per l'anno venturo; aggiungerò tre mesi di questo anno e perciò, prenderò 3 mesi di vacanze e tornerò fra poco a casa mia. Preferisco fare cosi di modo che abbia anche 9 mesi di borsa l'anno prossimo.

Dunque non rimarrò a Zurigo come avevo domandato e come Le avevo spiegato nelle mie ultime lettere. Per lo meno non lo farò con la borsa - ma forse per conto mio.

Ho ricevuto tutte le bozze del Suo libro, la prima parte in pagine e la fine in seconde bozze - ma non so il Suo indirizzo. Può darsi che Lei abbia lasciato Parigi molto tempo fa e che le mie ultime tre lettere non siano arrivate.

Manderò questa lettera a Roma; seguirà, spero. Ma per il pacco delle bozze, preferisco aspettare il mio ritorno a Roma - fra qualche giorno - e domandare a casa Sua, il Suo indirizzo.

Come Lei vedrà sulle bozze ho aggiunto qualche osservazione in Nota per dire che, quando fra le specie considerate alcune spariscono, i limiti delle altre hanno i valori che avrebbero senza le prime. Nel Suo corso Lei aveva negletto le prime dopo un certo momento; nei casi per i quali faccio l'osservazione, si può stabilire che la conclusione é valida conservando le prime specie (infinitamente piccole) - non do dimostrazione. Indico solo il fatto in una riga. Questa proprietà si stabilisce facendo certi ragionamenti facili che ho usato nella mia piccola memoria. È poco importante ma però credo che sia abbastanza interessante dal punto di vista teorico per farne l'osservazione in Nota.

La ringrazio infinitamente per la parte che Lei ha presa nell'ottenere la borsa e Le mando tanti rispettosissimi saluti.

M. Brelot fra quindici giorni à Boisseaux (Loiret) Francia

Se Lei non ha ricevuto ancora le mie lettere, vorrebbe Ella dirmelo? perché Le avevo scritto cose sulla prefazione - aggiungo che il libro é stato scritto per chi conosce press'a poco il corso di "Mathématiques Générales" in Francia.

4.34 Brelot to Volterra, Capri, 29.7.1930 *Cat. no. 35*

HÔTEL PENSION REGINA
CAPRI
Telefono N. 25

Capri, 29/7/30

Cher Monsieur

Je reçois votre paquet au moment où j'allais mettre à la poste une lettre que je joins à celle ci.

J'ai tout de suite regardé p. 39 et j'ai constaté que G. V. a oublié une page entière. En ce qui concerne le mot biologique, je l'avais supprimé bien des fois et je regrette de l'avoir oublié en certaines endroits; j'y veillerai. Enfin pour ce qui est de l'omission relativement à votre mémoire, je ne puis trop rien vous dire maintenant. Je n'ai pas ici ce mémoire. Je ne me souviens pas si j'ai mis ailleurs ou non ce que vous me dîtes. Il y a si longtemps que j'ai oublié. S'il y a vraiment omission grave, peut-être y a-t il une excuse à mon ignorance de l'italien il y a un an.

Je ferai de mon mieux pour réparer l'omission si elle existe.

Je vais donc rentrer prochainement à Rome. Si c'était utile peut être irais-je vous voir dans une dizaine de jours en rentrant en France. Je crois que ce me serait facile.

Veuillez croire à mes sentiments les plus respectueux

M. Brelot

4.35 Brelot to Volterra, Rome, 2.8.1930 *Cat. no. 36*

(Same heading of letter no. 34)

Rome, 2 août

Cher Monsieur

Je suis rentré hier soir à Rome et j'ai regardé tout de suite votre mémoire de Venise[21]. Il y a deux pages seulement dont je n'ai pas parlé (P. 74–75). Vous y étudiez un système impair dont une espèce est très peu nombreuse et les autres forment un système voisin d'un état d'équilibre.

Je ne me rappelle plus si j'avais vu ou non cela. Mais je me permettrai de vous faire remarquer que cette courte étude est tout à fait contenue dans celle plus générale (p 102) d'un système voisin d'un état d'équilibre et qu'on perturbe par l'introduction de nouvelles espèces. La condition $L < 0$ n'est autre que celle que le coefficient virtuel d'accroissement de la premier espèce (p 74–71) soit < 0.

De sorte que cette omission n'enlève rien, je pense aux résultats. Mais il peut y avoir utilité à traiter d'abord le cas particulier, peut être?

Ce serait tout différent si dans cette omission on ne se plaçait pas dans le cas de: 1ère espèce infiniment petite et autres espèces inf^{t} voisines d'un état d'équilibre.

Je rédigerai bien tout de suite ces 2 pages dont je pense que Gauthier-Villars ne ferait pas trop de difficulté à accepter, en refaisant une mise en pages. Mais, je pense que (p 102) dans le cas général, vous avez abandonné le raisonnement basé sur le fait de négliger le 2e ordre; vous avez substitué un raisonnement de calcul des variations, qui a des bases plus rigoureuses ou au moins plus nettes.

S'il faut intercaler le passage aussi, faut il conserver exactement la forme de votre mémoire avec ce procédé de négliger les infiniment petits, ou changer, pour répéter, somme toute, dans un cas particulier le nouveau procédé avec variations?

---

21 Volterra 1927d.

Je vous envoie le passage de votre mémoire. Je vous prierai de me les renvoyer car je n'ai qu'un exemplaire de ce mémoire. J'y joins les passages utiles de l'ouvrage en cours d'impression.

D'autre part je vous envoie aujourd'hui même la fin des placards, que j'ai corrigé ces jours derniers; mais comme le paquet ira moins vite que cette lettre, je mets dans cette lettre quelques feuilles en double intéressantes pour le point en question.

Je vais tâcher de faire parvenir à M. [not complete]

4.36 Volterra to Brelot, Saint Gervais-Les-Bains, 5.8.1930 *Cat no. 360*[22]

SAINT GERVAIS-LES-BAINS *5 Août 1930*
HÔTEL MONT JOLY (HAUTE SAVOIE)

Cher Monsieur

Je vous avais écrit hier la lettre ci-jointe lorsque j'ai reçu votre lettre de Rome du 2 Août avec les placards et les preuves en pages.

Je trouve inutile de changer les épreuves en pages 49–64 en ajoutant ce qui se trouve dans les pages 74–76 du mémoire Italien de Venise, parce que les résultats sont inclus dans l'étude que l'on fait après d'un système voisin d'un état d'équilibre et que l'on perturbe par l'introduction de nouvelles espèces (placards 19–20). Je trouve plutôt que l'on pourrait ajouter, après l'étude de ces perturbations, et comme application le cas où le système d'un nombre impair d'espèces se réduit à un nombre pair par épuisement de la première en remarquant que l'on peut obtenir le même résultat par une autre voie (celle de pages 74–76 du mémoire de Venise) que l'on pourrait indiquer très succinctement.

Le cas traité dans le §4 de la page 105 du mémoire de Venise me semble au premier abord qui ne rentre pas dans l' étude des perturbations des pages françaises 49–64 parce que les conditions sont un peu différentes, la racine q_1 de la page 105 n'étant pas un coefficient d'accroissement virtuel. Vous pourrez vérifier ce point mieux que moi.

Je vous renvoie ci inclus les épreuves en pages les placards et les pages détachées du mémoire de Venise.

C' est bien ce que vous dites par rapport à la série de Fourier.

4.37 Brelot to Volterra, Boisseaux, 11.8.1930 *Cat. no. 37*

Boisseaux, 11 Août 1930

Cher Monsieur

J'arrive et je trouve toutes vos lettres. Je vous envoie aujourd'hui la mise en pages corrigée – que vous pouvez expédier directement à G. V. – et aussi la préface

22 Draft.

avec des modifications au crayon. J'ai changé bien des choses pour que les tournures (grammaticalement correctes) soient plus appropriées ou simplement plus élégantes.

J'ai précisé pour les Notes mathématiques qu'il s'agirait des deux premières; je pense que vous pourriez dire que, dans une 3e Note vous avez cru devoir mettre quelques démonstrations d'existence qui auraient alourdi le texte; on aurait pu croire à lire votre préface que cette 3e Note m'était personnelle alors qu'il n'y a qu'une page où j'ai rappelé des résultats classiques sur les équations de 1ère et 2e espèce.

Vous pouvez alors envoyer directement aussi à G.V. ces quelques pages de préface.

J'envoie à M. D'Ancona le début du livre qu'il me réclame et son manuscrit italien qu'il veut comparer à ma traduction.

Je vais ces jours ci ajouter aux placards quelques indications sur le cas impair comme application de l'étude des perturbations et aussi lire l'extrait du travail allemand de M. D'Ancona[23].

Après avoir pensé à m'arrêter quelque temps à Chamonix, j'y ai renoncé, surtout à cause du mauvais temps. Dans le Nord de l'Italie la saison est bien mauvaise. Ce n'est plus Capri.

4.38 Brelot to Volterra, Boisseaux, 12.8.1930 *Cat. no. 38*

Boisseaux (Loiret), 12 Août 1930

Cher Monsieur

Vous me demandiez quand je pense partir en Allemagne. Vers la fin d'Octobre je crois. Plusieurs de mes camarades iront aussi à Berlin avec une bourse Rockefeller. D'autre part ira à Rome auprès de MM. Severi et Enriques, un de mes amis, Dubreil, qui a achevé de rédiger une thèse sur les courbes algébriques. Il s'est marié récemment avec la célèbre élève femme de l'Ecole, Melle Jacotin, agrégeé de mathématiques, préparant elle aussi une thèse - et d'ailleurs fort jolie[24].

M. d'Ancona voulait le début du livre, pour mieux rédiger la fin, et la notice biologique. Il me semble que contrairement à ce qu'il m'a laissé croire, il n'y a pas d'inconvénient à ce que soient répétées diverses références. Dans l'introduction et conclusion on donne l'essentiel surtout au point de vue mathématique. La notice spéciale doit être complète donc comprendre les indications de l'Introduction. C'est à ce sujet que je vous écris aujourd'hui, avant que vous ne receviez les épreuves annotées de M. d'Ancona.

Le Cercle de Palerme m'a reculé en septembre l'envoi des épreuves de mon mémoire (promises pour mai). Par contre je viens de recevoir celles du mémoire

---

[23] D'ANCONA 1927.

[24] This was the mathematician Paul Dubreil (n. 1904), who became well known for his work on lattice theory.

de l'Ist. Lombardo, pleines de fautes d'ailleurs. La typographie est loin d'être aussi bonne que celle des Lincei.

Veuillez croire à mes sentiments très respectueux et très reconnaissants.

M. Brelot

4.39 Brelot to Volterra, Boisseaux, 15.8.1930 *Cat. no. 39*

Boisseaux, 15 Agosto 1930

Carissimo Professore

Ho ricevuto oggi tutte le Sue bozze. Le mando oggi il foglio num. 21 coll'addizione sul sistema conservativo e dispari e anche il foglio n° 16 che era perduto; questo non è quello che G.V. dice di aver mandato col foglio delle prime bozze (che non ho mai ricevuto); è un altro, simile, credo, poiché c'è il n° 2 scritto nel angolo.

Lei potrà mandare a Parigi queste bozze e dare l'ordine del mettere in pagine. Dalla mia parte mando oggi le altre bozze. Lei potrà precisare di mettere in pagine soltanto fino alla conclusione; mi pare necessario di avere altre bozze della conclusione e dell'ultimo capitolo del Prof. d'Ancona.

Sono sicuro che la signorina Freda ha letto tutte le prime bozze; ho tenuto conto delle sue osservazioni per le seconde bozze che Lei ora sta leggendo.

Aspetto le osservazioni del Prof. d'Ancona che deve mandarle prima a Lei; dopo correggerò e aggiungerò qualche linea sopra il libro tedesco; poi quando saranno fatte le nuove bozze le manderò al signor d'Ancona.

Dalla mia partenza di Capri e Roma, non ho più trovato il vero bel tempo. C'era molta folla a Viareggio; ma dicono che l'anno scorso c'era doppia gente; doveva dormire nei camerini da bagno.

Può darsi che andrò in Belgio fra una settimana, ma per qualche giorno solo. Avrei molto da fare per il mio lavoro. Ma sono stanchissimo dei due ultimi mesi e del soggiorno a Roma di giugno-luglio. Avevo molto dimagrito e comincio solo a stare quasi bene. Spero scrivere una delle due o tre Note "nel aria" il mese prossimo.

La prego di salutare per me i figli; tanti auguri per la migliore salute di Enrico. Le mando, come alla Sua signora, i miei sentimenti rispettosissimi.

M. Brelot

Spero che l'addizione del sistema dispari basterà così; nel caso contrario, aspetto le Sue osservazioni e modificherò subito.

4.40 Brelot to Volterra, Boisseaux, 20.8.1930 *Cat. no. 40*

Boisseaux, 20 Août 1930

Carissimo Professore

Ho letto, questi giorni, le pagine estratte della opera del Friederichs[25], che mi aveva mandate il Prof. d'Ancona. I Suoi lavori sono citati e riconosciuti; ci si vedono anche alcune messe in equazioni. L'autore approva le Sue viste e idee. Osserverò che lascia credere, come la memoria del Prof. d'Ancona, che i sistemi di specie si dividono in conservativi e dissipativi. Nella redazione del libro Suo, ho scritto in evidenza che queste due sorte di sistemi sono distinti e non esauriscono tutti i casi; può darsi che sarebbe utile insistere, osservare che i sistemi conservativi sono improbabili (condizioni di uguaglianza) riguardo ai dissipativi (condiz. di diseguaglianza).

Ho dimenticato un'osservazione nell'ultima lettera; nella prefazione Lei non ha parlato del ultimo capitolo o piuttosto della Nota biologica e della contribuzione del Prof. d'Ancona; questo potrebbe essere utile per un libro che si rivolge ai naturalisti, essendo tale il professor d'Ancona. Come le Note matematiche, questa Nota è piuttosto fatta per i naturalisti.

Spero ricevere presto le ultime bozze e che fra due mesi il libro potrà essere stampato. Penso oggi che durante la prima settimana di settembre Lei potrà, se desidera guadagnare qualche giorno, scrivermi:

chez M. Houy, Cépoy (Loiret)

Mi scuso di parlarLe con tanta libertà ma è nell'interesse della Sua opera.

Le mando per tutta la famiglia e Lei stesso i miei rispettosissimi sentimenti e auguri

M. Brelot

4.41 Brelot to Volterra, Boisseaux, 21.8.1930 *Cat. no. 41*

Boisseaux, 21 Août 1930

Carissimo Professore

Ho ricevuto stamane la Sua lettera del 19, con le bozze. Ho fatto tutte le correzioni, poi ho mandato a Gauthiers-Villars, pregandolo mandarmi i fogli 1–6 corretti. Penso che Lei gli ha mandato le altre bozze e che G. V. potrà dunque mettere in pagine fino alla conclusione come glielo ho scritto.

Ho creduto bene di non aggiungere la citazione del mio nome per la piccola dimostrazione del placard 16. Infatti è insignificante, senza importanza e per tali cose bisogna fermarsi prima che la citazione divenga un po' ridicola.

Sono felice che il tempo sia un po' migliore nelle Alpi e che la Sua famiglia sta bene.

La prego di credere ai miei rispettosissimi sentimenti.

M. Brelot

---

[25] Friederichs 1930b.

4.42 Brelot to Volterra, Boisseaux, 22.8.1930 *Cat. no. 42*

Boisseaux, 22 Août 1930

Cher Monsieur

Je reçois les épreuves annotées et corrigées de la fin de l'ouvrage et je réponds tout de suite à vos remarques diverses.

1) Je ne vois aucun inconvénient, au contraire, à réunir la conclusion et la bibliographie mais je ne puis ne pas critiquer un peu le choix du titre conclusion, que M. d'Ancona a proposé. Il me paraît difficile que l'intérêt principal et la plus grande partie du chapitre "conclusion" d'un ouvrage mathématique consiste dans un historique et une bibliographie de travaux biologiques. Pourquoi ne pas réunir les deux titres et écrire "Conclusion – historique et bibliographie des associations biologiques"? On pourrait ajouter au moins quelque chose à "Conclusion". Je vous suggère cela sans plus. Sur les autres épreuves, il sera toujours temps de modifier. Je ne toucherai à rien en attendant selon le désir de M. d'Ancona.

2) Les difficultés de traduction que vous me signalez m'avaient déjà fort embarrassé car souvent la traduction littérale n'est pas usitée en français, ou bien n'a pas je crois le même sens, parce qu'elle laisse sous entendre quelque chose que n'existait pas. Il peut aussi n'y pas avoir de mot unique correspondant au mot italien. Enfin je ne connais pas le langage technique biologique et cela m'a beaucoup gêné.

Par ex: "espèces qui se défendent moins" M. D'Ancona voudrait "meno difese" et vous "moins bien défendues". Le passif français sous entend que la défense serait faite par d'autres individus; aussi avais-je employé la forme réfléchie "se défendre" parce que je pensais que le sens était tel, les espèces se défendant elles mêmes

inermi – sans armes. Cette expression française me semblait étrange et j'avais mis "passives", voulant dire que les espèces ne résistent pas, se laissent attaquer et dévorer. Le sens est différent, un peu seulement. Peut-être pourrait on dire "sans défense" en changeant et allongeant la phrase qui se prêtait seulement à un adjectif "au préjudice des espèces les plus passives"?

agrumi Dans mon dictionnaire il y a en regard "fruits acides, oranges, citrons etc". Je viens de chercher dans le gros Larousse en 7 vol. le mot agrume; il signifie une espèce de prune d'Agen. Je ne sais s'il existe un mot technique traduisant exactement. J'avais seulement regardé dans mon dictionnaire et je ne sais trop que faire de plus.

filet à traîne Je ne trouve rien ici dans le gros Larousse. Je tâcherai de me renseigner. A priori "à traîne" a une tournure plus français que "traînant" et il me semblait que c'était une expression qui s'emploie.

A propos de certaines de vos notes – vous supprimez, quand il s'agit d'un accroissement trop grand modifiant les conditions de vie, tout un membre de phrase que je souligne:

ce fait empêche pratiquement la croissance infinie qui résulte dans certains cas des équations adoptées jusqu'à présent.

Ce qui reste ne peut plus aller; je n'en vois plus guère le sens et je ne comprends pas votre suppression. Il est exact que le calcul ait fourni un accroissement $\rightarrow +\infty$ dans certains cas et c'est pour éviter cela que l'on modifie la mise en équations. Je voulais rappeler ce fait. Pourquoi supprimer la proposition et laisser "empêche la croissance" qui ne correspond pas sûrement à votre pensée. Que dire donc?

Apport unique d'individus. Je veux dire un seul apport, afin de bien marquer la distinction avec un apport continu qui, comme une destruction continue aurait pour but de modifier $\underline{\varepsilon}$ artificiellement, comme on a eu l'occasion de le faire. Il me paraît utile dans la phrase d'éviter la confusion "non plus une destruction continue, mais un apport unique d' individus ...".

Et "apport unique" me semblait mieux que "un seul apport", qui pourrait peut être interprété être "seulement un apport", "seulement" portant non sur le nombre des apports, mais sur la qualité; apport et non destruction. Je ne crois pas qu'en français unique puisse laisser croire qu'on n'apporte qu'un individu et d'ailleurs "individus" est au pluriel.

Je voulais déjà vous proposer de citer votre dernière note aux Lincei comme vous le faites.

Je vais tâcher de tout concilier sur ces divers points que je me permets de discuter, en général au point de vue de la langue française.

D'ailleurs il y a peu de telles difficultés et il sera possible d'y retoucher sur les secondes épreuves.

J'ai retardé mon court voyage in Belgique; je serai assez souvent absent pendant une quinzaine mais vers le 30 Août et le 6 ou 7 sept. je passerai chez mon grand père à Cépoy (Loiret) - adresse que je vous ai donnée. Je n'irai en Belgique qu'au début de septembre, mais avant je vais aller quelques jours chez un ami dans l'Yonne. Si vous voulez m'écrire et gagner 2 jours en évitant un renvoi de lettres, écrivez donc directement jusqu'au 8 ou 10 septembre a Cépoy (chez M. Houy).

Je vous souhaite une excellente fin de séjour à St. Gervais et vous prie de croire à mes sentiments de reconnaissance et de respect.

M. Brelot

4.43 Brelot to Volterra, Paris, 24.8.1930 *Cat. no. 43*

Paris, dimanche 24 Août 1930

Cher Monsieur

J'ai achevé la correction des derniers placards. C'est un travail assez difficile car après les diverses suppressions et additions de M. d'Ancona, le texte est devenu d'un français critiquable par endroits à cause des tournures, des phrases un peu déséquilibrées et des absences de liaisons. Sur plusieurs points je crains que le sens après correction ne diffère de la pensée de M. D'Ancona. Je me suis à peu près toujours tenu à conserver le modifications de M. D'Ancona; j'ai très peu varié mais peut être faudra-t-il faire quelques retouches plus faciles sur des nouvelles épreuves. Quand j'enverrai les épreuves à M. D'Ancona je soulignerai et commenterai les modifications que je crois devoir apporter. Il y a bien des difficultés de langage et de traduction, plus que vous ne pensez sans doute. J'ai pu améliorer certains points que vous m'avez signalés. J'ai remplacé "qui se défendent moins" correspondant à "meno difese" par "de moindre défense" - par ex. La traduction littérale est impossible comme en bien d'autres endroits.

J'ai été un peu surpris que M. D'Ancona supprime les quelques titres partiels que j'avais introduit pour faciliter les lecture d'un chapitre que me semblait bien compact mais je m'en tiens à cette réflexion, voulant respecter l'appréciation de M. D'Ancona.

Quant à la demi page dactylographiée où se trouvaient les références de Thompson, je l'avais collée à ma traduction au lieu de la transcrire. Je la demanderai à Gauthier-Villars et pourrai contrôler moi même les citations - d'ailleurs l'envoyer à M. D'Ancona s'il veut aussi faire le contrôle.

En ce qui concerne ces modifications de références, je vous dirai que j'avais à peine jeté un coup d'œil sur toutes ces Notes quand je vous ai envoyé les placards corrigés. Je me réserverai d'y revenir en détail, après modifications de votre part ou de celle de M. d'Ancona. Seulement M. d'Ancona m'a réclamé son manuscrit dans l'intervalle ce qui m'aurait fort gêné. De sorte que je suis très heureux qu'il ait fait cette vérification.

Je crois bien vous avoir demandé si vous avez expédié à G.V. la plus grande partie des placards 16–30. Au milieu de ces divers envoi, j'ai dû faire une confusion de souvenirs et hier envoyer directement à G.V. le gros paquet de ces placards. De sorte qu'ils sont tous à l'imprimerie maintenant.

Je ne pense pas qu'il y ait à ajouter quelque chose à propos du traité de Friederichs; M. d'Ancona en a parlé; cela suffit je pense. Mais M. D'Ancona désire-t-il les pages dactylographiées extraites de cet ouvrage?

Veuillez croire à mes sentiments les plus respectueux.

M. Brelot.

4.44 Volterra to Brelot, Saint Gervais-Les-Bains, 26.8.1930 *Cat no. 36*[26]

MONT - JOLY PALACE *26 Août 1930*
ST. GERVAIS - LES BAINS
MONTJOLY PALACE - ST. GERVAIS LES BAINS
pth 8

Cher Monsieur

Merci de vos lettres du 22 et du 24 courant. Je m'empresse à vous répondre.

1) Moi aussi je pensais de mettre comme titre du dernière chapitre: Conclusion - Historique - Bibliographie. C' est pourquoi je suis d'accord avec vous en prenant comme titre les trois mots précédents.

2) Il me semble qu' il ne faut pas à accepter la phrase de moindre défense et qu'il soit préférable d'adopter l'autre: moins bien protégés

3) inermi. Je crois que l'on pourrait traduire dépourvus d'armes. Êtes vous d'accord?

4) agrumi. Je vous ai déjà écrit que fruits acides va très bien;

[26] Draft.

5) Filet à traîne. J'ai écrit à M. D'Ancona qui doit voir sous peu M. Joubin de lui demander le terme exact technique Français.

6) Je trouve que vous avez parfaitement raison de conserver les mots:.... la croissance infinie qui résulte dans certains cas des équations adoptées jusqu'à présent.

7) Apport unique d'individus; unique va bien. M. D'Ancona aussi était très préoccupé des difficultés que vous avez rencontré à rendre en bonne langue française et dans un bon style ce qu'il avait écrit. Je vois que vous vous êtes donné beaucoup de peine pour le faire. Je ne doute pas vous y serez réussi et je vous en suis très reconnaissant.

Nous partirons Jeudi pour Genève. Le temps depuis quelques jours est très beau et il fait chaud. Si vous m'écrivez veuillez diriger les lettres à Rome.

Veuillez accepter, cher Monsieur, les meilleures salutations et l'expression de ma sincère affection et de mon estime.

Vito Volterra

Si vous m'écrivez veuillez diriger les lettres à Rome.

4.45 Brelot to Volterra, Bruxelles, 2.9.1930 *Cat. no. 44*

Bruxelles, 2 septembre 1930

Cher Monsieur

J'ai bien reçu vos deux dernières lettres (l'une pour Boisseaux, l'autre pour Cépoy). A propos d' "inermi" comme il y a, je crois, une comparaison, j'ais mis "celles qui sont le moins bien armées". Quant à "les moins bien protégées" c'est beaucoup mieux que "----défendues", et cela peut aller; mais pourquoi trouvez vous inacceptable "de moindre défense"; vous ne donnez pas de raison. Pour "agrumi" j'ai mis une périphrase "certains fruits tels que ...". Tout cela était fait et expédié à G.V. avant que votre réponse n'arrive. Vous pourrez y retoucher.

Je suis passé chez G.V. ce matin alors qu'il se disposait à m'envoyer les mises en pages ci-incluses que je viens de corriger. Il y a peu de corrections et vous pouvez peut être signer. Sinon je puis les revoir encore une fois, surtout si vous avez de nouvelles modifications à faire.

On m'a envoyé ces jours derniers chez moi, où je n'étais pas, les nouvelles mises en pages du début. Comme vous me l'avez dit, je m'assurerai des rectifications et je donnerai à votre place le bon à tirer; car je crois inutile de vous faire l'envoi.

On ne m'a pas donné la préface. Je pense que nous l'aurons bientôt; et vous pourrez y ajouter quelques lignes sur M. d'Ancona à coté de ce que vous dites de Melle Freda.

Dans une semaine je retournerai chez mon grand père à Cépoy, puis après le 15 septembre je serai chez mes parents à Cépoy.

J' envoie mes meilleurs souhaits pour toute votre famille et je vous prie de croire à mes sentiments les plus respectueux.

M. Brelot

4.46 Brelot to Volterra, Paris, 9.9.1930 *Cat. no. 45*

Paris, 9 septembre

Cher Monsieur

A mon retour de Belgique je passe chez Gauthier-Villars où je trouve diverses épreuves à corriger dont la préface. Je rentre chez moi demain où je trouverai d'autres épreuves. Je vous envoie tout de suite cette préface que j'ai annotée au crayon; je pense qu'on en peut améliorer le style. Voici quelques observations par exemple, entre autres:

1) a propos de vos recherches commencées en 1915 vous dites "elles continuèrent dans les 1ers mois ..." On a l'impression qu'elles sont indépendantes de vous; je crois qu'il est préférable qu'il y ait dans la phrase le sujet "je". Le "dans les 1ers mois" n'est pas très usité je crois. De même "dans l'automne"; on peut le supprimer ou mettre autre chose pendant ou au cours de

2) au lieu de renseignements n'est il pas mieux d'écrire "propriétés"?

3) A propos de la collaboration de Mlle Freda la tournure du texte semble dire que la façon dont Mlle Freda nous a aidé à corriger les épreuves consistant dans une collaboration du texte qu'elle seule aurait faite; aussi je vous propose autre chose.

4) J'intercale quelques lignes pour M. D'Ancona indiquant son titre de professeur a Sienne; il me semble utile de le dire (prof. de sciences naturelles).

Je crois que l'on peut lier deux ou trois paragraphes sans aller à la ligne.

Je vous soumets ces observations que vous apprécierez. Vous verrez si vous croyez devoir tenir compte de certaines d'entre elles. Je serais heureux si mes suggestions pouvaient vous être utiles.

Je vous enverrai très prochainement d'autres épreuves, et la page que me [illeg.] M. D'Ancona. Je serai à Cépoy chez M. Houy jusqu'au 15, ensuite chez moi à Boisseaux.

Je vous prie de croire à mes sentiments les plus respectueux.

M. Brelot

P.S. efficace me semble devoir être remplacé par un autre mot. p. ex. précieuse, si utile.

aider pourrait être remplacé par seconder à la cause de la proximité du mot aide. On pourrait dire "et seconder M. Brelot et moi même dans la correction des épreuves et l'amélioration du texte".

Ce serait plus correct et meilleur que l'emploi de l'infinitif "pour corriger etc." On n'a pas en français votre emploi si heureux de l'infinitif italien.

4.47 Brelot to Volterra, Boisseaux, 13.9.1930 *Cat. no. 46*

Boisseaux, 13 septembre 1930

Carissimo Prof. V. Volterra

Ho ricevuto le Sue bozze; ho corretto secondo gli spiegazioni che ho scritto su un foglio a parte, nel senso che Lei desiderava e ho già mandato a G.V. che dovrà spedire d'altre bozze. Sarà dunque possibile ancora di tener conto delle Sue nuove osservazioni.

Prima di ricevere la Sua lettera, Le avevo gia mandato le bozze in placards della fine. Lei potrà modificare i due o tre punti in discussione. In quanto allo stile, vi è sempre difficoltà, essendo due persone molto competenti non sempre d'accordo. E poi, l'uso di un'espressione può dipendere molto dal contesto.

Ho fatto per il meglio e ho qualche volta insistito nella discussione senza aver ragione. Spero che Lei ne accetterà le scuse.

Ho dato l'ordine di stampare le 96 prime pagine dopo le piccole correzzioni da fare. Ho cambiato soltanto per la citazione di Marchi in Nota le parole fra parentesi; era dato un riferimento incompleto; ho preferito rinviare all'ultimo capitolo dove sono dati due riferimenti precisi.

Mando i miei rispetti a tutta la Sua famiglia e a Lei stesso - senza sapere esattamente quale sarebbe in italiano la migliore formula finale.

M. Brelot

Sono tornato qui (Boisseaux) un po' più presto che pensavo - ieri sera -

4.48 Brelot to Volterra, Boisseaux, 14.9.1930 *Cat. no. 47*

Boisseaux (Loiret), 14 septembre 1930

Cher Monsieur

Je ne m'y reconnais plus dans les épreuves qui me sont arrivées de divers côtés. J'ai renvoyé corrigé à G.V. toute la fin de la mise en pages jusqu'à la conclusion non comprise en le priant de me renvoyer une seconde épreuve, sauf pour le début dont j'ai donné selon votre ordre le bon à tirer.

Mais je ne sais pas si vous avez vu ou non la mise en pages de 160 à la fin. Je vous envoie le double. Et j'attendrai de savoir si vous avez bien vu toute la mise en pages jusqu'au chapitre conclusion avant de donner le bon à tirer.

Je n'ai pas noté ce qui m'arrivait de vous et de G. V. et je ne suis plus sûr de rien, une fois tout réexpédié.

Veuillez bien m'excuser, rien ne sera retardé en tout cas. Je pense que vous avez vu la mise en pages jusqu'à 160.

Veuillez croire à mes sentiments de profond respect.

M. Brelot, Boisseaux (Loiret)

4.49 Brelot to Volterra, Boisseaux, 15.9.1930 *Cat. no. 48*

Boisseaux (Loiret), 15 septembre 1930

Cher Monsieur

Je reçois vos indications sur la préface; j'en tiens compte pour corriger. J'ai de petites choses à vous dire encore au sujet de l' “Ac. des Sciences” et autres.

Je crois qu'on gagnera du temps en ne faisant pas modifier maintenant à G.V. Vous pouvez achever vous même la correction, en lisant la note ci-incluse puis envoyer directement à G.V. qui fera toutes les corrections à la fois. Vous jugerez s'il est utile que vous relisiez encore vous même ou s'il suffit que je m'assure de l'exactitude de la correction typographique. De toutes façons je pense qu'on évite une correction de l'éditeur.

Je pense que tout sera fini avant votre départ et que l'ouvrage pourra paraître fin Octobre.

Je vous envoie mes meilleurs vœux et vous prie de croire à mes sentiments très respectueux.

M. Brelot

4.50 Brelot to Volterra, Paris, 17.9.1930 *Cat. no. 49*

IMPRIMERIE GAUTHIER - VILLARS & CIE *Paris, 17 septembre 1930*
Quai des Grands Augustins,

Cher Monsieur

Je vous écris de l'imprimerie; on a déjà refait les épreuves en pages de 97 à 196; je viens de les voir et de les corriger; j'ai demandé qu'on tienne compte de ces corrections (insignifiantes) et qu'on attendre l'ordre du bon à tirer. En effet:

1°) D'une part vous pouvez avoir quelques observations à faire sur la façon dont j'ai tenu compte de vos remarques relatives aux pages 97–160; je vous ai expliqué récemment ce que j'ai fait; je vous envoie aujourd'hui le feuillet qui contient les principales modifications; j'ajoute seulement que pour l'énoncé que vous critiquez:

"Pour des valeurs déterminées des individus d'une association dissipative (*), on peut trouver un nombre B tel que la valeur $\sum_r \alpha_r N_r$ de l'association ne saurait demeurer supérieure à B à partir d'un certain moment quel qu'il soit"

J'ai intercalé (*) "quelconque" comme on le voit p 116–117, seule chose importante du feuillet que suit celui que je vous envoie et d'ailleurs des autres jusqu'à la page 160.

Je vous ai déjà expliqué que j'avais mis "déterminées" parce que il faut fixer la valeur de l'individu de chaque espèce avant de pouvoir parler de la valeur de l'association.

2°) D'autre part vous n'avez pas vu la fin (p 160–196) pour laquelle vous aurez sans doute des observations à faire.

Quand j'aurais votre avis sur ces pages 97–196 je pourrai faire les corrections que vous m'indiquez et enfin seulement donner l'ordre de tirer.

Dans une semaine j'aurai sans doute votre réponse et j'écrirai à ce moment à G. V. Tous les bons à tirer donnés, il faudra 15 jours pour la parution de l'ouvrage.

Comme faudra t il faire la table des matières. Je pense que, vu les canevas détaillés en tête de chaque chapitre, il suffirait de mettre le titre des chapitres dans la table. Voulez vous plus de détail?

Enfin je pense que dans le contrat que vous avez passé avec G.V., il n'y avait rien concernant le nombre d'exemplaires qui me revendraient. Je pense qu'il vous faudra en aviser finalement la Direction. M. Picard m'avait donné 10 exemplaires. Ce m'est largement suffisant. Seulement comme ils seront pris sur les vôtres je crains que ce nombre vous réduise par trop le vôtre; de sorte que si vous êtes gêné, vous pourriez me laisser ce que vous voudrez bien.

Je vous prie de croire à mes sentiments très respectueux

M. Brelot

4.51 Brelot to Volterra, Boisseaux, 20.9.1930 *Cat. no. 50*

Boisseaux, 20 settembre 1930

Carissimo Professore

Mi pare che sia un malinteso a proposito della frase: "Pour des valeurs déterminées des individus d'une association.." Perché "iniziali"? Si tratta del valore attribuito al "individuo medio" di ogni specie e non del numero iniziale degli individui di ogni specie. Come Le ho già detto, quando un sistema è dato per mezzo dei coefficienti delle equazioni differenziali corrispondenti, il valore che sia possibile attribuire ad ogni individuo medio non è determinato, per un sistema dissipativo. Questi valori devono solamente essere tali che la $\sum_r \alpha_r p_{rs} N_r N_s$ sia una forma definita positiva; per l'enunciato considerato nel quale si parla del valore $\sum_r \alpha_r N_r$ dell'associazione, bisogna, mi pare, fissare i valori corrispondenti ad ogni specie; infatti si paragona $\sum_r \alpha_r N_r$ a un certo numero fisso B. Naturalmente si potrebbe pensare che la proposizione "Pour des valeurs déterminées des individus" è evidente e inutile; ho pensato però che sia meglio dirlo. Sarebbe meglio forse dire, invece di "déterminées", "choisies à l'avance"?

Prossimamente Lei deve scrivermi per le bozze della fine. Potrà dirmi se tutto va bene per la messa in pagine e dopo darò l'ordine di stampare dalle 97 alle 160 o anche 196.

Non ho ancora ricevuto le bozze della mia memoria di Palermo; l'ho mandato in gennaio e ho già ricevuto due o tre volte l'annuncio di un ritardo; l'ultima volta le bozze mi erano promesse per settembre. In compenso, ho ricevuto gli estratti delle memorie del R. Ist. Lomb. e Le ho mandato uno esemplare come a Suo figlio Enrico che aveva avuto l'amabilità di correggere parecchi sbagli.

Ho visitato rapidamente il Belgio in autocarro e sono rimasto due giorni in Namur presso un amico mio. Ricomincio a studiare la matematica; fra poco andrò vedere il professore Bouligand che potrà darmi diverse indicazioni pratiche ed altre, per la redazione della tesi. Prima di redigere, vorrei fare molte cose che si presentano naturalmente. Ma tutto diviene complicato.

Nel principio del mese prossimo, rimarrò a Parigi almeno una settimana, e farò allora, ogni giorno, della conversazione tedesca col lettore della Scuola Normale, studente tedesco. Poi verso la fine di Ottobre, partirò a Berlino.

Spero che tutta la Sua famiglia sta bene, in particolare Suo figlio Enrico che fu tanto malato in Luglio.

La prego di credere ai miei rispettosissimi sentimenti.

M. Brelot

4.52 Brelot to Volterra, Boisseaux, 24.9.1930 *Cat. no. 51*

Boisseaux, 24 settembre 1930

Cher Monsieur

Je viens de recevoir les feuilles de mise en pages que je vous avais expédiées de l'imprimerie. Je ne vous envoyais pas les épreuves précédentes parce que je m'assurais moi-même que les corrections indiquées étaient faites et qu'il était donc inutile que vous compariez à votre tour le deux épreuves. Quant à la compréhension du texte, il me semble que vous possédez un exemplaire complet de mise en placards de la 2e série (jusqu'à la conclusion); de sorte, si je ne trompe, que vous pouviez consulter ces placards pour situer dans l'ouvrage les pages que je vous expédiais. Je me demande maintenant si vous aviez bien ces seconds placards.

Je retrouve moi même des fautes d'impression, différentes d'ailleurs de celles que vous corrigez; il est bien difficile de n'en pas laisser et j'ai déjà été tout surpris d'en retrouver beaucoup dans le livre de M. Picard. J'ai lu plus attentivement le vôtre, sauf peut être le chapitre conclusion, et j'en ai laissé encore.

Vous me parlez des feuilles 113–160. Vous en aviez lu le 1er tirage en pages en m'indiquant quelques corrections en me disant de donner ensuite le bon à tirer. Il n'y avait d'important et de sujet à discussion que la phrase sur laquelle nous sommes d'accord maintenant. Aussi ai je trouvé inutile de vous réexpédier ces feuilles, puisque je vérifie le travail du typographe.

Quant à la table des matières, je pense qu'il convient, en répétant le canevas, de séparer les paragraphes en mettant en regard le n° de la page où il commence.

Je vais donner l'ordre d'imprimer jusqu'au dernier chapitre, en relisant pour trouver, encore s'il y en a des erreurs d' impression. Quant au dernier chapitre, désirerez vous voir la mise en pages ou bien suffit il que je m'assure que toutes les corrections se fassent bien exactement, si vous jugez définitif le texte actuel corrigé des seconds placards. Si vous êtes parti au début d'octobre, il sera difficile et long de vous attendre. M. D'Ancona voudra t il voir la mise en pages?

Enfin je croyais que vous aviez un double du contrat; je vais prier l'éditeur de vous renseigner à ce sujet et de vous écrire tout de suite. Vous pourrez alors juger et peut être l'éditeur consentirait-il à me donner quelques exemplaires non prévus et indépendants des vôtres.

Je pense bientôt me mettre à rédiger ma thèse. Si j'ai suffisamment de matière, j'ai avantage à ne point tarder; et je pourrai toujours d'ailleurs travailler, et publier des recherches consécutives dont j'ai une idée imprécise maintenant. Je vais voir M. Bouligand et après, s'il juge que je puis le faire, j'en parlerai à M. Picard.

Je vous prie de croire à mes sentiments de sympathie et de profond respect.

M. Brelot

4.53 Brelot to Volterra, Boisseaux, 27.9.1930 *Cat. no. 52*

Boisseaux, 27 settembre 1930

Carissimo Professore

Ricevo stamane le bozze di Suo genero. Non mi dice nella sua lettera, se desidera o no le bozze in pagina. Le ho già parlato di questo l'altro giorno, domandando se anche Lei vorrebbe leggere la messa in pagine.

È arrivata la risposta della ditta G.V.; mi rispondono che nel contratto è preveduto il numero di 40 esemplari di omaggio all'autore; essendo questo numero maggiore del numero ordinario, aggiungono che sarà dunque impossibile darmi esemplari a parte. Lei potrebbe forse, se non ha da utilizzare quasi tutti questi 40 volumi, dire a G.V. di riservarmi alcuni esemplari. La ringrazierei infinitamente.

Ho scritto ieri, dopo ricevuta Sua lettera, per l'aggiungere qualche parola a proposito della frase non chiara. No ho qui il doppio completo ora e devo aspettare che G.V. mi mandi il foglio di 16 pagine dove farò l'addizione. Questo a parte, ho dato l'ordine di stampare e di mandare tutto ciò che uscirà definitivamente man mano.

Penso che fra poco Lei partirà a Praga. Le mando i miei auguri di buon viaggio. In quanto a me, dopo la visita al Prof. Bouligand, andrò a Parigi e ci rimarrò qualche tempo; penso vedere i professori Vessiot e Picard per la questione della tesi e dell'entrata nel insegnamento superiore.

Le mando, come a Sua famiglia, i miei più rispettosi sentimenti di gratitudine e simpatia.

M. Brelot

4.54 Brelot to D'Ancona, Paris, 4.10.1930 *Cat. no. 2*

Paris, 4 ottobre 1930

Carissimo Signor D'Ancona

Ho l'onore di mandarle le bozze in pagine della parte che Lei ha redatta. Che Nota mancava? Non lo so e il tipografo ha semplicemente riguardato le bozze n° 2 invece delle bozze n° 1. Per andare più presto Le mando subito tutto ma se Lei non ha la Nota da aggiungere potrò andare io stesso alla casa G.V. per fare l'addizione.

Lei potrà dopo verificazione mandare le bozze rapidamente alla casa Gauthier-Villars. Credo che, dopo, basterà che io abbia nuove bozze senza che sia necessario per Lei di vederle.

La prego di credere ai miei rispettosissimi sentimenti

M. Brelot, Boisseaux (Loiret)

Imprimerie Gauthier- Villars
55 quai des Grands Augustins
Paris 6e

4.55 Brelot to Volterra, Boisseaux, 16.10.1930 *Cat. no. 53*

Boisseaux (Loiret), 16 octobre 1930

Cher Monsieur

Je viens de recevoir votre lettre et les épreuves. J'ai vite fait un paquet pour l'éditeur en le priant de vous les retourner après corrections et cela directement.

Je n'avais mis nulle part la citation anglaise. Si je l'avais fait, je ne l'aurais pas reproduite à la fin sans vous le dire au moins. Vous ne m'en avez jamais parlé. Il serait sans doute préférable que la phrase anglaise figurât tout de même, mais j'ai transmises exactement vos corrections. D'ailleurs la traduction classique française peut suffire. A ce propos n'y a-t-il pas une faute de copie dans votre texte; je vois "Il semble hors de doute que la quantité de perdrix dépend principalement ..." n'y a-t-il pas dépende ou ne dépende càd l'emploi du subjonctif qui me semble plus naturel. Je suis un peu surpris aussi que "grouses" ne soit pas traduit alors que "coqs de bruyère" me semblait bon. Enfin "tête de gibier" m'étonne un peu aussi; c'est la traduction littérale; mais si l'on dit "tête de bétail", je croyais que l'on disait seulement "pièce de gibier".

Je ne sais ce que vous pensez dans la table des matières; j'ai vérifié les phrases indiquées; ce sont les mêmes que dans le canevas. Peut être ne les trouvez vous pas claires. C'est du style un peu abrégé peut-être?.

Je vous envoie par le même courrier ce que je possède du tirage; la moitié des épreuves non définitives, mais les dernières, différant peu du texte qui paraîtra. Je n'ai pas tout car j'ai envoyé des corrections successives à faire d'après vos lettres et j'ai utilisé les épreuves en double.

Je vous souhaite un prompt rétablissement dont je ne doute pas; je suis moi même assez enrhumé et le début de cette saison est toujours mauvaise.

J'ai vu récemment M. Picard qui m'a proposé d'être rapporteur de ma thèse en même temps que président du Jury. Je pense que M. Julia acceptera de m'interroger sur le second sujet (la représentation conforme); M.M. Vessiot et Montel très pris par d'autres thèses préfèrent s'abstenir. Je ne sais encore qui sera le 3e membre du jury.

J'ai enfin reçu les épreuves de mon mémoire de Palerme et j'espère que je pourrai avoir les tirages à part assez rapidement.

Je vais donc me remettre à mon travail dont quelques points seraient à étudier; puis je rédigerai et je pense finir avant le 1er Janvier. M. Picard fera imprimer dans les Annales de l'Ecole Normale et je passerai peut être la thèse en Mai.

Croyez à mes sentiments de reconnaissance et de très respectueuse sympathie

M. Brelot

4.56 Brelot to Volterra, Boisseaux, 16.10.1930 *Cat. no. 54*

Boisseaux, 16 octobre 1930

Carissimo professore

L'altro giorno, leggendo per caso la prefazione, ho visto la parola "divulgant". Si deve scrivere "divulguant" con un u. Ho verificato nel dizionario e ho scritto all'editore. Avevo dimenticato di parlarne e come trovo ancora un osservazione da farLe, Le scrivo ora. Lei ha messo "Au cours de l'hiver 1929, M. Borel..." Non si sa cosi, se si tratta dell'inverno 1928–29 oppure dell'inverno 1929–30: credo che questo significherebbe piuttosto "inverno 1929–30", e sarebbe inesatto.

No ho niente altro da osservare e non so se sarebbe tempo ancora di modificare.

Le mando i miei auguri di buona salute e i miei rispettosissimi sentimenti di simpatia

M. Brelot

4.57 Brelot to Volterra, Berlin, 7.11.1930 *Cat. no. 55*

Berlin, 7 Novembre 1930

Carissimo professore.

Arrivato da cinque giorni, sono ora presso a poco installato. Ho visto l'Università, il professore Schmidt, la biblioteca del seminario. Infelicemente capisco poca cosa; sto per cominciare a prendere lezioni di tedesco e fare scambi regolari di conversazione. Vi é un circolo per studenti molto comodo e del quale la segretaria è ... italiana.

Sto redigendo la tesi ma non è rapido perché trovo ad aggiungere molte cose più o meno importanti, qualche volta abbastanza interessanti. Il soggetto non è esteso. Però si potrebbe approfondare molto, toccando a molte cose.

Vi è in Berlino una mezza dozzina di antichi allievi della "Ecole Normale", tanto in scienze quanto in lettere. Se voglio, non sono dunque isolato. D'altrove preferisco rimanere generalmente solo per avere qualche facilità di imparare la lingua. Molto difficile il tedesco! Non posso ancora seguire corsi; non potrei capire.

Spero che Lei ha corretto le ultime bozze del Suo libro e che questo uscirà fra breve.

Il prof. Julia mi ha dato il secondo soggetto di tesi. Dovrò studiare gli ultimi capitoli del t. II del Traité d'Analyse del Picard.

Berlino è un città molto uniforme dove le strade sono tutte uguali. Ma ci sono concerti splendidi dove andrò qualche volta. Ho tanto da fare, questi mesi che non mi preoccupo troppo dell'ambiente.

Le mando tutti i miei auguri e La prego di credere ai miei sentimenti di gratitudine e rispetto.

M. Brelot

Pension Boschkant - Rankestr 27 - Berlin W-50

4.58 Brelot to Volterra, Boisseaux, 10.8.1931 *Cat. no. 56*

Boisseaux (Loiret), 10/8/31

Carissimo Professore

Ho finito il mio soggiorno a Berlino e sono tornato a casa mia ieri dopo qualche giorno di vaggio in Bremen, Amsterdam, Anvers. Avevo già visitato Basilea, Monaco, Nürenberg e Dresden. Così conosco un po' la Germania.

La discussione della tesi è già una vecchia cosa; Le ho mandato un esemplare del lavoro alcune settimane fa. L'anno prossimo vado di nuovo a Berlino - senza borsa Rockefeller ma come allievo dell'Istituto francese, analogo all'Istituto Thiers à Parigi e creato un anno fa. Avrò anche una borsa francese di "chercheur" che hanno creato poco fa. Così potrò vivere press à poco come questi anni. Se è possibile andrò in Inghilterra due mesi fra i due semestri dell'Università. Ho molto piacere a vivere all'estero, imparare diverse lingue e viaggiare, senza dimenticare di studiare la matematica. Quest'inverno ho molto lavorato per redigere la tesi; dopo ho studiato poco a poco la seconda tesi, ma ho potuto anche uscire, studiare la lingua, conoscere i tedeschi e andare all'Università.

Ho fatto un Analisi del Suo libro e Le mando alcuni esemplari. Non dimentico che le devo la borsa Rockefeller colla quale ho potuto fare tante cose interessanti - tesi, viaggi, studio di lingue straniere. Le devo tutto il mio avvenire e Le mando tutta la mia gratitudine. La prego di presentare a tutta la Sua famiglia i miei rispettosissimi saluti e omaggi.

Il suo devotissimo allievo che Le presenta la sua rispettosa simpatia

M. Brelot à Boisseaux (Loiret)

4.59 Brelot to Volterra, Cepoy, 24.9.1931 *Cat. no. 57*

Cepoy (Loiret), 24 settembre 1931

Carissimo professore

Sto ora nella nuova casa del padre, nel villaggio di Cepoy vicino a Montargis, a 100 kilometri da Parigi e 40 da Fontainebleau. Il paese è piacevole ma lo sarebbe ancora di più se fosse bel tempo. Non piove più ma fa un freddo d'inverno.

Alcune settimane fa ho fatto un viaggio nella Haute-Savoie e molti giri nella macchina di un amico mio. In particolare sono passato vicino a Saint-Gervais che Lei conosce tanto bene.

Questi tempi mi occupo di diverse cose, matematica e altre. Partirò fra poco per Berlino, in una quindicina. Domani vado dal Professore Griaud che, malato, ha lasciato l'Università e sta vicino a Gien. Devo anche vedere il Villat. Spero che sarò iscritto senza difficoltà sulla lista per l'insegnamento superiore alla seduta di ottobre.

Lei si ricorda che ho fatto in maggio 1930 una piccola memoria a proposito del Suo libro. Ne ho corretto le bozze in Febbraio – e non ho più sentito parlare di niente. Alcune settimane fa, dal professore Cartan, che sta ora a Dolomieu (Isère) ho visto per caso un analisi della memoria in una rivista tedesca. Dunque la memoria è uscita negli Annali di matematica. Avevo già scritto in agosto per avere notizie. Ho scritto di nuovo, due settimane fa, alla stamperia Zanichelli. Nessuna risposta. Avrebbero dimenticato di farmi degli esemplari? E perché questo silenzio? Sarò obbligato di scrivere alla direzione degli Annali di matematica?...

Sono dunque ora "chercheur", pagato "dalla Caisse nationale de recherches" per quest'anno scolare. Sarò anche "pensionnaire" dell'Istituto francese di Berlino Landhaustr 14. È un po' come l'Istituto Thiers a Parigi. Spero che non ci saranno troppe difficoltà a vivere a Berlino quest' inverno. D'altronde il nuovo ambasciatore è un antico allievo della Scuola Normale. Forse che sarà l'occasione di avvicinare l'ambiente tanto chiuso delle ambasciate.

Le mando tutti i miei auguri per la fine delle vacanze e i miei sentimenti rispettosissimi. Rimango il suo allievo devotissimo

M. Brelot, Cepoy (Loiret), Francia.

4.60 Brelot to Volterra, Berlin, 24.10.1931 *Cat. no. 58*

M. Brelot
Franzosisches Akad. Hauy
Berlin - Wilmersdorf
Landhaus str. 14

Berlino, 24 ottobre 1931

Carissimo Professore

Prima di tutto le auguro un completo e rapido ristabilimento della salute e La ringrazio della Sua amabile lettera. Due settimane fa ho scritto al Levi-Civita a proposito della mia memoria e ho già complete risposte e chiarimenti. Avevano dimenticato di spedirmi gli esemplari, rimasti da alcuni mesi in magazzino. Sono spediti ora, forse arrivati a casa mia; la famiglia deve mandarLe alcuni numeri per Lei, il prof. d'Ancona e la sig[a] Freda. Potrei inviarLe altri esemplari se lo desidera. Ne faccio spedire due al prof. inglese Thompson e alcuni alla Rockefeller Foundation per le biblioteche americane.

Sono ora installato all'Istituto francese di Berlino. Ho una bella camera nella villa che, altre volte, era quella di Van Kluck dicono. Sono completamente tranquillo e ho alcune buone relazioni a Berlino. Quest'anno voglio seguire alcuni corsi all'Università perché conosco abbastanza (?) bene il tedesco. Infelicemente dimentico terribilmente l'italiano; proverò di avere ancora regolari rapporti con italiani studenti. Avrei bisogno di fare un giro di due o tre giorni in Italia per ritrovare le mie conoscenze della lingua.

Ho molto da fare quest'inverno; bisogna assolutamente che pubblichi qualche cosa di buono prima del 1° Maggio. Ho diversi soggetti interessanti di ricerca, ma non sono facili.

La ringrazio delle sue pratiche per la mia memoria e Le mando i miei migliori auguri per Lei e per la Sua famiglia.

Rispettosamente

M. Brelot

4.61 Brelot to Volterra, Berlin, 26.12.1931 *Cat. no. 59*

Berlino, 26/12/31

Illustrissimo professore

Le mando come pure a tutta Sua famiglia i miei migliori auguri per l'anno nuovo. Spero che la salute sarà sempre buona e che Lei avrà occasione di fare viaggi all' estero, senza difficoltà e collo stesso piacere che Lei ci trova sempre.

In quanto a me, ho lavorato abbastanza bene questi mesi e scritto tre Note per il Bull. Sc. Mat. e i Rend. Ist. Lomb. Le manderò estratti come quelli per la nota biologica e altri. Non ho ancora novelle della vendita della Sua opera presso Gauthier-Villars, ma sarò avvertito fra poco, come è scritto nel contratto e le scriverò. Ora devo fare la traduzione francese dell'opera del Prof. Levi Civita "Caratteristiche dei sistemi differenziali e propagazione ondosa"; ci sono commissioni francese e italiana che scelgono libri da tradurre e il Picard mi ha proposto al sig. Gauthier - direttore della Revue Scientifique - che si occupa dell'impressione.

Le scrivo a Roma dove Lei rimarrà forse, nonostante la decisione del Governo italiano che ha sorpreso tutti all'estero e della quale è meglio non parlare[27].

La prego di credere ai miei sentimenti rispettosissimi.

M. Brelot

Berlin - Wilmersdorf - Landhaus str. 14

4.62 Brelot to Volterra, Berlin, 10.1.1932 *Cat. no. 60*

Berlino, 10/1/32

Illustr^{mo} professore

La ringrazio della Sua lettera e dei Suoi auguri. Ho già parlato al professor Schmidt a proposito della traduzione possibile del Suo libro. Non conosce i

[27] Reference to Volterra's exclusion from the University and from all academic posts after his refusal to swear allegiance to the fascist regime. See Goetz 1993.

professori tedeschi di biologia o entomologia ma mi ha detto che avrà fra poco l'occasione di parlare del soggetto a qualche persona competente. Fra alcune settimane farò nel Colloquium una conferenza sopra queste ricerche e così avrò sicuramente l'occasione di conoscere tutti quelli che si interessano alla Sua opera. Secondo il Schmidt - e questo è anche il mio parere - siccome c'è tempo, tutto sarà così per il meglio. Avevo da molto tempo pensato a questa conferenza; aspettavo perché non parlavo tanto bene il tedesco. Ma ora, se mi permetto di leggere come ho fatto sul Seminario di Roma, potrò presentarmi abbastanza bene. Basta solamente un po' di lavoro.

D'altra parte ho ricevuto un avviso della Casa Editrice Gauthier-Villars. Fino al 30 settembre, hanno venduto 151 esemplari del Suo libro. Spero che l'anno prossimo, essendo le Sue ricerche più diffuse, la vendita aumenterà.

Lei può essere sicuro che farò il possibile per diffondere le Sue ricerche e che mi occuperò della traduzione possibile in tedesco. Penso tornare a questi studi che ho lasciato dacché ho scritto la memoria degli Annali di matematica. Ci sono soggetti interessanti di ricerche: per esempio il caso delle tre specie coll'eredità, oppure la dimostrazione dell'impossibilità di una periodicità nel caso ereditario generale (Lei non l'ha stabilito che per le piccole variazioni).

Indicherò simili cose nella mia conferenza ma me ne occuperò anche probabilmente io stesso. Avevo scritto al prof. inglese che proponeva altre equazioni, ma non ho ricevuto di risposta. E ho dimenticato cosa diceva. Forse che mi risponderà fra poco.

Ora debbo occuparmi della traduzione del libro del Levi-Civita.

La prego di gradire i miei sentimenti rispettosissimi

M. Brelot

Berlin - Wilmersdorf - Landhaus str. 14

4.63 Brelot to Volterra, Berlin, 26.2.1932 *Cat. no. 61*

Berlino, 26/2/32

Carissimo professore

Le mando le bozze della conferenza che ho fatto qualche tempo fa all'Università di Berlino e che sarà pubblicata nel Jahresbuch[28]. Non avrò gli esemplari prima di parecchi mesi; perciò le mando il testo delle bozze che ho poco modificato. Lei troverà strano che certe parole siano scritte in italico. Avevo sottolineato nel manoscritto col lapis queste parole o solamente una sillaba di esse, perché dovevano essere accentuate nella pronuncia. Ma è una altra cosa per la scrittura.

Non ho molti esemplari di bozze e non posso mandarle al professor Picard come lo desidererei. Forse che Lei avrà per caso l'occasione di mostrargli questa conferenza?

Mi sono dunque occupato della traduzione possibile in tedesco. L'assistente del Professor Bieberbach ha già parlato con un prof. di biologia e con un editore.

---

28 Brelot 1932.

Mi ha trovato anche un giovane dottore - Richard Rado - che queste ricerche interessano e che sarebbe pronto a far la traduzione. Ho scritto alla casa Gauthier-Villars per sapere chi può autorizzare la traduzione - cioè se tutto dipende da essa o solamente dall'autore. Aspetto la risposta che mi permetterà di aggiustare il contratto coll'editore tedesco. Ma finora niente di definitivo è stato detto. Le scriverò di nuovo se la traduzione verrà finalmente decisa, come è probabile.

Lei sa forse che ho tradotto recentemente il volumetto del prof. Levi-Civita sulle caratteristiche e la propagazione delle onde. L'autore ha esaminato il manoscritto che ho mandato a Parigi appunto ieri. Il Levi Civita mi ha detto che Lei è da poco a Parigi. Perciò Le scrivo al Hotel Lutetia dove Lei sta probabilmente colla Sua Signora.

Ho visto nei Comptes Rendus che il Suo figlio Enrico studia avanti l'elasticità e mi rallegro dei suoi successi.

Le mando - siccome a tutta la Sua famiglia - il mio grato ricordo e i miei sentimenti rispettosissimi.

M. Brelot

Berlin - Wilmersdorf - Landhaus str. 14

4.64 Brelot to Volterra, Berlin, 3.6.1932 *Cat. no. 62*

Berlino, 3/6/32

Carissimo professore

La casa editrice non ha ancora risposto a proposito del Suo libro. I tempi sono difficili e gli editori sono più che prudenti. Non c'è niente di urgente e si può aspettare e anche consultare altri editori. La mia conferenza sul libro Suo uscirà solamente in Luglio. Ne manderò esemplari dappertutto in Germania; forse che ciò aiuterà.

Frattempo il giovane Dott. R.Rado ha studiato detagliatamente la Sua opera e abbiamo già discusso sopra qualche errore di stampa o piccoli miglioramenti possibili. Rimarrà sempre pronto a far la traduzione e sono convinto che la farebbe benissimo. Lei si ricordi il suo nome. Ci sono punti per i quali il Rado vorrebbe confrontare colla memoria italiana originale. Perciò La prego di mandarne un esemplare al suo indirizzo per es. Hern Dr. Rado - Mathematisches Seminar Universität - Unter der Linden - Berlin.

Sto ora preparando due conferenze sul problema di Dirichlet (ricerche moderne) sui propri lavori. Ritornerò in Francia verso il 1° Agosto. L'anno prossimo avrò probabilmente la stessa borsa di “chercheur”. Ma non so dove starò.

Spero che Lei ha fatto un buon viaggio in Spagna. Non so se Lei è ora a Parigi o a Roma ma spero che la mia lettera Le giungerà presto.

Le mando i miei rispettosissimi auguri per Lei e la Sua famiglia

M. Brelot

Landhaus str. 14 - Berlin - Wilmersdorf

4.65 Brelot to Volterra, Berlin, 19.7.1932 *Cat. no. 63*

Berlin - Wilmersdorfs Landhausstr. 14

Indirizzo in Francia dopo il I° Agosto
Cépoy (Loiret)

Berlino, 19/7/32

Carissimo professore

Ho ricevuto la Sua lettera qualche tempo fa e Le mando per il lutto Suo le mie "condoléances". Ho dato le bozze delle Sue ricerche biologiche al Dott. Rado. Se la traduzione venisse decisa, gli manderei il mio esemplare di questa memoria. Ma l'editore non è entusiasta. Manca il denaro come dappertutto. Anche la traduzione della "Moderne Algebra" del Van der Waerden che era in questione per me è ritardata.

Le mando oggi a Parigi un esemplare del libretto, traduzione di una piccola opera del Prof. Levi-Civita. Spero che La raggiungerà presto, siccome questa lettera, dove Lei sta ora. Fra una diecina di giorni tornerò in Francia facendo probabilmente un piccolo viaggio per "Copenhague". In settembre andrò a Zürich per il congresso di matematici e avrò probabilmente così l' occasione di vederla.

La prego di credere ai miei sentimenti rispettosissimi.

M. Brelot

4.66 Brelot to Volterra, London, 3.1.1933 *Cat. no. 64*

UNIVERSITY OF LONDON CLUB
21, Gower Street, W.C.1.

Londra, 3.1.33

Carissimo professore

Sono arrivato qui da qualche giorno; il viaggio su un cattivo mare non mi è riuscito e sono stato ammallato qualche tempo. Ora sto meglio e Le scrivo per mandarLe come alla sua famiglia i miei migliori auguri per il 1933. Non dimentico tutto ciò che Le devo nel ottenere della borsa Rockefeller colla quale ho cominciato a lavorare per conto mio. E un piacere per me, nell'occasione del Capo d'Anno, di ringraziarLa ancora per questa sollicitudine che fu decisiva per il mio avvenire. Spero avere un posto nell'Università nel prossimo Novembre; ho dunque ancora sei mesi di libertà durante il periodo degli studi; quindi sono venuto in Inghilterra per imparare un po' la lingua del paese della quale si potrebbe dire che tutti la leggono, ma che nessuno la parla. Andrò principalmente a Cambridge dove sono i migliori matematici inglesi e una buonissima biblioteca.

Continuo naturalmente a pubblicare, come Le devo anche poiché sono sempre "chargé de recherches". Ora mi occupo delle funzioni subarmoniche, generalizzando molto diversi risultati della mia tesi che si rapportano al caso particolare degli integrali o dell'equazioni $\Delta u = cu$, $c \geq 0$. Precisamente sarà interessante parlarne col Hardy del quale teorema di media è una conseguenza di una proprietà del Riesz sulle funzioni subarmoniche.

Non ho notizie di traduzioni possibili del Suo libro sulla lotta per la vita. So solamente che l'editore ne ha venduto 69 esemplari quest'anno (la metà di più degli altri libri della collezione che sono usciti contemporaneamente)

Col mio rispettosissimo ricordo

M. Brelot.

4.67 Brelot to Volterra, Bordeaux, 26.11.1938 *Cat. no. 65*

FACULTÉ DES SCIENCES DE BORDEAUX
INSTITUT DE MATHÉMATIQUE

Bordeaux, le 26 Novembre 1938

Cher Monsieur

Je vous dois bien des excuses pour ne vous avoir plus donné de nouvelles depuis longtemps. Je suis depuis le début de 1938 professeur à Bordeaux, dans une chaire. Comme partie intéressante d'enseignement, je fais la moitié d'un cours d'Analyse supérieure à programme libre. Depuis plus d'un an je suis revenu à des questions touchant au potentiel et aux fonctions subharmoniques et ce sujet a l'air fécond. Je me rencontre sur bien des points avec M. de la Vallée Poussin, ce qui est un grand honneur.

M. Montel est passé l'autre jour par ici in allant au Portugal faire des conférences, et à mon collègue Milloux et à moi, il nous parlait des intellectuels italiens et spécialement de vous. Je suis donc mieux encore votre isolement et la situation de vos fils et de M. D'Ancona[29]. Puissent ces quelques mots de respectueuse sympathie vous assurer que vos amitiés internationales ne changent pas et que le temps, qui a depuis de si longues années consacré votre réputation scientifique dont l'Italie s'est enorgueillie, se chargera de faire de votre nom le symbole de la dignité intellectuelle.

La civilisation est en régression puisqu'on mêle à la science de questions de sous et de politique; le directeur de Zentralblatt, Neugebauer vient de démissionner car on voulait le contraindre à distribuer ses analyses selon des directives extra scientifiques[30]; la vague continue à déferler...

Je vous prie de me rappeler au souvenir de Mme Volterra et de votre famille. Je songe à mon heureux séjour à Rome, à une époque où les charges étaient encore fixes et où en France on ne craignait ni guerre, ni grèves générales.

---

[29] Obviously a reference to the consequences of the anti-Jewish laws introduced by the fascist regime, starting in Autumn 1938. It should be noted however that the reference to D'Ancona is unfounded as he (who was not Jewish) was never banned from teaching.

[30] Reference to the resignation of Otto Neugebauer actually due to the ousting of Tullio Levi-Civita from the editorial staff of reviews journal *Zentralblatt für Mathematik* because he was of "Jewish race". For these events and their consequences see ISRAEL, NASTASI 1998.

J'espère que ce mot ne sera pas intercepté et je vous prie de croire à mes sentiments de vielle reconnaissance et de respectueuse sympathie

M. Brelot

32 bis Rue de Patay, Bordeaux

4.68 Brelot to Virginia Volterra, Grenoble, 21.3.1947 *Cat. no. 66*

UNIVERSITÉ DE GRENOBLE
FACULTÉ DES SCIENCES
MATHÉMATIQUES

Grenoble, le 21 Mars 1947

Chère Madame

J'ai reçu avec beaucoup de plaisir et d'émotion la brochure nécrologique que vous avez bien voulu m'envoyer et je vous en remercie. La cérémonie d'Octobre a été assez restreinte, je crois et, peut-être sans présence étrangère. Nous nous en sommes un peu étonnés en France.

Récemment Pise a fait une cérémonie anniversaire de la mort de Tonelli et avait envoyé des invitations. La Faculté de Grenoble m'y a envoyé et j'ai fait le voyage récemment, ce qui m'a remis en contact avec les collègues italiens et permis d'avoir des nouvelles de votre famille.

Je me souviens avec une émotion multiple de mon séjour à Rome auprès de M. Volterra et de votre si bon accueil. Je vous prie d'adresser mon meilleur souvenir à vos enfants que j'ai connus et croire à mes sentiments très respectueux.

M. Brelot

4.69 Brelot to Volterra, Paris, n.d. *Cat. no. 7*

Paris, samedi matin

Monsieur

Je vous ai écrit jeudi que je n'avais point trouvé dans votre recueil de mémoire d'hydrodynamique. Je n'avais point vu la table des matières et en feuilletant l'extrait des Comptes Rendus de l'Académie des Sciences (1er octobre 1923) (Mouvement d'un fluide en contact avec un autre et surfaces de discontinuité) m'avait échappé.

Je vous prie de m'en excuser. Mais ce n'est point le mémoire en anglais dont vous m'aviez parlé je crois. Cependant veuillez me faire dire (j'attends en bas) si je dois vous laisser le livre que j'ai apporté, où si je puis le garder.

Avec mes excuses, veuillez accepter l'expression de tout mon respect.

M. Brelot

4.70 Brelot to Volterra, Paris, n.d. *Cat. no. 8*

Paris, lundi

Cher Monsieur

D'après mes camarades naturalistes le "Confused Flour Beetle" est un coléoptère, sorte de charançon, le "Tenebris Molitor" sans doute.

Excusez ma traduction, très mauvaise comme français. Mais j'ai du aller très vite et je suis fort déprimé ces jours ce de mon attaque de grippe. D'après l'Anglais, lecteur à l'Ecole, d'ailleurs, le style du texte n'est pas bon.

Veuillez croire à tout mon respect.

M. Brelot

4.71 Brelot to Volterra, n.p., n.d. *Cat. no. 10*

Dimanche matin

Cher Monsieur

Voici la rédaction du premier paragraphe du dernier chapitre. Je m'empresse de vous le porter avant votre départ; si vous pouvez jeter un coup d'œil d'ici demain matin, au moins sur la fin, sous paragraphe traitant de la notion la plus générale d'hérédité, nous pouvons en parler demain.

Croyez à tout mon respect.

M. Brelot

5 Letters between Royal N. Chapman and Vito Volterra

Royal Norton Chapman was born in Morristown (Minnesota, U.S.A.) on 17 September 1889. After completing his secondary schooling at the Pillsbury Academy (Owatonna, Minnesota) he entered the University of Minnesota in 1910, where he embarked upon entomology studies, graduating in Arts in 1914. He receive his MS degree the year after with a work on the life cycle of the two-lined chestnut borer *Agrilus bilineatus*, under the direction of A. G. Ruggles. He was immediately awarded a fellowship at Cornell University and in 1917 received his Ph.D. He then returned to the University of Minnesota where, for the years 1917 and 1918 he was teaching fellow at the Department of Animal Biology. For a number of years he had addressed his research to stored-product insects and the dynamics of insect populations, an interest that was increasingly incorporated in the general context of quantitative animal ecology.

In 1918 Chapman was appointed Assistant Professor and Assistant Entomologist at the Minnesota Experiment Station and, in 1923, Associate Professor. He became Full Professor in 1925, the following year taking over also the post of Chief of the Division of Entomology and Economic Zoology from W. A. Riley. The same year he published the first version of a book that was to become his best known work, *Animal Ecology* (Chapman 1925). In the years 1926 and 1927 he was awarded also a Guggenheim fellowship as well as a fellowship from the Rockefeller Foundation to carry out research in Europe. In 1928 he published an important article (Chapman 1928) in which, as well as presenting experimental results on *Tribolium* populations (see the following letter no. 4), he formulated his biotic potential-environmental resistance principle. This principle was derived from Ohm's law governing the transmission of electric current in conductors and was based on the formula $C = Bp/R$, where C was the concentration of insects, Bp the "biotic potential" of the species and R the physical and biotic resistance of the environment to population growth. Chapman justified a similar physical analogy by claiming that ecologists should "consider whether it would not be an advantage if it were possible, by a few principles briefly stated, to make a great variety of facts, which are known in our branch of science, appear as unity to the mind". This approach, which was aimed at transforming entomology and, more generally, animal population dynamics, from qualitative and descriptive sciences into quantitative sciences, found its full expression in the second expanded version of the book *Animal Ecology* (Chapman 1931, see also the following letter no. 5), which had a rapid circulation and became a university reference text in the United States for many years after. His research influenced the work of his pupils John Stanley and Samuel A. Graham.

Meanwhile, in 1930, Chapman had become director of the Experiment Station of the Hawaiian Pineapple Growers Association and Dean of the Graduate School of Tropical Agriculture in Hawaii. His authority in the field of entomology and animal ecology was now undisputed and he was also appointed chairman of the Entomological Society of America and director of the series *Ecological Monographs*. He remained at Hawaii until July 1939 and, on his return to the University of Minnesota, he was appointed Dean of the Graduate School. He however he contracted a disease that led him to a premature death in Minneapolis on 2 December 1939.[1]

* * *

5.1 Chapman to Volterra, St. Paul, Minnesota, 9.8.1928 *Cat. no. 67*

UNIVERSITY OF MINNESOTA
DEPARTMENT OF AGRICULTURE
UNIVERSITY FARM, ST. PAUL
DIVISION OF ENTOMOLOGY AND ECONOMIC ZOOLOGY

Professori Vitio Volterra *August 9, 1928*
Facoltà di Scienze
fisiche matematiche e naturali
ROMA, ITALIA

Dear Sir:

I have been very much interested in your recent publications on the fluctuations of organisms in natural systems. It happens that I have been doing experimental work in very much the same line. I am sending you under separate cover, a reprint of one of my recent papers. You will notice that I have been able to develop a technique whereby I can produce under experimental conditions, the same phenomena which you have been calculating.

So far I have been unable to obtain the paper which you published in 1926 with the following citation, "Mémoires de l'academie dei Lincei, 1926. Sev. VI. Vol II. foac III. pages 31–112"[2]. If it would be possible for you to send me a reprint of this paper I should very much appreciate it.

Unfortunately I was unable to meet you while I was in Italy a year ago. I am very much interested in the entire subject and hope that I may some day be able to meet you and discuss the subject with you personally.

Very truly yours,

Royal N. Chapman
RNC-b Chief of Division

[1] See GRAHAM 1941 and course notes by A. C. HODSON on the history of the Department of Entomology, Fisheries and Wildlife of the University of Minnesota (published by the university).

[2] VOLTERRA 1926a.

5.2 Volterra to Chapman, Napoli, 1928 *Cat. no. 362*[3]

GRAND HOTEL
NAPLES

Cher Monsieur

Je vous remercie beaucoup de votre aimable lettre et des intéressants mémoires que vous avez bien voulu m'envoyer.

J'ai publié des extraits des mémoires de l'Accademia dei Lincei e du Comitato Talassografico mais il n'y a rien outre ce qui a paru dans les dits mémoires[4]. Dès mon retour à Rome j'aurai l'honneur de vous les envoyer.

J'ai fait l'année passée à la Sorbonne des conférences sur le même sujet. On les a recueillies et elles sont sous presse chez Gauthier-Villars[5]. Dès qu'elles paraîtront je vous en férai envoyer un exemplaire. Il y a là quelques developpements nouveaux qui pourrant vous intéresser. Vos travaux y sont cités.

J' ai commencé des nouveaux calculs sur quelques cas particuliers se rapportant aux Insectes mais malheureusement j'ai été pris par d'autres occupations qui m'ont fait interrompre ces recherches. Je compte dès que je pourrai de pouvoir les reprendre.

Je suis heureux d'apprendre que vous continuez à vous intéresser de ce sujet et que vous avez tous le moyens pour les pursuivre d'une manière efficace.

Veuillez accepter, cher Monsieur, l'expression de ma haute estime et de tout mon dévouement

5.3 Chapman to Volterra, St. Paul, Minnesota, 26.9.1928 *Cat. no. 68*

(Same heading of letter no. 1)

Dr. Vito Volterra *September 26, 1928*
Via in Lucina, 17
ROMA, ITALY

Dear Sir:

I want to thank you very much for the reprints which you have sent me. I am greatly interested in the work which you have done. I believe that it is of great importance to the field of biology. In fact, I think it is comparable to what Dr. Willard Gibbs did for the field of physical chemistry and I hope that unlike his work, yours will come immediately to the attention of the biologists. In the next few days I hope to send on to you some of the results which I have recently gotten in experimental work which substantiate your calculations and I want

---

[3] N.d. but certainly written in 1928.

[4] VOLTERRA 1926a, 1927b.

[5] Clearly VOLTERRA 1931a. However, the editing of these lessons by M. Brelot actually began in 1929 (cf. relative correspondence and following letter no. 5.6, Cat. no. 361).

then to take up with you further the possibility of cooperating with you in this study.

Thanking you again and with best regards, I am
Very sincerely yours

Royal N. Chapman
Chief of Division

RNC-b

5.4 Chapman to Volterra, St. Paul, Minnesota, 28.1.1929 *Cat. no. 69*

(Same heading of letter no. 1)

Dr. Vito Volterra
Via in Lucina, 17
ROMA, ITALY

January 28, 1929

Dear Sir:

I have received your letter asking if I can give you some information with regard to qualitative and quantitative results of population fluctuations. I am sending you some reprints of my publications and also a rough manuscript describing some experiments the results of which have not yet been published, and I will be very glad if you can use any of these results in any way.

I think that you will be able to understand the possibilities of such experimental work as we are doing here at the University of Minnesota. One very interesting possibility in this experimental work is that we are able to duplicate experiments and get essentially the same results with very little variations as you can see by looking at the curves which I am sending to you. The present work represents only a very small beginning, but we have plans for extending it.

One possibility which I think would be interesting to you is to start a culture similar to the one shown in my Graph 1 and at the time that the pupae begin to appear to remove them from the culture. We would then have an adult population which would act as the origin of the egg and larval population. The eggs and larvae then should fluctuate as in the case which you have calculated for two species one of which feeds upon the other. The egg life would end when a larva hatched from it and the larval life would end when it emerged as a pupae. We would then take the pupae out of the culture. The egg population should then come up again, then the larval population would come up and these fluctuations should then continue as you have calculated.

I have a student here, Mr. John Stanley, who is taking work in mathematics and who is setting up a series of experiments in which we will have Tribolium confusum competing with another species very similar to it, Tribolium ferruginium and we hope from this that we can get the substantiation of your calculations for the cases where two species compete for the same food.

I have the rather ambitious hope that we may be able in time to set up experiments to substantiate practically all of the cases which you have calculated.

You may be familiar with the book entitled "Animal Ecology" published by Elton[6] of Oxford University in which he takes up in a qualitative way animal fluctuations. I presume you are also familiar with the publications of Dr. Raymond Pearl of John Hopkins University on the fluctuations of Drosophila culture[7]. These experiments, however, I think are open to a considerable amount of doubt and I question myself whether it is worth while spending much time in making any calculations on the basis of experiments in which there is question, as to the dependability of the data.

I hope that these data will arrive in time to be of some aid to you and shall be very glad to hear from you as to what to think of the results which I have obtained.

Very truly yours

Royal N. Chapman
RNC-p Chief of Division

P.S. I am interested in finding a method of calculating the chance of a larva or adult beetle finding an egg when it is known how many eggs are present per gram of flour and also how many adults and larval beetles are present. We must, of course, first obtain some exact experimental data on the rates of eating eggs at various concentrations. R. N. C.

5.5 Chapman to Volterra, St. Paul, Minnesota, 5.8.1929 *Cat. no. 70*

(Same heading of letter no. 1)

Dr. Vito Volterra *August 5, 1929*
Via in Lucina, 17
ROMA, ITALY

My dear Dr. Volterra:-

At the present time I am engaged in rewriting my text book on Animal Ecology[8] with special reference to insects and I wish to include a reference to your work on the fluctuations of the number of animal species living together. I do not feel that I am competent to give a complete treatment of the subject and one possibility which has occurred to me is that of obtaining permission from you to reprint the English translation by Miss Mary Evelyn Wells of your paper on "Variations and Fluctuation of the Number of Individuals in Animal Species Living Together". In event that you give me permission to reprint this, I would set it apart as a special chapter giving your name as the author of it and Miss Wells as the translator and state that it was an extract of the Journal Due Conseil International Pour l' Exploration De La Mer.

---

[6] ELTON 1927.

[7] Here Chapman is probably referring to PEARL, PARKER 1922. Another subsequent work by Pearl specifically dedicated to Drosophila is PEARL 1932. Cf. also PEARL 1925, 1927.

[8] The first version referred to is CHAPMAN 1925 and the rewritten version is CHAPMAN 1931. The first volume consisted of 183 pages while the second is a good 464 pages long.

It is possible that you have something more recent, more comprehensive and, possibly, a little briefer treatment of the general subject in English. If so, I should be glad to use this if you cared to give me permission to do so. I think one advantage of doing as I have suggested would be that it would bring your work more to the attention of the English readers of the subject of Ecology. I realize, however, that I am asking a great favor of you and if you consider that it would not be wise to use it in this way will you please tell me so?

Hoping that I may hear from you in the nearfuture, I am

Very truly yours

Royal N. Chapman

RNC:A Chief of the Division

5.6 Volterra to Chapman, n.p., n.d. *Cat. no. 363*[9]

M. Brelot qui a rèdigé le cours de mes leçons à la Sorbonne me donne la note ci-incluse[10]. Je serais heureux si vous voudrez bien la lire et m'écrire un mot à ce sujet.

J'espère que vous avez reçu ma lettre et ma note.

Volterra

5.7 Chapman to Volterra, St. Paul, Minnesota, 29.1.1930 *Cat. no. 71*

(Same heading of letter no. 1)

Dr. Vito Volterra *January 29, 1930*

Via in Lucina, 17

ROMA, ITALY

Dear Sir:

I want to thank you for your kind letter giving me permission to reprint your paper[11] and also for the more recent letter with the note from Paris. I hope very shortly to have some more of my data in shape so that I can send them over to you. I have been trying to make some definite determinations of the various factors which have to do with the rate of increase of these beetles, to see whether under the same conditions the rates of increase are always the same. I think you will find these data very interesting.

Thanking you again, I am

Very truly yours,

Royal N. Chapman

RNC-b Chief of Division

---

[9] Draft.

[10] Cf. letter no. 4.16 in the Brelot-Volterra correspondence (Cat. no. 23) and the introductory essay.

[11] VOLTERRA 1928a was published as an appendix to CHAPMAN 1931. See the later letter no. 5.10 (Cat. no. 365).

5.8 Volterra to Chapman, Roma, n.d. *Cat. no. 364*

Se voi potete mandarmi qualche memoria, o anche la notizia di qualche risultato quantitativo o qualitativo che avete ottenuto voi stesso o che siete a conoscenza che altri abbiano ottenuto il quale possa essere in connessione colle mie ricerche sulle fluttuazioni ve ne sarei molto grato. In modo particolare poi sarei lieto se potessi inoltre avere il permesso di citare le dette ricerche ed i detti risultati in alcune conferenze che debbo fare recentemente a Parigi sopra questi soggetti.

5.9 Chapman to Volterra, Honolulu, Hawaii, 4.4.1931 *Cat. no. 72*

Experiment Station of the Association of Hawaiian Pineapple Canners
University of Hawaii
Honolulu, Hawaii

Dr. Vito Volterra *April 4, 1931*
Via in Lucina, 17
Roma, Italy

Dear Professor Volterra:

I want to thank you for your pubblication from the Sorbona at Paris. I have gone over it with a great deal of interest. Mr. John Stanley, a student of mine, is just completing a thesis on the mathematics of the Tribolium population. He has also gone over this publication with a great deal of interest. I have been comunicating with Professor Evans of Rice Institute, Houston, Texas, with regard to the possibility of having it translated into English. The difficulty with publishers seems to be that they feel that most people who are interested in this publication can read French, and they would not, therefore, have any advantage in having it translated into English. I am investigating the possibilities, however, and shall be glad to let you know as soon as I find out what the situation is.

You may be interested to know that I am reading the galley proof on a book on "Animal Ecology with Special Reference to Insects", Which will be published by the McGraw-Hill Book Company very shortly. I shall be glad to send you a copy of it as soon as it is out.

Thank you again for your kindness, I am

Very sincerely yours,

Royal N. Chapman
Director

5.10 Volterra to Chapman, Roma, n.d. *Cat. no. 365*[12]

Roma 110
Via Giambattista Vico 1
Tel. 20.728

Dear Professor Chapman,

I have just received your publication on "Animal Ecology" & I want to thank you at once, very heartily and sincerely indeed for your kind thought. Besides I am very grateful to you inserting in your volume the appendix which includes my studies about the fluctuations in the number of animals.

You know how interested I am in these studies! I will read, therefore, with a great deal of interest, your book which is most valuable for me. I hope to find in it the ideas for some new researches.

Very sincerely yours

12 Draft.

6 Letters between Umberto D'Ancona and Vito Volterra

Umberto D'Ancona was born on 9 May 1896 at Fiume where he attended secondary school. He enrolled at the Faculty of Natural Sciences at the University of Budapest and from 1916 on continued his studies at the University of Rome. He interrupted studies during World War I to go and fight as an artillery officer on the Carso, where he was wounded and decorated. He graduated in Rome in 1920, defending a thesis on zoology under the guidance of Giulio Cotronei (1885–1962), a disciple of the eminent zoologist Giovanni Battista Grassi (1854–1925), who together with Sir Ronald Ross, had discovered the life cycle of the malaria parasite. In 1926 he married Luisa Volterra, the daughter of Vito Volterra, who also became his assistant.

He began his academic career as assistant at the *Regio Comitato Talassografico Italiano*. Subsequently, and until November 1929, he became Grassi's assistant and then deputy at the chair of comparative anatomy. After Grassi's death in 1925, he was appointed director of the Comparative Anatomy Institute and when, in 1929, Cotronei was called to take over Grassi's chair, he was given a lecturing position at the University of Camerino. He then transferred to the University of Siena, where he took over the chair of zoology and comparative anatomy. In 1936 he transferred to the chair of zoology at the University of Pisa and in 1937 to the same chair at the University of Padua, where he remained until his death.

In 1940 D'Ancona founded the hydrobiological station of the University of Padova at Chioggia. He was, among other things, also director of the Centro di Studi Talassografici of the National Research Council in Venice, President of the Società Internazionale di Limnologia and of the Consiglio Generale della Pesca nel Mediterraneo, of the Standing Committee for international zoology congresses, and of the NATO Oceanography sub-committee. He was a member of the Accademia Nazionale dei Lincei and of the Académie des Sciences of Paris, as well as of many other Italian and foreign scientific societies.

D'Ancona wrote more than three hundred scientific articles and important textbooks that were reprinted many times, such as *Biologia e zoologia generale* (Padova, 1947) and the *Trattato di zoologia* (Torino, 1953). The starting point for his research was a classical theme treated by Grassi and Cotronei, namely the study of eels and of the comparative morphology of the Muraenoidae. He also further developed Grassi's research on the sex of eels, making a fundamental contribution to the definition of the structural unity and delayed sexual differentiation of the Teleostei.

D'Ancona's field of interest was extremely vast and ranged from physiology, to embryology, hydrobiology, oceanography, and evolutionary theory. His research was marked by an initial interest in marine biology problems, especially ichthyology and fishery. In the 1920s he began publishing the results of his research on the effects of the stagnation of fishing during the wartime period 1914–18 on the biological equilibria of the marine fauna in the Upper Adriatic. The results of his statistical studies aroused Vito Volterra's interest in these topics and provided a starting point for the formulation of his mathematical theories on the "struggle for existence". D'Ancona helped his father in law in the development of this research at the biological level and later, encouraged by Ludwig von Bertalanffy, wrote a book on the struggle for existence (D'ANCONA 1939, 1954).

D'Ancona held fundamentally positivist scientific views and was convinced of the importance of the experimental method and of physical and mechanical analogy, and above all of the value of the mathematical method in the study of biological processes. However, his writings show that he was also convinced of the limitations of a mechanistic and reductionist approach aimed at reducing life processes to physico-chemical processes and to interaction among elementary constituents and claimed that it was necessary to devote considerable attention to the aspects of "complexity" characterizing biological phenomena. In his own words, he was convinced of the usefulness of an approach aimed "at identifying phenomena specific to life that elude physico-chemical analysis, that is, in opposition to the physical theory of life, at creating a true biological theory". This explains his prudent interest in organicist and holistic theories and in the new globalistic tendencies in animal ecology.

Umberto D'Ancona died at Marina Romea (Ravenna) on 24 August 1964[1].

* * *

6.1 D'Ancona to Volterra, Roma, 21.6.1926 *Cat. no. 81*

ISTITUTO DI ANATOMIA E FISIOLOGIA COMPARATE
DELLA R. UNIVERSITÀ DI ROMA

Roma, li 21–VI–1926

Caro Professore

Luisa mi ha parlato della risposta del Prof. Larmor. Thompson[2] è indubbiamente persona molto competente in questioni di matematica biologica, s'è occupato specialmente dell'accrescimento delle specie animali e delle leggi che regolano l'accrescimento in rispetto alla forma. Sarà bene mandargli la Sua

---

1 See also SERMONTI 1986 and the biographical note in VOLTERRA, D'ANCONA 1995, pp. XLIX-LI.

2 D'Arcy W. Thompson.

memoria oppena stampata, però credo che non sia il caso di fargliela rimaneggiare dandole una veste più biologica. A me pare che ciò possa essere fatto da noi.

Per la nota per il Nature[3], se Lei crede, si potrebbe ridurla e magari farci qualche aggiunta onde renderla più chiara ad uso dei biologi. Se Lei non ha nulla in contrario, potrei vedere io la brutta copia dell'articolo e poi fargliela vedere a Lei al Suo ritorno.

In quanto alla ripubblicazione della memoria al Talassografico[4], come già Le proposi, sarà pure bene metterla in forma più piana ad uso dei biologi. La memoria per i Lincei[5] va bene anche così perché essa sarà letta anche da matematici, invece quella del Talassografico sarà letta principalmente da biologi.

Questo lavoro però può esser fatto da noi. Io potrei fare benissimo la parte biologica, perché, per quanto non abbia la coltura matematica del Thompson posso sempre avere le necessarie spiegazioni da Lei.

Avrò molto piacere se Lei mi farà sapere che cosa ne pensa.

Le auguro un buon viaggio e spero presto di salutarla al ritorno.

Con affettuosi saluti

Suo Umberto

Caro Papà, sono stata dalla Freda[6] che ha detto che ha già le bozze e poi me le darà per rimandarle nel caso vadano bene. Non ti preoccupare per la Nota per il Nature che basta che tu dica cosa ne vuoi fare che si stamperà presto.

Dacci sempre tue buone notizie. Un abbraccio da Luisa.

6.2 D'Ancona to Volterra, Roma, 20.10.1926 *Cat. no. 82*

Roma, 20-X-1926

Caro Papà.

Come Le dirà Enrico i disegni non sono ancora pronti. Glieli porterò io Sabato sera. Quelli che dovevano essere pronti da mio fratello sono pronti. Il disegnatore di Enrico ne ha copiati due e deve copiarne ancora uno.

Farò fare le linee tratteggiate nelle Figg. 1 e 2 come Lei mi indicò.

Va bene in quanto alla nota sono d'accordo che convenga dire che si è giunti per via indipendente allo stesso risultato.

Qui è arrivato un pacco di libri da Catania. Tutto il resto l'abbiamo spedito.

Tanti saluti a Lei, a Mamma e ai ragazzi.

Umberto [...]

---

3 Volterra 1926b.

4 Volterra 1927d.

5 Volterra 1926a.

6 Elena Freda (1890–1978), Volterra's collaborator, who also made a contribution to biomathematics (see bibliography and the introduction).

6.3 D'Ancona to Volterra, Roma, 28.1.1929 *Cat. no. 83*

Roma, 28 Gennaio 1929

Carissimo papà.

Ho ricevuto ieri la Sua lettera e l'ho letta con molto interesse.

Mi fa piacere di sentire che ha cominciato le lezioni che sono certo interesseranno molto sia i matematici che i naturalisti.

Certamente l'osservazione di Lapique[7] è giusta perché il senso in cui Lei usa la parola Eredità non é quello di uso corrente in campo biologico. Noi intendiamo per eredità il passaggio di qualcosa da padre in figlio e in genere ammettiamo il passaggio di qualche cosa di materiale; anche per il carattere o per le singole proprietà ammettiamo infatti che siano trasmesse da dei fattori che si trasmettono attraverso gli elementi riproduttivi maschili e femminili.

Che d'altra parte determinate parole siano e siano state usate nelle nostre scienze in significati diversi non è cosa nuova.

Qui d'altronde non mi pare che ci possa essere tema di equivoci. Etimologicamente e logicamente non mi pare che possa essere considerato errato l'uso da Lei fatto. Come noi chiamiamo eredità ciò che un individuo lascia ai discendenti, così possiamo chiamare eredità ciò che un fenomeno fisico lascia sulla materia su cui ha agito. Per quanto poi riguarda gli scambi nutritizii tra le specie conviventi; mi pare che Lei abbia inteso di chiamare ereditarie l'azione del cibo che precedentemente è stato preso dagli individui della stessa specie e non dagli stessi individui. Ora ciò si può ammettere anche come ereditato dagli individui di una generazione da quelli della precedente.

Certamente a voler sottigliare si trova che non c'è una perfetta concordanza col significato della parola eredità che in biologia ha assunto un valore ben preciso. Però credo che purché si spieghi bene che cosa si intende dire non ci possono essere equivoci.

La parola Nachwirkung va forse meglio, ma non so se essa non abbia piuttosto il significato di azione ritardata, il che forse non è precisamente quello che ci vuole in questo caso. Forse la traduzione in postazione potrebbe corrispondere, ma non so se Le piace[8].

In quanto ad una parola greca non dovrebbe essere difficile trovarla purché le si dia un significato preciso. Tanto il valore etimologico delle parole greche non corrisponde mai al loro significato scientifico. Plancton per esempio non ci dice niente di quello che vuole indicare.

Volendo però adottare una parola greca sarà bene però sentire qualche letterato. Io però lascerei le cose come stanno mettendoci soltanto una esplicazione sul significato della parola eredità.

---

[7] Probably a reference to the physiologist, L. Lapicque, well known for his work in the field of the general physiology of the nervous system, in which he introduced the concept of chronaxy (or excitability characteristic).

[8] Reference is made in this letter to heredity *à la Volterra* and its relations with the analogous concept in biology. The concept of *Nachwirkung* is due to Ludwig Boltzmann. See ISRAEL 1991a, IANNIELLO, ISRAEL 1993.

Luisa e Silvia stanno bene. Io sono ritornato sabato da Camerino, vi ritorno ancora da Mercoledì a Sabato e poi per 15 giorni ho le vacanze di Carnevale. Oggi siamo stati a colazione dalla Mamma.

La prego di salutarmi Joubin e Le Danois. Riceva tanti affettuosi saluti

da Umberto

6.4 D'Ancona to Volterra, Roma, 10.2.1929 *Cat. no. 84*

Roma, 10 Febbraio 1929

Carissimo Papà.

Ho letto con molto interesse la Sua lettera del 2 e ho appreso con piacere che pubblicherà le lezioni.

Anch'io penso che il lasciare la parola eredità non dia gran noia ai biologi. Converrà però metterci una nota spiegando cosa Lei intenda per eredità e dicendo che usa questa parola in un significato alquanto diverso da quello comune.

Fra i tre titoli da Lei proposti penso che il secondo, Théorie mathématique de la lutte pour la vie, sia il migliore. Anche il primo potrebbe andare, Principes mathématique de la lutte pour la vie. Non mi piace invece il terzo, Principes mathématiques de biologie. La parte della biologia che il Suo lavoro riguarda è la ecologia o etologia, cioè lo studio dei costumi e dei rapporti coll'ambiente. Allora forse sarebbe preferibile accennare addirittura a che parte dell'ecologia si riferisce e precisamente allo studio delle associazioni biologiche o biocenosi, o dei rapporti tra specie.

Nel caso che di tre specie la 1ª mangia la 2ª e la 3ª contemporaneamente, non credo che possa separatamente prendersi la 1ª e la 2ª e poi la 1ª e la 3ª. Quanto più la specie 1ª mangia della 2ª tanto meno ne mangia della 3ª e viceversa. D'altronde il valore nutritivo della 2ª e della 3ª specie sarà diverso. Credo però che si potrebbe calcolare questo caso prendendo la somma del numero di individui della specie 2ª e 3ª, moltiplicati ciascuno per un coefficiente. Non le pare.

Quanto mi accenna dei parassiti delle piante mi pare che si accordi pienamente con la Sua teoria. Infatti l'Icerya parassita delle piante (quella che si trova anche in Ariccia sulle mimose) non è stata distrutta dal Novius, però la presenza di questo tiene quella in proporzioni tali da non riuscire tanto dannoso.

Ho visto in questi giorni la citazione dei due lavori riguardanti gli equilibri tra le specie. Uno è di un tale Bodenheimer dell'Università di Gerusalemme, è pubblicato in uno degli ultimi numeri del Biologisches Zentralblatt[9]. Non conosce il Suo lavoro, dice che in un caso da lui studiato non sono le specie che stabiliscono con la loro convivenza uno stato di equilibrio, ma che questo è determinato dalle condizioni ambientali (temperatura, ecc.). Bisognerà che io legga accuratamente questo lavoro di cui ho visto solo le conclusioni.

---

9 BODENHEIMER 1928.

L'altro è un lavoro di Pictet pubblicato nella Revue Suisse de Zoologie, che non si trova a Roma, ma c'è a Napoli. Se ne avessi trovato l'indirizzo, avrei chiesto il lavoro direttamente all'autore.

In quanto alla pubblicazione delle Sue lezioni, se il libro deve essere destinato anche ai naturalisti converrà, io penso, distribuire la materia in modo che i naturalisti possano leggere senza essere intralciati da troppo calcolo. Forse si potrebbe disporre la materia alternando capitoli o brani in carattere diverso o aggiungendo note in modo che chi vuole possa saltare la parte di calcolo usata per trovare una data formula e possa prendere senz'altro la formula stessa. Per questi lettori non matematici converrà invece aggiungere qualche spiegazione elementare sul modo di applicare la formula stessa presentando anche degli esempi.

Forse in questo modo si potrebbe fare un libro utile per matematici e per naturalisti. Altrimenti temo che i naturalisti trovandosi di fronte ad una lunga serie di calcoli perdano il filo del ragionamento e in mezzo alle numerose formule non sappiamo quale vada applicata né come vada applicata.

Converrebbe forse scegliere in letteratura qualche esempio concreto in cui ci siano dei dati statistici e veder di applicare agli stessi i Suoi calcoli e vedere anzitutto se dal tentativo di applicazione degli stessi si può riconoscere se si tratta di fluttuazioni di natura interspecifica oppure causate dai fatti ambientali. Perché non si può mai nel caso di esempi naturali escludere a priori che le fluttuazioni possano essere date da variazioni periodiche delle condizioni ambientali.

In quale modo si potrebbe fare questo esame avendo una serie di dati statistici per un certo numero di anni? Questo mi pare sarebbe il primo dato pratico da risolvere.

Sono contento di sentire che Lei sta bene e che si trova bene. Mamma la raggiungerà fra pochi giorni. Oggi siamo stati giù a colazione.

Noi stiamo tutti bene.

Tanti saluti affettuosi

da Umberto [...]

6.5 D'Ancona to Volterra, Siena, 25.11.1929 *Cat. no. 85*

Siena, 25-XI-1929

Carissimo Papà.

Le rimando quanto Lei mi ha mandato con un breve sunto delle ricerche di Chapman, dal quale credo potrà avere la risposta a tutto quanto chiede il sig. Brelot. Nella conclusione mi pare che le cose di Chapman non sono riferite nella parte essenziale che comprende appunto le fluttuazioni tra uova e larve. Per il resto va bene. Le cose di Marchi sono da prendere con riserva perché non mi pare portino una documentazione rigorosa[10]. Egli dice che porta una conferma alle Sue leggi e alle mie statistiche. Perciò non spetta certamente a noi smentirlo; ma in quanto a crederci è un'altra cosa.

---

[10] See Marchi 1929a, b.

Converrà che Lei risponda all'ultima lettera di Chapman che chiede di poter ristampare in un suo libro le memorie del Journal di Copenhagen.

Se ha bisogno di qualsiasi altra cosa io sono sempre a Sua disposizione. Mi scriva se quanto Le ho mandato va bene.

Tante grazie per l'offerta di tenere la bambina durante l'assenza di Luisa. Credo che Luisa verrà giovedì e che si tratterrà una decina di giorni.

Mi è dispiaciuto di sentire che la Mamma stava male. Spero che oramai sia passato tutto e che si sia ristabilita.

Tanti affettuosi saluti a Lei, a Mamma, Enrico e Gustavo.

Umberto

6.6 D'Ancona to Volterra, Siena, 10.3.1930 *Cat. no. 86*

ISTITUTO DI ZOOLOGIA E
ANATOMIA COMPARATA
DELLA R. UNIVERSITÀ DI SIENA
Piazza S. Agostino, 4 – SIENA

Siena, 10–III–1930

Caro Papà.

Le rimando le bozze del libro sulle fluttuazioni che ho riletto.

Lei mi chiede se non sarebbe il caso di mettere nell'introduzione anche qualche altra notizia biologica, quelle americane. Esse, come Lei sa, sono state citate nella conclusione e a me pare che stiano meglio lì perchè sono venute dopo la Sua memoria. Se Lei vuol ricordarle nell'introduzione bisognerebbe toglierle dalla conclusione, oppure accennarle semplicemente nell'introduzione rimandando alla conclusione. Mi pare che le ricerche di Thompson non siano state per nulla ricordate. Forse non sarebbe male citarle in base al riassunto di Pérès[11].

Sul primo foglio, subito sotto il paragrafo 2 è detto "les variations dans le développement". Non è preciso, bisognerebbe dire "variazioni di aumento, di moltiplicazione". Developpement è usato nel senso di sviluppo embriologico e simili e può dar luogo a confusioni.

Il capoverso che viene subito sotto la mia tabella statistica non è molto bello. Quel periodo "Or celle ci ....." sarebbe bene se fosse un po' modificato. Riunito in un periodo il fatto ove i Selaci sono voraci e che durante la guerra la pesca è stata meno intensa fa confusione. Io direi invece: Ciò prova che durante il periodo 1915–20, quando a causa della guerra la pesca è stata molto meno intensa, si è avuto un aumento relativo di Selaci che sono particolarmente voraci e si nutrono di altri pesci. Poi potrebbe continuare "Les statistiques.....".

Poco più giù è detto "En agronomie ...." Va bene? o non sarebbe preferibile dire: in agricoltura?

---

[11] PÉRÈS 1927.

In fine dello stesso foglio dice: “sans avoir connaissance de ces travaux”. Dopo tanto tempo non mi pare che sia il caso di ripeterlo, io direi piuttosto: indipendentemente da questi lavori, oppure per altra via.

Cosa significa nella nota “presentée”?

Una cosa che non capisco bene è quella della “sparizione passeggera” di una specie, vedi foglio 4. Forse converrebbe darci qualche spiegazione. Se ho ben compreso va inteso in questo senso. Una specie può diminuire fino a scomparire quasi completamente; però vi sussistono alcuni esemplari, una coppia almeno, che possono moltiplicarsi appena le condizioni ridivengono favorevoli. È così? Converrebbe allora dire che si intende, una scomparsa quasi completa, con cessazione della riproduzione. Però dato che alcuni individui si conservano, essi possono nuovamente moltiplicarsi. Naturalmente in questo caso avrà importanza la durata di questa apparente scomparsa. Se essa dura poco gli individui sopravvissuti possono ritornare a riprodursi, se dura molto a causa d'invecchiamento o di morte non più.

Sempre per questa questione ho fatto dei segni anche al foglio 11. Invece di “Deux individus” converrà dire “una coppia”. Nel caso di specie inferiori può servire a questo scopo anche una semplice sospensione della riproduzione, come effettivamente avviene in molti casi, oppure un incistamento o una conservazione di nova, una “vita latente” insomma. Certe volte questi periodi di vita latente (ne sono noti a causa di sfavorevoli condizioni fisiche dell'ambiente) possono durare anche relativamente a lungo.

Sullo stesso foglio non mi persuade invece il postulato biologico che “una specie non può resistere a un numero abbastanza grande di sparizioni passeggere”. Io dal punto di vista biologico ritengo invece più importante la durata delle sparizioni e la loro frequenza. Se sono molto frequenti allora possono essere più efficaci perchè nei periodi intermedi la specie non può sufficentemente riprodursi. Se invece sono piuttosto distanziate possono essere anche molte senza che la specie scompaia definitivamente. Ci sono delle specie che ogni anno a causa di condizioni sfavorevoli scompaiono una o più volte.

Al foglio 13 invece di “développement” mettere aumento o moltiplicazione.

Al foglio 15 c'è nuovamente un altro interrogativo per il solito motivo.

In altri punti ho fatto dei segni per richiamare la Sua attenzione su errori tipografici o in punti che mi ponevano dubbi.

Ho ripensato sull'opportunità di presentare la nota dell'Académie. Ritengo che ora non sia il momento; data la prossimità del concorso potrebbe venir interpretato come esibizionismo e nel complesso mi sarebbe più dannoso che utile. Penso che sarà meglio attendere e presentare la Nota dopo il concorso.

Per ora ho preparato invece una nota che presenterò alla locale Accademia dei Fisiocratici. Le manderò domani il testo con preghiera di volermela leggere per quelle solite questioni fisiche di interpretazione del funzionamento.

La ringrazio per l'invio del Bollettino della Facoltà di Oceanografia. Peccato che alla posta si sia un po' sciupato. È preferibile metterlo in una busta o disteso sotto una fascia più grande.

Silvia ringrazia per il giornalino che ha avuto stamane.

Luisa e Silvia mandano tanti saluti. Anche da parte mia i più affettuosi saluti a Lei, alla Mamma, Nella, Enrico, Edoardo, Gustavo.

Umberto

Le rimando anche la lettera e la nota del sig. Van der Pol, al quale converrà forse che Lei risponda.

11-III-1930

Le mando il manoscritto della mia nota. Mi farà molto piacere se la vorrà rileggere e se mi farà sapere il Suo parere. Io conto di presentarla alla prima seduta dell'Accademia dei Fisiocratici in modo da poterne avere gli estratti per il concorso.

6.7 D'Ancona to Volterra, Siena, 15.3.1930 *Cat. no. 87*

(Same heading of letter no. 6.6)

Siena, 15-III-1930

Caro Papà.

La ringrazio tanto per l'interessamento per la mia nota e per l'attenzione con cui l'ha voluta leggere e per le correzioni e le indicazioni consigliate.

A pag. 1 io riferisco l'idea del Münch che nella fibra non tutta la sostanza contrattile sia anisotropa, ma che sia una sostanza anisotropa birifrangente disposta a spirale. È il Münch che si esprime così. La concezione non è però precisa.

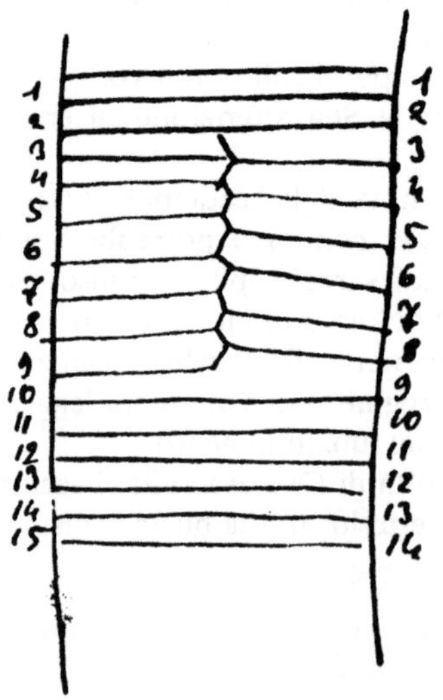

A pag. 2 va bene fibre stirate. Si tratta di fibre che il Tiegs ha stirato per vedere meglio questa disposizione.

I "periodi a nonio", così chiamati dal Heiderhain, sono dei tratti della fibra in cui tra destra e sinistra da una parte si nota una stria in più che dall'altra, come in un nonio.

Nella gran maggioranza dei casi non si vede un asse centrale attorno a cui si avvolge l'elicoide. È come una scala a chiocciola che non avesse un asse al centro.

L'ordinamento a zig-zag indica una elicoide semplice perché se noi facciamo una sezione longitudinale passante per l'asse di una elicoide semplice abbiamo una figura come quella qui indicata a sinistra, se invece, facciamo una simile sezione lungo una doppia elicoide abbiamo una figura come quella a destra. Ora nei preparati si ha l'ordinamento a zig-zag come si vede nella figura di sinistra[12].

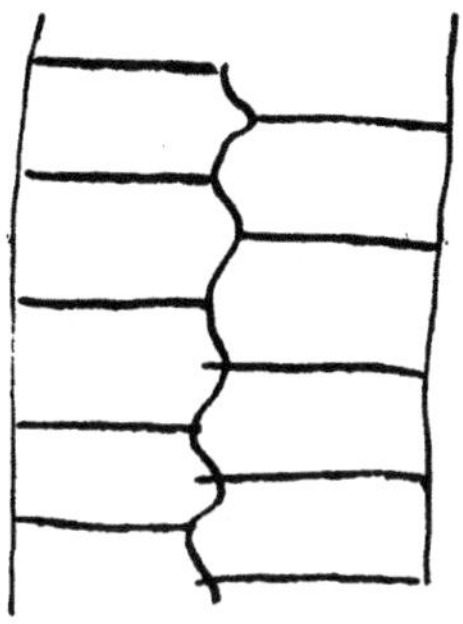

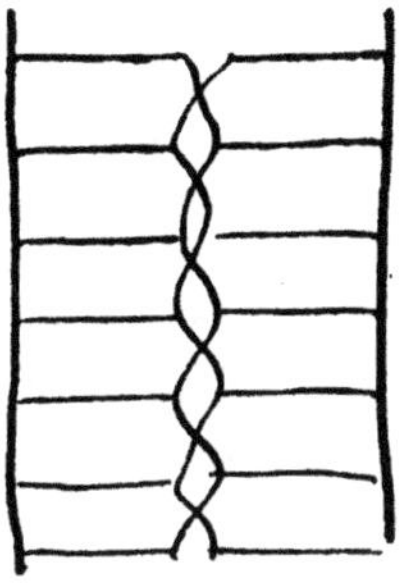

Ho detto stazionarie le contrazioni elementari perché esse si formano sempre in punti fissi e non progrediscono lungo la fibra come fa l'onda di contrazione. Mentre questa procede, le contrazioni elementari sono stazionarie come delle onde stazionarie. Può andare così questa espressione?

Correggerò "potrebbe determinare lo sfondamento" invece di dovrebbe. Però ritengo che tale sfondamento si dovrebbe avere per il fatto che le membrane trasversali sono molto più esili e verosimilmente meno resistenti della parete del cilindro. Ora sotto l'effetto delle pressioni interne che si sviluppano durante la contrazione la parete del cilindro si dilata. Se una pressione così forte si esercitasse sulle membrane trasversali, esse a mio avviso dovrebbero essere sfondate.

Correggerò "maniera continua" invece di "forma". È più preciso.

Perché non va bene "continuità dell' onda di contrazione". Se non ci fosse continuità tra i tratti elicoidi formati da due piani elicoidali completamente indipendenti?

Farò volentieri degli altri modelli, però il sistema seguito da Trabacchi[13] non si presta bene. Con la carta non si possono fare elicoidi a passo più stretto perchè la carta è poco cedevole. Bisognerebbe farle con qualche sostanza più cedevole, forse con della gomma dura. Quando verrò a Roma sarà forse bene che ne parli anch'io a Trabacchi. Forse si potrebbe fare con qualche sostanza plastica e che poi si possa indurire. Ne parlerò anche a mio fratello per fargli fare dei disegni.

Tanti saluti affettuosi a tutti

da Umberto.

---

[12] Reference to the left-hand figure in the group of two figures below.

[13] Reference to Giulio Cesare Trabacchi (1884–1959), who was director of the physics laboratory of the Istituto di Sanità Pubblica in Rome (later Istituto Superiore di Sanità) from 1922 to 1958 and collaborated with the University of Rome physicists in nuclear research.

6.8 D'Ancona to Volterra, Siena, 28.3.1930 *Cat. no. 88*

Siena, 28/III/1930

Carissimo Papà

La ringrazio tanto per tutto quanto mi ha scritto a proposito delle elicoidi.

Ora ho capito come si originano le varie elicoidi. Di alcune altre cose a questo proposito ne riparleremo a voce tra pochi giorni quando verremo a Roma.

Mi dispiace che Lei non sia soddisfatto della redazione del libro francese. Io credevo che quelle note di matematica le avesse fatte mettere Lei per uso dei matematici. A dire il vero io non credo che un naturalista troverà più chiara questa redazione francese che le precedenti memorie italiane. Quelle erano comprensibili nei risultati anche per i naturalisti. Certamente noi non abbiamo una preparazione matematica sufficiente per comprendere tutto l'andamento delle operazioni. Perché i naturalisti potessero capire tutto lo svolgimento delle operazioni bisognerebbe premettere degli elementi di calcolo e allora si dovrebbe fare un trattato e non più una memoria. Oppure bisognerebbe esporre tutto in forma più elementare il che non mi pare sia possibile.

Perciò forse come aveva fatto Lei in italiano era il meglio di tutto. Caso mai conveniva mettere qualche nota in calce per spiegare meglio qualche passaggio o qualche concetto matematico non adatto per i nostri denti.

Ma come ha fatto il Sig. Brelot non pare nemmeno a me che sia il modo migliore per rendere la memoria accessibile ai naturalisti.

Però noi naturalisti abbiamo in genere una preparazione matematica molto elementare, perciò non ritengo che molti di noi possano applicare o seguire dei calcoli di questo genere.

La maggior parte di noi può servirsi della matematica applicando semplicemente delle formule bell'e pronte come si fa se non sbaglio anche in statistica. Perciò a facilitare la diffusione del lavoro fra i naturalisti converrebbe al caso precisare delle equazioni da applicarsi in singoli casi e dare degli esempi di tali applicazioni. Forse ciò converrà fare in tempo successivo. Applicare per esempio i Suoi calcoli ai dati numerici di Chapman e vedere di far corrispondere l'andamento teorico a quello empirico dell'aumento della popolazione.

Tanti saluti affettuosi a Lei, alla Mamma, Enrico e Gustavo,

Umberto. [...]

6.9 D'Ancona to Volterra, Siena, 15.6.1930 *Cat. no. 89*

(Same heading of letter no. 6.6)

Siena, 15-VI-1930

Carissimo Papà

Le invio un foglietto con alcune citazioni che converrà aggiungere a quelle già date al Signor Brelot.

A Firenze ho avuto notizie del libro di Friederichs, recentemente uscito[14]; pare che in esso i Suoi calcoli sono citati abbastanza dettagliatamente. Vedrò se mi sarà possibile farmelo inviare in prestito.

Rivedrò molto volentieri le bozze che mi invierà il Signor Brelot. Sarà bene che egli coordini le varie parti che io in diversi tempi ho mandato.

Preferirei che il capitolo figurasse assieme agli altri del libro senza alcuna indicazione che io ho contribuito alla sua elaborazione. Dall'indicazione che l'ho fatto io non mi viene alcun vantaggio, perchè si tratta di una compilazione di poca importanza. Perciò mi farà piacere se non metterà alcuna indicazione.

Brunelli ha risposto abbastanza violentemente alla mia ultima nota. Gli sto preparando la risposta.

Tante grazie per i biglietti speriamo di rivedervi prima della partenza. Tanti saluti affettuosi a tutti

Umberto [...]

6.10 D'Ancona to Volterra, Siena, 26.7.1930 *Cat. no. 90*

Scriverò con piacere quella prefazione per Papà.

Anch'io vi invio tanti ringraziamenti per gli affettuosi auguri.

Ho letto la prefazione che Papà ha scritto. Mi pare che vada bene. C'è soltanto in due punti che farei una piccola modificazione di forma. Quando Luisa l'avrà ricopiata ne manderò una copia a Papà con le modificazioni a lapis da me proposte. Porterò l'altra copia al signor Brelot. Prego di farmi vedere le bozze del capitolo conclusivo bibliografico, perché bisognerà forse farci qualche aggiunta.

Speriamo di avere spesso vostre notizie e di sentire che costì vi trovate bene. Tanti affettuosi saluti.

Umberto [...]

6.11 D'Ancona to Volterra, Roma, 28.7.1930 *Cat. no. 91*

Roma, 28-VII-1930

Carissimo Papà.

Luisa ha ricopiato la prefazione del libro. Le mando una copia unitamente al manoscritto, l'altra la porto oggi stesso al Signor Brelot.

La prefazione stessa mi pare che vada bene. Però alla riga 3ª, dove sta scritto "sur les variations des associations biologiques" io vedo che sarebbe preferibile mettere: variazione nella composizione delle associazioni biologiche.

Alla fine del primo capoverso, dove Lei cita Friederichs, mi pare che messo così dia l'impressione che questo autore abbia fatto un riassunto completo del Suo lavoro, il che non è. Sembra anche che Friederichs abbia citato le differenti riviste in cui sono comparse delle recensioni. Io credo che converrebbe modi-

---

14 FRIEDERICHS 1930b.

ficare quelle righe dicendo: fra l'altro le mie ricerche sono state riportate nella recente operà di Friederichs, in cui ne sono rilevate le possibili applicazioni allo studio della zoologia agraria.

Se ha bisogno di altro mi scriva che io sono sempre lieto di poterle essere utile.

Ieri è partito Gustavo, dal quale avrà nostre notizie. Noi contiamo di partire giovedì.

Tante grazie per gli auguri per l'anniversario del nostro matrimonio.

Della mia commissione non sono molto soddisfatto. Non sono persone della serenità di giudizio di Russo. Il lato debole, nei miei riguardi, di questa commissione è che tre dei commissari hanno allievi da mettere a posto, per cui c'è molto pericolo che giungano ad un accordo. D'altra parte io non ho alcun sostenitore energico, come avrei invece avuto se nella commissione ci fossero stati Cotronei o Enriques[15]. Fra tutti faccio maggior assegnamento su Pierantoni[16].

Spero di aver presto loro notizie. Tanti affettuosi saluti a lei e alla Mamma. Tanti saluti anche da Luisa che ora non è in casa e baci da Silvia.

Umberto.

6.12 D'Ancona to Volterra, Fiume, 16.8.1930 *Cat. no. 92*

Fiume, 16 Agosto 1930

Carissimo Papà.

Brelot mi ha mandato tutto quanto mi serviva e così ho potuto correggere.

Le rimando ora sia le bozze dell'introduzione che quelle della conclusione con le correzioni e modificazioni che riterrei conveniente apportarvi. Veda Lei che cosa Le pare.

Anzitutto io non farei quelle note biologiche a parte perché riesce meno organica l'esposizione e perché si dà alla nota stessa una importanza che non ha e non deve avere. Perciò io ho fuso la nota biologica stessa nella conclusione. Così ho tolto le indicazioni dei singoli paragrafi della nota e le ho elencate all'inizio della conclusione come il Brelot ha fatto in genere per tutti i capitoli.

Il Brelot ha preso semplicemente i fogli che io gli avevo inviato traducendoli, ma senza coordinarli col resto come io avevo raccomandato. Ne sono venute così delle ripetizioni e delle sproporzioni che ho dovuto togliere.

---

[15] Reference to Giulio Cotronei (1885–1962), researcher in the field of experimental embryology and the physiology of development, who was director of the Zoology Institute of the University of Siena (see also D'Ancona biography). Paolo Enriques (1878–1932), brother of the mathematician Federigo, lectured in zoology, anatomy and comparative physiology at Padova, was also interested in genetics and eugenics and is known for his law on the "independence of variability".

[16] Umberto Pierantoni (1876–1959), a geneticist, was professor of genetics and the biology of race at the University of Rome and in 1938 joined the Higher Council for Demography and Race (see Israel, Nastasi 1998).

Ho fatto anche alcune aggiunte di citazioni bibliografiche. Le parti che ho scritto a lapis devono essere tradotte in francese. Converrà poi che Lei riveda lo stile della conclusione perché mi pare che ci siano parecchie ripetizioni di lingua.

Anche nella Introduzione ci sono alcuni errori che vanno corretti.

Bisogna sostituire sempre paludisme a malarie.

Nella conclusione dove dice "ce fait empêche pratiquement la croissance infinie qui résulte dans certains cas des équations adoptéés jusqu'à présent", io toglierei le parole dopo "croissance" perché non è molto preciso.

Fondendo la nota biologica con la conclusione basta che al paragrafo 5 sia messa l'indicazione in una nota in calce che le indicazioni bibliografiche sono state fornite da me. Ciò se Lei lo ritiene necessario, io per conto mio non ce lo metterei.

Le citazioni sono per la maggior parte piene d'errori. Perciò ho fatto molte correzioni controllando il testo. Però Brelot non mi ha mandato un pezzetto di manoscritto con le citazioni di Thompson. Converrà per maggior sicurezza che esse siano accuratamente controllate sul mio manoscritto che deve essere presso Brelot.

Le citazioni di Chapman vanno spostate come da me indicato.

Come si dice in francese agrumi? Va bene "fruits acides"?

La citazione delle mie ricerche era prolissa. L'ho abbreviata tenendo anche conto che se ne fa parola già nell'introduzione.

"Pesca a strascico" credo si dica "pêche au filet traînant" e non "pêche au filet a traîne". Forse Lei potrà chiederlo a qualcuno.

"Specie meno difese" non mi pare equivalga a "qui se défendent moins".

Inermi o inoque e non "passives".

Credo che Lei troverà giuste le correzioni da me fatte. Mi farà piacere se mi manderà un'altra bozza corretta e se con essa mi restituirà per facilitare il confronto questa con tutte le correzioni.

Ho ricevuto una lettera da Joubin, che mi dice che andrà a Padova.

Abbiamo stamane avuto dopo lungo silenzio una vostra lettera. Luisa era preoccupata della mancanza di notizie.

Noi stiamo benissimo. Peccato che anche qui i tempi siano poco favorevoli ai bagni. Però possiamo uscire tutti i giorni. Silvia anche se il tempo non consente i bagni, può pigliare tutti i giorni l'aria di mare.

Riceva tanti saluti affettuosi e auguri di passare bene la villeggiatura. Saluti affettuosi alla Mamma e Enrico e Gustavo.

Umberto [...]

6.13 D'Ancona to Volterra, Napoli, 9.10.1930 *Cat. no. 93*

Napoli, 9-X-1930

Caro Papà

Il signor Brelot mi ha inviato le bozze. Le ho riviste e gliele invio qui unite. C'è ancora da aggiungere una citazione che la tipografia aveva saltato. Credo che si potrà fidarsi di far fare l'ultima revisione a Brelot, raccomandandogli di stare attento. Per il resto va bene.

La ringrazio tanto per la lettera sulle elicoidi. Domenica ne potremo riparlare in Ariccia.

Ha poi deciso per i libri di Casella?

Tanti saluti affettuosi a Lei, alla Mamma, Edoardo, Enrico e Gustavo.

Aff.mo Umberto

6.14 D'Ancona to Volterra, Siena, 18.11.1930 *Cat. no. 94*

Siena, 18-XI-1930

Carissimo Papà.

La ringrazio per il Suo continuo interessamento per le mie ricerche. Avevo già da tempo pensato di vedere se le fibre muscolari sono sinistrorse o destrorse e conto di compiere questa ricerca non appena sarò libero da altri lavori. Ritengo anch'io che possa esserci una qualche regola sull'orientamento destrorso o sinistrorso delle fibre muscolari. Chissà che non ci sia differenza fra i vari muscoli, o tra quelli di una metà e quelli dell'altra metà del corpo? oppure tra le diverse specie? Ciò è anche verosimile per il fatto che tra le conchiglie dei molluschi alcune specie sono destrorse e altre specie sono sinistrorse, costantemente, con eccezioni rarissime. Perciò penso che questa ricerca possa essere fruttuosa di osservazioni interessanti.

Per quanto riguarda il passo sulle elicoidi ho già visto che ci sono differenze tra specie e specie e talvolta tra i diversi muscoli della stessa specie. Ciò può essere misurato direttamente col micrometro.

La ringrazio anche per la discussione del Senato sulla Stazione Zoologica. Mio fratello me ne ha già procurato i Rendiconti e li ho inviati a chi me li aveva richiesti.

Ho ricevuto anch'io una lettera dal Dottor Tirrazzo che mi dice che sta facendo una rivista sintetica sulle Sue ricerche sulle fluttuazioni e mi chiede perciò la mia memorie su questo argomento. Non so chi sia questo dottore, che sento nominare per la prima volta. Cosa voleva da Lei?

Sono contento di sapere che stanno tutti bene. Spero che anche Edoardo si sia completamente ristabilito. Io sto cercando casa; Luisa mi scrive che fra giorni verrà qui. Così potremo decidere come sistemarci quest'anno.

Riceva tanti affettuosi saluti, unitamente alla Mamma, Enrico e Gustavo.

Umberto

6.15 D'Ancona to Volterra, Siena, 7.12.1930 *Cat. no. 95*

7-12-1930

Vi ringrazio tanto per gli auguri che avete mandato per la festa di Luisa. Luisa era un po' "bizzosa" perché non trovava casa e perché mancava di vostre lettere; dopo la telefonata s'è però calmata.

Ringrazio Papà per il Suo interessamento sugli elicoidi. Ho fatto nuove osservazioni e ritengo che in tutti i muscoli sia degli insetti che delle [illeg.] siano mescolate fibre sinistrorse e fibre destrorse. Però l'osservazione è molto delicata e facili sono le illusioni ottiche perché si tratta di strutture molto minute non rigide, per cui un piccolo spostamento può far ritenere che le condizioni siano diverse da quelle che sono. Non saprei davvero con quale altro mezzo si potrebbe avere una maggior fiducia nelle osservazioni. Ne riparlerò a Papà a voce e gli farò vedere alcuni disegni che ho fatto eseguire.

A voi tutti tanti affettuosi saluti. Baci a Silvia

Umberto [...]

6.16 D'Ancona to Volterra, Siena, 27.1.1931 *Cat. no. 96*

Siena, 27-I-1931

Carissimo Papà

La ringrazio per ma Sua lettera del 25. Ho letto con interesse il Feuilleton del Journal des Debats. È molto interessante e scritto con competenza.

Io credo che da Thompson non Le sarà difficile avere il suo libro. Si tratta di pubblicazione a quanto sembra ufficiale che Lei potrebbe chiedergli. È strano che egli non abbia risposto.

Lei non gli ha scritto, a quanto io mi ricordi, mandandogli il libro. Gli ha scritto ultimamente, mi sembra, da Siena. Anche Luisa dice di non avergli scritto.

Non so se valga la pena di mandarne una copia a S[illeg.]. Mi pare che egli sia oramai completamente dedito alla politica.

Con Luisa Le do le nostre ottime notizie. Stiamo tutti bene. La casa è oramai completamente sistemata e ci troviamo abbastanza bene. Oggi abbiamo avuto la visita di zio Piero.

Riceva tanti affettuosi saluti unitamente alla Mamma, Enrico e Gustavo.

Umberto.

6.17 D'Ancona to Volterra, Siena, 11.2.1931 *Cat. no. 97*

Siena, 11-II-1931

Carissimo Papà.

Ieri sera al telefono non ho potuto esporre completamente quanto desidero farle sapere in merito alla questione con Brunelli[17].

Le ho già detto che i nostri rapporti sono completamente rotti dopo quello che egli ha fatto nei miei riguardi.

---

[17] Gustavo Brunelli (1881–1960), researcher in development morphology and embryology, in the 1930s he founded the Royal Central Laboratory of Hydrobiology in Rome and was director of the Naturalists Society of Naples (see also Israel, Nastasi 1998).

Circa un anno fa Issel[18] mi aveva chiesto se ero disposto di elaborare il capitolo sui pesci della laguna per l'opera che la Commissione del Mediterraneo sta preparando sulla laguna Veneta. Accettai, ma quando si venne in discussione al Talassografico Brunelli disse che doveva fare lui tutto.

Precedentemente nel lavoro della Monografia di Napoli io avevo chiesto di fare anche un capitolo generale e la Stazione Zoologica era disposta a farmelo fare, ma poi Brunelli si oppose.

Altre volte io dovevo ricevere del materiale di studio dalla Stazione di Piscicultura di Roma, ma lui si oppose ed non ebbi nulla.

E ci sarebbero anche altri esempi. In tutti i modi sia a voce che per iscritto egli cercò poi di mettere in ridicolo e di screditare appunto il mio lavoro sui pesci dell'Adriatico. Metteva in ridicolo che io avevo trovato che dopo la guerra si aveva un aumento di pescicani.

Sarebbe troppo lungo dirle tutto.

In seguito alla nostra recente polemica, egli terminò dicendo come un nume sdegnato e annoiato da un misero mortale che "la mia posizione nella sperimentazione non era così elevata" da poter discutere con lui.

Ora i nostri rapporti sono tali che io incontrandolo non lo saluto. Perciò non posso nemmeno accettare di fare alcun lavoro in cui lui c'entri. Avrei desiderato andare a Parigi al congresso di Pesca, ma sapendo che Brunelli è il delegato italiano non ci andrò perchè non voglio avere contatti di alcun genere con lui. Perciò se anche lui mi chiedesse di fare la relazione rifiuterei.

La prego quindi non solo di non proporre che io collabori con lui nell'elaborazione della relazione, ma di non chiedere neppure che un biologo in genere collabori, perchè questo biologo non potrei essere io. Potrò naturalmente collaborare con Lei per aiutarla a trovare i dati biologici necessari, ma in nessun caso desidero che il mio nome figuri.

A Lei che non conosce tutta la situazione ciò potrà sembrare eccessivo, ma non lo è quando si considerino tutti gli elementi che sono a mia cognizione. Perciò avrò piacere se lei vorrà fare la relazione da solo. Lei può benissimo rispondere che farà solo una relazione d'indole generale sugli equilibri biologici nell'economia del mare soffermandosi particolarmente sulle Sue leggi matematiche. Dica che preferisce che la relazione venga divisa e venga affidata a un biologo (che in nessun caso potrà essere io) la parte più strettamente biologica. È naturale che questa sia fatta da un biologo dato che ci sarà probabilmente discussione e che è necessario che il relatore sia a giorno di tutta la enorme letteratura che ha attinenza col problema della produttività del mare.

Sono certo che Lei comprenderà la ragioni che mi inducono a rispondere in questo senso alla Sua gentile offerta e che in questo senso Lei vorrà rispondere al Brunelli.

La prego di gradire i nostri più affettuosi saluti unitamente alla Mamma. Spero che abbiano fatto un buon viaggio e che costì abbiano trovato tempi buoni.

Aff. mo Umberto [...]

---

[18] Reference to Raffaele Issel (1878–1936), marine biologist and son of the geographer and geologist Arturo Issel.

6.18 D'Ancona to Volterra, Siena, 18.2.1931 *Cat. no. 98*

Siena, 18-II-1931

Carissimo Papà.

Ho ricevuto stamane la lettera Sua e della Mamma e rispondo subito in merito alla questione Brunelli.

Le ho già esposto in parte quali sono i rapporti col Brunelli e ciò non è conseguenza di spirito litigioso da parte mia o di ripicchi, ma è la conseguenza di una lotta sorda e ingiustificata che lui ha cominciato a farmi fino dai primi tempi ch'io ho cominciato a lavorare. E non è dipeso nemmeno da divergenze di vedute nella questione dell'inesauribilità del mare, ma forse questo è stato un pretesto nella campagna ostile che lui mi ha fatto.

In tutte le occasioni e di più recentemente lui ha voluto trattarmi come un ragazzino che non va preso sul serio.

Perciò torno a ripeterle che non ho alcun desiderio di avere nulla a che fare con quell'individuo. Non è che io sfugga di sostenere le mie idee in sua presenza. Però l'andare a Parigi con lui a capo della delegazione italiana vuol dire averlo fra i piedi più di quanto non sia necessario e vuol dire anche il non poter avere quella indipendenza di azione che sarebbe necessario. D'altra parte la sua posizione ufficiale è superiore alla mia per cui io in una discussione mi troverei sempre in una condizione d'inferiorità.

Queste sono le mie ragioni.

Se però lei desidera fare differentemente io non mi ci oppongo, però La prego di tener sempre presente i miei rapporti col Brunelli e la necessità da parte mia di mantenere un tono di rapporti consono all'atteggiamento che lui ha avuto nei miei riguardi.

Se Lei vuole assolutamente che io collabori alla Sua relazione, lo metta come condizione al Brunelli e non lo chieda come concessione. La Sua posizione nella scienza è talmente superiore che Lei può benissimo condizionare al Brunelli la Sua collaborazione e non fargli vedere che lui ha fatto a Lei una gentilezza invitandola. È certo che se il Brunelli si è rivolto a Lei vuol dire che a Parigi egli ha bisogno di qualcuno più autorevole che gli dia autorità.

Perciò mi rimetto a Lei.

Ho tanto piacere che abbiano fatto un buon viaggio e che si trovino bene anche col tempo. Quando inizia le lezioni? Riceva unitamente alla Mamma tanti affettuosi saluti da

Umberto [...]

6.19 D'Ancona to Volterra, Napoli, 4.4.1931 *Cat. no. 99*

Napoli, 4-IV-1931

Carissimo Papà

Domani consegnerò a Luisa le pagine introduttive e gliele spedirà Lunedì mattina. Conviene che Lei le riveda, perché può darsi che Lei preferisca fare qualche modificazione. Io ci ho messo quelle indicazioni che Lei desiderava.

Ho potuto avere appena ieri delle pubblicazioni da Silvestri e oggi ho potuto completare i dati qui in Biblioteca. Prima non mi è stato possibile.

Sto preparando la relazione per il Congresso di Pesca. Appena l'avrò stesa gliela manderò pregandola di restituirmela subito col suo parere e con le eventuali modificazioni per poterla presentare nel termine utile.

Silvestri mi ha incaricato di salutarla. Egli avrebbe piacere di esporle un caso di lotta biologica affinché Lei vedesse possibilmente di trattarlo matematicamente[19]. Si discute tra gli entomologi se è utile o dannoso l'introdurre in un paese più di una specie di insetti che distruggono gli insetti dannosi di una pianta. Siano questi insetti dannosi della specie A. Questa è parassitata da due specie B e C indipendenti e di costumi diversi che tra loro non si combattono ma contendono lo stesso nutrimento. Sono specie che depongono le uova sulle larve della specie A. Dall'uovo nasce una larva della specie B o C che si divora la larva A e quando l'ha divorata completamente si metamorfosa e diventa un individuo adulto. Se un solo uovo della specie B o C viene deposto su una larva A la larva B o C che sguscia riesce a svilupparsi perché ha nutrimento sufficiente. Se invece sulla stessa larva A vengono deposte due o più uova di B o C, sia della stessa o delle due specie non trovano nutrimento sufficiente e finiscono per morire. Le due specie B e C avendo costumi diversi avranno naturalmente diversi coefficienti di accrescimento e in diverso modo saranno efficaci nella lotta contro A. Ora alcuni entomologi dicono che B e C avendo costumi diversi e distruggendo la specie A in modo e condizioni diverse si coadiuvano e completano la lotta naturale. Altri invece dicono che le due specie contendendosi lo stesso nutrimento si combattono a vicenda e finiscono col diminuire l'efficacia della lotta esercitata dalle singole specie.

Silvestri è della prima opinione, egli però desidererebbe sapere se applicando i Suoi calcoli si può trarne qualche conclusione.

Mi è dispiaciuto di non averla potuto vedere prima di partire, ma sono molto occupato perché oltre al lavoro che faccio qui devo completare per il 1 maggio il rapporto e la relazione per il Congresso di pesca. Perciò non mi è possibile interrompere il lavoro per più di una giornata. La prego di salutarmi tanto la Mamma e di gradire i miei più affettuosi saluti.

Umberto [...]

6.20 D'Ancona to Volterra, Napoli, 12.4.1931 *Cat. no. 100*

Napoli, 12-IV-1931

Carissimo Papà,

Luisa mi ha dato vostre notizie, dalle quali apprendo che state bene.

Il giorno di Pasqua Luisa e Silvia sono venute a Formia e sono rimaste con me fino alla sera. La giornata era bella e la gita è riuscita molto bene.

---

19 Filippo Silvestri made important contributions to systematic zoology, ethology and to the theory of the fight against noxious animals in agriculture. For the problem raised here by Silvestri, see Falcone, Israel 1985.

Io conto di rimanere qui fino a venerdì, poi da Roma ripartiremo tutti per Siena lunedì 20.

Luisa Le ha mandato copia della relazione che ho preparato per il Congresso di Pesca. Ho ritenuto preferibile unire la parte biologica a quella matematica. Ho cercato di esporre quest'ultima in modo molto semplice, senza dimostrazioni matematiche perché ho tenuto conto che si tratta di un congresso di pesca. Credo che sia meglio così di mettere come è esposto in quella memoria di Milano. Già così nella lettura converrà lasciare gli accenni più strettamente matematici.

Mi dica come Le pare e se ha delle correzioni da fare La prego di rimandarmi il manoscritto corretto in modo che io lo possa rivedere e preparare definitivamente per poterlo presentare in tempo utile prima della fine del mese.

Le ho fatto mandare da Luisa anche la prefazione per quell'altra memoria. Andava bene?

Io cercherò di passare uno di questi giorni da quell'antiquario a vedere se ha la Divina Proporzione.

Fino a quando rimangono a Parigi? Spero che dopo ritornati verranno a farci una visita a Siena. Anche Luisa ne avrebbe molto piacere.

Se mi deve mandare manoscritti corretti, li spedisca a Roma, che qui forse non arriverebbero più in tempo.

Tanti affettuosi saluti a Lei e alla Mamma

da Umberto

6.21 D'Ancona to Volterra, Roma, 18.4.1931 *Cat. no. 101*

Roma, 18-IV-1931

Carissimo Papà.

Sono ritornato a Roma ieri sera e ho trovato tutti bene. Martedì alle 3 del pomeriggio Luisa, Silvia ed io ripartiamo per Siena.

Ho trovato qui la Sua lettera con la relazione per il congresso di pesca. Sono contento che nel complesso l'abbia approvata.

Farò le correzioni della parte matematica come da Lei indicato.

Per quanto riguarda quella osservazione sulla disponibilità di energia solare, bisogna tener conto che per gli animali non si verificano regolarmente dei fenomeni di migrazione verso i paesi caldi come si può verificare nelle migrazioni umane. Le migrazioni animali hanno in genere carattere periodico stagionale. Migrazioni d'altro genere avvengono soltanto se cambiano le condizioni ambientali oppure se la moltiplicazione di una specie avviene in proporzioni eccessive per un dato ambiente. L'energia solare è sempre in eccesso per le necessità del mondo vivente, è la parte di energia solare trasformata in energia chimica nella materia vivente che può essere deficiente. Perciò anche le migrazioni umane si possono far rientrare in quest'ordine di fenomeni. Le grandi migrazioni storiche verso il Sud sono state dovute forse più al desiderio di conquistare nuove terre fertili e ricche di nutrimento vegetale che non al solo desiderio di venire a godere il bel sole meridionale. In ogni modo sebbene per l'uomo può essere un'attrattiva anche questo, non lo è che scarsamente per gli animali.

È vero che anche gli animali vengono attratti dalla luce, ma questi fenomeni sono dei riflessi indipendenti del bisogno di maggior o minor energia luminosa o termica. In generale le specie animali tendono a occupare tutto lo spazio disponibile in cui si verificano le condizioni fisiche ambientali sufficienti alla loro esistenza. In questa area di diffusibilità la limitazione è data dalla concorrenza vitale che si manifesta soltanto nella lotta per la conquista del nutrimento e dello spazio.

Le specie nordiche in genere hanno un'area di diffusione in cui si hanno determinate condizioni di ambiente che esse non desiderano né possono sostituire con quelle che si hanno in altri ambienti.

Perciò nel complesso mi pare che quell'accenno alla disponibilità di energia possa andare.

Quella frase dello scetticismo la modificherò.

Devo fare anche alcune altre modifiche per alcune citazioni nuove, ma nel complesso la relazione rimane come Lei l'ha letta.

Perciò appena pronta, senza più rimandargliela, la invierò a Brunelli.

Qui unita Le mando una copia della Prefazione della Sua memoria che Luisa ha ricopiato.

Quando conta di essere nuovamente a Roma? Facciamo assegnamento su una visita a Siena.

Gradisca unitamente alla Mamma i più affettuosi saluti

Umberto [...]

6.22 D'Ancona to Volterra, Siena, 1.6.1931 *Cat. no. 102*

Siena, 1 Giugno 1931

Carissimo Papà,

La ringrazio per le bozze della relazione di Parigi che ho ricevuto e rispedite corrette al congresso. Alcune delle correzioni di stile da Lei indicate non ho potuto eseguirle perché portavano come conseguenza una modificazione dell'impaginazione. La nostra relazione supera di 2 pagine il limite massimo concesso, con le correzioni si sarebbe aumentato ancora, con pericolo pure di far fare nuovi errori tipografici. Poiché le modificazioni stesse non cambiavano il senso, ho creduto preferibile lasciare il testo come stava. Anche Luisa è di questa opinione. Ho chiesto una nuova bozza perché gli errori tipografici erano molti. Ho chiesto gli estratti da Lei indicati.

La ringrazio per le informazioni che mi dà sull'Accademia dei XL. La medaglia per la biologia è stata assegnata a Pasquini[20]. Desideravo perciò sapere chi erano i commissari che gliel'avevano assegnata. Chi è il presidente?

Spero che avrete fatto un buon viaggio e che il soggiorno a Vienna e a Praga sia stato piacevole. Quando fa le lezioni?

---

[20] Pasquale Pasquini (1901–1977) was a zoologist and experimental embryologist.

Noi stiamo tutti bene. Silvia nei giorni passati ha avuto qualche disturbo causato da acetone. Ora sta bene. Il medico ha prescritto una dieta da seguire nei primi tempi. È stata cosa di brevissima durata, 2 giorni, e così per fortuna non si è sciupata.

Al ritorno dalla Cecoslovacchia speriamo di vedervi qui a Siena, ma non soltanto per mezza giornata.

Tanti affettuosi saluti a Lei e alla Mamma.

Umberto [...]

6.23 D'Ancona to Volterra, Siena, 24.7.1931 *Cat. no. 103*

Siena, 24-VII-1931

Carissimo Papà

Spero che abbiano fatto un buon viaggio di ritorno e che costì abbiano trovato tempo buono e non troppo caldo.

Qui dopo la Loro partenza il tempo è migliorato, così che abbiamo potuto girare Parigi. Abbiamo visto parecchio approfittando dei giorni in cui i lavori del Congresso erano di minore interesse.

Lunedì pomeriggio ho letto la relazione scusando la Sua assenza. C'è stata qualche discussione con [illeg.] e con Georgevitsch, ai quali ho fornito degli schiarimenti. Si sono interessati alla relazione. Ho parlato con Brunelli, che è stato gentile e che ha detto che si trattava di un malinteso, che riteneva eliminato. Ho visto anche Le Danois, che ha fatto una relazione sull'impoverimento dei mari. Ieri siamo stati a fare la gita a Fontainebleau. Domani terminano i lavori del Congresso e lunedì partiamo per la Brettagna. Speriamo che ci concedano il ritardo del biglietto.

Se ci concedono di prolungarne la validità saremo a Roma probabilmente il 4, altrimenti già il 2.

Sono stato a trovare Joubin e Gravier, ma tutti e due sono assenti da Parigi.

A Loro tutti invio tanti affettuosi saluti e nuovi ringraziamenti per i gentili auguri.

Umberto [...]

6.24 D'Ancona to Volterra, Siena, 26.7.1933 *Cat. no. 104*

Siena, 26-VII-1933

Carissimo Papà.

Ho esaminato le tabelle e i diagrammi del Signor Régnier. Essi sono in sé stessi interessanti dal punto di vista batteriologico e credo che varrebbe la pena pubblicarli come nota preliminare. Però dal punto di vista della dimostrazione delle Sue leggi, gli elementi portati dall'esperienza del Régnier sono pochi e tali da portare una scarsa prova. Qui più che di due specie che si contendono lo stesso nutrimento mi pare che si tratti di specie che si danneggiano reciprocamente. Forse anche qui si manifestano delle fluttuazioni, ma bisognerebbe portare le colture più avanti. Capisco però che la cosa non è facile e temo che una vera dimostrazione sarà difficile a ottenersi a causa delle difficoltà sperimentali.

In ogni modo però i risultati che il Régnier potrà ottenere saranno sempre interessanti e conviene che egli li continui.

Riceva con la Mamma, Edoardo e Gustavo tanti affettuosi saluti

da Umberto [...]

6.25 D'Ancona to Volterra, Siena, 28.10.1933 *Cat. no. 105*

Siena, 28-X-1933

Carissimo Papà

La ringrazio per l'invito a venire a Roma, ma in questo momento ciò è impossibile perché abbiamo molto da fare.

Qui unita le mando la Nota dei lavori Suoi e miei e quella dei lavori biologici e matematici apparsi dopo il 1931. La Nota dei lavori Suoi credo sia completa; quella dei lavori apparsi dopo il 1931 è invece certamente incompleta, specialmente per quanto riguarda i lavori matematici. Lei deve avere ricevuto certamente parecchi di cui io non ho l'indicazione; Luisa mi ha parlato di alcuni lavori che sono stati portati da Levi Civita[21].

Se Le serve altro mi scriva.

Mi farà molto piacere se manderà i Suoi lavori al Prof. Negri.

Riguardo alle riserve sull'applicazione biologica non so se è stato pubblicato nulla. Sentii da Cantelli[22] che qualcuno aveva fatto delle riserve sull'applicazione dei calcoli; perciò lui mi disse che se io avevo degli esempi lui li avrebbe pubblicati. Però gli dissi che per ora non avevo dati nuovi oltre a quelli pubblicati a suo tempo sulla pesca nell'Adriatico.

Perciò non so se siano soltanto state fatte delle osservazioni verbali a Cantelli oppure se sia stato pubblicato qualcosa nel suo periodico.

Di obbiezioni a me constano soltanto quelle di Bodenheimer[23], al quale probabilmente farò una risposta.

Riceva con la Mamma e Gustavo tanti affettuosi saluti

Umberto. [...]

---

21 Reference to the mathematician Tullio Levi-Civita (1873–1941).

22 Probably refers to the mathematician (and specialist in probability theory) Francesco Paolo Cantelli (1875–1966).

23 See BODENHEIMER 1932b.

6.26 D'Ancona to Volterra, Siena, 3.11.1933 *Cat. no. 106*

Siena, 3-XI-1933

Carissimo Papà.

Ho ricevuto anch'io la memoria di Jensen[24] e l'ho scorsa un po' superficialmente. Egli osserva fluttuazioni nell'abbondanza di alcune specie di pesci della Danimarca e discute le cause di tali fluttuazioni biologiche perché trova che le specie predatrici (Gadidi) hanno un ciclo del tutto diverso da quello delle specie suddette prese in esame; propende perciò ad attribuire tali fluttuazioni a cause fisiche. Naturalmente il trovare cicli diversi in più specie conviventi non è una buona ragione per ammettere le cause fisiche ed escludere quelle biologiche; quando ci sono due sole specie una divorante e l'altra divorata si avranno cicli di uguale ampiezza sfasati, ma quando l'associazione è di molte specie tra le quali corrono rapporti non ben precisati non si può sapere come si comporteranno i singoli cicli.

Nel caso delle fluttuazioni del Jensen mi pare in complesso che sia difficile dire se si tratti di fluttuazioni dovute a cause fisiche o a cause biologiche. Certamente in natura esistono le une e le altre, ma nella maggior parte dei casi osservati in natura è difficile o addirittura impossibile precisare quanto spetta alle une o alle altre.

Il Jensen però cita le Sue ricerche ed ammette l'esistenza di fluttuazioni dovute a fattori biologici di convivenza.

Noi stiamo bene, Silvia pure. Lunedì ricominciano le lezioni. Mandateci vostre notizie.

Tanti affettuosi saluti

da Umberto [...]

6.27 D'Ancona to Volterra, Siena, 30.11.1933 *Cat. no. 107*

Siena, 30-XI-1933

Caro Papà.

Ho esaminato i lavori di Teissier e Kostitzin[25], che sono interessanti.

Il lavoro di Kostitzin mi sembra fatto bene dal punto di vista biologico, per quanto non porti alcun elemento nuovo. Tutta l'impostazione dei problemi mi pare ci sia già nei Suoi lavori; egli applica il calcolo a un maggior numero di casi specifici. La trattazione matematica lascio a Lei a giudicarla, però mi pare che il Kostitzin segua il metodo da Lei ideato. Perciò mi sembra che egli trascuri un po' troppo i Suoi lavori e che attribuisca a Lotka più di quanto a questi spetta. Veda a pag. 6.

Domani le manderò i due fascicoli unitamente a quello del Journal of animal ecology.

---

[24] See Jensen 1932 as well as Jensen 1933.

[25] It is not clear what works are being referred to, but they are almost certainly V. A. Kostitzin 1934 and Teissier 1934.

Le rimando la lettera del Professor Adolf Meyer. Egli vuol pubblicare una serie di monografie di biologia teorica e chiede che Lei autorizzi a mettere il Suo nome fra quelli dei co-editori. Non conosco questo Prof. Meyer. Non so in quali termini Lei vorrà rispondergli.

Lei mi aveva scritto che c'era una recensione da fare di un Suo lavoro; ma non ho mai ricevuto la lettera.

Ho piacere di sapere che Lei e la Mamma stanno bene. Anche qui il tempo è brutto; ha nevicato un po' e la temperatura si è rapidamente abbassata. Speriamo che poi l'inverno sia buono.

Riceva con la Mamma tanti affettuosi saluti.

da Umberto [...]

6.28 D'Ancona to Volterra, Siena, 7.2.1934 *Cat. no. 108*

Siena, 7-II-1934

Carissimo Papà.

La ringrazio per la Nota di Gause[26], che Lei mi ha mandata. Ne terrò conto per il lavoro che sto scrivendo. Spero di poterlo ultimare prima di Pasqua in modo da poter completare durante quelle vacanze le consultazioni bibliografiche. Poi le darò il manoscritto affinché Lei lo riveda e ne potremo parlare assieme. Ho molto piacere che Lei abbia continuato a lavorare e che abbia fatto delle parti nuove.

Oggi vi scrivo io solo, perché Luisa da domenica sera ha la mano destra immobilizzata. Domenica sera usciamo per andare al cinematografo e Luisa salendo il marciapiede è caduta e si è battuta il polso. L'ho subito fasciata e lunedì siamo andati alla Clinica chirurgica, dove il Professore ha riconosciuto trattarsi di una leggera frattura del radio. Oggi siamo ritornati e le hanno fatto una piccola gessatura, che dovrà portare per una quindicina di giorni. Perciò per questo periodo e per una settimana successiva in cui dovrà riattivare la mano, Luisa non vi potrà scrivere. Sta per altro benissimo ed esce tutti i giorni. Luisa è un po' contrariata di non poter lavorare per casa, ma forse tutto il male non viene per nuocere perché in questi giorni che la ragazza ha la mano ammalata, Luisa stava lavorando troppo. Così è un poco obbligata a starsene tranquilla.

La mano della ragazza è quasi guarita, così ha potuto riprendere almeno in parte il lavoro; abbiamo poi una donna a mezzo servizio.

Silvia sta benissimo; s'è molto divertita alla lettera del Nonno, a cui risponderà quanto prima perché in questi giorni è stata occupatissima con i compiti per la scuola, che piglia molto sul serio.

Luisa si sta interessando per le tovaglie di Gustavo. Mandatele commissioni perché così ha una occupazione.

Tanti saluti affettuosi da noi tutti a Lei, alla Mamma e Enrico e Gustavo.

Aff.mo Umberto [...]

---

26 This could refer either to Gause 1932a or to Gause 1934c.

6.29 D'Ancona to Volterra, Siena, 8.6.1934 *Cat. no. 109*

Siena, 8-VI-1934

Carissimo Papà.

Con Luisa Le ho mandato il manoscritto. Oggi ho riveduti i disegni che le rimando con le seguenti osservazioni.

1) Per tutte le figure è necessario togliere le lettere x e y dagli assi e sostituirle con N e t. Ciò è facile anche se si deve bucare la carta lucida perchè tanto questa dovrà essere montata su un fondo bianco e per il cliché va bene anche se il disegno è stato corretto.

2) La figura che rappresenta una specie sola in cui il coefficiente di accrescimento diminuisce con l'aumento del numero di individui va bene, però perchè è stato messo l'asse verticale nel mezzo della figura anzichè fuori della stessa più a sinistra?

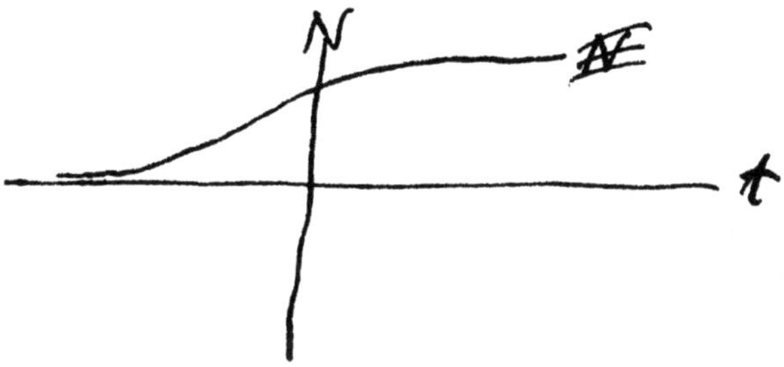

3) Va bene la figura che rappresenta una sola specie che tende a un esaurimento.

4) Va pure bene la figura che rappresenta una sola specie che tende asintoticamente a un limite q

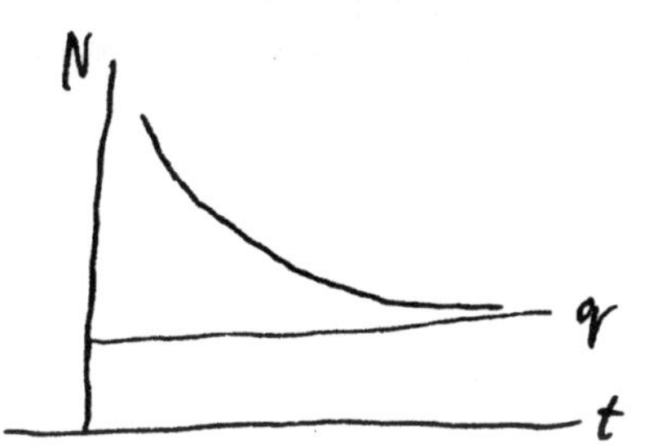

5) Va pure bene la figura che rappresenta una specie che presenta fluttuazioni smorzate

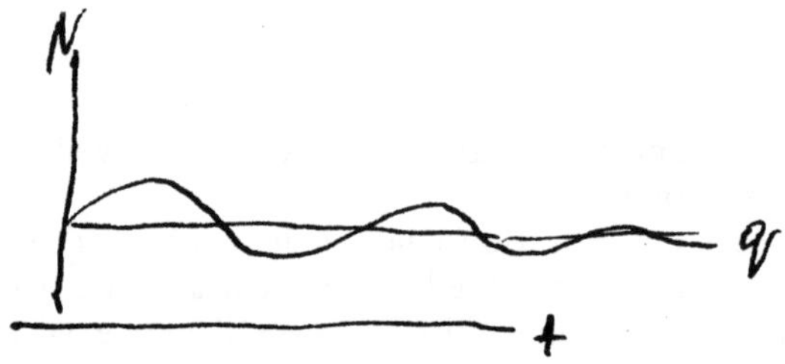

6) Non è stata invece fatta la figura che rappresenta una specie con fluttuazioni non smorzate non periodiche (oppure anche periodiche)

7) Non è stata fatta neppure la figura che rappresenti la sovrapposizione di fluttuazioni proprie e di fluttuazioni forzate

8) Manca la figura che rappresenti nell' associazione di tre specie il primo caso (vedi memoria [illeg.]) in cui una specie (carnivori) si esaurisce, mentre le altre 2 presentano fluttuazioni non smorzate. Se questo caso è rappresentato dalla figura che ho indicato con X, non mi pare che essa sia dimostrativa, anzitutto perchè negli altri casi dell'associazione di tre specie le tre curve sono state disegnate una sopra l'altra, poi perchè le tre curve dovevano essere fatte tutte a tratto continuo. Io credo che andrebbe fatta così questa figura[27].

[27] On the left side, it is written in Volterra's own hand that: "It is the first that is damping and the two other are undamped oscillations and so it would be as indicated in the sketch"; and on the right side: "This is damped". "Non smorzato" means "undamped".

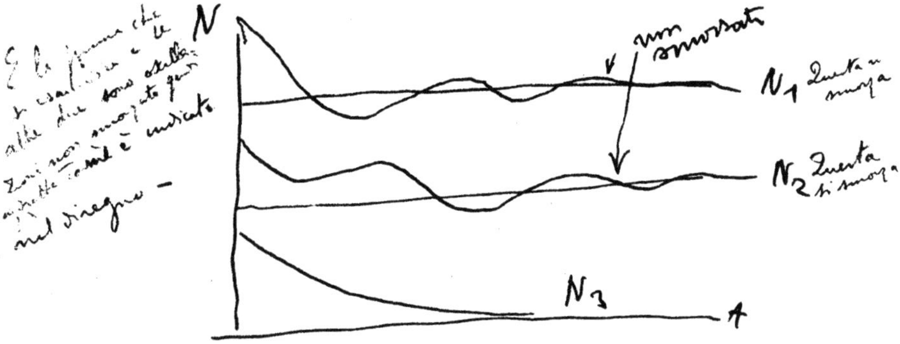

9) Non è stata fatta la figura che rappresenti l'altro caso in cui due specie tendono ad esaurirsi e l'altra tende a un valore costante.

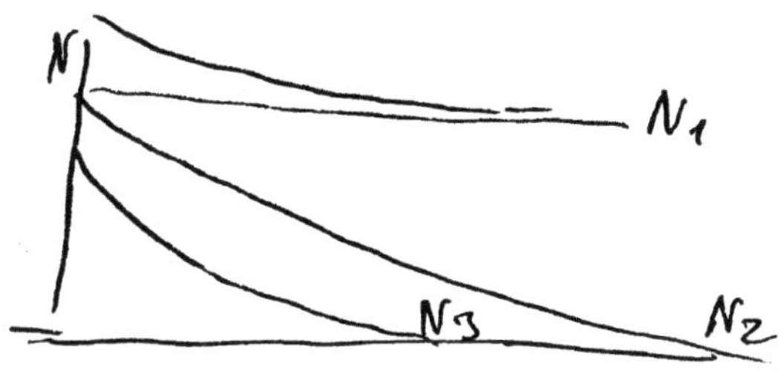

10) Manca pure la figura che rappresenti il caso in cui una specie tende ad esaurirsi e le altre due presentino fluttuazioni smorzate e infine tendono ad uno stato stazionario.

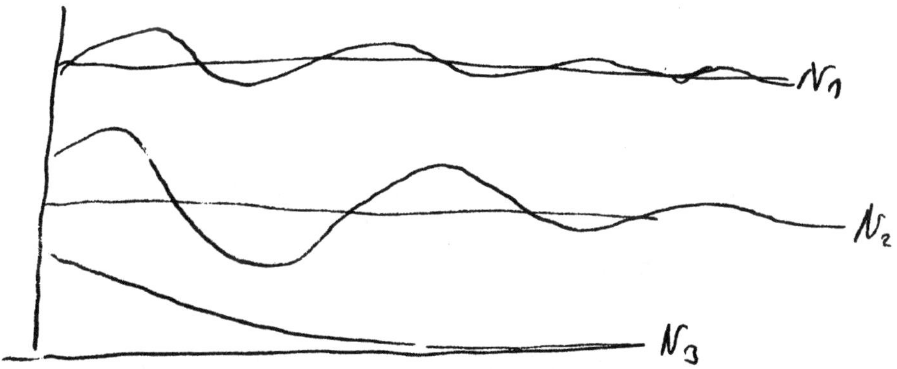

11) Sarebbe bene che la figura che rappresenta il caso in cui tutte e tre le specie hanno fluttuazioni smorzate presentasse le onde più piccole o che esse andassero smorzandosi maggiormente verso destra. Così è poco dimostrativo che si tratta di fluttuazioni che tendono ad uno stato stazionario.

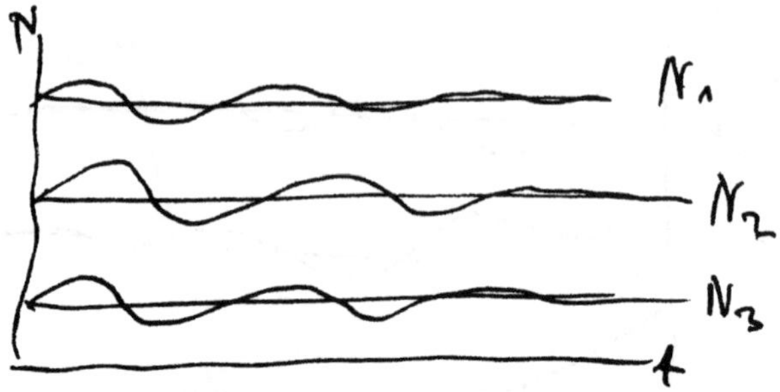

12) Va bene la figura che rappresenta che tutte e tre le specie tendono asintoticamente a uno stato stazionario.

13) Rimangono alcune figure delle quali non comprendo lo scopo. Se vanno bene La prego di volermi spiegare che cosa rappresentano[28].

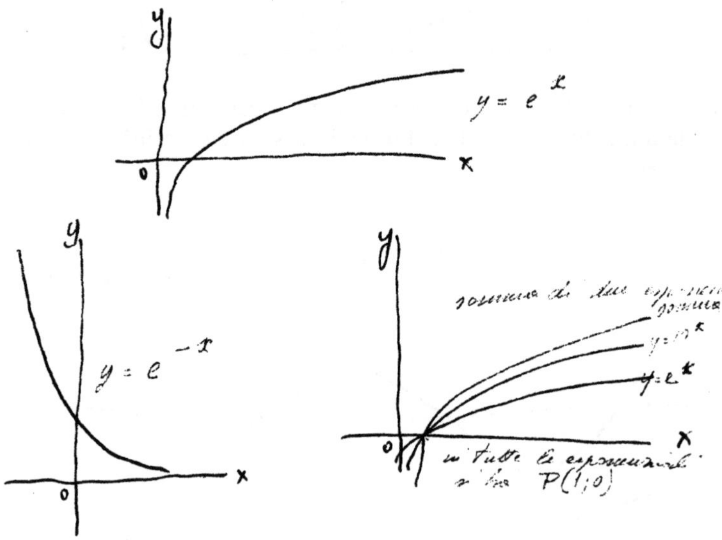

14) Ho riguardato la fig. di Gause, ma ho visto che già così è disegnata molto male per cui la riproduzione diventerebbe anche peggiore. Perciò io farei ridisegnare la figura senza scritte ingrandendola al doppio in modo che poi riproducendola ridotta aumentasse di esattezza. Non le pare che sia meglio fare così.

[28] On the left of the figure it is written in Volterra's own hand: "These figures are useless".

Trattengo la nota della figura, che copierò e poi Le rimanderò quando avrò avute le figure corrette.

Quando avrò avute tutte le figure corrette penserò io a sistemarle.

Luisa Le avrà portato i miei saluti. Ringrazio la Mamma per la lettera da mandare alla Banca.

Mi farà un piacere se mi scrive a proposito del lavoro. Riceva con la Mamma ed Enrico tanti affettuosi saluti

da Umberto.

6.30 D'Ancona to Volterra, Siena, 29.6.1934 *Cat. no. 110*

Siena, 29-VI-1934

Carissimo Papà

Stasera speravo di potere parlare con Lei, ma non mi è stato possibile. Le faccio i migliori auguri di un buon soggiorno a Parigi, perché la mia presente la raggiungerà a viaggio ultimato.

Qui unito Le mando il manoscritto corretto secondo le Sue indicazioni.

Soltanto a pag. 3 non ho capito la correzione e a pag. 57 non ho compreso le ultime parole dell'aggiunta da Lei fatta a riguardo di Bailey. Credo che Lei potrà fare queste piccole correzioni. A riguardo di quanto Lei ha indicato a pag. 3, bisogna tener presente che temperatura e umidità sono i fattori che più frequentemente agiscono sugli animali terrestri, ossigenazione e composizione dell'ambiente quelli che agiscono sugli animali acquatici.

Credo che così Lei potrà far tradurre il manoscritto. Qualche piccola aggiunta potrà essere apportata al manoscritto della traduzione.

Fra pochi giorni Le manderò i disegni che sto facendo completare.

Riceva con la Mamma tanti affettuosi saluti.

da Umberto

Come mi ha detto la Mamma questa sera al telefono sarà difficile combinare di trovarci in Svizzera. Noi stiamo qui fino alla fine di luglio. Gli ultimissimi di questo mese andremo per un paio di giorni a Roma, da dove devo fare una gita a Nemi. Poi se fosse possibile trovarci con Voi verremmo in Svizzera per una decina di giorni, per andare poi a Fiume per un mese.

Se invece i vostri progetti sono diversi e vi è impossibile trovarci, allora forse Luisa ed io andremo forse ai primi di Settembre in Jugoslavia in un Congresso di Limnologia.

Per prendere una decisione attendiamo perciò una vostra risposta.

U. [...]

6.31 D'Ancona to Volterra, Siena, 3.7.1934 *Cat. no. 111*

Siena, 3-VII-1934.

Caro Papà.

Le mando tutti i disegni che ho fatto completare. Le figg. 1-2, 5-9 sono da riprodurre dal Suo lavoro di Scientia[29] e da quello del Comitato Talassografico[30]. La fig. 3 fu fotografata e Lei deve avere la relativa stampa. Tutte le altre figure sono nel presente plico. L'unito foglio contiene le spiegazioni da porre sotto le figure. Dove non c'è alcuna spiegazione ci si riferisce al testo.

Le figg. 10–14 rappresentano le definizioni: esaurimento (10), fluttuazione periodica (11), fluttuazione non periodica (12), fluttuazione smorzata (13), tendenza a un limite (14). Le figg. 17–21 rappresentano i vari casi che si possono verificare nell'associazione di tre specie. Ho rappresentato tanto il caso in cui tutte e tre le specie tendono assintoticamente a un limite quanto quello in cui si presentano fluttuazioni smorzate.

Vedrà che io ho fatto fare alcune figure diversamente da quanto aveva indicato Lei. Mi pare che nel caso delle tre specie mettere tutte e tre le curve assieme, anziché una sopra all'altra rende l'idea più chiara. Così pure per le fluttuazioni smorzate proprie e forzate.

Le allego in ogni modo anche i disegni fatti fare da Lei e le Sue indicazioni. Se qualche cosa non va bene le farò riffare.

Affettuosi saluti a Lei e alla Mamma

Aff.mo Umberto

Le figure vanno tutte ridotte della metà.

6.32 D'Ancona to Volterra, Siena, 9.7.1934 *Cat. no. 112*

(Same heading of letter no. 6.6)

Siena, 9-VII-1934

Carissimo Papà.

Le mando subito la figura di Pearl che Lei non trova più e che probabilmente è rimasta a Roma. Non è stata presa dalla memoria di Pearl che ha Lei ma da un libro di questo autore[31].

Sono contento che le figure fatte fare da me siano andate bene.

Desideriamo vostre notizie; Luisa desidera sapere come state.

Tanti affettuosi saluti a Lei e alla Mamma.

da Umberto

---

29 Volterra 1927g.

30 Volterra 1927d.

31 Pearl 1925.

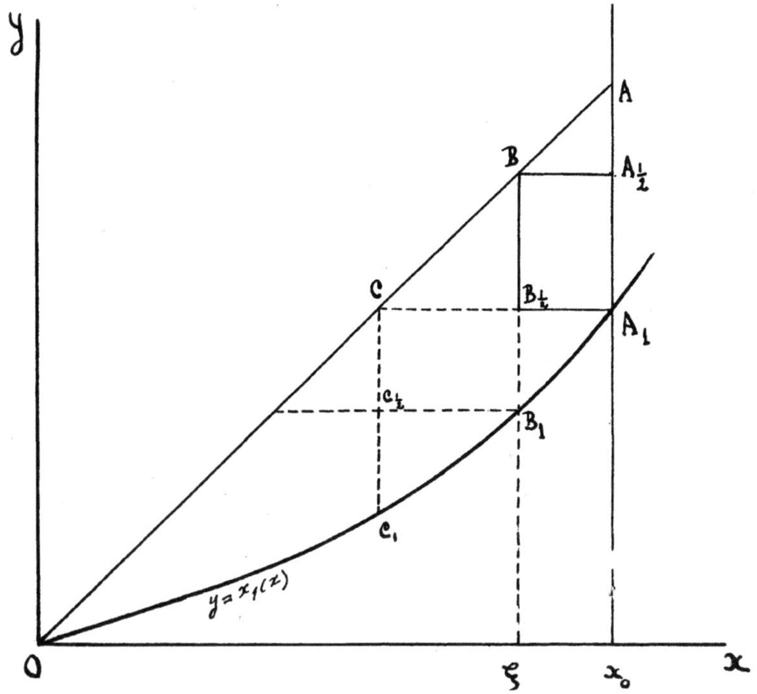

6.33 D'Ancona to Volterra, Fiume, 19.8.1934 ***Cat. no. 113***

Fiume, 19-VIII-1934

Carissimo Papà.

Abbiamo appreso con piacere le vostre buone notizie. Noi avevamo mandato una lettera a Stresa il giorno successivo alla vostra partenza. L'avete ricevuta?

Qui unita Luisa Le manda la lettera inglese. Mi sono informato per le scarpe Bata, ma le indicazioni che Lei mi ha mandato non sono sufficienti. Manca il numero della scarpa, che deve essere scritto nella scarpa stessa. Se Lei me lo manda potrò subito acquistare le scarpe.

Sono contento che Hermann abbia mandato la traduzione. Rileggerò con cura il manoscritto e lo confronterò col testo italiano. Però io dovrei fare qualche piccola aggiunta al capitolo bibliografico citando alcuni lavori visti di recente. Questo completamento lo potrei fare soltanto per la fine di settembre, perché prima devo finire una mia relazione. Credo che Lei non avrà difficoltà a ritardare l'invio alle stampe fino alla fine di settembre.

Riceva con la Mamma tanti affettuosi saluti.

Umberto [...]

6.34 D'Ancona to Volterra, Fiume, 4.9.1934 *Cat. no. 114*

Fiume, 4-IX-34

Carissimo papà.

Ho consegnato ieri alla Freda la prima parte del lavoro e oggi mando la seconda. Sulla traduzione ho fatto delle correzioni a lapis. Qui unite Le mando alcune osservazioni che desidererei Lei leggesse quando riavrà la traduzione.

Quando Lei avrà rivisto tutto, mi mandi la traduzione a Siena, che vi farò ulteriori aggiunte. Mi pare che basterà copiare a macchina le parti aggiunte e corrette, le altre mi pare possano andare anche così, dato che sono scritte chiaramente e che così possono andare in tipografia.

Riceva con la Mamma tanti affettuosi saluti.

Umberto [...]

6.35 D'Ancona to Volterra, Siena, 2.12.1934 *Cat. no. 115*

Siena, 2-XII-1934

Carissimo Papà

Mi scusi se ho tardato alcuni giorni a rispondere alla Sua del 26. Volevo però rileggere con attenzione quanto scriveva Régnier e la sua lettera e ciò non mi è stato possibile nei giorni passati.

Luisa ha copiato a macchina le note di Régnier e gliene manda unita una copia assieme a quella manoscritta.

Vedo che dalle ricerche di Régnier risulta una azione dannosa allo Stafilococco, non al Cobacillo. Régnier dice che non si può trattare del caso in cui una specie si nutre dell'altra e ciò mi pare giusto. Non so però se si può trattare soltanto di due specie che si disputano lo stesso nutrimento. In questo caso la convivenza dovrebbe essere dannosa a ambedue le specie. Piuttosto si tratta, come Régnier osserva, di una azione nociva del Colibacillo verso lo Stafilococco. Forse il primo produce sostanze cataboliche o tossiche dannose al secondo.

L'andamento delle curve di sviluppo delle singole specie è simile a quello delle curve già date dal Régnier e delle quali una è stata riprodotta. Mi pare che in questa nota il Régnier concreti un fatto che già era stato trattato nelle tabelle manoscritte che a suo tempo egli ci aveva mandate.

L'accrescimento a S di una popolazione è stato verificato sperimentalmente da Pearl; ma in tutti i casi considerati si trattava di una popolazione vivente in un ambiente uniforme, ma che andava man mano saturandosi di individui che trovavano anche sempre minor disponibilità di nutrimento. Si ha la curva a S quando lo spazio e il nutrimento diminuiscono con l'aumentare del numero di individui, ma la qualità dell'ambiente e del nutrimento rimane sempre uguale.

Invece nel caso di una coltura batterica quando il terreno di coltura non si rinnova, si verifica una modificazione del terreno, che agisce sia come ambiente circondante, almeno in parte, il batterio, sia come nutrimento dello stesso. Quali modificazioni intervengono è difficile precisare; certamente nel terreno nutritivo si accumulano i prodotti catabolici dei batteri, che a loro volta sono tossici. Si potranno verificare anche modificazioni della natura colloidale e della composizione chimica del terreno. Sono tutte condizioni difficili a precisare, ma che certamente vengono rivelate dal modificato accrescimento della coltura. Ciò è noto in batteriologia e infatti per allevare dei batteri è necessario trapiantarli periodicamente, rinnovare cioé il terreno di coltura.

Conviene perciò tener distinti i due casi: 1) quello in cui diminuisce il nutrimento e lo spazio disponibili, ma la qualità dell'ambiente e del nutrimento rimane invariata, 2) quello in cui oltre alla diminuzione di spazio e nutrimento si verifica anche una alterazione dell'ambiente e del nutrimento.

Al primo caso si applicano le considerazioni da Lei fatte precedentemente e la figura a S della curva, al secondo le considerazioni che Lei fa ora e la curva, corrispondente a quella empirica di Régnier.

Sarà interessante vedere i risultati che Régnier otterrà nelle colture e ambienti rinnovati. Probabilmente avrà una curva a S.

Trovo giusto il ragionamento che Lei fa, penso che sarà bene fare una aggiunta al libro. Potremo fare questa aggiunta sulle prime bozze e mettere pure una nuova figura.

Non so se questo caso di curve d'accrescimento sia stato già studiato. Certamente la curva empirica sarà stata ottenuta in tutti i casi (batteri, protozoi) nei quali l'ambiente viene progressivamente alterato. Se sia stato esaminato matematicamente questo non lo so. Bisognerebbe vedere i lavori di Lotka e di Pearl, nei quali probabilmente queste cose sono esaminate. Specialmente Pearl si è occupato dell' accrescimento delle popolazioni ed ha pubblicato molti lavori, fra gli altri un libro riassuntivo, che io ho consultato e che si trova da Cotronei a Roma. Ma probabilmente c'è qualcosa di più recente. Penso che sia un bene che Lei vedesse i libri di questo autore americano. Io, a quello che ricordo, ho visto nel suo libro soltanto il caso della curva a S, la così detta curva logistica, che egli trova nella Drosophyla (un moscerino) e che applica anche alle popolazioni umane.

Se ha bisogno di altre indicazioni mi scriva che sarò ben lieto d'indicarle quel poco che so.

Riceva con la Mamma tanti affettuosi saluti

Umberto [...]

6.36 D'Ancona to Volterra, Siena, 20.1.1935 *Cat. no. 116*

20-I-1935

Carissimo Papà.

La ringrazio dell'invio del libro di Gause[32] che leggerò con molto piacere. Non l' ho ancora ricevuto.

[32] This is almost certainly GAUSE 1934a (see Cat. no. 179 and 180). It is not however clear what the note is referring to immediately after this, while the work published in an English journal must be GAUSE 1932b.

Non trovo nemmeno io la nota di Gause, che la Mamma aveva copiata. Non ricordo se rimase da me o restò in Francia. Però si tratta di una nota copiata da "Nature", che lei potrà facilmente trovare in originale sul fascicolo del periodico.

Ricordo poi di aver avuto l'estratto di un altro lavoro di Gause, stampato in un giornale inglese. Ma ricordo di averglielo restituito.

Ho piacere di sapere che state tutti bene. Riceva con la Mamma e Enrico tanti affettuosi saluti.

Umberto

6.37 D'Ancona to Volterra, Siena, 23.1.1935 *Cat. no. 117*

Siena, 23-I-1935

Carissimo Papà

La ringrazio dell'invio del pacco con il bel regalo per Silvia, i fascicoli per me e i lavori di Gause.

Tra i fascicoli che mi ha mandato ci sono due uguali della Revue des Travaux de l'Office des Pêches. Uno dei due fascicoli è dedicato a Lei da Le Danois, perché tutto il fascicolo contiene un suo lavoro sulle trasgressioni oceaniche. Glielo rimando per posta; sarà bene che Lei lo ringrazi.

Ho visto il libro di Gause, che mi sembra interessante. Sarà necessario esaminarlo attentamente in occasione delle correzioni delle bozze del libro francese. Se Lei non ha nulla in contrario io tratterrò qui il libro di Gause e farò le necessarie correzioni e aggiunte quando verranno le bozze francesi. Poi indicherò a Lei di rivedere la parte matematica, che a me sarà riuscita più difficile.

Ho visto anche gli altri lavori di Gause, ma mi pare che sono tutti lavori già compresi nel libro. Le due traduzioni russe sono fatte malissimo e sarà necessario riscriverle completamente. Però mi pare che sia un lavoro inutile fare tutte queste correzioni; basterà servirsi della traduzione per comprendere che cosa dice e vedere se è necessario servirci per il libro francese. Se però Lei desidera che io faccia queste correzioni delle due traduzioni le farò ben volentieri.

Se Lei non ha difficoltà tratterrò qui tutto in attesa delle bozze.

Luisa sta copiando gli elenchi di lavori.

Riceva con la Mamma tanti affettuosi saluti.

da Umberto.

6.38 D'Ancona to Volterra, Siena, 31.1.1935 *Cat. no. 118*

Siena, 31-I-1935

Carissimo Papà

Sono lieto di aver trovato il fascicolo di Scienze che Le ho mandato assieme agli altri 3 fascicoli che avevo e che mi ero dimenticato di aver riposto assieme ai manoscritti del libro.

Trovo giustissimo quanto Lei dice di Lotka e penso che sarà bene dirlo nel libro citando questo suo nuovo lavoro. Se Lei mi vorrà mandare questo lavoro del Lotka mi farà molto piacere. Sarò molto contento di ricevere una copia della nota dell'Académie des Sciences, che suppongo Lei avrà ricevuta.

Per quanto Lei mi dice dei lavori di Régnier e Gause non so quanto sia utile aggiungere un intero capitolo dedicato a questi due autori. A me sembra che ciò sia esagerato. Essi con i nuovi lavori non hanno portato un contributo molto maggiore di quello dato con i lavori già citati. In fondo si tratta di rifacimenti di cose che sono state già da noi citate. Converrà certamente aggiornare il capitolo citando anche i nuovi lavori e aumentando se sarà necessario qualche figura.

Però mi sembra che il Suo lavoro abbia importanza già in se stesso come impostazione teorica del problema della convivenza anche indipendentemente dalla eventuale verifica sperimentale delle Sue leggi. Certamente queste acquisteranno valore, qualora saranno verificate empiricamente. Ma a me non pare che né Régnier né Gause abbiamo dato la dimostrazione della parte essenziale della Sua trattazione, quella che per effetto della convivenza si determinano fluttuazioni. Tutto quanto essi hanno trovato può essere spiegato con la Sua teoria, ma non si può dire che essa venga dimostrata dai lavori di Régnier e Gause. Perciò penso che a valorizzare troppo i lavori di questi Lei finisca collo sminuire il Suo lavoro. Non mi pare conveniente che Lei leghi la sua teoria ai lavori sperimentali di nessuno. Qualora venga una precisa elaborazione sperimentale converrà metterla in valore. Ma per ora mi pare che si tratti soltanto di tentativi, in gran parte concordanti, che conviene citare, ma non sopravalutare.

In ogni modo se Lei pensa di dare una elaborazione un po' più estesa dei lavori di questi due autori, sono ben disposto a farlo, ma dedicare ad essi un capitolo mi pare esagerato.

Auguro a Lei e alla Mamma un buon viaggio. Con tanti saluti affettuosi.

Umberto.

6.39 D'Ancona to Volterra, Siena, 6.2.1935 *Cat. no. 119*

Siena, 6-II-1935

Carissimo Papà.

Abbiamo ricevuto la lettera da Roma, la cartolina da Parigi e oggi le lettere del 4. Vi ringraziamo delle notizie che ci date e ci fa piacere di leggere che state bene e che il tempo è favorevole. Anche qui non è male; è coperto ma la temperatura é mite.

La ringrazio per la Sua comunicazione all'Académie[33]. È molto interessante. Ho visto poi il lavoro di Lotka e trovo giusto quanto Lei dice.

Oggi mi sono giunte da Roma le bozze del lavoro francese. Sono venute due copie delle bozze e il manoscritto originale. Hermann non ha però mandato nessuna figura. Io ho le prove che egli aveva mandato a suo tempo, però allora

[33] It is not clear what publication is being referred to.

alcune figure mancavano. Sarebbe bene che me le mandasse e così io potrei collocare a posto le figure.

Ho visto la figura della nuova esperienza di Régnier. Si tratta effettivamente di una curva simile a quella di Pearl. Non so perché Régnier non accetti questa assomiglianza.. La curva di Pearl non dipende da queste o da quelle condizioni. Io, se Lei ricorda, prevedevo che nel caso di colture in ambiente rinnovato si sarebbe avuta una curva di Pearl. Ciò concorda con tutto quanto Lei ha scritto finora. Perciò sul capitolo dell'accrescimento di una singola specie si potrà tener conto di queste nuove ricerche di Régnier.

Io nella correzione delle bozze terrò conto delle pubblicazioni di Régnier e Gause aggiungendo quanto è necessario.

Vedo dalla Sua precedente lettera che Lei pensa che io veda con gelosia le nuove ricerche di questi due autori e che io tenga a far risaltare le mie ricerche. Sono ben lontano dal pensare così e sarei ben lieto che ci fossero delle precise dimostrazioni sperimentali delle Sue teorie matematiche. Le ricerche di Régnier a mio giudizio non danno una dimostrazione delle Sue leggi; studiano più che altro l'accrescimento di una singola specie e questa era cosa già nota. Lei può spiegare questo accrescimento con le Sua teoria ma questa non viene dimostrata dalle suddette esperienze perché lo studio di una singola specie era soltanto il punto di partenza della Sua trattazione. Il trovarla concordante con le esperienze di Régnier non è più dimostrativo che il trovarla concordante con quelle precedenti di Pearl.

In quanto alla convivenza di due specie mi pare che Régnier non abbia portato dei risultati molto dimostrativi.

Più interessanti mi sembrano le esperienze di Gause perché trattano di due specie conviventi, ma anche esse non dimostrano (almeno per quello che ho visto finora) la fluttuazione come effetto della convivenza.

Certamente le mie osservazioni sulla pesca nell'Alto Adriatico dovrebbero dare un sostegno più sicuro alla Sua teoria perché esse ne dovrebbero dimostrare un punto essenziale, la 3ª legge. Purtroppo però le mie osservazioni statistiche possono essere interpretate come ho fatto io, ma possono essere interpretate anche in senso diverso e di questa opinione sono Pearson, Bodenheimer, Gause. Perciò anch'io riconosco che queste critiche un fondo di serietà ce l' hanno. Le mie osservazioni possono essere interpretate nel senso della Sua teoria ma ciò non è un fatto assolutamente indiscutibile, è soltanto una interpretazione.

È questo che io volevo dire. Non creda che io voglia sminuire le ricerche sperimentali che appoggiano la Sua teoria, ma io ritengo che bisogna essere molto cauti nell'accettare come dimostrazioni queste ricerche sperimentali. A accettarle troppo facilmente si va incontro alla possibilità di vederle smentite da altri.

La Sua teoria in tutte questa questione non viene toccata per niente. Essa è una teoria impostata logicamente e in modo verosimile, concordante con molti dati noti e verosimili. Perciò essa rimane come ipotesi di lavoro che può essere fonte di nuove indagini e che rimane anche se non è appoggiata da prove empiriche. Certamente essa da queste può acquistare maggior autorità. Ma bisogna esser cauti nell'accettare queste prove, che siano sicure e dimostrative altrimenti è meglio che Lei non leghi la Sua teoria a una base sperimentale, che è certamente meno solida della teoria stessa.

Mi scusi se io Le scrivo tutto ciò, ma avrò molto piacere di sapere il Suo parere su quanto Le dico. Apprezzo quanto hanno fatto Régnier e Gause, che certamente hanno lavorato seriamente, ma appoggiarsi troppo a loro mi pare indebolisca la Sua teoria, che non deve appoggiarsi nemmeno alle mie ricerche, ma che in tutte deve vedere soltanto delle conferme più o meno parziali.

Riceva con la Mamma tanti affettuosi saluti

Suo aff.mo Umberto.

6.40 D'Ancona to Volterra, Siena, 17.2.1935 *Cat. no. 120*

Siena, 17-II-1935

Carissimo Papà.

Ho ricevuto le bozze che Lei mi ha mandato. Le vedrò e le confronterò con le mie che ho già letto e corretto. Una copia di bozze da me corrette le ho mandate alla signorina Freda, con preghiera di rimandarmela.

Vedrò ora le correzioni e aggiunte che Lei ha fatto e aggiungerò quanto è necessario per gli ultimi lavori di Gause e Régnier.

Vedrò come si comporta l'ultima grafica di Régnier, qualora ai logaritmi si sostituiscano i numeri. Però già ora, a mio avviso, si tratta di una curva simile a quelle di Pearl.

Avrò piacere se Hermann mi manderà le figure. Tutte quelle che avevo le ho messe a posto. Mancano però due. Forse sarà necessario aggiungere qualche figura [illeg.] gli ultimi lavori di Régnier e Gause. In questo caso farò fare le figure e poi le rimanderò unitamente alle bozze. Spero di rimandarle tutto quanto prima. In questi giorni ho dovuto un po' ritardare perché avevo dei lavori urgenti da ultimare.

Quelle questioni riguardanti l'esaurimento di una singola specie andrebbero secondo me citate dove si parla della curva di Pearl. Le farò vedere come farei le aggiunte quando Le manderò le bozze.

La ringrazio per quanto Lei mi scrive, ma vedo che Lei attribuisce un eccessivo valore probativo alle mie statistiche. Se io Le consigliavo di non sopravvalutare le ricerche di Gause e di Régnier altrettanto devo dire e a maggior ragione per le mie. Ma di ciò avremo occasione di riparlare.

Ho letto con interesse l'articolo di "Science et vie", che Lei ha inviato a Silvia.

Tanti affettuosi saluti a Lei e alla Mamma.

Umberto.

6.41 D'Ancona to Volterra, Siena, 21.2.1935 *Cat. no. 121*

Siena, 21-II-1935

Carissimo Papà.

Ho confrontato la bozza Sua con quella da me corretta. Nei fogli che allego metto le mie osservazioni a quanto Lei ha scritto. La maggior parte di correzioni le ho trascritte sulla copia mia. In questa, confrontandola col manoscritto ho trovato parecchi altri errori.

Trattengo ancora la mia copia per fare le aggiunte in base agli ultimi lavori di Gause e Régnier. Attendo poi le bozze della Freda.

Non ho ricevute le altre figure da Hermann.

Se Lei ha altro da aggiungere o da replicare alle mie osservazioni La prego di rimandarmi la Sua bozza per poter precisare i riferimenti.

Mi farà molto piacere se mi indicherà quanto altro ha da dire.

Spero che continui costì il bel tempo e che insieme alla Mamma si trovi sempre bene. Quando contano di ritornare?

Tanti affettuosi saluti a Lei e alla Mamma

Umberto

Le sarei grato se volesse tradurre le spiegazioni delle figure, contenute nel foglio scritto a macchina allegato. Saranno le spiegazioni da mettere sotto alle figure.

6.42 D'Ancona to Volterra, Siena, 23.2.1935 *Cat. no. 122*

Siena, 23-II-1935

Carissimo Papà.

Ricevo la Sua lettera del 20 in cui trovo la figura presa da Gause. Questa va bene. Manca però sempre la figura dell'accrescimento d'una singola specie, che era stata fotografata da Pearl. Io mandai la fotografia, ma Hermann non inviò mai il cliché. Bisognerebbe sollecitarla.

Per le figure di Régnier, mi pare che quella che è stata copiata sia la più comprensiva. Le altre non dicono nulla di più. Anche le nuove osservazioni di Régnier possono essere passate sotto silenzio se lui non ha pubblicato nulla. Certamente la curva che lui ora trova è una curva alla Verhulst-Pearl. Caratteristica di questa è di avere una concavità superiore iniziale, poi una inflessione e in seguito una convessità. Tutto il resto dell'andamento varia molto.

Che si dovesse trovare una curva di questo tipo mantenendo il terreno invariato mediante rinnovo era prevedibile.

Le Sue nuove ricerche matematiche io le metterei subito dopo la trattazione dell'accrescimento di una sola specie. Si direbbe prima dell'accrescimento a coefficiente invariato, poi dell'accrescimento secondo Pearl, poi del nuovo caso che si accorda alle osservazioni di Régnier. Per quest'ultimo caso io metterei una nuova curva non sperimentale, ma teorica tratta dalla formula che Lei trova nella nota dei CR. Io però non so costruire la curva in base a quella formula, che è alquanto complessa. Forse anche ci potrei riuscire, ma con notevole difficoltà. Penso perciò che sarà più facile che Lei mi dia una piccola tabellina numerica in base alla quale farò la figura.

Poi vedrò cosa conviene aggiungere dalle figure di Gause.

Quando avrò riavuto le bozze dalla Freda e quando Lei mi avrà mandato la figura mancante, i dati per quella da fare ed eventualmente le Sue osservazioni su quanto ho scritto sulle bozze, completerò tutto e Le manderò la mia bozza così aggiornata.

Ho piacere di sentire che con la Mamma si trova bene costì. La ringrazio per gli auguri che ha voluto mandare ai miei genitori ed a me.

Mi dispiace di sentire che Joubin sia gravemente ammalato. Che cos'ha?

Conosco le opinioni di Caullery sull'evoluzione[34] perchè ho sentito e letto il discorso che egli ha fatto a Padova. Trovo che è molto obbiettivo.

In merito all'evoluzione non credo che oggi nessuno zoologo possa obbiettivamente dire di essere darwinista. Oramai questa è una fase superata. Si può essere evoluzionista, ma non più darwinista.

In quanto poi all'evoluzione essa è entrata su un terreno obbiettivo di sperimentazione e soltanto su quest'ultimo si può discutere. Effettivamente i risultati finora non sono molto promettenti, per cui lo scetticismo della maggior parte degli zoologi è ben giustificato.

Si può riconoscere però che l'evoluzione è sempre l'unica ipotesi plausibile di cui noi disponiamo e che fino a che non abbiamo di meglio dobbiamo ammetterla come ipotesi di lavoro.

Ma non più. Certamente si andrà avanti e si formuleranno nuove ipotesi più aderenti ai fatti. Probabilmente l'evoluzione ritornerà sotto altra forma. Ma non credo sotto la forma darwiniana perché sarebbe un ritorno indietro.

In ogni modo non si può dire d'essere darwinista o no; tale espressione sarebbe antiscientifica. È soltanto sul terreno sperimentale che si può discutere e questo è molto diverso da quello che era ai tempi di Darwin.

La ringrazio per tutto quanto mi scrive. La prego di gradire, unitamente alla Mamma tanti affettuosi saluti.

Umberto [...]

6.43 D'Ancona to Volterra, Siena, 24.2.1935 *Cat. no. 123*

Siena, 24-II-1935

Carissimo Papà

Le mando alcuni manoscritti di aggiunte, che penserei di fare al libro. Glieli mando affinché Lei li veda e se vanno bene me li voglia tradurre e rimandare.

La Sua nota dei C. R. ho pensato di dividerla in due parti, una per quanto riguarda una sola specie e una per quanto riguarda 2 specie. Nel libro la trattazione è sempre separata e perciò mi pare conveniente dividere anche in questo caso. Una parte la aggiungerei al Cap. IV e una al Cap. V.

[34] See CAULLERY 1931.

Per la nota di Régnier c'è poco da aggiungere. Più che altro si tratta di modificare quanto noi avevamo detto. Ho modificato conservando quasi completamente le parole di Régnier e Lambin. L'aggiunta va fatta all'ultima pag. La figura già fatta è perfettamente aderente al testo.

Non so se Lei crede di mettere il capoverso finale da me aggiunto al Cap. IV. Se Le pare indelicato perché accenna alle ricerche inedite di Régnier lo tolga. A me pare che possa andare.

Le mando pure un'aggiunta da fare per Gause. Per questi dovrò però fare anche altre aggiunte. Intanto Lei potrebbe vedere queste. Non ho ancora letto tutto il suo libro. Tutte le altre memorie sono comprese in questo. Farò poi fare una o due altre figure, copiate da Gause, oltre a quella teorica Sua di cui già Le scrissi.

Domani conto di mandarle le altre eventuali aggiunte da fare in base al libro di Gause.

Se quando avrà visti questi manoscritti, me li vorrà tradurre mi farà molto piacere. Dove ho potuto ho lasciate le parole francesi. Poi La prego di rimandarmi tutto che copierò a macchina le aggiunte e le inserirò nelle bozze.

Tante grazie.

Spero di sentire sempre buone notizie Sue e della Mamma. Tanti affettuosi saluti.

Umberto [...]

6.44 D'Ancona to Volterra, Siena, 25.2.1935 *Cat. no. 124*

Siena, 25-II-1935

Carissimo Papà

Ho finito di esaminare il libro di Gause ed ho fatto un'altra aggiunta, di cui Le allego il manoscritto.

La prego di volerla rileggere e se va bene di rimandarmela tradotta.

Farò aggiungere altre 3 figure prese dal libro di Gause, oltre a quella teorica sui calcoli Suoi.

Spero che tutto vada bene.

Riceva con la Mamma tanti affettuosi saluti da noi tutti. Luisa e Silvia sono già a letto, che è tardi. Perciò Luisa non scrive.

Aff. mo Umberto

6.45 D'Ancona to Volterra, Siena, 28.2.1935 *Cat. no. 125*

Siena, 28-II-1935

Carissimo Papà.

Ricevo oggi la Sua lettera del 26 con la figura che mancava.

Qui unite Le mando le tre figure nuove che io ho fatte copiare dal libro di Gause. Esse vanno rimpicciolite come indicato. Quando Lei mi manderà le prove di queste e della Sua le metterò a posto.

La signorina Freda mi ha mandato le bozze con le sue correzioni. Le ho aggiornate sulla bozza mia, salvo alcuni punti sui quali ero incerto e che vorrei sottoporre al Suo giudizio. Quando Le manderò le bozze da me corrette invierò anche quelle della Freda sulle quali ho segnato i punti sui quali desidero richiamare la Sua attenzione.

Ora non manca che le traduzione delle aggiunte che Le inviai. Quando le avrò le copierò a macchina e le metterò a posto. Quindi Le manderò le bozze con tutte le correzioni.

Non ho ancora fatto la curva di Régnier con i numeri anzi che con i logaritmi, perché in questi giorni sono stato molto occupato, però lo farò nei giorni prossimi.

Lei mi chiede se non sono d'accordo con la prefazione di Pearl? Non completamente. Non che io voglia discutere gli enormi meriti di Darwin, che è stato veramente geniale nella sua teoria e che ha portato il più grande contributo che mai sia stato dato alle ricerche zoologiche. Indubbiamente delle sue ricerche, oltre allo stimolo che hanno dato, molto rimane ancora, fra altro nel caso specifico la teoria della lotta per l'esistenza nei suoi riguardi all'economia della natura. Ma che la lotta per l'esistenza abbia importanza per l'evoluzione non ci crede più nessuno. Non ci crederebbe più nemmeno Darwin se vivesse.

Ho piacere di leggere le buone notizie Sue e della Mamma. Tanti affettuosi saluti

da Umberto

6.46 D'Ancona to Volterra, Siena, 9.3.1935 *Cat. no. 126*

Siena, 9-III-1935

Carissimo Papà

Le mando la figura che ho fatto fare. Ho fatto riunire in una tutte e tre le curve. Così mi sembrano più dimostrative.

Però ho dovuto fare delle modificazioni perché la curva della figura A Sua non corrisponde a quella della B. Suppongo che il disegnatore abbia calcolato diversamente e che a ciò siano dovute le differenze. Mi pare che come sono state fatte le curve vadano bene. Se c'è qualche cosa che non va mi scriva che farò rifare il disegno.

Sullo stesso non ho messo numeri né dei tempi, né del numero di individui perché essi non sono stati mai messi sulle curve teoriche. Ho indicato con A, B, C le tre curve. Poi sulla didascalia metteremo le loro definizioni.

Sulle figure ho omesso pure i punti o le linee riguardanti i dati empirici perché al punto in cui è stata inserita la nuova figura non si è ancora parlato di dati empirici. I dati del Régnier sono stati d'altronde compresi in un'altra figura alla fine del volume.

La differenza tra curva empirica e teorica credo sia accidentale perché nella figura di Régnier da noi pubblicata, che si riferisce evidentemente ad un'altra coltura il ramo discendente della curva ha forma convessa e non concava.

Le unisco la curva dei nuovi calcoli di Régnier fatta con i numeri accanto a quella con i logaritmi. Quest'ultima mi pare caso mai più dimostrativa.

In ogni modo mi pare che si tratti di una curva a S di Verhulst-Pearl.

Attendo le traduzioni per inserirle nelle bozze. Se avrò le prove delle nuove figure metterò a posto anch'esse.

Riceva con la Mamma tanti affettuosi saluti

Suo Umberto.

6.47 D'Ancona to Volterra, Siena, 13.3.1935 *Cat. no. 127*

(Same heading of letter no. 6.6)

Siena, 13-III-1935

Carissimo Papà.

Ho ricevuto ieri il libro del Fauré Fremiet e la memoria di Ludwig[35]. La ringrazio poi tanto per il bel libro che ha mandato a Silvia. Ne è rimasta molto entusiasta.

Il libro del F. F. lo conosco e l'ho qui all'Istituto. Perciò, se sapevo di cosa si trattava, Le avrei detto di non mandarlo. Non ritengo di nessun interesse citarlo perché non tratta dell'accrescimento delle popolazioni, ma dell'accrescimento degli individui e delle colonie animali considerate come unità individuali. È un bellissimo libro, ma non riguarda per niente il nostro argomento.

Conoscevo anche la memoria di Ludwig, che ho ricevuta dall'autore. Non avevo ritenuto necessario citarla; però, ora, dato che Lei ha dato più sviluppo alla trattazione di una singola specie, si può tenerne conto. Ho fatto una breve aggiunta, che dopo tradotta in francese (se va bene) andrà aggiunta in fine del Cap. IV.

Le mando tutte e tre le bozze, quella vista da Lei, quella corretta dalla Freda e la mia. Su quest'ultima sono state trascritte tutte le correzioni e fatte le aggiunte. Essa è la copia aggiornata. Avrò piacere se Lei vorrà darle ancora un'occhiata. Guardi specialmente le aggiunte per lo stile.

Troverà poi in alcuni punti delle delle note a lapis che si riferiscono alle correzioni della Freda. Sono punti sui quali desidero richiamare la Sua attenzione perché non ho saputo decidere se quanto notava la Freda era giusto o meno.

Nelle nuove aggiunte ho visto che Lei tanto citando se stesso che Régnier ha messo l'indicazione M. (M. Volterra, M. Régnier, Mlle Lambin). Altrove sono state messi soltanto i nomi, come in genere nel citare i vari autori. Soltanto nella prefazione era stato usato il M.

Vedrà che ho modificato la citazione di Gause, così credo che sarà più chiara. Come aveva messo Lei non spiegava abbastanza.

Troverà uniti alcuni fogli con altre osservazioni che sottopongo al suo esame.

Per tutto il resto va bene.

---

35 Fauré-Fremiet 1925 and Ludwig 1928.

Spero che abbia ricevuta l'ultima figura e che stia bene. Vedrà che ho indicato il posto anche per le nuove figure così che nelle seconde bozze si potrebbero mettere a posto tutte le figure.

Se ci sono altre modificazioni o aggiunte da fare, La prego di scrivermi.

Le unisco una mia memoria sull'accrescimento, che forse potrà interessarla.

Gradisca con la Mamma tanti affettuosi saluti.

Umberto

6.48 D'Ancona to Volterra, Siena, 15.3.1935 *Cat. no. 128*

(Same heading of letter no. 6.6)

Siena, 15-III-1935

Carissimo Papà.

Ricevo oggi la Sua lettera del 13 e rispondo subito in merito alla denominazione di curve di Pearl o di Verhulst-Pearl.

Non c'è pericolo di confusione perché anzi la maggior parte degli autori usano la denominazione di Verhulst-Pearl. Credo che anche lo stesso Gause faccia così. D'altra parte è giusto nominare Verhulst che per primo ha riconosciuto l'andamento a S delle curve d'accrescimento delle popolazioni.

Perciò se anche Lei nella nota dell'Ac. des Sc. ha nominato soltanto Pearl penso che ora sia preferibile chiamare le curve a S curve di Verhulst-Pearl. Non c'è nessun pericolo di confusione perché ripetutamente e in genere i due nomi sono uniti.

Ho piacere che Lei trovi ben fatta l'ultima figura. Ora avrà ricevute le bozze con tutte le correzioni.

Se ci sono altre cose che non vanno, mi scriva.

Abbiamo saputo ieri sera a mezzo di una telefonata di Enrico che è morto improvvisamente Giovannino D'Agostini. Suppongo che anche loro saranno stati informati.

Luisa e Silvia stanno bene. Io vi scrivo dal laboratorio.

Riceva con la Mamma tanti affettuosi saluti

da Umberto.

6.49 D'Ancona to Volterra, April 1935 *Cat. no. 7*[36]

Carissimi.

Ho piacere di sentire le vostre buone notizie. Sono contento che Papà abbia rivedute le bozze e che abbia trovate giuste le correzioni. Capisco benissimo che Hermann si sia spaventato delle modificazioni e aggiunte e che non ne voglia altre per le seconde bozze. Quando verranno queste le correggerò subito.

---

36 N.d. but clearly written on April 1935.

Noi restiamo qui fino al 14. Il 13 finiscono le lezioni e io devo andare subito a Venezia per le mie ricerche lagunari. Mi tratterrò lì due settimane facendo una breve gita a Fiume per Pasqua. Al ritorno da Fiume verrò a Roma per solo due giorni perchè poi dobbiamo ritornare a Siena perché c'è un congresso. Durante la mia assenza Luisa e Silvia verranno a Roma, dove speriamo di trovarvi. Probabilmente allora sarete già tornati e non sarete andati ancora in Ariccia.

Io conto di vedervi in quei brevi giorni che verrò prima di tornare a Siena.

Auguro buon lavoro a Papà, a tutti e due tempo buono in questi giorni che ancora vi tratterrete a Parigi.

Affettuosi saluti

da Umberto

6.50 D'Ancona to Volterra, Siena, 13.4.1935 *Cat. no. 129*

(Same heading of Letter n. 6)

Siena 13-IV-1935

Carissimo Papà.

Le rimando le bozze con le mie osservazioni. La prego di volerle vedere e poi rispedirmele. Io riporterò le correzioni sulla copia buona e la manderò a Hermann.

Sul frontespizio è stato omesso etudiées, che invece c'è sul titolo.

Da Pag. 21 a Pag. 23 Lei parlando di se stesso dice M. Volterra. Si può premettere il M., dato che Lei é l'autore?

Per il resto veda il foglietto unito, su cui ci sono le altre osservazioni mie.

Io parto domani per Venezia. La prego di scrivermi all'Albergo Manin. Io avrò con me le bozze per fare le necessarie correzioni. Mi mandi pure la continuazione quando la avrà rivista.

Luisa arriverà a Roma lunedì. Io conto di venire per uno o due giorni il 28 o il 29.

Riceva con la Mamma tanti affettuosi saluti

da Umberto

Scrivo dal laboratorio, perciò Luisa e Silvia non possono aggiungere i loro saluti.

6.51 D'Ancona to Volterra, Venezia, 17.4.1935 *Cat. no. 130*

Venezia, 17-IV-1935

Carissimo Papà.

Ricevo stasera la Sua lettera del 15 e le bozze. Le rispondo subito e appena avrò tempo, spero domani, riscriverò tutte le correzioni sull'altra copia di bozze che ho e gliela rimanderò.

Per tutte le correzioni va benissimo. Soltanto non mi piace ancora la correzione di pag. 6. Capisco benissimo quanto Lei mi dice, ma appunto io avevo cercato di mettere il testo in modo da distinguere tra ricerche quantitative e ricerche matematiche. Ora Lei giustamente osserva che nel testo tale distinzione non è chiara. Però non mi piace come Lei ha corretto; penso che si dovrebbe dire in altro modo.

Io volevo dire "... hanno fatto largamente impiego di metodi quantitativi e per conseguenza si sono serviti delle applicazioni matematiche". Forse in francese si potrebbe dire "... ont largement eu recours aux methodes experimentales quantitatives et par consequence aux application mathématiques."

La correzione che Lei ha fatto mi sembra che lasci un equivoco, così invece viene chiarito quello che si voleva dire. In sostanza io accenno ai metodi quantitativi perché essi hanno portato per conseguenza l'uso di applicazioni matematiche. Credo che Lei troverà giusto e che vorrà senz' altro correggere sulle bozze che le manderò.

Per quanto riguarda Verhulst non si trattava di rivendicare la priorità di questi. Io avevo soltanto osservato che in alcuni punti dello stesso testo si diceva in un modo, in altri nell'altro. Ciò non stava bene. Ora va benissimo.

Spero che presto arrivino le altre bozze e che si possano correggere definitivamente. Mi mandi tutto qui che provvederò subito.

Luisa mi dà Loro buone notizie. Io spero di passare uno o due giorni con Loro a Roma. Qui ho molto da fare con i pesci della Laguna. Dovrei girare, ma il tempo è brutto.

Andrò a passare il giorno di Pasqua a Fiume, che tanto qui non potrei fare nulla quel giorno.

Riceva con la Mamma e tutti i fratelli tanti affettuosi saluti.

da Umberto.

6.52 D'Ancona to Volterra, Venezia, 18.4.1935 *Cat. no. 131*

Carissimo Papà.

Ho riportato su queste copie delle bozze tutte le correzioni, meno quella di pag. 6, per la quale mi riferisco alla lettera di ieri.

Trattengo l'altra copia delle bozze in attesa della continuazione.

Affettuosi saluti

Umberto, Venezia, 18-IV-1935

6.53 D'Ancona to Volterra, Venezia, 22.4.1935 *Cat. no. 132*

Venezia, 22-IV-1935

Carissimo Papà.

Al mio ritorno da Fiume ho trovato le bozze. Le correggerò e appena pronte gliele spedirò.

La correzione per la pagina 6 va benissimo secondo il testo che Lei ora mi invia: ont largement en recours aux méthodes quantitatives qui ont conduit aux applications mathématiques.

Della prima parta che Le ho già inviato io credo che si possa dare il bon à tirer. Una nuova bozza mi pare superflua. Però bisognerà raccomandare che eseguiscano accuratamente le correzioni fatte.

Ho avuto anche i lavori di Chapman e di Stanley[37]. Li leggerò e gliene scriverò.

Ho comperato a Fiume le scarpe che porterò al mio ritorno. Conto di tornare domenica in giornata.

Qui il tempo è buono e posso girare la Laguna che in questa stagione è molto bella.

Riceva con la Mamma e i Fratelli tanti affettuosi auguri.

Umberto.

6.54 D'Ancona to Volterra, Venezia, 25.4.1935 *Cat. no. 133*

Venezia, 25-IV-1935

Carissimo Papà.

Le rimando le bozze corrette. Ho trovato ancora qualche piccolo errore tipografico o qualche virgola mancante. Ho poi riportato tutte le correzioni sull'altra copia di bozze. Veda Lei quale le sembra più chiara per spedire a Parigi.

Le aggiungo un foglio su cui ho riportato alcune questioni dubbie. Queste vanno poi indicate sulle bozze, perché sulle stesse non ho fatto alcuna correzione in attesa che Lei decida.

Mi pare che si possa dare il bon à tirer. Basterà raccomandare che facciano con cura tutte le correzioni e che qualcuno le riveda.

Le rimando unitamente anche i lavori di Chapman. Essi mi sembrano fatti bene. Riconfermano le sue cose precedenti, ma portano anche nuovi contributi. Trova che nella convivenza in ambiente omogeneo di larve e adulti di cui questi mangiano quelle si stabiliscono delle fluttuazioni. È vero che sui suoi diagrammi queste ultime non appaiono sempre molto evidenti, però mi pare giusta la sua conclusione e che egli dimostri con ciò che in ambiente uniforme si verificano delle fluttuazioni dovute ai rapporti di convivenza, confermando con ciò la Sua seconda legge.

Spero che così tutto andrà bene. Io verrò a Roma domenica. Spero di trovarla completamente guarito. Ho saputo da Luisa della laringite che Le dà noia. Le faccio i migliori auguri e La saluto affettuosamente, unitamente alla Mamma e ai Fratelli

Umberto.

---

[37] Probably referring to several works cited in the references.

6.55 D'Ancona to Volterra, Siena, 29.6.1935 *Cat. no. 134*

Siena, 29-VI-1935

Carissimo Papà

Ho ricevuto le tre copie del libro. Pare anche a me che la stampa sia riuscita benissimo. La ringrazio tanto per le copie che mi ha mandato.

Io penso che per parecchio tempo non verrò a Roma. Penso di venire dopo il 15–20 di Luglio. Perciò sarà forse preferibile metterci d'accordo per iscritto dei nomi cui inviare il libro. Se Lei mi manda la nota di coloro cui Lei desidera inviarne una copia, io potrei completarla e rimandargliela.

Spero che Emilia sarà in via di guarigione. Le notizie della sua malattia ci hanno fatto molto dispiacere e ci hanno tenuto in preoccupazione. Spero che anche Lei sarà più tranquillo e riposato.

Riceva con la Mamma e tutti di casa tanti affettuosi saluti

da Umberto. [...]

6.56 D'Ancona to Volterra, Siena, 4.11.1935 *Cat. no. 135*

Siena, 4-XI-1935

Carissimo Papà.

Le mando qui unito il foglietto con i titoli dei due Suoi nuovi lavori. Spero che vada bene, altrimenti la prego di rimandarmelo corretto e lo ricopierò.

Ringrazio tanto Lei e la Mamma per tutte le premure usate a noi durante questo breve soggiorno a Roma. Sono stato molto contento di aver potuto fare con Lei la passeggiata di domenica pomeriggio; mi dispiace soltanto che Lei abbia creduto che io non mi sia trovato bene a Roma. Le assicuro che mi sono trovato benissimo e che mi sarebbe piaciuto soltanto passare con Loro qualche giorno di più.

Sono arrivato a Siena dopo un buon viaggio. Ora le comunicazioni con Roma sono molto più rapide; si impiega quasi un'ora meno di prima.

Qui ho trovato tutto in ordine; la ragazza è arrivata questa mattina. Le lezioni non cominciano anche perché non sono stati ancora confermati gli incarichi e siamo soltanto pochi titolari. Ne approfitto per lavorare.

Spero di avere presto Sue buone notizie. La prego di salutare a mio nome la Mamma e i ragazzi. Riceva nuovi ringraziamenti e affettuosi saluti.

da Umberto

Tanti saluti a Luisa e Silvia

6.57 D'Ancona to Volterra, Siena, 17.4.1936 *Cat. no. 137*

17-IV-1936

Carissimi,

Vi ringrazio per tutte le affettuose premure avute da tutti durante il nostro breve soggiorno romano. Ci siamo trovati molto bene e ci è dispiaciuto di dover ripartire così presto.

Se Papà mi manderà il manoscritto del Suo articolo lo rivedrò molto volentieri per quanto riguarda le notizie zoologiche. Se ha anche bisogno di qualsiasi altra informazione sono sempre a sua disposizione. Spero di potergli mandare in esame il fascicolo della Rivista di Biologia.

Affettuosi saluti.

Umberto.

6.58 D'Ancona to Volterra, 1.5.1936 *Cat. no. 138 and 139*[38]

1-V-1936

Carissimo Papà.

Le faccio anch'io i migliori auguri per il Suo compleanno. Come scrive Luisa, lo festeggeremo anche noi da lontano.

Ho letto con interesse quanto Lei mi ha scritto nell' ultima Sua lettera. Io penso però che l'opinione di Caullery, che d'altra parte non è sua soltanto, ma rispecchia l'opinione della grande maggioranza degli zoologi in tutto il Mondo, sia perfettamente obbiettiva. Certamente molti non sogliono rinunciare a una teoria molto seducente e in apparenza perfettamente logica e giustificata. In Inghilterra poi, un po' anche per ragione di sentimentalismo nazionale, prendono ancora le parole di Darwin come fatti sicuri.

Invece bisogna riconoscere che dal 1859 a oggi, per merito di Darwin anzitutto, le conoscenze zoologiche hanno fatto grandissimi progressi. Quanto allora sembrava sicuramente documentato oggi non ci sembra più così sicuro. Perciò dobbiamo cercare prove nuove, fatti nuovi. Non si può quindi riesumare la teoria darwiniana senza appoggiarla su fatti nuovi. I fatti vecchi sono stati esaurientemente esaminati e discussi. Di fatti nuovi però finora non se ne vedono.

Al momento attuale l'ultima evoluzione che abbia avuto una sicura dimostrazione sperimentale è quella delle mutazioni. Ma queste non dimostrano l'evoluzione darwiniana.

Ho piacere che Lei legga il libro di Jennings[39]. Non è quanto di meglio si possa desiderare, specialmente poi non è un libro di divulgazione, perché presuppone già delle conoscenze sull'argomento. Non saprei però cosa altro consigliarle.

---

[38] They are separated in the Volterra Archive but they form a unique letter.

[39] The reference is to a book of Herbert Spencer Jennings (1868–1947), almost certainly JENNINGS 1935.

Enriques ha introdotto il termine italiano di genidi, ma tutti usano invece il nome corrispondente a quello in uso in altre lingue di gene, plurale geni. Enriques volle modificare in gerundio per evitare le confusione con genio. Ma una tale confusione non è possibile, almeno in genere. Perciò è meglio dire geni.

Per il volume di [illeg.] La ringrazio tanto. Desideriamo però sapere quanto ha speso. Era da anni che cercavamo questo libro, che abbiamo sempre dovuto prendere in prestito ai Lincei.

Le rinnovo i migliori auguri e La prego di gradire con la Mamma e i Fratelli i più affettuosi saluti.

Aff. mo Umberto [...]

6.59 D'Ancona to Volterra, Siena, 12.5.1936 *Cat. no. 140*

Siena, 12-V-1936

Carissimo Papà.

Ho letto il manoscritto del Suo articolo. Mi sembra che nel complesso vada molto bene; soltanto sarebbe bene dare un'ultima limatina al testo, che in qualche punto sembra un po' slegato. Potrebbe forse far apparire meglio la distinzione della parte spettante al primo da quella in cui parla del secondo libro.

Il titolo poi non mi pare molto aderente al contenuto. Anzitutto Lei nomina nel titolo uno solo dei due libri, mentre poi parla ugualmente di tutti e due. Per quanto riguarda la "nuova scuola di biologia sperimentale" veda i fogli scritti a macchina nei quali faccio le mie osservazioni.

Faccia l'uso che crede delle osservazioni stesse. Se ha bisogno di altre indicazioni sono naturalmente sempre a Sua disposizione. Più che altro ho proposto qualche piccolo cambiamento per questioni di nomenclatura. In Zoologia, come d'altra parte in tutte le scienze, alcune parole hanno assunto un determinato significato e non possono essere sostituite da altre che letterariamente sembrano sinonimi.

Per quanto riguarda l'opinione di Pearl, io non la condivido, ma è questione d'opinioni e dato che essa viene riferita al suo [illeg.] non pregiudica nulla.

Come sta Gustavo? Spero che si sia ristabilito. Noi stiamo benissimo.

Riceva con la Mamma, Enrico, Emilia e Gustavo tanti affettuosi saluti.

da Umberto [...]

6.60 D'Ancona to Volterra, Siena, 20.5.1936 *Cat. no. 141*

Siena, 20-V-1936

Carissimo Papà.

Appena ricevuto il manoscritto[40] lo leggerò con molto piacere. Sono certo che sarà fatto benissimo e che non ci sarà bisogno di modificarlo. Però se per qualche questione biologica vorrà sentire il mio modesto parere, vedrò volentieri di rendermi utile.

---

[40] Perhaps VOLTERRA 1936f.

Dove pubblicherà il lavoro? Su Scientia?

Le rinnovo i miei ringraziamenti per gli affettuosi auguri per il mio compleanno. Riceva con la Mamma e i fratelli e il nipote i più affettuosi saluti.

da Umberto

Scusi se scrivo in modo così indecente, ma è domenica e siamo rimasti senza carta.

6.61 D'Ancona to Volterra, Siena, 14.7.1936 *Cat. no. 142*

Siena, 14-VII-1936

Carissimi.

Spero che a quest'ora sarete già in Ariccia dopo aver fatto un buon viaggio ed avervi sistemati pel ritorno

Se ciò non vi da disturbo, Luisa e Silvia verrebbero a Roma giovedì prossimo per raggiungervi in Ariccia. Luisa si tratterrebbe per alcuni giorni e poi verrebbe qui a sistemare la casa. Silvia invece, col vostro gradimento, rimarebbe in Ariccia fino alla fine del mese, quando verrei anch'io per un paio di giorni prima d'andare tutti a Fiume.

Però prima di mandare Silvia desideriamo avere notizie sulla poliomelite. A Roma ci sono dei casi; desideriamo sapere se ce ne sono anche ad Ariccia perché in questo caso non sarebbe prudente che Silvia si trasferisse da un paese immune in uno in cui c'è la malattia. A Velletri credo ci sia qualche caso, ma di ciò non ci sarebbe da preoccuparsi se non ce ne sono in Ariccia dato l'isolamento del villino. Vi sarò grato se al più presto ci darete delle notizie precise e attendibili.

Papà ultimamente mi ha scritto di osservazioni fatte in Brettagna su un caso di concorrenza tra animali marini e alghe. Io non ne so nulla e così mi è difficile cercare notizie se non ho dati più precisi. Vi prego di volermi dare qualche indicazione più particolareggiata.

Con la Mamma, Enrico, Gustavo e Emilia, e tanti baci al nipote, vi invio i più affettuosi saluti

Umberto [...]

6.62 D'Ancona to Volterra, Siena, 23.7.1936 *Cat. no. 143*

Siena, 23-VII-1936

Carissimo Papà.

La ringrazio tanto per i fascicoli e il libro di Douvier che mi ha mandato. Quest'ultimo specialmente è molto interessante e mi riesce molto utile.

Ieri ho subito letto le bozze e le ho rispedite. Credo che avrà visto le mie correzioni; c'era però ben poco da correggere.

Luisa è arrivata dopo un buon viaggio e mi ha portate le Loro buone notizie. Ho piacere di sentire che Silvia si trovi bene; grazie per tutte le premure.

Fra pochi giorni contiamo di essere a Roma e avremo il piacere di rivederla.
Riceva, con la Mamma, Emilia, Gustavo e Vito, tanti affettuosi saluti
da Umberto.

Ringrazio tanto Lei che la Mamma per gli affettuosi auguri per il 10° anniversario del nostro matrimonio.[...]

6.63 D'Ancona to Volterra, n.p., 1936 *Cat. no. 136*[41]

Carissimo Papà.
La ringrazio vivamente per il gentile invio della Sua nuova nota[42]. L'ho letta, ma vedo che è molto strettamente matematica. Quando uscirà l'ultima?
Io ho scritto al Professor van der Klaauw di Leiden, che è il direttore degli Acta Biotheoretica. Appena ne avrò una risposta gliela comunicherò.
Ho molto piacere di sentire che Lei fa nuovi lavori; spero al Suo ritorno da Parigi di avere il piacere di poter parlare con Lei di questi nuovi svolgimenti della Sua teoria.
Per ora non mi è possibile di venire a Roma, perché sono ingolfato nella relazione, che devo fare qui dove ho accumulato tutto il materiale.
Luisa e Silvia verranno fra giorni a Roma.
Riceva tanti affettuosi saluti
da Umberto
Prof. C. J. van der Klaauw
Zoologisch Laboratorium
Leiden

6.64 D'Ancona to Volterra, Siena, n.d. *Cat. no. 74*[43]

Carissimo Papà.
La ringrazio tanto per l'invio dei 3 fascicoli, che sono molto interessanti. Specialmente mi ha fatto piacere di ricevere il fascicolo 10 della crociera di Schmidt, che è seguito a quelli che mi ha dato precedentemente.
Ho letto quanto mi ha scritto a proposito delle fluttuazioni. Credo anch'io che possa avere importanza la lotta tra specie nello stabilirsi di un equilibrio tra batteri. È noto che in certi casi la completa sterilità non è vantaggiosa alla vita degli organismi come la presenza di alcune specie di batteri innocui.
Forse tale fatto è applicabile anche ai batteri [illeg], però è difficile vedere fino a quale punto ciò può essere applicato nella pratica.

---

[41] N.d. (except for the subsequent indication, 1936), and there is insufficient information to date it. It has thus been left in the position it occupies in the Volterra Archive ordering.

[42] Obviously VOLTERRA 1937a, which was in fact published by Van der Klaauw's journal.

[43] N.d. and insufficient information to date it. It has thus been left in the position it occupies in the Volterra Archive ordering.

Sono molto contento di sentire le Sue buone notizie, che ci sono state confermate oggi da zio Piero.

Verso la metà di Dicembre contiamo di venire a Roma. Stiamo benissimo e pigliamo in santa pace la pioggia che il grigio cielo senese ci manda in abbondanza.

Le faccio tanti auguri e con nuovi ringraziamenti Le mando tanti affettuosi saluti.

Umberto

6.65 D'Ancona to Volterra, Siena, 18.9.1936 *Cat. no. 144*

Siena, 18-IX-1936

Carissimo Papà.

Le mando le lettere di risposta del Prof. van der Klaauw. Come vedrà accetta con molto piacere di pubblicare la Sua memoria e mi pare che anche la data di pubblicazione sia conveniente[44].

Immagino che gli risponderà Lei direttamente. In ogni modo anch'io lo ringrazierò.

Ho piacere di sapere che Luisa ha trovato tutti bene. Anche qui il tempo è brutto. Quest'anno l'autunno è cominciato presto.

Riceva con la Mamma, Edoardo, Nella, Gustavo tanti affettuosi saluti

da Umberto

6.66 D'Ancona to Volterra, Siena, 24.9.1936 *Cat. no. 145*

Siena, 24-IX-1936

Carissimo Papà.

Le mando la nota delle persone alle quali fu inviato il libro; non me la rimandi che è una copia.

Io credo che il direttore del giornale olandese quando vedrà il manoscritto lo pubblicherà. Al caso Lei potrebbe un po' modificarlo; corredarlo di qualche spiegazione più abbondante che non nei C. R. Trattandosi di un giornale non matematico qualche maggior chiarimento sarebbe forse utile.

Nel caso che non volessero pubblicarlo ci sarebbe sempre la rivista di Polimanti[45].

Però penso che sia meglio attendere la risposta.

Riceva con la Mamma e i fratelli tanti affettuosi saluti

da Umberto

---

[44] See preceding letter no. 6.63 (Cat. no. 136).

[45] Reference to the zoologist, and hydrobiology specialist, Osvaldo Polimanti, the founder and director of the Lake Trasimeno hydrobiological station. Polimanti founded and directed, from 1919 on, the *Rivista di Biologia*, in which Volterra's mentioned work was published, VOLTERRA 1937c.

6.67 D'Ancona to Volterra, Siena, 23.10.1936 *Cat. no. 146*

Siena, 23-X-1936

Carissimo Papà.

Le mando qui subito un capitolo della mia relazione di Tripoli, sul quale gradirei molto il Suo parere. In particolare mi farà piacere se vorrà rileggere quanto riguarda i Suoi lavori. Non so se ho riferito bene sulle ultime Sue comunicazioni all'Accademia di Parigi, che non erano molto facili ad essere citate.

Mi farà anche piacere se mi dirà il Suo parere sulla parte introduttiva e su quella riguardante l'articolo del Baldi.

Se non ha tempo non me lo rimandi subito, che mi basta sentire quanto Lei dice per la stampa della relazione. Altrimenti se dovesse rispedire il capitolo prima del 26, me lo mandi direttamente a Roma, Via in Lucina.

Riceva con la Mamma i più affettuosi saluti e tanti ringraziamenti

Aff.mo Umberto [...]

6.68 D'Ancona to Volterra, Siena, 10.11.1936 *Cat. no. 147*

Siena, 10-XI-1936

Carissimo Papà.

Le invio i miei più vivi rallegramenti per la nomina ad accademico pontificio. Ne ho letto l'annunzio sul giornale mentre ero a Tripoli. Vercelli e Toniolo fecero le congratulazioni per la Sua nomina. Ho visto e parlato anche con Somigliana che mi ha chiesto di Lei. Scorza poi mi ha detto che aveva letto i Suoi lavori sulle associazioni.

La ringrazio per aver voluto leggere e correggere il manoscritto della mia relazione. Esso mi è giunto quando io ero già partito; ne terrò conto nella correzione delle bozze.

Io ho letto il Suo manoscritto che mi pare vada benissimo[46]. Esso potrà fare la base dell'articolo per la rivista di Polimanti. Sarà bene che l'articolo per la parte matematica sia scritto in forma piana o meglio che le varie questioni matematiche siano tutte spiegate.

Io sono ritornato oggi a Siena, questa notte ho dormito a Roma e ieri sera ho visto Enrico che sta bene. Ringrazio tanto Lei e la Mamma per avermi voluto ospitare a Roma, sono sempre stato a mangiare a Via in Lucina e mi sono servito di Arrigo per fare le pescate e per andare alla stazione.

Se mi vorrà comunicare il giorno e l'ora del passaggio da Pisa ne avrò piacere perchè se sarò li potrò venirvi a salutare.

Il Congresso di Tripoli è stato interessante. Abbiamo visto bene tanto la città che i dintorni, che sono molto interessanti.

---

[46] See previous note.

Ringrazio tanto per la premura nel ricercare il libro sulle anguille.
Silvia e Luisa stanno benissimo. Riceva con la Mamma tanti affettuosi saluti.

Umberto [...]

6.69 D'Ancona to Volterra, Pisa, 10.3.1937 *Cat. no. 150*

Pisa, 10-III-37

Carissimo Papà

Ricevo dal Prof. Bertalanffy[47], noto cultore viennese di biologia teorica, la lettera che qui Le accludo.

Come vedrà il Bertalanffy vuol iniziare la pubblicazione di un giornale di biologia esatta, che dovrà comprendere anche argomenti di biologia teorica e matematica. Accanto a questo giornale vuol cominciare la pubblicazione di una serie di monografie in forma di volume e propone di inaugurare questa serie con un volume dedicato alle Sue teorie matematiche della lotta per l'esistenza. A tale scopo propone di fare una traduzione tedesca di una fusione dei due Suoi volumi francesi, aggiornata in base alle pubblicazioni più recenti.

Il Bertalanffy chiede il mio parere prima di rivolgersi a Lei. Io però prima di dargli una risposta desidero sapere cosa Lei ne pensa.

Dalla lettera e dal prospetto del giornale Lei potrà farsi un'idea delle intenzioni del Bertalanffy. Questi è uno studioso serio; l'editore offre pure affidamento.

Unitamente alla Sua risposta Le chiedo di rimandarmi lettera e programma affinché io possa scrivere al Bertalanffy.

Riceva con la Mamma e Enrico tanti affettuosi saluti.

da Umberto [...]

6.70 D'Ancona to Volterra, Pisa, 1937 *Cat. no. 148*[48]

Carissimi.

Ringrazio anch'io per lo cheque e per il regalo per la festa di Luisa. Luisa vi da nostre dettagliate notizie.

Avrei anch'io piacere di poterci incontrare una domenica, però la distanza tra Roma e Pisa e abbastanza grande.

Sarebbe interessante vedere le rovine di Populonia e le rovine etrusche di Cosa e dell'Ansedonia. Però tanto Grosseto che Orbetello sono abbastanza lontane tanto da Roma che da Pisa. Più giù non sarebbe possibile venire in una giornata.

Ringrazio tanto per il bollettino di Oceanografia.

---

[47] Ludwig von Bertalanffy (1901–1972), now considered one of the founders of modern system theory.

[48] The letter bears the subsequent indication 1937.

Immagino che Polimanti avrà già risposto a Papà. In ogni modo non tarderà molto a farlo perché ci teneva molto a quel lavoro.

Con Bertalanffy io non mi sono messo d'accordo per fare il lavoro da solo, anzi gli dissi che lo facevo assieme a Papà. Però egli ora scriverà direttamente e così Papà gli potrà rispondere dicendo quali sono i suoi desideri. Bertalanffy alla mia prima e unica lettera rispose che avrebbe scritto direttamente a Papà.

In ogni modo Papà non faccia i complimenti e dica direttamente a Bertalanffy se desidera che il lavoro sia fatto e come.

Avremo molto piacere se ci scriverete spesso.

Tanti affettuosi saluti a tutti

Umberto [...]

6.71 D'Ancona to Volterra, n.p., 1937 *Cat. no. 149*[49]

Carissimi.

Ho letto con piacere le vostre buone notizie.

Spero che Papà avrà già avuto risposta da Polimanti. Può darsi però che durante le vacanze egli sia stato a Roma.

Ho ricevuto una lettera da Bertalanffy che dice di essere contento che io abbia accettato di fare il lavoro, per la sua collezione. Dice che è contento che il lavoro stesso compaia a nome di Papà e mio. Promette di scrivere direttamente a Papà. Quando riceva questa lettera Papà si regoli come preferisce che il lavoro sia fatto. Io mi rimetto completamente alla Sua decisione.

Tanti affettuosi saluti

da Umberto. [...]

6.72 D'Ancona to Volterra, Pisa, 14.3.1937 *Cat. no. 149*

Pisa, 14-III-1937

Carissimo Papà.

Le restituisco il copione della cinematografia sulla lotta per la vita[50]. L'ho letto e ne sono rimasto inorridito. Se si tratta di fare una parodia può forse andare, ma come cosa seria non di certo.

Si passa dalle fluttuazioni alle leggi di Mendel, alle mutazioni e infine all'evoluzione dei cavalli. Roba da far rizzare i capelli. È uno di quegli esempi di volgarizzazione che ha screditato la teoria dell'evoluzione. Fortunatamente però oggi è passata la moda di simili divulgazioni e si è convinti tutti che se si vuol parlare di evoluzione bisogna essere più cauti e più seri. Il signor Kostitzin è evidentemente rimasto al periodo immediatamente post-darwiniano.

---

[49] The letter bears the subsequent indication 1937.

[50] Reference to the film Kostitzin was preparing for the Palais de la Découverte. See MIILLÁN GASCA 1993.

Io per conto mio ritengo che se ci si associa a una simile opera si corre il rischio di coprirsi di ridicolo. Per parte mia la prego perciò di non far includere il mio nome e Le consiglio di non far mettere neppure il Suo.

Le Sue ricerche sulle associazioni biologiche non sono oggetto adatto a una divulgazione. Dire che il pesce grosso mangia il pesce piccolo è cosa ovvia che tutti sanno da tempo. L'aver messo in forma concreta i rapporti interspecifici è il merito Suo, ma questi rapporti interspecifici non possono essere tradotti in espressioni elementari, tantomeno poi essere raffigurati nella forma banale che si può ottenere in un cartone animato. Perciò si cade nel grottesco, come ha fatto il signor Kostitzin.

Io credo invece che queste ricerche, che sono cosa seria, debbano essere mantenute entro l'ambito dell'indagine seria e non essere date in pasto a persone superficiali.

Mi scusi se Le scrivo in forma così categorica, ma desidero, che Lei sia persuaso del vero valore di questo pasticcio confezionato dal Signor Kostitzin.

La ringrazio per averlo voluto far vedere a me e per aver pensato di includere il mio nome.

Riceva tanti affettuosi saluti

da Umberto.

6.73 D'Ancona to Volterra, Pisa, 18.3.1937 *Cat. no. 152*

Pisa, 18-III-1937

Carissimo Papà.

Ho piacere che Lei sia d'accordo nell'accettare l'offerta del Bertalanffy. Gli risponderò perciò subito aderendo alla proposta di fare un volumetto sulle associazioni e gli equilibri biologici.

Io per parte mia lo metterò assieme ben volentieri. Naturalmente avrò bisogno del Suo consiglio e della Sua revisione. Se poi Lei vorrà far rivedere la parte matematica anche alla signorina Freda io non ne avrò nulla in contrario, anzi ne avrò piacere perché l'accurata revisione della Signorina potrà evitare eventuali errori che dovessero sfuggire alla mia attenzione.

Come propone il Bertalanffy io penserei di fare una traduzione ampliata e aggiornata del volumetto di Hermann. Conserverei anche lo stesso titolo; non mi pare che con Hermann ci siano contratti che impediscano traduzioni, specialmente poi se queste non sono esatte ripetizioni.

Poiché vedo dalla lettera del Bertalanffy che questi non parla di limitazioni di spazio, anzi desidera una trattazione estesa, penso che il lavoro potrà essere ampliato in confronto a quello francese.

Il libro è destinato a naturalisti. Perciò io penserei di dare una più estesa esposizione del lato zoologico delle associazioni biologiche. Per la parte matematica aggiungerei le parti nuove che Lei ha fatto di recente e quelle che ci sono nel libro di Gauthiers-Villars e che furono omesse nel volumetto di Hermann. Non farei però trattazioni matematiche eccessive come ha fatto Brelot, che si rivolgeva a matematici. Dove sia conveniente dare lo svolgimento matematico di qualche dimostrazione, esso potrebbe essere inserito in carattere più piccolo in modo che anche saltando queste dimostrazioni il lettore non matematico possa rendersi conto del ragionamento e vederne le conclusioni.

Come nel volumetto di Hermann metterei il mio nome in collaborazione al Suo. Va bene?

Quando verrò a Roma Le sarò grato se Lei vorrà prestarmi tutto il materiale che potrà servirmi a questo lavoro, che inizierò subito dopo il mio ritorno a Pisa.

Per quanto riguarda Kostitzin mi dispiace di non essere d'accordo con Lei. Io ritengo che il soggetto non si presti a una divulgazione cinematografica senza cadere nel banale e senza ridursi a voler dire delle cose ovvie. Questo è quanto ha fatto a mio parere il Kostitzin, che avendo poi mescolato Mendel, mutazioni, evoluzione dei cavalli ha fatto un pasticcio piuttosto ridicolo. Io perciò non ritengo opportuno associare una ricerca scientifica seria a una così grottesca parodia.

Che il Signor Kostitzin sia un buon matematico non lo nego, ma per la parte biologica mi pare che si limiti a fare delle divulgazioni, che nel caso presente non dimostrano neppure di avere ben assimilate le questioni.

Io non lo conosco, perciò non credo opportuno di fare una critica precisa al suo copione. Per parte mia direi che non ritengo opportuna una divulgazione del genere.

Ho piacere di sentire che Lei lavora e fa delle cose nuove. Fra pochi giorni contiamo di essere a Roma e ne potremmo parlare. Io sono stato due giorni all'Elba e conto di andare prossimamente al Giglio.

Riceva con la Mamma tanti affettuosi saluti.

Umberto

6.74 D'Ancona to Volterra, Pisa, 21.4.1937 *Cat. no. 153*

Pisa, 21-IV-1937

Carissimo Papà.

La ringrazio per la due lettere che mi manda. Dato il Suo desiderio di non voler figurare come autore dell'opera che mi chiede il Bertalanffy, mi atterrò senz'altro allo stesso. Farò perciò il lavoro sotto il mio solo nome, dando sviluppo alla parte biologica unitamente a quella matematica che riferirò in base ai Suoi lavori. Spero che Lei vorrà anche così darmi il Suo aiuto e consiglio, che altrimenti non potrei fare un lavoro privo di gravi deficienze ed errori.

In quanto a informare di tale decisione il Bertalanffy io temo che a scrivergli oggi, la mia lettera si incroci con quella che egli mi disse Le avrebbe scritto. D'altra parte penso che sia meglio attendere e vedere che cosa egli scrive.

Da quanto mi pare egli tiene ad avere un lavoro sugli studi matematici Suoi, ma immagino che voglia che il lavoro sia volto ad uso e portata dei naturalisti. Penso perciò che egli avrebbe preferito un lavoro in collaborazione. Ma di fronte alla Sua precisa decisione credo che sarà d'accordo che il lavoro sia fatto da me soltanto.

Però penso che le ragioni che Lei ha comunicato a me, e che mi sembrano tali da giustificare il desiderio di non prendere parte attiva al lavoro, Lei potrebbe scriverle direttamente al Bertalanffy in risposta alla lettera che questi Le manderà. Un Suo rifiuto messo in questa forma non è tale da offendere il Bertalanffy, né da provocare una più lunga discussione epistolare. D'altra parte io immagino che il Bertalanffy avrebbe piacere di sapere da Lei che ha piacere che il lavoro sia fatto da me.

Io perciò penso che sia meglio attendere per vedere se egli scrive e cosa scrive e rispondergli in conseguenza. Se caso mai la lettera del Bertalanffy dovesse tardare allora gli scriverò in conformità alla Sua.

Non pensa anche Lei che sia meglio fare così?

Mi ha scritto Polimanti che ha ricevuto il suo manoscritto, cosa di cui sembra molto contento.

Per quanto riguarda la gita che Lei intende fare per incontrarci a metà strada tra Pisa e Roma ho piacere di sentire che Lei desidera combinarla o presso Grosseto o presso Orbetello.

Sarebbe interessante andare a vedere Cosa (Ansedonia) subito a sud di Orbetello dove ci sono mura etrusche, il così detto bagno della regina e la tagliata etrusca. Da Orbetello si potrebbe andare per Porto S. Stefano sulla strada panoramica che gira per un tratto l'Argentario.

Più a nord sono interessanti le rovine di Populonia.

Per fare colazione sarebbe meglio fermarsi a Grosseto. Perciò si potrebbe vedere in mattinata Cosa e nel pomeriggio Populonia o viceversa.

Noi si potrebbe partire da Pisa alle 7.41 ed essere a Campiglia alle 9.43 o partire da Pisa alle 9.05 e essere a Campiglia alle 10.54. In questo caso si potrebbe vedere subito Populonia, poi proseguire per Grosseto e poi per Orbetello.In serata noi ripartiremmo da Orbetello alle 20.06 oppure alle 17.01.

Qualora invece per non fare un viaggio troppo strapazzoso Lei preferisse lasciare Populonia, si potrebbe trovarci alle 11.02 a Grosseto, visitare la città e far colazione e poi proseguire per Orbetello, dove visitare nel pomeriggio Cosa e l'Argentario.

Sarà forse difficile combinare per la domenica prossima anche a causa dei tempi incerti. Si potrebbe invece fissare la gita alla domenica successiva.

In ogni modo noi ci rimettiamo a Lei per la decisione.

Gradisca con la Mamma e Enrico tanti affettuosi saluti.

da Umberto [...]

6.75 D'Ancona to Volterra, Pisa, 25.4.1937 *Cat. no. 154*

Pisa, 25-IV-1937

Carissimo Papà.

Riceviamo oggi la Sua lettera del 24 alla quale rispondiamo subito.

Dato il Suo desiderio che io scrivessi al Bertalanffy gli ho mandato oggi stesso la lettera di cui Le accludo copia. Ho creduto preferibile trascrivergli alcuni passi della Sua lettera precedente anziché mandargliela. Così mi pare che sia chiaro che il Bertalanffy vedrà qual'è il Suo punto di vista e Suoi desideri in merito alla pubblicazione.

Io credo però che sarebbe stato preferibile attendere la promessa lettera del Bertalanffy prima di scrivergli. Dato che egli aveva l'intenzione di scriverLe, conveniva anche veder cosa diceva e in base a ciò regolarsi nel rispondergli. Non credo che questa attesa avrebbe portato la necessità di fare delle discussioni dato che i motivi da Lei addotti erano chiari e persuasivi.

In ogni modo ho fatto come Lei ha desiderato. Se però viene una lettera del Bertalanffy penso che sarà meglio che Lei gli risponda.

Per quanto riguarda la gita ritengo che sarà preferibile incontrarci a Grosseto da lì poi andare a Cosa o a Populonia. A quale ora Loro potrebbero essere a Grosseto, verso le 11 o prima? Noi ci regoleremo poi in base al Loro arrivo.

Riceva con la Mamma e tutti tanti affettuosi saluti

da Umberto

Ho riveduto le bozze che vanno bene.

Carissimi, abbiamo ricevuto la Vostra lettera nella quale parlate sempre del Sig. Bertalanffy e poco di voi. [...]

Umberto ha scritto al Sig. Bertalanffy per mettervi tranquilli. Non scriveva perché quello gli aveva detto che non poteva dargli ancora nessuna istruzione per il volume né parlargli della combinazione editoriale. Così Umberto non voleva tornare a scrivere, sembrava che avesse gran fretta di combinare il volume. [...]

Affettuosi saluti e baci a tutti

Luisa [...]

6.76 D'Ancona to Volterra, Pisa, 9.5.1937 *Cat. no. 155*

Pisa, 9-V-1937

Carissimi Papà e Mamma.

Vi ringrazio tanto unitamente a Enrico per i vostri affettuosi auguri. Ho piacere di sapere che state tutti bene e che anche la tosse di Vito sia leggera e in corso di guarigione.

Gradite i miei più affettuosi saluti e ringraziamenti.

Umberto

Carissimo Papà

La ringrazio per avermi mandato in visione la lettera del Bertalanffy. Questi mi ha scritto anche direttamente. Gli dispiace che Lei non abbia voluto far uscire il libro in collaborazione, ma trova giusti i motivi per cui Lei ha rinunziato a tale idea. Bertalanffy però desidererebbe che Lei premettesse al mio lavoro una breve prefazione, dicendo qualche cosa sul significato della biologia matematica, mettendo in rilievo l'importanza dei rapporti tra matematica e biologia. Trascrivo per precisione le parole del Bertalanffy.

"Wenn Prof. Volterra auch an dem geplanten Werke selbst nicht teilzunehmen gedenkt, so wäre ich ihm doch ausserordentlich dankbar, wenn er in einer kurzen Vorrede zu Ihrem Werke über die Bedeutung der mathematischen Biologie spräche; da ja ein Hauptziel der neuen Abhandlungen die Arbeit gerade auf diesem Gebiete sein soll, so wäre es besonders zu begrüssen, wenn Volterra als führender Mathematiker diese Verbundenheit von Biologie und Mathematik hervorheben würde. Eine grosse Arbeit für Volterra wäre ja damit wohl nicht verbunden, und es würde vielleicht ein solches Einführungswort der Verbreitung des Buches in mathematischen Kreisen förderlich sein."

Non so quali sono i Suoi desideri a questo riguardo; se Lei pensa di poter fare o no questa prefazione dato il luogo dove il libro dovrà essere pubblicato[51]. Se Lei però fosse dell' idea di accettare la proposta del Bertalanffy penso che non si dovrebbe trattare di un lavoro lungo e faticoso perché forse basterebbe ripetere alcune delle cose dette nel Suo discorso inaugurale del Università di Roma.

Attendo la Sua risposta prima di scrivere al Bertalanffy. Naturalmente ritengo che non sarà il caso di fare complimenti; io comunicherò quindi quanto Lei mi dirà.

Con nuovi affettuosi saluti

Umberto [...]

6.77 D'Ancona to Volterra, Pisa, 15.5.1937 *Cat. no. 156*

Pisa, 15-V-1937

Carissimi Papà e Mamma.

Vi ringrazio vivamente per il dono che avete voluto inviarmi per la mia festa. Esso mi è giunto molto gradito, per pagare un nuovo conto di libri che avevo in sospeso. Vi ringrazio ancora molto per gli affettuosi saluti.

Avrò molto piacere di sapere cosa pensa Papà in merito alla richiesta di Bertalanffy di una prefazione. Naturalmente non è il caso di far complimenti, ma Papà può rispondere come meglio preferisce che io riferirò al Bertalanffy. Se non desidera fare la prefazione si potrebbe forse riportare come prefazione semplicemente qualche brano del discorso inaugurale in cui Papà disse dell'importanza dell'applicazione della matematica alla biologia.

Non è per ora il caso di pensare a come scrivere questa prefazione o quale citazione riportare in sostituzione di una prefazione. Per ora basta che Papà mi dica cosa desidera che io risponda al Bertalanffy.

Che tempi avete trovato in Ariccia? Immagino buoni. Anche qui il tempo si sta mettendo al bello e comincia a fare caldo.

Ricevete con Enrico tanti affettuosi saluti. Come sta il piccolo Vito?

Umberto [...]

---

51 This passage – "given the place where the book is to be published" – is a clear indication that the reason for the refusal was not scientific, but linked to the place of publication – Berlin. It was thus most probably due to the German political situation, in particular, the politics of racial persecution.

6.78 D'Ancona to Volterra, Pisa, 20.5.1937 *Cat. no. 157*

Pisa, 20-V-1937

Carissimo Papà.

La ringrazio vivamente per il fascicolo dell'Ufficio pesca e per il libro di P.M. Savielli che mi ha voluto mandare. La ringrazio particolarmente per quest'ultimo, che mi interessa come ricordo dei Fisiocritici.

Noi Le abbiamo mandato il volume di geometria. La Mamma ci ha detto che Lei già lo aveva. Però lo abbiamo avuto per poco e abbiamo pensato che Le potesse essere utile per fare cambi con altre opere.

La ringrazio anche per quanto mi ha scritto a proposito della prefazione; ho scritto in questo senso al Bertalanffy.

Spero che avranno buoni tempi in Ariccia. Mi fa molto piacere che Vito piccolo sta bene.

Riceva con la Mamma tanti affettuosi saluti.

da Umberto [...]

6.79 D'Ancona to Volterra, Pisa, 2.6.1937 *Cat. no. 158*

Pisa, 2-VI-1937

Carissimo Papà.

Abbiamo letto sui giornali dell'inaugurazione dell'Accademia Pontificia. Immagino che la cerimonia, anche senza la partecipazione del Papa, sarà riuscita molto solenne.

Conosco i lavori di Fontaine, che per mezzo di ormoni ha provocato la maturazione sessuale delle anguille. Contemporaneamente e indipendentemente dal Fontaine lo stesso risultato era stato raggiunto a Milano da Schreiber. La cosa è certo molto interessante ma non segna un fatto nuovo, perché si sapeva già che con determinati ormoni in altri animali poteva essere ottenuta la maturazione anticipata. Sarebbe interessante se ora si potesse determinare la fecondazione di individui così maturati. Per ora però si è avuta la maturazione dei soli maschi.

Non conosco invece i lavori di cui mi parla della fluorescenza dovuta ad azioni nervose. Le cellule nervose in genere presentano fenomeni di fluorescenza. Vengono essi aumentati o modificati durante l'attività funzionale?

Del Congresso di Parigi non so nulla per cui non so neppure quanto esso sarà interessante. Esso sarà forse un buon pretesto per andare a Parigi a vedere l' esposizione.

Abbiamo visto domenica il Giuoco del Ponte. Il corteo è bello, ma il giuoco non vale molto.

Comincia a far parecchio caldo e le zanzare diventano insopportabili.

Immagino che in Ariccia farà una temperatura più piacevole.

Riceva tanti affettuosi saluti

da Umberto [...]

6.80 D'Ancona to Volterra, Siena, 22.9.1937 *Cat. no. 159*

Siena, 22-IX-1937

Carissimo Papà.

Sono ritornato stamane da Bologna, dove ero andato per il Congresso di Zoologia, che è riuscito bene. Al mio ritorno ho trovato la lettera della Mamma, in cui scrive che Lei sta correggendo il manoscritto francese.

Avrò molto piacere se vorrà vedere i punti da me messi in rilievo in quella lettera che Le inviai da Firenze. Sarei contento che Lei, se ciò Le è possibile, mi mandasse il manoscritto entro i prossimi giorni. Così io potrei vedere le aggiunte da fare per citare i nuovi lavori. Se il lavoro verrà qui presto, procureremo di ricopiarlo in tutto o in parte prima della nostra partenza per Roma. Noi contiamo di partire domenica 30 e di essere in Ariccia lunedì.

Riceva con la Mamma, Edoardo e Enrico tanti affettuosi saluti.

Umberto [...]

6.81 D'Ancona to Volterra, n.p., n.d. *Cat. no. 75*[52]

Ha ricevuto Papà un fascicolo pubblicato da Oxford, sugli equilibri biologici? In esso c'è una estesa bibliografia sull'argomento; si é citato pure il libro francese ed è detto che Papà è un pioniere in questi studi.

Termino inviandovi i più affettuosi saluti.

Umberto

6.82 D'Ancona to Volterra, n.p., October 1937 *Cat. no. 76*[53]

Carissimo Papà.

Ho piacere di sentire che ha fatto nuovi calcoli. Cosa ha ottenuto?

La ringrazio tanto per i fascicoli che mi ha mandato.

Ieri Le ho spedita la prima serie delle fotografie del villino. Sarà bene che attenda anche le altre prima di scegliere le migliori. Per le altre bisogna che completo prima il rotolo.

Alcune mi sembrano buone. Converrà però farle ritoccare prima di riprodurle.

Le mando unita una cartolina di un mio collega che è stato a Parigi. Sarà contento di sapere che il film di Kostitzin è riuscito bene. Anch'io ne ho piacere e mi ricredo.

Le unisco pure una bozza del frontespizio del fascicolo che pubblicherò per Bertalanffy. Io veramente volevo mettere in forma un po' diversa il solo attuale

---

[52] Fragment.

[53] N.d. but certainly written on October 1937. Incomplete

sottotitolo. Hanno però insistito per farci entrare anche la lotta per l'esistenza. Così mi pare che vada bene.

Riceva con la Mamma e i Fratelli tanti affettuosi saluti da

Umberto [...]

Enclosure: *Cat. no. 1* (postcard of Baldi to D'Ancona, Paris, 7.10.1937)

Parigi, 7 Ott. 1937

Ho visto proiettato al Palais de la Découverte un buon film Volterra-Kostitzin sulla lotta per l'esistenza. Cordiali saluti

Baldi[54]

6.83 D'Ancona to Volterra, Pisa, 24.10.1937 *Cat. no. 160*

Pisa, 24 ottobre 1937

Carissimo Papà. La ringrazio per quanto mi scrive nella Sua lettera.

Ho piacere di sentire che le Sue nuove ricerche continuano e che danno risultati interessanti. Penso che sarà molto bene se Lei le svilupperà anche indipendentemente da altri eventuali lavori del genere, perché sono certo che Lei darà loro uno sviluppo nuovo e originale. In seguito sarà però bene vedere se c'è già qualche cosa del genere. Per ora io sono molto occupato con lavori da ultimare prima di partire, ma appena a Padova mi sarò un po' sistemato vedrò di cercare cosa è stato fatto su questi argomenti.

Oggi ho potuto vedere sul libro di Lotka, che Pearl e Parker (Proc. Nat. acad. sci. Vol. VIII, p. 21, 1922)[55] hanno trattato dell'influenza della densità della popolazione sull'accrescimento della stessa. Hanno sperimentato con la Drosophila. Si tratta di un lavoro che Lei potrà facilmente vedere; svolgono però il problema da un altro punto di vista. Trovano che la mortalità aumenta con la densità della popolazione, la natalità invece diminuisce con l'aumento della densità. La durata della vita invece aumenta fino a un certo punto con l'aumento della densità, poi diminuisce. C'è un ottimo di densità in cui la durata della vita è massima.

Sono contento che Le siano piaciute le fotografie. Non è il caso di parlare di pagare le spese. Sono stato lieto di renderle questo piccolo servizio. Io credo che Lei possa dare a Alinari o a qualche altra ditta le cartoline che io ho mandato e che da quelle possono trarre la riproduzione. In ogni modo Le darò anche le pellicole e se vuole gliele manderò subito. Per la riproduzione credo sia bene se sente Alinari e che si faccia consigliare e mostrare qualche campione.

---

[54] Edgardo Baldi (1899–1951) was a physiologist and zoologist, director of the Istituto Italiano di Idrobiologia "Dott. Marco de Marchi" of Pallanza. He founded the journal of the Institute (*Memorie dell'Istituto Italiano di Idrobiologia Dott. Marco de Marchi*).

[55] Pearl, Parker 1922.

Noi speriamo di trasferirci presto, ma finora non sappiamo ancora nulla di preciso.

Speriamo di vederli al Loro passaggio da Pisa.

Riceva con la Mamma tanti affettuosi saluti

Umberto [...]

6.84 Volterra to D'Ancona, London, 5.11.1937 *Cat. no. 367*[56]

CUMBERLAND HOTEL
MARBLE ARCH,
LONDON. W. 1

5 Novembre 1937

Carissimo Umberto

Ho trovato qui il Prof. Pearl (senza Verhulst) col quale mi sono intrattenuto e anzi ho assistito alla sua ultima lezione alla Scuola d'Igiene[57] ove ho conosciuto anche il Prof. Grenwood direttore dell' Istituto e che si occupa specialmente di statistica. Il Sig. Pearl mi ha detto di pubblicare la memoria in Inglese sul suo giornale di America.

Dopo l'ultima estensione di calcoli ho ceduto alla tua insistenza che la memoria sia fatta a mio nome.

Però per mettere le cose a punto e per la verità la memoria comincia con le parole: Il Prof. D'Ancona mi ha chiesto di stabilire le condizioni per cui una popolazione cresce o si esaurisce valendosi degli incontri. Il Pearl mi ha detto che darà a questo principio la forma più inglese ma questa è la sostanza. Ti prego di dirmi se la cosa ti va così o se desideri modificazioni ma mi sembra che questo corrisponda a come sono andate le cose.

6.85 D'Ancona to Volterra, Pisa, 11.11.1937 *Cat. no. 161*

Pisa, 11-IX-1937

Carissimi.

Abbiamo letto con molto piacere le vostre buone notizie. Sono contento di sapere che avete fatto un buon viaggio e che Papà abbia avuto utili scambi d'idee con Pearl. Immagino che Pearl sia persona che si interessi delle ricerche di Papà; è forse la persona che più si occupa di applicazioni di matematica alla biologia.

Sono anche molto contento che Papà abbia combinato di pubblicare il lavoro sul periodico di Pearl[58]. Non è nemmeno il caso però di spiegare che lo pubblichi

---

[56] Draft.

[57] Cf. letters nos. 3, 4 and 5 in the Pearl-Volterra correspondence.

[58] VOLTERRA 1938b.

a nome Suo. Io insistevo sempre in questo senso perché non ritenevo possibile fare diversamente. Il mio nome non c'entrava affatto; anzi non vedo nemmeno la ragione perché mi debba citare. Il mio suggerimento non ha certamente nessun valore perché era un'indicazione molto banale di applicare il metodo degli incontri agli accoppiamenti dei due sessi. L'importante era trovare il modo di studiare la cosa dal punto di vista matematico. Mi fa molto piacere che Papà si sia interessato alla cosa e abbia trattato in modo così brillante questo problema. Ma io non credo di entrarci per nulla. Mi fa soltanto piacere d'aver provocato che Papà facesse un nuovo interessante lavoro.

Di questi nuovi risultati terrò conto nel libro per Bertalanffy che sto portando avanti un po' lentamente a causa delle noie di dover fare una nuova prolusione

Noi contiamo di partire per Padova verso la fine della prossima settimana. Voi quando tornerete?

Vorrei pregare la Mamma di farmi due pagamenti a Parigi. Le unisco due fatture con preghiera di volermele saldare se non le da troppo scomodo.

Vi ringrazio tanto, vi auguro un buon soggiorno costì e vi invio tanti affettuosi saluti.

Umberto [...]

6.86 D'Ancona to Volterra, Abbazia, 22.8.1938 *Cat. no. 162*

Abbazia, 22-VIII-1938

Carissimo Papà.

Ho avuto oggi la risposta da Bertalanffy, che è sempre in America. Dice che il lavoro va bene e che lo pubblicherà presto. Ha incaricato della traduzione il libero docente di matematica Holzer di Vienna. Bertalanffy ritornerà in Europa in Ottobre e si interesserà di sollecitare la traduzione. Prima di stampare il libro mi manderà il manoscritto della traduzione.

Spero che anche la stampa possa essere presto ultimata.

Abbiamo ricevuto con piacere vostre buone notizie. Qui abbiamo tempo brutto. Speriamo che faccia presto nuovamente bello, però certamente grandi calori non ci saranno più.

Silvia si diverte, sta bene.

Fra poco più di una settimana ritorneremo a Padova.

Io conto di venire in Ariccia un paio di giorni alla fine di settembre o ai primi di ottobre prima di andare a Napoli.

Ho visto giorni fa Maria, che è a Fiume in vacanza.

Riceva unitamente alla Mamma e ai Fratelli tanti affettuosi saluti.

da Umberto [...]

6.87 D'Ancona to Volterra, October 1938 *Cat. no. 163*

Carissimo Papà.

Questa mattina il Prof. Levi Civita mi ha riportato il manoscritto del Suo lavoro[59]. Mi dice che Le ha scritto in proposito.

Per la traduzione Luisa e io saremmo ben lieti di poterla fare, ma purtroppo Luisa non ha una padronanza tale dell'inglese da fare una traduzione soddisfacente. È necessario che questa sia fatta in un buono stile e con conoscenza dei termini tecnici.

Io non so se forse a Roma non sarebbe più facile trovare persona adatta a tale scopo. Altrimenti cercheremo di trovare qui un buon traduttore; per i termini tecnici sarà però necessario rivolgersi in seguito a qualcuno che li conosca..

Se Lei crede provvederemo così, ma non è possibile fare da noi.

Per la ricopiatura Luisa è naturalmente a Sua disposizione.

Abbiamo piacere di sentire che tutti stanno bene. Riteniamo che Gustavo non avrà noie in seguito agli ultimi provvedimenti. Per il resto speriamo bene[60].

Luisa non scrive di più perché io ho preso la lettera in laboratorio, da dove scrivo. Luisa doveva andare oggi a fare visita alla Signora Libera.

Affettuosi saluti a Lei, alla Mamma e a tutta la famiglia.

Umberto

6.88 D'Ancona to Volterra, 4.1.1939 *Cat. no. 164*

Padova, 4-I-1939

Carissimo Papà

La ringrazio per la pagina di formule che mi ha mandato. Essa mi sarà molto utile per stendere la nota.

Ho risposto a Luisa su quanto lei mi chiedeva a proposito delle mutazioni. Credo che il libro più adatto per mettersi al corrente su queste questioni sia il volume di Trinofeeff-Ressousky, recentemente tradotto anche in italiano e edito da Hoepli.

Bertalanffy mi ha scritto che ha riveduto la traduzione e che a giorni la manderà in tipografia.

Io cercherò di scrivere al più presto la nota, che però mi richiederà un po' di tempo.

Ringrazio Lei e la Mamma per tutte le gentilezze usate a me, a Luisa, a Silvia durante il nostro soggiorno a Roma.

Le faccio i migliori auguri e la saluto affettuosamente con la Mamma, Enrico, Emilia e Gustavo.

Umberto.

---

[59] Probably VOLTERRA 1939b.

[60] Another reference to the racial laws.

6.89 D'Ancona to Volterra, 28.1.1939 *Cat. no. 165*

Carissimo Papà.

La ringrazio per la revisione del mio riferimento sul lavoro di Wright. Cercherò di controllare i punti che Lei dice non chiari.

Io penso che sarebbe molto interessante se Lei potesse riprendere in esame la questione e cercare di trattarla più estesamente.

Ho ricevuto i due lavori che Lei mi ha mandati. Tante grazie.

Ho appreso con piacere la notizia che Lei ha cominciato a uscire. Spero che il tempo sia buono.

Gradisca con la Mamma e i fratelli tanti affettuosi saluti.

Umberto. [...]

6.90 D'Ancona to Volterra, 22.5.1939 *Cat. no. 78*

Padova, 22-V-1939

Carissimo Papà.

Ho avuto ieri sera molto piacere di sentire la Sua voce al telefono. Spero di continuare ad avere sue buone notizie e di miglioramenti del Suo stato di salute.

Oggi ho avuta una lettera da Bertalanffy che mi dice che il lavoro è in stampa e che a giorni mi manderà una prima serie di bozze che lui ha già rivedute. Sono stati fatti anche i clichés in modo che il volume potrà essere stampato fra breve.

Bertalanffy mi chiede di mandargli in prestito il volume di Gause per riprodurre alcune figure che io desidero riportare nel libro. Se Lei non ha nulla in contrario manderò in prestito la Sua copia del volume stesso, che io ho già qui.

Io ho scritto la prefazione al libro; gliela manderò fra giorni affinché Lei la legga e mi dica se va bene.

La ringrazio e invio a Lei e alla Mamma tanti affettuosi saluti.

Umberto. [...]

6.91 D'Ancona to Volterra, n.p., May 1939 *Cat. no. 79*[61]

Carissimo Papà.

Ecco la prefazione per il libro. Le sarò grato se vorrà leggerla e se mi dirà se così può andare o se ha bisogno di qualche correzione.

Le faccio tanti auguri e La saluto affettuosamente insieme alla Mamma.

Umberto [...]

61 N.d. but clearly written on May 1939.

6.92 D'Ancona to Volterra, Padova, 29.5.1939 *Cat. no. 166*

Padova, 29-V-1939

Carissimo Papà.

Ho letto quanto mi scrivono Lei e Zio Roberto. Ho riletto anche la mia prefazione, ma non trovo giustificata la preoccupazione di Zio Roberto. Il Bertalanffy ha voluto un lavoro sui Suoi studi ed è quindi naturale che ciò risulti dalla prefazione. Caso mai se il Bertalanffy farà delle eccezioni, se ne potrà riparlare, ma se lui non dice nulla io lascio così. Si tratta di un lavoro puramente scientifico nel quale è naturale che le cose siano esposte in modo del tutto obbiettivo.

Per evitare la possibile impressione (non giustificata però) che io abbia voluto insistere intenzionalmente al Suo nome, nominerò questo 2 sole volte invece di 3 come era finora. Ma il resto lascerò come sta[62].

Manderò il libro di Gause, come Lei mi autorizza.

Sento con piacere che Lei sta alzato in poltrona; spero che il miglioramento continui regolarmente.

Noi stiamo benissimo.

Il tempo è regolarmente brutto.

Tanti affettuosi saluti a Lei e alla Mamma. A Zio Roberto rispondo separatamente; Vi prego di consegnarli l'unita lettera.

Umberto [...]

6.93 D'Ancona to Volterra, Padova, 10.7.1939 *Cat. no. 167*

Padova, 10-VII-39

Carissimo Papà.

Ho appreso con piacere da Luisa le Sue migliori notizie; sono contento di sentire che presto andrete in Ariccia, dove probabilmente si troverà meglio per il clima e per la possibilità di stare in giardino.

Qui fa parecchio caldo; oggi per fortuna abbiamo avuto un temporale, che ha un poco rinfrescato.

Ho visto con interesse le memorie che mi ha mandato. Non potrò più tenerne conto nel libro, che è già impaginato; delle Sue due nuove pubblicazioni avevo però già parzialmente tenuto conto in base ai manoscritti che mi aveva dato.

Ho visto anche la memoria di Fellner. Come avrà notato anche Lei, questi Suoi studi destano sempre interesse.

La prego di salutarmi anche la Mamma e di gradire i miei migliori auguri di una buona villeggiatura in Ariccia.

Umberto

Affettuosi auguri a Lei e alla Mamma per l'Anniversario di Matrimonio

---

62 These passages are clear confirmation that Volterra's refusal to write the book was linked to the situation caused by the racial persecution.

6.94 Volterra to an unkown correspondent, n.p., n.d. *Cat. no. 425*

Monsieur,

Je vous remercie beaucoup des exemplaires de l'ouvrage que vous m'avez envoyés.

Je pense d'en envoyer quelques uns à des Journaux scientifiques qui pourront en faire des comptes rendus, Scientia, Rivista di Biologia, Berichte v. Wess Biol.

M. Picard m'a écrit qu'il en a reçu un exemplaire. Pouvez vous lui en envoyer un autre pour le présenter en hommage à l'Académie des Sciences?

Je desire envoyer un exemplaire à mes Collègues de l'Istitut: M. M. Hadamard, Goursat, Borel, Lebesgue, Perrin, Joubin, Gravier, Caullery.

Pouvez vous vous en charger?

Je vous serai aussi très reconnaissant si vous voudrez bien envoyer un exemplaire à la Revue Générale des Sciences, à Nature et à American Mathematical Society, New York qui en [illeg.] un compte rendu.

Veuillez bien m'annoncer les autres journaux auxquels vous l'envoyez pour ne pas faire double emploi.

6.95 Letter written by D'Ancona, n.p., n.d. *Cat. no. 80*[63]

Ho letto con molto interesse la sua pubblicazione "Zur Frage der Eigengesetzlichkeit der theoretischen Biologie" e trovo che Lei ha impostato il problema su basi molto precise.

La possibilità di considerare i fenomeni biologici anche da un punto di vista diverso da quello fornito dalle leggi fisiche è promettente di vasti sviluppi suscettibili di portare a una più completa conoscenza dei fenomeni stessi.

La ringrazio vivamente per l'invio di questa Sua pubblicazione che porta un così utile contributo agli studi sui problemi generali della scienza.

Gradisca i miei distinti saluti.

6.96 D'Ancona to Volterra, n.p., n.d. *Cat. no. 77*[64]

Carissimi.

Rimando i contratti che vanno benissimo. Mi direte poi come posso fare.

Ieri sono stato a Pavia, dove abbiamo avuto una piccola riunione famigliare per discutere i problemi evolutivi. C'era, oltre a vari zoologi, anche Rasetti, che si interessa molto di questioni zoologiche. A Pavia il tempo era anche più brutto di qui. [...]

---

63 The Volterra Archive catalogue indicates that it is a translation by D'Ancona of a letter the addressee and the sender of which are both unknown.

64 The Volterra Archive catalogue indicates 1938, although this date, in view of the contents, cannot be considered reliable.

Fra breve contiamo di rivedervi. Avrei da sottoporre a Papà alcuni problemi sulle associazioni. Egli ha trattato il caso in cui una specie nuova viene introdotta in una associazione e ha trovato che quando essa è rappresentata da pochi individui si esaurisce. Si deve però ammettere la possibilità che pur essendo rappresentata da pochi individui essa si conservi ed aumenti. In quali condizioni si esaurirà e in quali si conserverà? Se Papà avrà voglia di pensarci, ne avrò molto piacere.

Tanti affettuosi saluti a tutti

da Umberto.

7 Letters between Charles S. Elton and Vito Volterra

Charles Sutherland Elton was born in Liverpool (England) on 29 March 1900. He studied first at Liverpool College and then at New College, Oxford. Here, while still an undergraduate, he became assistant to Julian Sorell Huxley (1887–1975), the well-known biologist and philosopher who was to have a considerable influence on modern embryology and evolutionary studies (HUXLEY 1942). The latter organized an Oxford University expedition to Spitzbergen to study the animal communities in the area from the ecological point of view, in which Elton participated. There is no doubt that this experience had a strong influence on Elton's ideas; he also made a profound study of the work of the American biologist Victor Ernest Shelford (1877–1968), who had been one of the first to establish ecology as a separate branch of science (SHELFORD 1913). During the Spitzbergen expedition, Elton had the occasion to get to know another participant, the demographer Alexander Carr-Saunders (1886–1966), author of the book *The Population Problem* (CARR-SAUNDERS 1922), who also had a strong influence on his ideas.

Elton graduated in zoology at Oxford in 1922 and continued the research he had undertaken during the Spitzbergen expedition in three further expeditions in the Arctic area in 1923, in 1924 and in 1930. The interest in the ecological approach that these experiences aroused in him led him to make a strong criticism of the approach then in vogue in zoology. This was focused on comparative anatomy and tended increasingly to address the study of animal behaviour in the laboratory following a physico-chemical approach in which the animals were considered merely as physical mechanisms. Elton reappraised the naturalist approach by considering the life of animals in their natural habitat and their interrelations with the surrounding environment. These studies were gradually to progress from the prescientific phase of "natural history" to the scientific phase represented by the new branch of ecology, which Elton defined as "the sociology and economics of animals". According to Elton, one of Charles Darwin's greatest contributions – under the influence of the great German naturalist Alexander von Humboldt (1769–1859) – was to have shed light on "the tremendous intricacy and importance of plant and animal interrelationships".

The first book in which Elton expressed his point of view and his ecological theories in an organized fashion was *Animal Ecology* (ELTON 1927) in which he defined concepts that would later become basic to ecology, such as "food chains", "food cycle" and "niche". In 1930 he published another book called *Animal Ecology and Evolution* (ELTON 1930), in which he went further in his criticism of the reductionist approach, claiming that "the balance of nature does not exist and perhaps never has existed".

As a result of his Arctic experiences, Elton was taken on as a consultant by the Hudson's Bay Company, where he studied the fluctuations in certain animal populations, using the data contained in the company records concerning the number of animals caught. One of the most interesting results was the well-known statistics on the prey-predator ecosystem involving hares and lynxes in the area, which provided one of the main experimental confirmations of Volterra's theories (and to which reference is made several times in the correspondence contained in the present book). It was probably these experiments that led him to take an increasing interest in the problem of the fluctuation in numbers in animal populations. This led him to take an interest in Volterra's mathematical research.

In 1932 he founded the Bureau of Animal Population at Oxford, a sort of clearing house for data on the variation of the number of animal species, again within the framework of a general study of terrestrial ecology. The same year he founded the review *Journal of Animal Ecology* (mentioned in his correspondence with Volterra) which became the organ of the British Ecological Society.

In 1936, Elton was appointed reader in animal ecology at the University of Oxford and senior research fellow at Corpus Christi College.

At the outbreak of World War II, Elton's Bureau of Animal Population was given the job of finding methods for controlling rodent pests responsible for serious food losses. The results obtained by him and his coworkers were later collected and published in a book (ELTON 1954). Again during the war he published another book on the problems of population dynamics (ELTON 1942), the topics of which were further discussed in another postwar book dedicated to the problem of the effects due to the diffusion in a natural area of organisms extraneous to it (ELTON 1958). In 1953 he was elected fellow of the Royal Society of London and foreign member of the American Academy of Arts and Sciences. His final book on ecological issues was published in 1966 (ELTON 1966). He actually retired from active scientific and academic work in 1967, the year he was awarded the Darwin Medal of the Royal Society.

Charles S. Elton died in Oxford on 1st May 1991.

* * *

7.1 Elton to Volterra, Oxford, 6.1.1931 *Cat. no. 168*[1]

UNIVERSITY MUSEUM
OXFORD DEAUFORT HOUSE
KIDLINGTON
OXFORD

January 6th 1931

Dear Professor Volterra,

I beg to thank you very much for doing me the honour of sending me your very important work on animal associations[2], in which I am extremely interested.

It appears to me that the study of marine animals is indeed more likely to provide the first materials for the application of your theories, since they are probably less subject to the interference from external factors. In this connection I enclose an Abstract of a paper which may have escaped your notice, by being published in a small English Journal. I am sending by separate post some numbers of these "Ecology Abstracts".

I hope to publish shortly some works upon the fluctuation in numbers of Canadian wild animals (rabbits, linx etc.) upon which I have managed to get more light by examining the old records of the Hudson's Bay Company[3]. I do not know, however, whether these figures and facts will be sufficiently accurate or relevant, to provide the material you would wish.

I agree completely with your main thesis i.e. that external factors are not in themselves necessary to cause fluctuations in the populations of wild animals. In fact I have been led to postulate both the existence of an external climatic cycle[4], and also a process similar to the ones you have described from mathematical data, in explaining the rabbits-linx 10-year cycle in Canada. It will give me great pleasure to send you this report when it appears.

I shall probably be in a position to obtain a good deal of material concerning these mammalian cycles, during the next few years, and shall attempt to procure figures which may be of use to you in carrying further the analysis of these problems.

My only regret is that I am unable to follow the purely mathematical arguments, and must confine myself to the working out of an adequate technique for obtaining biological data which may be of use to the mathematician.

yours very truly,

Charles Elton

P. S. I am also sending a small book in which some arguments about annual associations are set forth.

---

1 Includes a short review (ELTON 1929b) of JOHNSTONE 1928.

2 VOLTERRA 1931a.

3 As we observed in the introductory essay and in the biographical note, Elton research on the ecosystem hares-lynxes of Hudson's Bay in Canada represents one of the few empirical verifications of the first law of Volterra, stating the periodicity of the oscillation of the number of preys and predators. See also ELTON 1924–5, 1925, 1927, 1931, 1946.

4 See also ELTON 1929a.

7.2 Volterra to Elton, 20.1.1931 *Cat. no. 369*[5]

January 20th 1931

Dear professor Elton,

I am very grateful to you for your letter and for the papers & indications you had the kindness to send me.

I am very glad to hear that your ideas agree perfectly with those I have expressed in a mathematical form. I have already read the first of your lectures on "Animal Ecology & Evolution"[6] with much interest and have been very pleased to see how well our ideas agree. Now I am reading your second lecture on "migration".

If you will be so kind as to send me your statistic and the result of your researches, I should be grateful to you as they would be most useful to me.

I should be very glad to publish again my researches in an english translation if I could be helped in finding an editor.

I should like to add to it all that part about the remarks & statistic measures much more developed than in the first edition. In this way it could be used as a mathematic's subsidy to the studies of Ecology

Yours very truly

7.3 Elton to Volterra, Oxford, 24.9.1931 *Cat. no. 169*[7]

DEPARTMENT OF ZOOLOGY AND COMPARATIVE ANATOMY,
UNIVERSITY MUSEUM,
OXFORD

24 September, 1931

Dear Professor Volterra,

I am enclosing a notice of a new Journal of Animal Ecology which we are trying to start here and which I shall be editing. It will contain a good deal of work on animal population and fluctuations in numbers, so I think it may interest you know about it.

Yours sincerely,

Charles Elton

Professor Vito Volterra
via in Lucina 17, Roma, Italy

---

[5] Draft.

[6] ELTON 1930.

[7] Contains a notice of presentation entitled "Proposed *Journal of Animal Ecology*".

7.4 Elton to Volterra, Oxford, 30.12.1931 *Cat. no. 170*

DEAUFORT HOUSE
KIDLINGTON
OXFORD

30/12/31

Dear Sir,

I beg to thank you for your kind letter of December 20th, enclosing your cheque for £ 1.5.0, as subscription to the new Journal.

The subscription of £ 1.5.0 applies to members of the British Ecological Society, and I shall be very glad to have the privilege of proposing your name as a member of the Society. There is no entrance fee, and no other dues. If you will kindly fill in the enclosed form and return it to me when you next write, and will see thet your name is put through.

I should much appreciate a paper dealing with your fluctuation theory, especially if you can write it without too much technical mathematical detail, which the ordinary biological reader like myself must, in any case, take "on trust"[8]. The new journal will be short at first, until the subscriptions increase in number, so that I would prefer your paper to be of not more than about 5000–6000 pages. Later I shall be able to increase the size of the journal, and therefore of the papers.

yours sincerely,

Charles Elton

7.5 Elton to Volterra, Oxford, 9.3.1932 *Cat. no. 171*

OXFORD UNIVERSITY
DEPARTMENT OF ZOOLOGY AND COMPARATIVE ANATOMY.
UNIVERSITY MUSEUM
OXFORD
BUREAU OF ANIMAL POPULATION

9/3/32

Dear Professor Volterra,

I must apologise for not writing sooner, in reply to your letter of Jan.6th. It was, of course, a mistake in my letter when I wrote "pages" instead of "words".

I was glad to see that Professor Chapman has included an appendix by yourself in his latest book on animal ecology.

Your name comes before the Council for election next week. The first number of our next journal will be out in May. Contribution for the second number should reach me before July, or if intended for the 1933 issue, before December of this year (1932).

yours very truly,

Charles Elton

---

[8] This work (see also the subsequent letter) was never written.

7.6 Volterra to Elton, 1932 *Cat. no. 368*[9]

Cher Professeur Elton

Vous voudrez bien m'excuser si je vous reponds avec un si long retard, mais j'ai été pendant ce temps toujours en voyage. Je me trouve maintenant en Espagne où j'ai fait à l'Université de Madrid quelques leçons sur la question des fluctuations biologiques. Dans quelques jours je retournerai à Paris (Hôtel Lutetia) d'où je me rendrai à Rome.

Je tiens à vous exprimer toute ma reconnaissance pour le grand honneur d'avoir été nominé membre de la British Ecological Society. Veuillez bien exprimer à mes Collègues mes remerciements les plus sincères.

On me demande quel journal je prefère recevoir des deux que la Société publie. Je pense que le nouveau journal que vous avez fondé est le "Journal of animal Ecology". Et c'est dans celui-ci que j'espère pouvoir collaborer. C'est donc ce journal que je choisis. Je pense qu'il suffit de vous en avoir averti par cette lettre pour le recevoir.

7.7 Elton to Volterra, Oxford, 28.9.1936 *Cat. no. 172*

OXFORD UNIVERSITY
BUREAU OF ANIMAL POPULATION

28th September, 1936

Dear Professor Volterra,

We are assembling a special library dealing with animal population studies, especially on the theoretical side. It would be of great service if you can possibly send us any of the papers which I enclose on a list. These we have no reprints of, and some of them are not available in Oxford.

The Bureau of Animal Population has now become a University Institution, and we are being given facilities for expanding our research work. I hope that we shall, at any rate, be able to collect all important literature together

Yours very sincerely.

Charles Elton

Professor Vito Volterra,
via in Lucina 17, Roma, Italy

VOLTERRA

1923. ... *J. Wash. Acad. Sci.* 13(8): ______
1920. ... *Proc. Nat. Acad. Sci.* 6(7): 410–415
1926. Lois de fluctuation de la population de plusieurs especes coexistant dans la meme milieu
Assoc. Fr. Avanc. Sci.,Lyons p. 96

[9] Draft, n.p., n.d. but clearly written in reply to letter no. 7.5, and therefore in 1932.

1927. Sulle fluttuazioni biologiche
Rend. R. Acad. Lincei ser 6,5: 3
1927. Leggi delle fluttuazioni biologiche
Rend. Acad. Lincei ser 6,5: 61
1927. Sulla periodicità delle fluttuazioni biologiche
Rend. Acad. Lincei ser 6,5: 465
1928. Variations and fluctuations of the number of individuals in animal species living together
J. Cons. Int. Expl. Mer 3: 3
1929. Sulle fluttuazioni biologiche (redazione a cura del Prof. A. Masotti),
Rend. Seminario Matem. Fiasco Milano 3
1931. Ricerche Matematiche sulle associazioni biologiche.
Giorn. Ist. Ital. Attuari 2(3):
1932. Proc. Nat. Acad. Sci. 18(2): 172–178.
1924. Remarques sur la note de M. Regnier et M.lle Lambin.
C. R. Acad. Sci. 199: 1682.

7.8 Elton to Volterra, Oxford, 2.10.1936 *Cat. no. 173*

(Same heading of letter no. 7.7)

2nd October, 1936

Dear Professor Volterra,

Just after writing to you I received the reprints of your latest papers, which you were kind enough to send us. We are very glad to have these and to know that you are continuing the analysis of this problem, which becomes every year more important in the work we are engaged upon.

Yours sincerely,

Charles Elton

P.S. Please send future communications to the address given above as I have left Kidlington and now I live in Oxford.

Professor Vito Volterra,
via in Lucina 17, Roma, Italy

7.9 Elton to Volterra, Oxford, 12.10.1936 *Cat. no. 174*

(Same heading of letter no. 7.7)

12th October, 1936

Dear Professor Volterra,

I beg to thank you very much for sending me the valuable set of papers, which I have placed in our library, and also for the further copies of those already received. I shall be publishing shortly the Annual Report of our research work and shall have pleasure in sending you a copy of it when it appears.

Yours sincerely,

Charles Elton

Professor Vito Volterra,
via in Lucina 17, Roma, Italy

7.10 Elton to Volterra, Oxford, 11.2.1937 *Cat. no. 175*

(Same heading of letter no. 7.7)

11th February, 1937

Dear Dr. Volterra,

I wish to thank you very much for sending me your recent paper on "Principes de biologie mathématique"[10].

Yours sincerely,

Charles Elton

Professor Vito Volterra,
via in Lucina 17, Roma, Italy

[10] Volterra 1937a.

8 Letter between Karl Friederichs and Vito Volterra

It is very difficult to find any detailed information concerning the life and scientific career of Karl Paul Theodor Friederichs, even though he is considered to be a zoologist and entomologist of great worth. He was born on 13 October 1878 at Wismar. He published a number of zoological observations (including Friederichs 1906, 1918), and became Privatdozent at the University of Rostock in 1919 and, in 1921, was appointed extraordinary professor at the same University, where he directed the Entomological Seminar in 1927. He carried on research in the Dutch Colonies from 1921 to 1924, and in 1928–1929 was guest professor at the University of Minnesota, Minneapolis. He was appointed professor at the University of Posen in 1940 and later transferred to the University of Göttingen where he remained until 1958. In the same year he was named professor emeritus. He died in 1969.

Friederichs received a number of awards including the Karl Escherich Medaille of the Deutsche Gesellschaft für angewandte Entomologie in 1954. In 1935 he was appointed member of the Deutsche Akademie der Naturforscher (Leopoldina), Halle (Saale), of the Zoologisch-Botanische Gesellschaft Wien (1960) and of the Deutsche Entomologische Gesellschaft (1963).

His research was focused on the field of entomology and ecology, although he also showed considerable interest in general topics of the philosophy of nature. He wrote some 140 scientific articles and several books, including in particular, *Die Grundfragen und Gesetzmässigkeiten der land- und forstwirtschaftlichen Zoologie insbesondere der Entomologie* (Friederichs 1930b), *Ökologie als Wissenschaft von der Natur* (Friederichs 1937) and *Die Selbstgestaltung des Lebendigen* (Friederichs 1955). In his treatment of the problem of the biological equilibrium of insects he adopted a position that supported Volterra's approach, in opposition to the criticism advanced by F. S. Bodenheimer (see Introduction).

* * *

8.1 Friederichs to Volterra, Rostock, 10.10.1936 *Cat. no. 176*[1]

ENTOMOLOGISCHES SEMINAR
ROSTOCK (MECKLBG.)
BLÜCHERPLATZ

Herrn Prof Dr.
Vito Volterra
Hotel Lutetia
Boulevard Raspail
Paris

Rostock, d. 10. Oktober 1936

Sehr geehrter Herr Professor und Kollege,

Ich habe Ihnen noch zu danken für das Buch, das Sie mir vor längerer Zeit zu schicken die Güte hatten; aber ich erfahr Ihre genaue Adresse erst durch Ihre neue Sendung, für die ich Ihnen ebenfalls bestens danke. Ich selbst habe in den letzen Jahren Veröffentlichungen die Sie interessieren könnten, leider nicht herausgebracht[2]. Sobald das einmal wieder der Fall ist, werde ich nicht verfehlen, sie Ihnen zu senden.

In vorzüglicher Hochachtung ergebenst

K. Friederichs

---

[1] Postcard.

[2] Almost certainly a reference to FRIEDERICHS 1929, 1930a, b.

9 Letters between Georgii F. Gause and Vito Volterra

Georgii Frantsevich Gause was born in Moscow on 27 December 1910. He attended the Faculty of Biological Sciences of the University of Moscow and was a pupil of the biologist Vladimir V. Alpatov, who transmitted to him his interest in Raymond Pearl's research on *Drosophila* and on quantitative and experimental ecology in general. In the late 1920s Gause worked intensely in this field (see GAUSE 1930, 1931), in particular on the topic of animal density as a function of habitat. He studied the implications of Pearl's logistics law, publishing a book on the topic in collaboration with Alpatov (GAUSE, ALPATOV 1931). He graduated in the same year. In this period Gause extended his interest to general population theory, collecting all the available information on the most recent research and consequently coming into contact with Volterra's work. Gause's interests were also stimulated by the holistic approach to ecology promoted by the socalled "Moscow school", the leader of which was Vladimir I. Vernadsky, the author of the widely known book, *The Biosphere* (VERNADSKY 1929). Gause also had the occasion in 1933 to work in Vernardsky's laboratory. Volterra's work was well known in this environment and Gause set himself the research goal of experimentally verifying the mathematical theory governing the struggle for life. This was the topic treated in a book published in 1935 (GAUSE 1935) in the series directed by Georges Teissier for the Paris publisher Hermann, that was preceded by another book devoted to a general exposition of the theory of the struggle for existence (GAUSE 1934a). Gause also takes credit for the general statement of the "principle of competitive exclusion" according to which, if two species coexist in the same ecological niche, the species with the greater capacity for environmental adaptation will tend to prevail and to eliminate the other.

Gause continued to carry out this research at the Faculty of Biology of the University of Moscow until 1940, first as lecturer in experimental biology and then, from 1936 on, as "research professor". That year the situation in the biological circles of the University of Moscow was worsened by the rising influence of the Lysenko school, which championed the principle of the inheritance of acquired characters and came into conflict with the evolutionist view that had widely been accepted in those circles. The University of Moscow was in fact a centre of opposition to lysenkoism and, in the late 1940s, suffered a purge. In 1949 it was rumoured that also Gause was one of the scientists purged and sacked from key posts. In actual fact, this was not true, but only because Gause had shifted his primary research interests and was now investigating topics with direct applications in microbiology, in particular in the field of antibiotics. For several more years he continued to publish in the field of ecology (GAUSE

1941, 1947, GAUSE, SMARAGDOVA, ALPATOV 1942, GAUSE, ALPATOV 1945) but then abruptly ceased all interest in these topics and in theoretical biology in general.

In 1940 Gause transferred to the Academy of Sciences of the USSR, heading the laboratory for procurement and cultivation of producers of Research Institute for New Antibiotics. He then became head of the antibiotics procurement laboratory of the Institute of Pharmacology and Experimental Chemotherapy and Chemoprophylaxis of the Academy of Medical Sciences of the USSR (of which he was appointed corresponding member). In 1946 he was awarded the Stalin Prize for his research on antibiotics. He also travelled abroad: in 1956 to Italy for an immunológy congress and, between 1957 and 1959, to the U.S.A., where he lectured on antibiotics at the Universities of Yale, Cornell, Chicago and New York. He was the author of numerous articles and books on antibiotics and on the ways and means of discovering new ones (GAUSE 1960, 1983). In 1949, in the USSR, he published *Lectures on Antibiotics* which was reprinted many times. He died in Moscow in 1989.

* * *

9.1 Gause to Volterra, Moscow, 12.11.1932 *Cat. no. 177*[1]

Dr. G. F. GAUSE
DOCENT
Of Experimental Biology
Malaia Bronnaia 12, kv. 33,
Moscow, U.S.S.R.

12th Nov. 1932

Dear Professor Volterra,

It was a great pleasure to get your letter of November forth as well as a copy of your recent article. I am continuing now the experimental analysis of the struggle for existence and several papers are now ready for publication. One among them deals with the experimental study of carnivorism, and the integro-differential equations obtained are based on the principles discovered by you[2]. The experimental research showed, however, that this case is a complicated one, and the quantitative analysis may be given at present for the simplest conditions only.

I would wish to point out that the simultaneous logistic differential equations which have been obtained by me in the experiments with yeasts and are in agreement with the observed data, have been solved only with the aid of the

[1] It also contains an Italian translation of the letter, probably done by one of Volterra's collaborators.

[2] See G. F. GAUSE 1932, 1934b, c.

approximate numerical method. At the same time there is now independent "chemical" confirmation of their correctness, and it is very important, therefore, to analyse them more closely. I am working in the field of experimental biology, and I am not qualified enough to analyse more closely the mathematical part of the problem and it would be very interesting if it will be possible for you to investigate these equations in the future, and to find their solution in the general form.

I hope you will favor me with any papers that you may publish in the future

Yours very sincerely

G. F. Gause

Caro Professore Volterra

Ho ricevuto col maggior piacere la sua lettera del 4 novembre e una copia del suo recente articolo. Io continuo adesso la analisi sperimentale per la lotta per la esistenza e diversi lavori sono pronti per essere pubblicati. Fra questi uno si occupa dello studio sperimentale del carnivorismo, e le equazioni integro-differenziali ottenute sono basate sui principi scoperti da lei. La ricerca sperimentale ha tuttavia mostrato che questo è un caso complicato e che la analisi quantitativa nel momento presente può essere data soltanto per le condizioni più semplici.

Desidero rilevare che le equazioni differenziali simultanee logistiche che ho ottenuto negli esperimenti con i lieviti e sono in accordo con i dati osservati, sono state risolte soltanto con l'aiuto del metodo numerico approssimato. Nello stesso tempo ora si ha una indipendente conferma "chimica" della loro esattezza ed è perciò molto importante di analizzarle più esaurientemente. Io lavoro in biologia sperimentale e non sono abbastanza competente per analizzare più esaurientemente la parte matematica del problema; sarebbe molto interessante se le fosse possibile di occuparsi nel futuro di queste equazioni e di trovare la loro soluzione nella forma generale.

Spero che vorrà favorirmi con i lavori che pubblicherà nel futuro.

9.2 Gause to Volterra, Moscow, 12.10.1933 *Cat. no. 178*

(Same heading of letter no. 9.1)

12 X 1933

Monsieur le Professeur!

I have recently written a book entitled "The Struggle for Existence"[3], and I am sending it now for publication in the United States of America. In this book I described various experimental verifications of your mathematical study of the struggle for life, and most of the experimental data very well harmonize with your theory.

[3] Gause 1934a. See also the following letter (no. 9.3, Cat. no. 179).

I am now writing to ask about your more recent publications on this subject. I carefully mentioned in the book all your articles beginning with 1926, but it is possible that some of your more recent papers (published since your french book of 1931) have been overlooked. I should therefore be very grateful if you could send me a complete set of all your publications since 1931. I remain

very sincerely yours

G. F. Gause

9.3 Letter of Gause to Volterra, Moscow, 25.11.1934 *Cat. no. 179*

(Same heading of letter no. 9.1)

25 November, 1934

Dear Professor Volterra,

I am delighted to be able to report you that I have recently confirmed experimentally several of your laws of periodic fluctuations in populations. These exist, evidently, when the condition of your equations are realised, and disappears as soon as the biological conditions of interaction are different.

These experiments are described in my recent book "The Struggle for Existence" which is being published by The Williams and Wilkins Company of Baltimore, U.S.A. The date of the issue of the book will be December 15th. Unfortunately I shall receive no author's copies and shall have no opportunity to send you the book.

Several my more recent experiments together with mathematics I am publishing in the form a small book in Paris[4]. As far as editors promise to give me the author's copies I shall be very glad indeed to send you this small book as it appears, probably, in the early Summer of 1935.

With best regards

yours truly

G. F. Gause

9.4 Letter of Gause to Volterra, Moscow, 1.8.1935 *Cat. no. 180*

(Same heading of letter no. 9.1)

August 1st, 1935

Dear Professor Volterra

I wish to thank you very much for sending me a copy of your excellent recent book on biological associations[5]. I want to express also my appreciation of the very favorable mention which you made of my researches.

---

[4] GAUSE 1935.

[5] V. VOLTERRA, U. D'ANCONA 1935.

I am publishing in the same series, Exposés de Biométrie, a small book describing the results of some of my new experiments[6]. It will be a pleasure to send you a copy of it as well as some new papers which I expect to get out shortly.

Very sicerely yours

G. F. Gause

9.5 Letter of Gause to Volterra, Moscow, 23.12.1935 *Cat. no. 181*

(Same heading of letter no. 9.1)

23 XII 1935

Dear Professor Volterra:

I am taking liberty of sending you under separate cover my recent french monograph on biological associations. I should greatly appreciate it if you could review it, together with my American monograph, in Scientia or elsewhere[7]

I remain

Very sincerely yours

G. F. Gause

9.6 Letter of Gause to Volterra, Moscow, 2.10.1936 Cat. no. 182

Dr. G. F. GAUSE
RESEARCH PROFESSOR
Of Experimental Biology
Malaia Bronnaia 12, kv. 33,
Moscow, U.S.S.R.

2 X 1936

Dear Professor Volterra:

I wish to thank you very much for your kindness in forwarding me the reprints of your interesting publications. I appreciate very much your favourable mention of my recent investigations in the field of mathematical biology.

With best regards

Very sincerely yours

G. F. Gause

---

[6] GAUSE 1935.

[7] The review was actually published in *Scientia* (VOLTERRA 1936f).

10 Letters between Samuel A. Graham and Vito Volterra

Samuel Alexander Graham was born in Salisbury (Maryland, USA) on 22 March 1891. He studied at the University of Minnesota where he obtained his Bachelor of Science in Forestry in 1914 and a Ph.D. on forest ecology topics in 1921. In the meantime, in 1916, he obtained a Master of Forestry at Cornell University. He remained at the University of Minnesota until 1927 as lecturer and supervisor of an entomological field station of the U.S. Department of Agriculture. In those years he collaborated with Chapman, who played an important role in his training and of whom he wrote a biographical note after the latter's death (GRAHAM 1941).

In 1927 he transferred to the University of Michigan (Ann Arbor) as Associate Professor of Entomology in the new School of Forestry and Conservation. In 1929 he became Professor of Economic Zoology and Research Associate in the Division of Insects of the Museum of Zoology. He remained at the University of Michigan until 1961, the year he retired, and on this occasion received the title of Professor Emeritus of Economic Zoology. In the years that followed he participated in the visiting scientists program of the Society of American Foresters and ran the summer field work in insect archeology at Mesa Verde, Colo. He died on 21 September 1967 in the hospital of the University of Michigan, Ann Arbor.

He was a fellow of the American Association for the Advancement of Science, of the Entomological Society of America and of the Wildlife Society, as well as a member of the committee on population dynamics of the International Union of Forest Research Organizations.

Graham was one of the leading figures in the field of the forest area studies and, in particular, of forest entomology. In this latter field, he dedicated particular attention to the issue of forest defoliation by insect fauna. In 1929, he published an important book on the topic which emphasized the ecological aspects of insects and concentrated on interrelationships between insects, protection, and forest management (GRAHAM 1929). It was extremely successful and underwent many re-editions with contributions by other researchers. He also made an important contribution to population dynamics and to the study of insect ecology in a book (GRAHAM 1956b) in which he stated the "law of natural compensations". According to this law if a species is to remain in relatively constant numbers over successive generations in spite of changes in the strength of different factors of environmental resistance, it is thus apparent that the changes must be compensating. In 1963 he published a book in collaboration in which he collected forty years of his research on the aspens of the Great Lakes (GRAHAM, HARRISON,

WESTELL 1963). Graham was involved in a number of classical studies on the biology and ecology of major forest pests (GRAHAM 1935, 1937, 1943, 1956a) and was one of the first to bring the principles of ecology in the field of management and resource development. He always considered research in the field as a primary factor in research and went as far as to publish a handbook of behaviour in wild environments (GRAHAM, O'ROURKE 1943).

* * *

10.1 Graham to Volterra, Ann Arbor, 10.1.1928 *Cat. no. 183*

UNIVERSITY OF MICHIGAN
SCHOOL OF FORESTRY AND CONSERVATION
ANN ARBOR

Dr. Vito Volterra, *January 10, 1928*
Via in Lucina, 17
Roma, Italy.

Dear sir:

I wish to thank you very much for sending me copy of your paper "Variation and Fluctuations of the Number of Individual Animal Species Living Together"[1].

This is an extremely interesting method of attack on this problem and is, I believe, a method that has a wide possible application. We have only recently ventured into the field of quantitative biology and all of us, I believe, are feeling about for satisfactory means of expression. This work of yours I feel is a real contribution to the subject and will be of great value to biologists.

Very truly yours,
S. A. Graham,
Associate Professor.

10.2 Graham to Volterra, Ann Arbor, 1.11.1928 *Cat. no. 184*

(Same heading of letter no. 10.1)

Professor Vito Volterra, *November 1, 1928*
Via in Lucina, 17
Roma, Italia

Dear sir:

For some time I have been interested in quantitative studies of forest insects and have frequently been faced with mathematical problems in connection with this work.

---

[1] VOLTERRA 1926a, 1927d.

A short time ago, Dr. Chapman of the University of Minnesota showed me two papers of yours discussing the variations and fluctuations of the number of individuals in animal species living together. I was very much interested in these papers and I should like very much indeed to have reprints of them. If you have any available for distribution I should be very thankful to receive copies. In return I shall be very glad to send you any of my entomological papers in which you may be interested[2].

Yours very truly,

S. A. Graham,
Associate Professor

10.3 Graham to Volterra, Ann Arbor, 30.1.1929 *Cat. no. 185*

(Same heading of letter no. 10.1)

Professor Vito Volterra,
Via in Lucina, 17
Roma

January 30, 1929

Dear sir:

Your letter of the second of January has been received and I shall be very glad indeed to give you what information I have that may be of help to you in your work on fluctuations. I am afraid, however, that the material is too fragmentary to be of much use to you. I know of no quantitative studies that have been made in America concerning the results of competition between two species contending for the same food although we all realize that this kind of competition does occur all about us. For instance we know that such barkbeetles as Dendroctonus ponderosae may, in some instances, be reduced in numbers as a result of competition with certain Cerambycidae. The Dendroctonus beetles by killing trees provide an abundant supply of food for those cerambycids which feed on the inner bark and cambium of injured and recently killed trees but cannot live successfully in healthy trees. These cerambycids come into direct competition with the barkbeetles since they are both contending for the same food. Under the favorable food conditions brought about by a Dendroctonus outbreak, the cerambycids will increase in numbers until they may consume such a large proportion of the cambium and inner bark that the barkbeetle broods will find insufficient food for development, with the result that many of the larvae will perish. Dr. M. W. Blackman of Syracuse University, Syracuse, New York, has studied this interrelationship on the Kaibab Forest, Utah. He has not as yet published his materials but I am sure that you can obtain information concerning his results by writing to him.

Competition between parasites is well illustrated by the work of entomologists of the U. S. Bureau of Entomology in Hawaii, especially that of Pemberton and

---

[2] The contents of this letter seem to suggest that it was written before letter no. 10.1 (Cat. no. 183). However, the (typed) date is unequivocal.

Willard concerned with the interrelation of Diachasma tryoni and Opius humilis both parasites that were introduced into Hawaii to help in the control of the Mediterranean fruit fly, Ceratisis capitata. This work appeared in a series of articles in the Journal of Agricultural Research, Government Printing Office, Washington, D. C. One of these articles entitled Interrelations of Fruit-fly Parasites in Hawaii, Vol. XII, No. 5, pp. 285 to 295, is especially interesting. It is pointed out that although Opius humilis is a more effective parasite than Diachasma tryoni the former is unable to compete successfully with the latter when double parasitism occurs, with the results that its effectiveness is much reduced. The end result is a reduction in the total parasitism to a value equal to that of the inferior parasite. Later articles show that Opius humilis has been reduced to a point where it is a negligible factor in fruit-fly control.

When we come to consider the association of two species, one of which feeds upon the other, we find more material available. For instance there is J. D. Tothill's rather extensive paper entitled "The Natural Control of the Fall Webworn (Hyphantria cunea) in Canada", Bul. 3, new series (technical) published by the Entomological Branch, Dominion of Canada, Department of Agriculture, Ottawa, Ontario, Canada. In this paper Tothill discusses the rise and fall of a webworm outbreak and presents quantitative data concerning the parasitic and predatory agents concerned.

Another possible source of information concerning the interaction of parasitic and predatory organisms upon a host is the Gypsy Moth Laboratory at Melrose Highlands, Massachusetts. At that place representatives of the Bureau of Entomology have been studying the rise and fall of gypsy moth numbers for more than twenty years. Also the amount of parasitic activity has been recorded. Up to the present little data of a quantitative nature have been published but I am sure there must be a great deal available. The number of egg masses per acre as determined in their scouting work could be used as an index of abundance of the moths from year to year and the percentage of eggs, larvae, and pupae parasitized by the various parasitic species together with a record of the number of individuals examined each year would serve as an index of parasite fluctuation. I should suggest that you write to Mr. C. W. Collins, who is now in charge of the laboratory, requesting the information you desire to obtain. I am sure, if Mr. Collins understands the purpose for which you expect to use the data, thet he will be more than glad to place them at your disposal.

I am afraid that my own material will not be of any use to you in its present stage of development. Most of my work in the past has been concerned with the influence of physical environmental factors upon the relative abundance of certain species and only in the last few years have I been collecting data bearing upon the interrelation of biotic factors.

One series of experiments, started under my direction, I am watching with great interest. There should in the course of a few more seasons produce valuable data regarding the influence of parasites upon a host. These experiments deal with the interrelations of pine tipmoth Rhyacionia frustrana and its parasites in an artificially estabilished forest in the naturally treeless sand-hill region of Nebraska. Each year since 1924 we have recorded the number of tips killed by this moth on a series of sample plots. This will serve as an index of moth abundance. We have also records of the numbers of parasites of a large series of species in samples taken from these plots. During the second year of the work a parasite that appears to be much more effective than any naturally present was introduced. Since that time this species has increased tremendously, raising the percentage of parasitism on tipmoth larvae in western yellow pine, Pinus ponderosa, from between 10 and 20 per cent to about 60 per cent. The condition of equilibrium has not as yet been reached and so we dont know what the end result will be, but whatever it is the information collected during the course of the study is bound to be interesting.

I have also been studying the larch sawfly quantitatively but the data concerning its interrelations with other organisms are still in the formative stage and are not adequate to serve as basis for reliable conclusions.

Although I am afraid they will not be of much use in connection with the immediate problem under discussion I am sending to you a few of my reprints in the hope that you may find something of interest in them.

With kindest regards, I am

Very truly yours

S. A. Graham

Associate Professor.

11 Letters between William O. Kermack and Vito Volterra

William Ogilvy Kermack was born in Kirriemuir (Angus, Scotland) on 26 April 1898. His mother died in 1904 and he was brought up by his father's sister. After attending the local Webster Seminary he was admitted to Aberdeen University in 1914, obtaining his Master of Arts in mathematics, natural philosophy and chemistry in 1918. For the next six years he served in the Royal Air Force. From 1919 to 1921 he worked for the Dyson Perrins Laboratory where he developed a strong interest in the chemistry of heterocyclic compounds, then embarking on a research career in the chemistry department of the Royal College of Physicians Laboratory in Edinburgh. But at the age of 26, he was completely blinded in a laboratory accident and it was only by virtue of his extraordinary will power that he succeeded in continuing in his scientific research career. In this he was helped by his assistants, who compensated for his inability to read – he never actually learned braille – and write, as well as by his exceptional memory.

In his work at the Royal College of Physicians Laboratory he supervised the research activities of the chemistry students and pursued his research interests. He developed methods for the synthesis of heterocyclic compounds with anti-malarial properties such mepacrine, which was produced a large scale in England and used in many countries during World War II. During the period 1927–40, Kermack, in collaboration with A. G. McKendrick, was interested in the mathematical theory of the spread of epidemics and, in particular, in the condition in which an infectious disease reaches a stationary endemic state. The basic equations governing the spread of an epidemic and the Kermack-McKendrick "threshold theorem" are considered starting points of modern mathematical epidemiology.

In 1949 Kermack was appointed as the first MacLeod-Smith professor of biological chemistry at Aberdeen University, a post that he kept until 1968. During this period, he participated in university activities to the full extent of his energy, continuing his research on anti-malarial drugs and occupying also the post of Dean of the Faculty of Science from 1961 to 1964.

Kermack was elected Fellow of the Royal Society of Edinburgh in 1924 and Fellow of the Royal Society in 1944. He was awarded the Freeland Barbour prize by the Edinburgh College of Physicians in 1925 and the Macdougall Brisbane prize of the Royal Society Edinburgh in 1929. He retired to private life in 1968 although he continued to take an interest in scientific matters until his death in Aberdeen on 20 July 1970[1].

---

[1] See Grant 1961–1970 and Davidson, Yates, McCrea 1971.

11.1 Kermack to Volterra, Edinburgh, 14.3.1938 *Cat. no. 186*

2, Forrest Road, *14th March 1938*
Edinburgh

Dear Professor Volterra,

Very many thanks for your three reprints which I have just received. I have not yet had time to read them but I am sure they will be very interesting and useful. I enclose one or two reprints of ours which you may care to have.

Yours faithfully,

W. O. Kermack

11.2 Letter of Volterra to Kermack, 11.9.1938 *Cat. no. 370*[2]

Dear Professor W. O. Kermack

Sono molto grato delle importanti memorie che si è compiaciuta inviarmi. Esse mi saranno molto utili e le studierò con gran piacere. La ringrazio della menzione che Ella fa delle mie formule e delle mie ricerche.

I remain

I am very grateful to you for your reprints you have kindly sent to me. I am sure they will be very important and useful. I shall read them with great interest.

I thank you for having made mention of my formulas and my works.

Mon cher Collègue

Merci de votre aimable lettre et de la Communication du Colloquium de St. Andrews.

La notice m'en était déja parvenue et je voudrais assister très volontiers à ce Colloquium ainsi qu'à la réunion d'Edinburgh à la même époque, si mes conditions de santé me le permettaient.

Veuillez agréer, mon cher Collègue, l'expression de mon sincère dévouement.

[2] Drafts in french, italian and english.

12 Letters between Vladimir A. Kostitzin and Vito Volterra

We still know little about the life of Vladimir Aleksandrovich Kostitzin. He was apparently born in 1882, although the date of his death is uncertain: it was probably around 1963 since – as revealed by the correspondence – he was in the habit of sending a greeting card to Virginia Volterra every year, a habit that was interrupted after the last card sent on 24 December 1962. Some biographical data were collected by Francesco M. Scudo, also on the basis of evidence provided by Luisa D'Ancona Volterra, Claudine Petit, Marcel Prenant and Yuri Shmiglevsky (see SCUDO, ZIEGLER 1976).

He studied in the city of Smolensk, although nothing is known about his higher education which was however clearly oriented toward geophysics, chemistry and mathematics. It is said that he participated in the revolutionary uprisings that took place in 1905 in Russia and, as a consequence of this participation, was exiled. He later returned to his country and was called up for military service in World War I. He took part in the October Revolution, in his military capacity. It was during the revolutionary period that his rapid scientific career took off, first as university professor of mathematics for chemists, and then as Director of the Moscow Geophysics Institute (during the 1920s).

As we learn from Poggendorf's biographical dictionary (*J. C. Poggendorffs biografisch-literarisches Handwörterbuch fur Mathematik, Astronomie, Physik mit Geophysik, Chemie und Kristallographie und verwandte Wissensgebiete*, Bd. VI, 1923 bis 1931, II Teil, Berlin, Verlag Chemie, p. 1388), much of Kostizin's work published in Russia prior to 1926 was dedicated to the theory of integro-differential equations, and he was apparently quite familiar with Volterra's work in this field. It also seems that he collaborated with the eminent Russian geochemist Vladimir Vernadsky (1863–1945), who championed a holistic and integrated view of the biosphere, according to which the natural processes of the planet must be situated in a global context of the transformation of matter and of biogeochemical cycles, lumping together phenomena related to the atmosphere, the earth's crust, the biosphere and the presence of man. The University of Moscow cultivated a biological research tradition that was strongly ecologically oriented and Darwin's ideas, cultivated in particular by Serguei Chetverikov (1880–1959), exerted a strong influence. This was the scientific environment in which, in addition to Kostitzin, also Georgii F. Gause was trained. The latter was also interested in Volterra's research and maintained close relations with international research circles. This milieu was criticized for these relations considered as unpatriotic activities and this ultimately led to the dissolution of the Moscow school.

It is not clear why Kostitzin broke with the Soviet regime, although it is clear (and supported by some evidence) that his faith in socialism remained untarnished throughout his life, while it is not clear to which opposition current to the left of the regime he belonged. Some mention of this situation is contained in letter no. 12.4 in which he declares his inability to do anything for the dissemination of Volterra's work in Russia as he was "sous l'interdit" in that country. In any case, in the 1920s (probably after 1926) he moved into exile in Paris, together with his wife Julie, who was a specialist in parasitology and subsequently worked in the comparative anatomy and histology laboratory of the Sorbonne. She died in 1950 (as emerges from the correspondence).

Academically, Kostitzin encountered considerable difficulty in finding a job in France. He did fill-in jobs and only in 1939 did he succeed in obtaining a permanent position as *chargé de recherches* at the Conseil national de recherches scientifiques which was set up that very year. In 1942 he was awarded the Montyon Prize of the Académie des Sciences for statistics. The outbreak of World War II marked the beginning of a dramatic period for him. After Germany declared war on the Soviet Union, he was arrested and interned in the German civilian prison camp of Compiègne, where he remained from 22 June 1941 to 23 March 1943. After his release from this camp, he joined the French resistance together with his wife Julie. He was again arrested by the Gestapo on 29 January 1944, but as luck would have it, succeeded in escaping and spending the rest of the war in hiding.

His wife's death in 1950 was a serious blow to him and marked him for the rest of his life. Until his death in 1963 he was unable to regain to recover the optimism, dynamism and joyfulness that had always been distinctive traits of his character.

Kostitzin combined the scientific and cultural tradition he had absorbed from the Moscow environment with the elements he later acquired in his relations with Paris scientific circles. In particular, he entertained fertile relations with distinguished French mathematicians like Paul Montel, Jacques Hadamard and Émile Borel as well as with scientists deeply interested in theoretical biology, such as Philippe L'Héritier and Georges Teissier, deputy director of the Roscoff Zoological Station. In the 1930s he began to take an interest in problems of theoretical biology and, in particular, in the application to it of integral and integro-differential equations. Kostitzin most probably met Volterra at one of the lectures given by the latter at the Institut Henri Poincaré. What is certain is that, starting in those years, Kostitzin began to consider himself a "pupil" of Volterra with whom he established increasingly close scientific relations.

His early work in the field of theoretical biology included statistics on hermit crab infestations (in collaboration with his wife, J. KOSTITZIN, V. A. KOSTITZIN 1931), a geological application for differential equations (V. A. KOSTITZIN 1932) and asymptotic solutions of the differential equations of the theory of the growth of organisms (V. A. KOSTITZIN 1933). In 1933, Georges Teissier started up the new series of publications entitled "Exposés de biométrie et statistique biologique", to which Kostitzin contributed two extremely interesting books (V. A. KOSTITZIN 1934, 1935a). In this way, Kostitzin became a full participant in the main fields of modern ecology – the dynamics of population and biological associations and the study of the biosphere and biogeochemical cycles - following an original interdisciplinary approach, as his background allowed him - in which he combined a solid mathematical training with biological, chemical and geological knowledge. The period from 1933 to 1940 was that of the most intense collaboration with Volterra, as emerges also in his correspondence (and as has been described in the introductory essay). It was marked by the publication of many works, including the significant book *Biologie mathématique* (V. A. KOSTITZIN 1937b), which was translated also into English (V. A. KOSTITZIN 1939a). After the war, Kostitzin ceased almost completely to take any interest in biomathematics (except for one work, V. A. KOSTITZIN 1956)[1].

* * *

12.1 Kostitzin to Volterra, Paris, 16.3.1933 *Cat. no. 187*

Paris, 16 mars 1933

Monsieur et cher maître,

je veux tout d'abord vous remercier de l'envoi de vos mémoires dont la lecture est toujours si attrayante et si profitable pour moi. Je veux vous prier ensuite de bien vouloir présenter la note ci-jointe à l'Académie des Sciences de Paris ou à la Accademia dei Lincei[2]. Cette note sera suivie d'autres publications sur le développement des animaux. En ce qui concerne le développement des organes il est très tentant de faire des analogies entre lui et celui des espèces se trouvant soit en cohabitation, soit en concurrence. Je crois pourtant que ces analogies sont plutot dangereuses et que malgré la similitude apparente il y a des différences profondes entre ces problèmes.

Veuillez agréer, cher maître, l'expression de mes meilleurs sentiments.

V. A. Kostitzin

3, rue Bellier-Dedouvre, Paris (13)

12.2 Kostitzin to Volterra, Paris, 6.11.1933 *Cat. no. 188*

Paris, 6 novembre 1933

Monsieur et cher Maître,

mon fascicule vient enfin de paraitre, et je suis heureux de pouvoir vous l'envoyer en même temps que celui de M. G. Teissier[3]. Je me propose dans le

---

[1] See ISRAEL, MILLÁN GASCA 1993 and SCUDO, ZIEGLER 1976, 1978 (the latter contains also a selection of Kostitzin writings).

[2] V. A. KOSTITZIN 1933.

[3] Kostitzin is referring to TEISSIER 1934 and to V. A. KOSTITZIN 1934.

fascicule suivant[4] de traiter les processus cycliques dans la nature. Je ne peux pas dire que ce travail m'ait donné entière satisfaction: la série de M. Teissier étant destinée aux biologistes j'ai été obligé de réduire la partie mathématique sans pouvoir naturellement l'éliminer: ce sont les démonstrations qui ont le plus souffert et je crains bien de ne satisfaire pleinement aucune de deux catégories de lecteurs (biologistes et mathématiciens). Sans compter qu'il y a des biologistes qui partagent le point de vue du professeur Bouasse[5] sur l'inutilité totale de la méthode mathématique en sciences physiques et naturelles.

J'étudie actuellement le problème de l'hystérésis élastique en me servant de quelques résultats expérimentaux obtenus au laboratoire de M. Bouasse. Qu'est ce que je vais prendre! A ce propos serait-il possible de publier ce mémoire en Italie?

M. G. Teissier me charge de renouveler auprès de vous sa demande de collaboration. Il serait infiniment content si vous consentiez à écrire un fascicule sur un des problèmes de biologie mathématique qui vous intéressent. M. Freymann, directeur de la maison Hermann et C promet la rapide parution de votre ouvrage.

Et maintenant, permettez moi, cher Maître, d'attirer votre attention sur mon problème personnel. J'ai derrière moi, ainsi que vous le savez peut-être, une experience scientifique, pédagogique et administrative assez importante; j'aime la science et le travail scientifique par dessus toutes autres choses et je peux fournir un rendement utile. Or, depuis mon expatriation je n'arrive pas (pour des raisons variées) à sortir de l'impasse matérielle qui m'oblige d'executer tous les travaux qui me tombent sous la main et de me considérer encore comme très heureux quand quelque possibilité de ce genre se présente; or, elle ne se présente pas tous les jours.

Mes désirs ne sont pas excessifs: je veux pouvoir vivre et travailler-exploiter rationnellement mon savoir et mon énergie. Et je m'adresse à vous en vous demandant de bien vouloir penser à moi si une occasion se présente qui permettrait de ne pas continuer à gacher inutilement des années précieuses.

Veuillez agréer, monsieur et cher Maître, l'expression de mes meilleurs sentiments et de ma considération la plus distinguée.

(same address of letter no. 12.1) V. A. Kostitzin

12.3 Kostitzin to Volterra, Paris, 14.12.1933 *Cat. no. 189*

Paris, 14 décembre 1933

Monsieur et cher maître,

j'ai été très heureux de pouvoir transmettre à M. G. Teissier votre consentement à écrire un fascicule pour sa collection. En ce qui concerne

[4] V. A. Kostitzin 1935a.

[5] Here Kostitzin refers to the French physicist P. M. Henri Bouasse (1866–1953), a researcher in the field of the theory of elasticity and thermodynamics.

l'exécution, puis je vous être utile à quelque chose. Si la distance entre Paris et Rome ne vous effraie pas trop, disposez de moi: cela me fera le plus grand plaisir.

Je ne manquerai pas dans mon prochain fascicule ou mémoire de prendre en considération les lois des fluctuations biologiques. Il me semble d'ailleurs que le mode de raisonnement conforme à ces lois est employé tout le long de mon fascicule paru. Malheureusement, la place me manquant énormement il m'a été impossible de donner tous les développements, toutes les démonstrations, toutes les indications nécessaires. Je n'ai utilisé qu'un tiers du matériel dont je dispose. Mon fascicule n'est qu'un resumé de trois mémoires non publiés.

Dans mon mémoire sur les phénomènes élastiques héréditaires la théorie héréditaire est confrontée avec quelques faits expérimentaux. Ma conclusion est que cette théorie + le principe du cycle fermé permet d'expliquer qualitativement un ensemble assez considérable de faits, mais l'accord quantitatif n'y est pas. Ou peut remédier à ce défaut en admettant que le noyau $K(t,u)$ possède quelques propriétés analogues à celles du noyau $K(t-u)$. Je crois cependant que la dépendance du coefficient d'hérédité du temps "absolu" et non du temps "relatif" $t-u$ contient quelque chose de choquant pour l'esprit. Peut-être faudrait-il considérer une forme non linéaire $K(t-u,$ déformation) ou $K(t-u,$ effort). Le mémoire est écrit en français et contient 20–25 pages du format 4 . Naturellement je lis assez couramment l'italien, mais je ne me risquerais pas à malmener votre belle langue.

Avez vous vu l'exposé très long, très detaillé mais très rebarbatif, de ces problèmes par M. Fromm (Handbuch der Phys. Mechanik, Bd. IV, p. 436–550 1931)[6]?

Je vous remercie bien sincèrement des bonnes intentions manifestées à mon égard. Je ne sais que trop bien combien il est difficile de trouver une situation pour un intellectuel hors de son pays. Mais il y a des périodes bien difficiles à passer et on consentirait à effectuer n'importe quel travail, même chez les éthiopes ou chez les lapons.

Veuillez agréer, monsieur et cher maitre, l'expression de mes meilleurs sentiments.

(same address of letter no. 12.1) V. A. Kostitzin

12.4 Kostitzin to Volterra, Paris, 11.7.1935 *Cat. no. 190*

Paris, 11 Juillet 1935

Monsieur et cher maître,

je suis très flatté de ce que mon mémoire ait trouvé en votre personne un lecteur bienveillant. Je ne manquerais pas d'étudier votre thèse qui contient certainement des choses précieuses pour moi. Dès qu'il me sera possible je m'occuperais de votre dernier mémoire des Annales de l'Institut H. Poincaré, qui est très riche en idées et en suggestions[7].

---

[6] FROMM 1931a, b.

[7] VOLTERRA 1934b.

En ce qui concerne votre fascicule, l'éditeur n'a aucune intention de me le donner, et je serai d'autant plus heureux de l'avoir de vous. Je vous remercie d'avoir pensé à moi.

Je vous enverrai le mien dès sa parution[8]: en principe il devrait déja être paru, mais étant donné la saison très avancée, les vacances proches etc, l'éditeur laisse trainer les choses et, tout en m'amadouant avec des promesses, louche du côté de la rentrée. Je vois très bien son jeu et ne le lui facilite pas. Qu'il ait au moins comme punition quelques inconvenients d'ordre moral!

Je ferai tout mon possible pour contribuer à la propagation de votre livre mais étant sous l'interdit en Russie je ne peux rien faire de ce côté-là. En revanche, je m'adresse à un de mes amis-redacteur scientifique dans une des revues françaises[9] - pour qu'il consacre un article special à votre livre. Il le fera avec d'autant plus de plaisir qu'il a déja publié un article sur votre "Théorie mathématique de la lutte pour la vie" (Le Mois n° 3 1931) et un autre sur mon fascicule "Symbiose et Parasitisme" (Le Mois n° 5 1934). Je verrai aussi du côté de l'Amérique.

J'espère que le séjour à la campagne vous sera profitable et que vos chères malades sont déja rétablies. J'espère aussi que l'année prochaine nous vous verrons et vous entendrons à Paris.

Veuillez agréer, mon cher maitre, l'expression de mes sentiments respectueux.

V. Kostitzin

12.5 Kostitzin to Volterra, Richelieu, 13.8.1935 *Cat. no. 191*

Richelieu, 13 Août 1935

Monsieur et cher Maitre,

je tiens tout d'abord à vous remercier, ainsi que M. d'Ancona, pour l'envoi de votre nouvel et si important ouvrage.

Je me permets de vous communiquer quelques reflexions à propos de l'intoxication d'un milieu par les produits cataboliques d'une population. En regardant de près votre equation (p. 22)

$$\frac{dN(t)}{dt} = N(t)\left[\varepsilon - hN(t) - c\int_0^t N(\tau)\,d\tau\right] \qquad (1)$$

je m'aperçois qu'on peut l'intégrer exactement. Posons, comme vous le faites,

$$n(t) = \int_0^t N(\tau)\,d\tau \qquad (2)$$

l'equation (1) devient

[8] V. A. Kostitzin 1935a.

[9] This is Jacques Delevsky (see letter no. 12.11, Cat. no. 196).

$$n'' = \varepsilon n' - hn'^2 - cn'n \tag{3}$$

ou bien

$$e^{hn}(n'' + hn'^2) = e^{hn}(\varepsilon n' - cnn')$$

ou bien

$$\frac{\mathrm{d}}{\mathrm{d}t}\left(e^{hn}n'\right) = \frac{\mathrm{d}}{\mathrm{d}t}\left[e^{hn}\left(\frac{\varepsilon}{h} + \frac{c}{h^2} - \frac{c}{h}n\right)\right].$$

On a par conséquent

$$M = e^{hn}n' - \left(\frac{\varepsilon}{h} + \frac{c}{h^2}\right)e^{hn} + \frac{c}{h}e^{hn}n \tag{4}$$

ou en tenant compte de ce que

$$n(0) = 0, \;\; n'(0) = N_0$$

$$N_0 - \frac{\varepsilon}{h} - \frac{c}{h^2} = e^{hn}n' - \left(\frac{\varepsilon}{h} + \frac{c}{h^2}\right)e^{hn} + \frac{c}{h}e^{hn}n \;. \tag{5}$$

On en tire ensuite

$$n' = \left(N_0 - \frac{\varepsilon}{h} - \frac{c}{h^2}\right)e^{-hn} + \left(\frac{\varepsilon}{h} + \frac{c}{h^2}\right) - \frac{c}{h}n \tag{6}$$

$$t = \int_0^n \frac{\mathrm{d}s}{\left(N_0 - \frac{\varepsilon}{h} - \frac{c}{h^2}\right)e^{-hs} + \left(\frac{\varepsilon}{h} + \frac{c}{h^2}\right) - \frac{c}{h}s} \;. \tag{7}$$

L' inversion de cette intégrale donne

$$n = \Phi(t) \tag{8}$$

et on trouve finalement

$$N(t) = n' = \Phi'(t) \tag{9}$$

La partie droite de l'expression (6) est positive. Elle s'annule pour une certaine valeur finie de n que nous appelerons n_∞:

$$0 = \left(N_0 - \frac{\varepsilon}{h} - \frac{c}{h^2}\right)e^{-hn} + \left(\frac{\varepsilon}{h} + \frac{c}{h^2}\right) - \frac{c}{h}n_\infty \tag{10}$$

Cette valeur est atteinte pour $t = \infty$ ce qui signifie que N tend vers 0 pour $t \to \infty$.

D'autre part, en supposant

$$N_0 < \frac{\varepsilon}{h} \tag{11}$$

on trouve que N croît pour les valeurs suffisamment petites de t. Il existe une valeur maximale de N que nous appelerons N_m et qui est atteinte au moment t_m. On peut montrer sans peine que

$$N_m = \frac{1}{h}\left[\varepsilon - \frac{c}{h}\log\frac{c+\varepsilon h - N_0 h^2}{c}\right] \qquad (12)$$

$$n_m = \frac{1}{h}\log\frac{c+\varepsilon h - N_0 h^2}{c}\,. \qquad (13)$$

En remplaçant n par cette valeur dans l'équation (7) on obtient t_m. De cette façon le problème se trouve entièrement résolu.

Je séjourne actuellement à Richelieu où il existe une station expérimentale appartenant à la Faculté de Médecine de Paris. Ma femme va y travailler pendant les vacances. Du point de vue touristique le pays n'est pas intéressant, mais la station est située dans un parc magnifique (400 hectares) un peu delabré, mais d'autant plus agréable.

Veuillez agréer, mon cher Maître, l'expression de ma très haute considération ainsi que les meilleurs souhaits pour vos vacances.

V. A. Kostitzin
Station Expérimentale, Richelieu (Indre et Loire)

12.6 Kostitzin to Volterra, Richelieu, 24.8.1935 *Cat. no. 192*

Richelieu, 24 Août 1935

Monsieur et cher maître,

en me conformant à votre conseil j'ai transformé la partie mathématique de ma lettre précédente en une note ci-jointe[10]. Je suis très heureux de l'intérêt que vous temoignez à mes travaux et des encouragements que vous me donnez.

Mon fascicule sur l'évolution de l'atmosphère a paru et je vous l'enverrai des que j'aurai des exemplaires moi-même.

J'écris en ce moment un livre sur la biologie mathématique pour la collection Colin et en l'écrivant je fais des constatations et des observations intéressantes et susceptibles d'être développées ultérieurement[11]. Ainsi, j'ai constaté que le terme intégral (héréditaire dans le sens mathématique) doit être introduit dans les équations de la croissance des organismes. Cela me parait tout indiqué en ce qui concerne le vieillissement, et j'ai réussi à le montrer dans le cas de la croissance embryonnaire. Peut-être dans quelques semaines vous prierai-je de bien vouloir présenter encore une ou deux notes à l'Académie des Sciences.

[10] V. A. Kostitzin 1935b.

[11] V. A. Kostitzin 1937b.

Nous revenons à Paris vers le 1 Octobre, et je serai infiniment content de vous révoir là. J'espère que le repos estival vous est profitable et que les spécialistes de Paris sauront remettre votre gorge en bon état.

Veuillez agréer, mon cher Maître, l'expression de mes sentiments les meilleurs.

V. Kostitzin.

12.7 Kostitzin to Volterra, Richelieu, 19.9.1935 *Cat. no. 193*

Richelieu, 19 Septembre 1935

Monsieur et cher Maître,

j'ai bien reçu et réexpédié les epreuves de ma note. Je vous remercie vivement de l'avoir présentée. Serait-il abusif de ma part de vous demander d'en presenter encore une, cette fois sur le sexe des parasites renfermés dans un hôte?[12] Cette question n'a pas de rapports directs avec l'analyse mathématique et se résout par des considérations probabilitaires élémentaires mais le résultat est net et palpable. Je dois dire que les biologistes, en général, ont une répulsion et une méfiance pour les méthodes de calcul "trop calées" et que cette méfiance est un peu fondée. On trouve d'ailleurs cet état d'esprit au début de la mathématisation de toutes les sciences dites rationnelles.

Je reviens à Paris vers le début d'octobre et j'espère vous y trouver encore.

Veuillez agréer, mon cher Maître, l'assurance de mes meilleurs sentiments et de ma très haute considération.

V. Kostitzin

12.8 Kostitzin to Volterra, Paris, 6.11.1935 *Cat. no. 194*

Paris, 6 novembre 1935

Monsieur et cher Maître,

je regrette infiniment de n'avoir pu vous voir avant votre départ. J'espère que ce séjour à Paris vous aura permis de rétablir votre santé malgré les nombreuses et fatigantes visites que vous avez dû recevoir. Je suis très heureux de votre approbation et de ce que ma note a pu vous être de quelque utilité. Je ne manquerai pas d'entrer en relations avec M. Régnier d'autant plus que M. Hadamard m'a chargé de faire dans son séminaire au Collège de France l'exposé de ce problème mathématico-biologique. Aussi vous serai-je particulièrement reconnaissant si, en prévision de ma conférence, vous vouliez bien me donner quelques détails complémentaires sur vos calculs.

Je vous ai envoyé il y a quelques jours ma dernière note parue aux CR. et mon mémoire "Sur les phénomènes élastiques héréditaires".

---

[12] V. A. Kostitzin 1935c.

Veuillez agréer, mon cher Maître, l'expression de mes meilleurs sentiments, et transmettez, je vous prie, à Mme Volterra mes hommages respectuex

V. A. Kostitzin
1, Square Vermenouze, Paris (5)

12.9 Volterra to Kostitzin, Roma, 13.11.1935 *Cat. no. 372*[13]

13 Novembre 1935

Cher Monsieur Kostitzin

Je vous ai envoyé ce matin par la poste la redaction de deux conférences que j'ai tenues il y a déja quelque temps. Elles pourront peut-être vous servir. Je pense que vous connaissez déjà tout ce qui a été publié sur le sujet. La bibliographie se trouve à la fin de l'opuscule publié par Hermann. Dernièrement a paru un mémoire de Royal N. Chapman sur les fluctuations des insectes (Proceedings Hawalian Entomological Society 8 279–92, 1933). Elle a été résumée dans The Journal of animal Ecology (Vol 4, N. 2, Novembre 1935). M. Chapman vérifie les lois des fluctuations que j'ai données (les trois lois) pour les insectes[14]. J'avais déjà parlé avec M. Hadamard de votre conférence au séminaire. Je suis heureux de voir confirmée cette notice.

Je tiens à vous repeter directement que lorsque j'ai publié mon mémoire des Lincei sur les fluctuations ainsi que mon mémoire du Comité talaxographique je ne connaissais aucun travail de M. Lotka. Je n'ai connu ses travaux que plus tard lorsqu'il m'a envoyé son livre de physique biologique. Mais c'est moi qui a donné les trois lois des fluctuations et a traité le cas de n espèces pour la première fois en introduisant la méthode des rencontres que j'ai établie en partant des postulats fondamentaux. C'est ainsi que j'ai donné pour la première fois les équations générales des fluctuations de n espèces et que j'en ai tiré les lois générales et leurs conséquences.

On n'aurait rien trouvé en écrivant seulement comme Lotka a fait que à chaque instant la variation du nombre des animaux des n espèces est une fonction de leur état actuel. Il fallait établir la forme de ces fonctions. C'est ce que j'ai fait pour la première fois. Je vous prie de lire à ce propos ce qui est imprimé dans l'opuscule publié par Hermann à page 11 et à page 36. Je pense que l'on ne peut pas dire que l'on a traité le cas de n espèces si l'on ne particularise la forme des équations c'est à dire si l'on ne donne les lois qui font dépendre les dérivées des nombres des individus du nombre de ces individus. En laissant ces lois indéterminées on ne pouvait rien bâtir dessus.

Je vous remercie de votre interêt pour ma santé. Elle est toujours telle qu'elle était à [illeg.].

Je lirai avec le plus grand interêt dans tous les détails le travail que vous avez eu l'obligeance de m'envoyer, sur l'hérèdité. J'attends d'être un peu moins fatigué.

---

[13] Two drafts. We reproduce the first (the second is almost identical).

[14] Chapman 1933.

Mais je vous remercie beaucoup des ce moment de l'intérêt que vous avez à des questions auxquelles j'ai consacré beaucoup d'années de travail.

12.10 Kostitzin to Volterra, Paris, 31.12.1935 *Cat. no. 195*

Paris, 31.12.1935

Monsieur et cher Maître,

je vous écris avec un énorme rétard, mais j'ai attrapé une très mauvaise grippe dont je ne suis pas encore tout à fait remis. C'est pour cette même raison que je n'ai pas fait ma conférence au seminaire de M. Hadamard et que je n'ai pas encore terminé l'étude du cas de deux espèces microbiennes.

Votre lettre m'a beaucoup attristé. Comment se fait-il que vous croyiez nécessaire de souligner exprèssement que les recherches de Lotka vous étaient inconnues à l'époque où vous commenciez les vôtres? Je ne vois pas qui pourrait avoir des doutes là-dessus. Ceux qui ont eu le plaisir de lire un des mémoires publiés par vous savent très bien avec quelle correction et exactitude scrupuleuse vous citez le plus petit travail publié sur le même sujet. Je crois savoir que M. Lotka lui-même n'a jamais émis de doutes de cette nature, au contraire. Si au début de mon livre sur la symbiose et la parasitisme je parle des travaux de Lotka, c'est par souci d'exactitude historique et aussi pour des raisons sentimentales. L'humanité scientifique (tout comme l'humanité en général) est bête, injuste et encline à frustrer les uns au profit des autres, et surtout les pauvres au profit des riches. Ainsi se créent des légendes. Exemple-légende de Fredholm, au profit du quel on frustre beaucoup de gens très méritants. Je crois que la gloire très méritée de Fredholm n'a aucun besoin de cette expropriation de ses précurseurs, de ses devanciers, je crois que votre gloire très réelle, incontestable, durable n'a aucun besoin de ce qu'on oublie les travaux de Lotka. Pour lui la biologie mathématique – c'est toute sa vie scientifique, sa seule raison d'être, la seule justification de son existence. Comment voulez-vous qu'il ne soit pas un peu courroucé, lorsqu'il voit ses travaux, pourtant si récents, négligés et oubliés injustement? Vous n'y êtes pour rien, mais le fait existe. Lotka sait très bien qu'il mérite mieux que ça, et il a raison. Son livre “Elements of physical biology” et ses mémoires de la même époque contiennent beaucoup de pensées, de remarques, de refléxions très intéressantes, de germes des recherches ultérieures. S'il n'est pas le premier qui ait introduit des équations de cette forme, il en donne une étude très convenable dans le cas de deux espèces avec des aperçus sur les problèmes plus généraux. Ses refléxions sur l'absence de contradiction entre l'existence de la matière vivante et la loi d'entropie ne sont pas encore suffisamment appréciées par les biologistes et par les philosophes de la nature. Je crois que si dans votre prochaine publication, en grand seigneur scientifique que vous êtes, vous le citerez avec sympathie, vous ferez une bonne action et mettrez fin à une situation qui n'a aucun besoin d'être envenimée.

J'ai reçu deux lettres très contradictoires de M. Régnier.... J'avoue n'y rien comprendre.

Veuillez agréer, mon très cher et très honoré maître, mes meilleurs voeux pour la nouvelle année, ainsi que mes hommages respectueux à l'intention de M.me Volterra

V. A. Kostitzin

12.11 Kostitzin to Volterra, Paris, 8.7.1936 *Cat. no. 196*

Paris, 8 juillet 1936

Monsieur et cher maître,

je vous envoie l'adresse et le téléphone de M. J. Delevsky (280, Bd. Raspail, Paris (14), tél. Odéon 71-27). La révue dont il dirige et rédige la section scientifique, est très repandue, et je crois que votre portrait romancé y serait à sa place.

Pour prolonger notre entretien d'hier, permettez moi de vous raconter une histoire:

Thèse Un mathématicien polonais bien connu M. Sierpinski[15] à l'époque de la guerre fut compromis dans les agissements anti-russes en Pologne. Il fut envoyé comme prisonnier civil dans un camp de concentration en Sibérie, alors que sa famille fut dirigé sur un autre camp. Situation terrible. Mes collègues de Moscou ne partageaient certainement pas les idées russophobes de Sierpinski, mais ils croyaient que les savants de tous les pays forment une famille. Résultat: Sierpinski fut extrait avec sa famille de son exil, a reçu l'autorisation de séjourner à Moscou; on lui a procuré une possibilité modeste de gagner sa vie, et il fut amicalement accueilli par les mathématiciens de Moscou.

Antithèse Un professeur de droit international (spécialité ridicule, n'est ce pas?) à la Faculté de Droit de Petrograd Gorovtsoff fut trouvé pendu à un arbre du bois de Boulogne. Cause: faim, absence de logis. Il a laissé une lettre explicative où il exprimait l'éspoir, que son suicide reveillera la conscience etc. Naïf!

Synthèse Je m'en abstiens.

Veuillez agréer, mon cher maître, avec mes hommages respectueux pour M.me Volterra, l'expression de ma très haute considération

V. Kostitzin

[15] The mathematician Waclaw Sierpinski (1882–1969) studied at Warsaw, where he was born, and taught at Lvov until 1914 and then at Warsaw. He founded a school of mathematics in Poland focused mainly on set theory and the foundations of mathematics. He was Vice President of the Polish Academy of Science.

12.12 Kostitzin to Volterra, Paris, 15.7.1936 *Cat. no. 197*

Paris, 15 Juillet 1936

Monsieur et cher maître,

je suis très confus en présence de ce résultat inattendu de ma mauvaise humeur. Je suis très touché de votre intervention spontanée auprès de MM. Borel et Perrin[16] et je vous remercie de votre grande bonté.

Je connais M. Perrin mais je n'ai mais eu et je n'ai jamais cherché l'occasion de l'approcher. J'essaierai de me mettre en rapport avec lui.

En ce qui concerne mon livre, la solution que vous proposez est la plus raisonnable, et je vous en enverrai les epreuves en automne[17].

Veuillez agréer, mon cher maître, avec mes hommages respectueux pour M.me Volterra, l'expression de ma réconnaissance et de ma très grande considération. Je vous souhaite des bonnes vacances

V. Kostitzin

12.13 Kostitzin to Volterra, Paris, 11.10.1936 *Cat. no. 198*

Paris, 11 Octobre 1936

Monsieur et cher maître,

j'espère que votre voyage s'est accompli dans les meilleures conditions, que vous n'êtes pas trop fatigué et que Paris vous accueille bien. Quant à moi j'ai eté accueilli à mon retour par une grippe féroce qui m'empêchera probablement de vous revoir lundi à l'Académie des Sciences.

Mon livre est sous presse, et l'éditeur m'a promis de vous envoyer à l'hôtel Lutetia un jeu de placards. Dans le cas où cela ne serait pas encore fait je vous transmettrai mon exemplaire personnel.

En ce qui concerne M. J. Perrin je lui ai envoyé suivant votre conseil une lettre accompagnée de toutes mes publications parues pendant ces années d'exil. Aucune réponse. Bien que ce silence puisse être interprété de différentes façons je me demande si il ne faut pas réfaire la demarche? Je serais heureux d'avoir et de suivre votre conseil à ce sujet.

Veuillez agréer, monsieur et cher maître, avec mes hommages les plus respectueux à l'intention de M.me Volterra, l'expression de mes meilleurs sentiments.

V. Kostitzin

---

[16] Émile Borel (1871–1956) was one of the most distinguished French mathematicians of the time and a great friend of Volterra's (see also *Dictionary of Scientific Biography*, sub voce). Jean Perrin was the well-known Nobel physics laureate (see note 10 in the Régnier-Volterra correspondence).

[17] This certainly refers to the introduction Volterra wrote for V. A. Kostitzin 1937b.

12.14 Kostitzin to Volterra, Paris, 14.10.1936 *Cat. no. 199*

Paris, 14 Octobre 1936

Monsieur et cher maître,

je suis à peu prés guéri; en tout cas je sors et je suis entièrement à votre disposition n'importe quel jour sauf vendredi. Je vous remercie de vos bonnes intentions et je vous prie d'agréer avec mes hommages respectueux à l'adresse de M.me Volterra, l'expression de ma très haute considération et de mes sentiments les plus affectueux

V. Kostitzin

12.15 Kostitzin to Volterra, Paris, 16.10.1936 *Cat. no. 200*

Paris, 16 octobre 1936

Monsieur et cher maître,

je ne sais pas comment de vous remercier de votre intervention.

Je serai très heureux de vous voir demain soir.

Veuillez agréer, monsieur et cher maître, avec mes hommages respectueux à l'intention de M. Volterra, l'expression de ma vive reconnaissance et de mes meilleurs sentiments

V. Kostitzin

12.16 Kostitzin to Volterra, Paris, 4.11.1936 *Cat. no. 201*

Paris, 4 Novembre 1936

Monsieur et cher maître,

j'ai reussi à être grippé la deuxième fois depuis mon retour à Paris. Maintenant je vais mieux, et j'ai pu voir M. Montel qui m'a montré votre préface sous sa forme définitive. Tout est parfait, et je vous remercie infiniment.

L'adresse de M. Lotka est:

Prof. Dr. A. J. Lotka
Statistical Bureau
Metropolitan Life Insurance
1, Madison Avenue
New-York (N. Y.) États-Unis

Ma femme va vous téléphoner aujourdhui ou demain à propos du rendez-vous.

Veuillez agréer, Monsieur et cher Maître, avec mes hommages respectueux pour Madame Volterra, l'expression de mes meilleurs sentiments

V. Kostitzin

12.17 Kostitzin to Volterra, Paris, 30.12.1936 *Cat. no. 202*

Paris, 30 decembre 1936

Monsieur et cher maître,

malgré toutes les prédictions pessimistes de charlatans et de gens serieux, de politiciens et d'économistes, d'astrologues et de magiciens qui nous promettent unanimement en 1937 tous les maux imaginables, je veux espérer que le sort sera un peu plus clément et que les orages en préparation seront dissipés. Je vous souhaite, et ma femme se joint à moi, à vous et à M.me Volterra une bonne et heureuse année et la réalisation de tous vos souhaits.

Veuillez agréer, mon cher maître, l'expression de mes meilleurs sentiments et mes hommages les plus respectueux à l'intention de M.me Volterra.

V. Kostitzin

12.18 Kostitzin to Volterra, Paris, 4.3.1937 *Cat. no. 203*

Paris, 4 Mars 1937

Monsieur et cher maître,

je vous envoie le scenario préalable tiré par M. Dufour (cinéaste) de ma notice et que je dois transformer en un scenario définitif, qui sera (théoriquement) réalisé vers le mois de Mai. Si pendant ce temps vous venez à Paris, vous serez irrevocablement filmé[18].

Mon livre doit paraître d'un jour à l'autre. J'attends ce moment avec une impatience et une appréhension, car je vois très bien tous les défauts de mon ouvrage, qui m'a valu beaucoup de travail, beaucoup d'ennuis, et qui m'attirera le courroux de biologistes: quand même, de tous mathématiciens, je suis plus près d'eux.

J'ai lu avec beaucoup de plaisir et d'émerveillement votre mémoire sur les "Principes de Biologie Mathématique"[19]. Mon âme de naturaliste n'est pas d'accord avec vous sur l'interprétation des coefficients vitaux, mais mon âme de mathématicien comprend très bien toute la partie philosophique de votre travail: parmi tous les phénomènes vitaux vous délimitez une région très étroite, mais dans ces limites vous construisez une biologie rationnelle, analogue à la mécanique rationnelle, holonome, canonique, vérifiant des conditions économiques, des principes extrémaux. C'est un progrès énorme.

---

[18] This is the scientific documentary *Images mathématiques de la lutte pour la vie* screened for the first time at the Palais de la Découverte during the Universal Exposition of Paris in 1937. The final version, presented by the Scientific Cinema Institute under the auspices of the the mathematical section of the Palais, benefited from the intervention of its director Jean Painlevé (1902–1989). In this connection, see also the D'Ancona-Volterra correspondence. For further details, see MILLÁN GASCA 1993.

[19] VOLTERRA 1937a.

J'ai vu l'article consacré à vous par M. Delevsky dans "le Mois" qui m'a appris beaucoup de choses que j'ignorai.

M. Lotka m'écrit qu'il sera à Paris vers la fin de juillet en vue du Congrès International de Population.

Veuillez agréer, mon cher maître, l'expression de mes sentiments respectueux et affectueux, avec mes hommages très respectueux à l'intention de M.me Volterra.

V. Kostitzin

P. S. Où dois-je vous envoyer mon livre?

12.19 Kostitzin to Volterra, Paris, 5.4.1937 *Cat. no. 204*

Paris, 5 Avril 1937

Monsieur et cher maître,

je trouve votre lettre en revenant d'un voyage de deux semaines dans les Vosges où nous avons vu la neige et le debut de printemps.

Le scenario est certainement defectueux, car il est extrêmement difficile d'exposer pendant 10–12 minutes un ensemble des problèmes compliqués; il est encore plus difficile de doser ce melange de façon que tout soit bien équilibré. Il me semble quand même que la question mendelienne, qui n'occupe d'ailleurs que 40 secondes, est à sa place, car du point de vue transformiste le sort des groupes mutés, apparus dans une population-mère, est au moins aussi important que le sort des associations des groupes non apparentés. Donc, si l'on peut montrer qu'ici aussi existe un mécanisme d'élimination, tout va bien. En ce qui concerne l'assistance mutuelle entre les espèces, ce problème était beaucoup discuté par Darwin lui-même qui, en anglais très conformiste, voulait à tout prix défendre ses théories contre le reproche d'immoralité. Un des darwinistes anglais, plus consécutif que Darwin, – Sutherland – a ecrit un ouvrage très intéressant sur les origines de l'instinct moral et de l'instinct social (à ce propos, la lecture de cet ouvrage m'a valu jadis, au lycée de Smolensk, quatre dimanches d'arrêts de rigueur)[20]. Il ne me semble pas qu'il soit possible de compléter le scenario par un exposé, même sommaire, de ces problèmes si compliqués, mais on peut ajouter quelques images montrant l'assistance mutuelle entre les espèces et quelques phrases expliquant l'importance biologique de ce phénomène. Demain je verrai M. Dufour – cinéaste scientifique – et lui en parlerai.

Veuillez agréer, monsieur et cher maître, l'expression de ma vénération et de mon affection, ainsi que mes hommages respectueux à l'intention de M.me Volterra. Ma femme me prie de transmettre à m.me Volterra et à vous l'expression de ses sentiments les meilleurs et inaltérables.

V. Kostitzin

---

[20] SUTHERLAND 1898.

12.20 Kostitzin to Volterra, Paris, 12.6.1937 *Cat. no. 205*

Paris, 12 Juin 1937

Monsieur et cher maître,

je vous remercie de l'aimable envoi de votre mémoire, comme toujours, important et intéressant. Je ne vous écrivais pas car j'espérais vous voir bientôt à Paris, et encore j'ai pensé que cette lettre ne vous trouve pas à Rome.

Mon scenario a passé par des avatars innombrables: il était tantôt raccourci, tantôt étiré, tantôt moralisé, tantôt démoralisé, tantôt démendelisé, tantôt rémendelisé; depuis trois semaines il se trouve en "execution" on en "réalisation" à l'antre de M. Jean Painlevé.

Je vous enverrai bientôt une note contenant une généralisation du cas d'une population handicapée par ses propres déchets[21]. Cette généralisation est motivée par l'existence des populations expérimentales disparaissant au bout d'un temps fini.

Mon livre me rapporte tantôt des joies, tantôt des chagrins.

Côté joie: j'ai reçu la proposition d'écrire un article sur la biologie mathématique pour Tabulae Biologicae - une sorte de l'Encyclopédie biologique allemande publiée en Hollande.

Côté chagrin: il y a des biologistes (comme il y a des physiciens) qui trouvent que la méthode analytique est beaucoup moins importante que la méthode statistique. Sans vouloir dénigrer le probabilisme, je crois que, par exemple, l'étude statistique des inondations à Paris, avec la démonstration de leur distribution Gaussienne, vaut beaucoup moins que l'étude de leurs causes avec la prédiction, même mauvaise, de ces calamités.

Veuillez agréer, monsieur et cher maître, avec mes hommages respectueux à l'adresse de M.me Volterra, l'expression de ma profonde vénération. Ma femme me prie de transmettre à M.me Volterra et à vous ses salutations très cordiales

V. Kostitzin

12.21 Kostitzin to Volterra, Paris, 24.6.1937 *Cat. no. 206*

Paris 24 Juin 1937

Monsieur et cher maître,

j'ai appris par M. Montel votre présence à Paris et je m'empresse de vous souhaiter la bienvenue. J'espère que tout va bien chez vous et que M.me Volterra et vous-même étes en bonne santé.

Je vous transmets simultanément une note contenant une généralisation de votre problème d'intoxication. Au prix d'un coup de pouce (équation 4) j'ai réussi à intégrer effectivement l'équation intégro-différentielle. Je prévois quelques ramifications et quelques applications possibles de ces résultats.

---

[21] V. A. Kostitzin 1937a.

Je serais bien content de vous revoir mais j'ai peur de vous importuner trop.

Veuillez agréer, monsieur et cher maître, avec mes hommages très respectueux à l'adresse de M.me Volterra l'expression de ma très haute considération.

V. Kostitzin

Ma femme me prie de transmettre à M.me Volterra et à vous ses meilleurs salutations.

12.22 Kostitzin to Volterra, Paris, 3.7.1937 *Cat. no. 207*

Paris, 3 Juillet 1937

Monsieur et cher maître,

je veux vous remercier (ainsi que M.me Volterra) de votre accueil toujours si cordial et de ce climat de bienveillance que je trouve auprès de vous. Je ne suis pas facile à apprivoiser et j'ai dans mon caractère beaucoup des pointes acérées, mais quand je suis dans votre société toutes ces armes intérieures n'existent pas*.

Je suis très content que mon petit livre vous plait, ce qui signifie pour moi que mon travail n'était pas tout à fait inutile, et vous savez, avec mon hypertrophie russe de l'instinct social, se voir à l'âge actif voué à l'inutilité, se voir tout à fait en dehors de la vie - c'est une des plus grandes souffrances qu'on peut concevoir.

Veuillez agréer, monsieur et cher maître, avec mes hommages les plus respectueux à l'adresse de M.me Volterra, avec les meilleurs salutations de la part de ma femme, l'expression plus haute de ma très grande considération et de mon affection

V. Kostitzin

* Parmi les mathématiciens français seul M. Montel me donne une sensation de securité intérieure.

12.23 Kostitzin to Volterra, Paris, 6.7.1937 *Cat. no. 208*

Paris, 6 Juillet 1937

Monsieur et cher maître,

voici le titre exact de ce livre allemand don je vous ai parlé l'autre jour:

<u>Ringleb</u> Die mathematischen Methoden der Biologie
Leipzig, B. G. Teubner 1937

C'est un livre exclusivement statistique consacré à l'étude de la variation, de la survivance des races etc. Tout ce qui est analytique en est sévèrement éliminé. Comme introduction aux méthodes statistiques c'est assez bien fait quoique incomplet.

Veuillez agréer, monsieur et cher maître, avec mes hommages très respectueux à l'intention de M.me Volterra, l'expression de mes sentiments les meilleurs.

V. Kostitzin

Mon adresse à partir de 15 Juillet: 6, Square de Port-Royal, Paris (13)

12.24 Kostitzin to Volterra, Paris, 10.7.1937 *Cat. no. 209*

Paris, 10 Juillet 1937

Monsieur et cher maître,

permettez-moi de vous communiquer une solution du petit problème sociologique posé hier par facétieux M. Lebesgue. Supposons qu'une population est composée de $\underline{x}$ individus virtuellement honnêtes et de $\underline{y}$ individus virtuellement criminels. On peut supposer (hypothèse fondamentale) que les éléments dangereux ne naissent pas tels et que, à cause des facteurs sociaux très variés, une certaine partie $\underline{\alpha x}$ dévient socialement malade. D'autre part, une certaine partie $\underline{\beta y}$ accomplit des actes criminals (assassinats), et la société en élimine régulierement $\underline{\gamma y}$. On peut resumer ces hypothèses par les équations différentielles très simples

$$\left.\begin{aligned} x' &= \varepsilon x - hx^2 - \alpha x - \beta x \\ y' &= \alpha x - \gamma y \end{aligned}\right\} \qquad (1)$$

Il est facile de voir que pour

$$e > \alpha + \frac{\alpha\beta}{\gamma} \qquad (2)$$

il existe un état stationnaire stable

$$\left.\begin{aligned} x_{\lim} &= \frac{1}{h}\left(\varepsilon - \alpha - \frac{\alpha\beta}{\gamma}\right) \\ y_{\lim} &= \frac{\alpha}{h\gamma}\left(\varepsilon - \alpha - \frac{\alpha\beta}{\gamma}\right) \end{aligned}\right\} \qquad (3)$$

et que dans ce cas

$$\frac{y_{\lim}}{x_{\lim}} = \frac{\alpha}{\gamma} \,. \qquad (4)$$

Donc, il est impossible par des repressions éliminer complètement la criminalité, car la repression accrue devient nécessairement *repression préventive*, et ceci n'est pas possible sans des dangers très graves pour la stabilité sociale. Il ne reste qu'une seule méthode rationelle – l'abaissement du coefficient α par l'education des masses et par l'amélioration des conditions d'existence.

Je suis très confus à cause de ma venue tardive dont je suis doublement puni, mais je ne savais pas du tout que c'était votre tour de parler.

Je n'ai pas oublié votre promesse aimable de venir chez nous, et ce serait pour nous une très grande joie. Malheureusement, notre vie est troublée par les préparatifs au démenagement prochain. Aussi permettez nous de repondre à l'invitation de M.me Volterra et de venir vous voir la veille de votre depart - dimanche ou lundi soir.

Veuillez agréer, monsieur et cher maître, avec mes hommages respectueux pour M.me Volterra, l'expression de ma très haute considération et de ma très vive reconnaissance.

V. Kostitzin

12.25 Kostitzin to Volterra, Paris, 22.7.1937 *Cat. no. 210*

Paris, 22 Juillet 1937

Monsieur et cher maître,

je sais que vous n'êtes plus à Paris, mais j'ignore complètement votre séjour actuel; aussi je vous écris un peu au hasard. Je vous remercie de tout mon coeur de votre demarche auprès de M. Perrin. C'est en effet lui qui doit décider maintenant après les avis favorables de la section mathématique et de la commission des recherches.

Je connais M. Bauer qui m'est très sympathique et qui a de la sympathie pour moi. C'est un physicien de grande mérite, mais un peu effrayé par les transformations kaléidoscopiques des théories physiques et désirant dans le secret de son coeur le rétour à la stabilité; et comme ce secret n'est secret pour personne, M. Bauer est assez mal vu par la jeunesse physique à Paris et en souffre.

Après demain nous partons en vacances. Ma chère femme en a un très grand besoin: pendant les derniers mois c'était un travail fou, absorbant et épuisant, nécessité surtout par l'instabilité de la vie autour de nous et de la notre situation. C'est comme la Cathédrale de St. Isaak à Petrograd construite sur le pilotis composé de 10000 chênes: chaque année la Neva en emporte une dizaine, mais il est impossible de déviner à l'avance ceux qui vont partir. Chez nous la proportion est encore moins avantageuse.

Veuillez agréer, monsieur et cher maître, nos meilleurs souhaits pour vos vacances, sourtout ceux de la bonne santé pour M.me Volterra et pour vous même.

Croyez en ma vénération affectueuse et reconnaissante.

V. Kostitzin

Ma nouvelle adresse à Paris:	6, Square de Port-Royal, Paris (13)
Mon adresse en vacances de 24 Juillet à 1 Octobre:	Villa les Fougères, Erquy, Côtes du Nord

12.26 Kostitzin to Volterra, Paris, 23.11.1937 *Cat. no. 211*

Paris, 23 Novembre 1937

Monsieur et cher maître,

nous regrettons vivement de n'avoir pu voir M.me Volterra et vous-même avant votre départ: nous espérions que votre séjour à Paris sera cette fois un peu plus long.

Je me permets de vous envoyer l'ouvrage de M. de Lur-Saluces sur le premier savant russe Lomonossof[22]. Actuellement, la science russe n'est plus une pauvre parente, et vous verrez qu'avec Lomonossof elle débuta d'une façon brillante. Quant à l'ouvrage lui-même, il est écrit par un moyen français qui a lu consciencieusement quelques ouvrages russes et qui a su en tirer l'essentiel sans trop le déformer. C'est déja quelque chose, quoique par endroits un lecteur russe peut rire et s'amuser énormement.

Veuillez agréer, mon cher et vénéré maître, avec mes hommages respectueux à l'intention de M.me Volterra, l'expression de mes meuilleurs et affectueux sentiments. Ma femme me prie de transmettre à M.me Volterra et à vous son souvenir cordial.

V. Kostitzin

12.27 Kostitzin to Volterra, Paris, 30.12.1937 *Cat. no. 212*

Paris, 30 Décembre 1937

Monsieur et cher maître,

je suis très content que le problème psycho-sociologique Lomonossoff vous ait intéressé. C'était une des réactions de l'organisme national russe aux procédés hardis et parfois brutaux de Pierre le Grand.

Veuillez agréer, monsieur et cher maître, mes meilleurs voeux pour la nouvelle année, pour Madame Volterra et pour vous- même.

V. Kostitzin

12.28 Kostitzin to Volterra, Paris, 16.3.1938 *Cat. no. 213*

Paris, 16 Mars 1938

Monsieur et cher maître,

je vous remercie de l'aimable envoi de votre si intéressante conférence génevoise[23]. En ce qui concerne la terminologie, j'emploie toujours action

---

22 MARQUIS DE LUR-SALUCES 1933. Mikhail Vasilievic Lomonosoff (1711–1765) was a scientist and poet. He taught chemistry at the University of St. Petersburg and was instrumental in the reform of the Russian language, contributing to structuring it as a means of literary expression.

23 VOLTERRA 1937c. This is the text of a lecture delivered by Volterra in June 1937 at the University of Geneva in the series of *Conférences Internationales des Sciences mathématiques*.

résiduelle pour désigner les termes intégraux dans les équations qui ont l'honneur de porter votre nom.

Je m'occupe actuellement du problème de la selection naturelle dans le sens Darwin-Mendelien de ces mots. Je réussi a perfectionner le modèle mathématique et il me semble qu'on peut reconstruire rationnellement les traits essentiels du processus de l'évolution[24].

Nous esperons tous de vous voir et de vous entendre bientot à l'Institut Poincaré.

Veuillez agréer, monsieur et cher maître, l'assurance de ma profonde vénération, ainsi que mes hommages très respectueux à l'intention de M.me Volterra

V. Kostitzin

12.29 Kostitzin to Volterra, Paris, 4.5.1938 *Cat. no. 214*

Paris, 4 mai 1938

Monsieur et cher maître,

je suis très content de vous savoir à Paris, et je serais heureux de venir chez vous aujourdhui, mais ma femme est grippée, et je crois qu'il vaut mieux un peu remettre notre entrevue, disons jusqu'à demain jeudi ou jusqu'à samedi 7 Mai (Vendredi 6 Mai est occupé).

Veuillez agréer, monsieur et cher maître, avec mes hommages respectueux à l'intention de Madame Volterra, l'expression de ma plus haute considération et de mes meilleurs sentiments

V. Kostitzin

12.30 Kostitzin to Volterra, Paris, 5.5.1938 *Cat. no. 215*

Paris, 5 Mai 1938

Monsieur et cher Maître,

je vous remercie de votre aimable lettre, et je ne manquerai pas de venir Samedi soir, peut être accompagné de ma femme, si l'état de sa santé le permet.

J'ai reçu votre mémoire sur la croissance des populations[25] et je tiens à vous signaler deux publications contenant la même idée de départ que votre mémoire. Tout d'abord, dans un livre destiné aux naturalistes, "Notions élémentaires de mathématiques pour les sciences expérimentales, p. 215, remarque, M. L. Brillouin (que j'apprécie beaucoup plus de son jumeau siamois scientifique) dit[26]: "Pour

---

[24] One first work by Kostitzin on this topic is V. A. Kostitzin 1936. In 1938 he published the notes: V. A. Kostitzin 1938a, b, c, d, e.

[25] Volterra 1938b.

[26] Brillouin 1935. The author is Léon Nicolas Brillouin (1889–1969), former professor of physics at the Collège de France, later renowned for his information theory studies. The twin

une espèce bisexuée, le nombre des naissances dépend du nombre des rencontres entre individus des deux sexes; si ceux-ci sont peu nombreux, le nombre des naissances suivra une loi en $\alpha N^2 \delta t$ et c'est seulement à partir d'une densité minima de population que les naissances deviendront proportionnelles au nombre N des individus. On pourrait d'ailleurs établir les règles théoriques satisfaisant à ces deux conditions extrêmes". D'autre part, dans ma note "Sur une généralisation des équations biologiques dans le cas d'une population intoxiquée par les produits de son activité chimique" (CR. 204, p. 1683, 1937) j'écris: "... le principe des rencontres qui régit les actions limitatives donne des termes rectangulaires ou quadratiques lorsque le nombre des rencontres est relativement petit, et des termes linéaires lorsque ce nombre est relativement grand, et les individus, ayant subi les rencontres, cessent, du moins provisoirement, de figurer dans les dénombrements"[27].

Mais nous n'émettons qu'une idée, alors que vous donnez une étude mathématique complète de ce cas curieux et qu'on peut surement justifier sur des population raréfiées.

Veuillez agréer, mon cher et vénéré maître, avec mes hommages respectueux pour M.me Volterra, l'expression de mes meilleurs sentiments.

V. Kostitzin

12.31 Kostitzin to Volterra, Paris, 15.5.1938 *Cat. no. 216*

Paris, 15 Mai 1938

Monsieur et cher maître,

les recherches de M. R. A. Fisher* ont le même objet que les vôtres ou les miennes, mais par la méthode, purement statistique, en diffèrent entièrement. Il déploie une somme énorme de travail pour obtenir une partie insignifiante de ce que la méthode analytique donne en quelques lignes. Il appartient à cette espéce des statisticiens qui consentent à utiliser la function exponentielle mais évitent de se servir des dérivées et des intégrales quoique l'erreur commise dans les deux cas est absolument la même. Il me semble que cet excès de la méthode statistique est une maladie de notre temps ou plutôt une mode. Il y a quelque temps j'ai eu la visite d'une jeune mathématicien qui désirait d'approfondir la théorie de la population. Voici sa méthode: soit f(p,q) la probabilité de ce que la population égale à p au moment t devient égale à q au moment (t+1); cette fonction ne dépend pas directement de t; il faut trouver la loi la plus probable du développement de la population. Vous voyez d'où lui pouvait venir cette idée saugrénue d'assimiler un processus organique à une suite d'épreuves du style de battage des cartes. Vraiment, ces analogies statistiques rendent la génération mathématique actuelle incapable de rémonter du plan moléculaire au plan molaire. Ce jeune homme m'a quitté très mécontent de mes remarques critiques

referred to here by Kostitzin is Marcel Louis Brillouin (1854–1948), also professor of mathematical physics at the Collège de France.

27 V. A. Kostitzin 1937a.

et il exerce ses talents du batteur des cartes dans un des meilleurs établissements scientifiques de Paris. Dommage, car il est doué.

Veuillez agréer, monsieur et cher maître, avec mes hommages respectueux à l'intention de M.me Volterra, l'expression de mes meilleurs sentiments

V. Kostitzin

* On trouve dans l'ouvrage de M. Ph. L'Héritier "Génétique et évolution" (Hermann 1934) un court exposé des travaux de Fisher.

12.32 Kostitzin to Volterra, Etelley par Samoens, 13.8.1938 *Cat. no. 217*

Etelley par Samoens (H. Savoie) France, 13 Août 1938

Monsieur et cher maître,

je vous remercie infiniment de votre aimable envoi et de votre lettre. Je lirai votre important mémoire avec d'autant plus de profit et de plaisir que je réfléchis en ce moment sur la méthode historique dans la connaissance scientifique. Je dis: "réfléchis" et non pas "travaille", parce que depuis 18 Juillet nous nous reposons dans un coin très tranquille et très sauvage de Savoie. J'aime beaucoup ce pays qui me rappelle mutatis mutandis ma patrie.

Je dois vous remercier beaucoup, car grace à vos conseils, j'ai fait connaissance epistolaire avec MM. Donnan et Van der Klaauw, et mon prochain mémoire[28] paraitra dans Acta Biotheoretica. En ce qui concerne le roman de J. H. Rosny, je serai très heureux s'il vous donne un peu de plaisir[29]. Je dois dire toutefois que ce roman, tout dédié, qu'il est, à MM. Borel et Perrin, contient à côté des idées ingénieuses, des développements purement enfantins et montre une certaine faiblesse logique. Comme logicien, Wells est beaucoup plus fort: il n'oublie jamais ses prémisses et il en tire des résultat non contradictoires. Comme "anticipateur" Rosny est beaucoup plus faible que, par exemple, Robida qui a su, dans son "Vingtième siècle", publié dans les environs de 1880, prévoir toutes les réalisations techniques de notre époque[30]. Malgré tout, j'aime beaucoup les romans simili-scientifiques de J. H. Rosny, peut-être grace à ses "Vamireh" et "Xipehus" que j'ai lu dans mon enfance et qui ont frappé mon imagination.

Veuillez agréer, mon très cher et très honoré maître, avec mes hommages très respectueux à l'adresse de Madame Volterra, l'expression de mes meilleurs sentiments. Madame Kostitzin me prie de transmettre à M.me Volterra et à vous son très cordial et affectueux souvenir

V. Kostitzin

---

[28] V. A. Kostitzin 1940a.

[29] J. H. Rosny is the pseudonym shared by the brothers Joseph-Henry and Séraphin-Justin Boëx, Belgian writers who were the authors of a number of novels, first written together, and then separately. In many of these novels science undergoes a fantastic transfiguration. The books cited are *Les Xipéhuz* (1887), *Vamireh* (1892), dedicated to prehistory. Kostitzin probably refers to the older of the two, Joseph-Henri Rosny (1856–1940).

[30] Albert Robida (1848–1926), French painter and writer, the author of *Le Vingtième siècle* (1883).

12.33 Kostitzin to Volterra, Paris, 12.10.1938 *Cat. no. 218*

Paris, 12 Octobre 1938

Monsieur et cher maître,

je viens d'apprendre par M. Régnier que la fin de l'été n'était pas très favorable pour votre santé, et nous sommes très inquiets. D'autre part, notre époque est riche en émotions; pendant ces jours si critiques pour le monde nous pensions souvent à vous. Il faut quand-même être optimiste et plus que jamais travailler quoique cela n'est pas toujours facile.

M. Régnier m'a apporté le manuscrit de sa future note avec vos remarques. Tout d'abord, je vous remercie infiniment de votre appréciation si flatteuse pour moi, et ensuite je veux vous présenter quelques réfléxions suggérées par les chiffres de M. Régnier.

1° En ce qui concerne la phase de latence occupant les deux premières heures de la culture, on peut l'expliquer soit par la structure hétérogène (granulée?) du milieu peptoné, soit par le fait que ce milieu n'est pas favorable et que presque tous les germes y périssent rapidement. C'est cette derniere hypothèse qui est acceptée par M. Régnier.

2° La concentration du milieu en peptone diminue pendant chaque expérience, et M. Régnier m'a promis de mesurer cette diminution.

3° Dans ces conditions il serait nécessaire de perfectionner un peu la théorie, soit en introduisant la concentration du peptone comme une nouvelle variable, soit en l'introduisant dans les coefficients vitaux. Il me semble qu'on peut obtenir de cette façon des formules plus générales et plus exactes que mes formules (9)[31].

4° A propos de ces formules je me permets de signaler un petit malentendu: ce que vous avez pris pour l'indice (') de dérivation est en realité la puissance l. C'est ma faute car j'ai mal choisi les notations.

Maintenant, du point de vue pratique, je vous demande respectueusement l'autorisation de compléter vos Remarques par les résultats de mes calculs que je vous enverrai bientôt. Il me semble qu'on peut concilier la théorie avec les expériences de M. Régnier.

Veuillez agréer, monsieur et cher maître, avec mes hommages respectueux à l'intention de M.me Volterra, l'expression de ma vénération affectueuse. Ma femme me prie de transmettre à vous et à M.me Volterra, les meilleurs voeux et les meilleurs souvenirs.

V. Kostitzin

---

[31] Reference to the formula that provides the population maximum, obtained in KOSTITZIN 1937a.

12.34 Volterra to Kostitzin, n.p., October 1938 *Cat. no. 371*[32]

Cher monsieur Kostitzine,

je vous suis très reconnaissant de votre bonne lettre et des paroles si amicales et si affectueuses que vous m'adressez.

Je ne me porte pas bien. Helas. Pendant ce temps j'ai eu trop de soucis et de souffrances.

Venant à la question de Monsieur Régnier voici ce que j'ai à vous dire.

1) Est ce que mon calcul est exact en corrigeant ma bévue, c'est à dire en remplaçant l'indice (') par la puissance 1 ?

Ma remarque n'est que négative et si vous pensez qu'elle devient formellement exacte après la correction on pourrait la publier après la note de monsieur Régnier et de Melle Lambin comme c'était entendu. Mais si mes calculs ne sont pas exacts il faut supprimer complètement la remarque. La remarque avait été faite uniquement en vue de la précédente note, et je désirais montrer que la seule hypothèse de l'empoisonnement, comme je faisais dans cette note, n'était pas suffisante. Or, d'après ce que vous m'écrivez, vous avez expliqué d'une manière complète le résultat de Régnier et Lambin. Cette explication par voie théorique est tout à vous et à mon avis une grande importance. C'est pourquoi vous devez publier une note à part sous votre seul nom. Ce serait un vol si je m'en appropriais d'une partie quelconque[33]. Si vous me l'envoyez je serais honoré de la présenter à l'Academie, mais j'ai aussi tellement confiance en vous que je vous autorise à la presenter directement au secretariat de l'Académie.

Veuillez m'écrire un mot si vous êtes favorable à tout cet arrangement. Veuillez presenter de la part de ma femme et de moi même nos meilleurs souvenirs à Mme Kostitzin et veuillez agréer, cher Monsieur, l'expression de ma plus haute estime et de mon dévouement

12.35 Kostitzin to Volterra, Paris, 27.10.1938 *Cat. no. 219*

Paris, 27 Octobre 1938

Monsieur et cher maître,

votre lettre m'a profondément ému. Je voudrais vous savoir à l'abri de tous les soucis et jouissant tranquillement de votre gloire bien méritée. Je sais que le sort est aveugle et frappe sans choisir mais quand-même!... En tout cas il y a une chose que vous devez savoir: vous avez partout des amis que vous considérent non seulement comme le plus grand mathématicien de notre époque, mais aussi comme un homme bon, noble et courageux, cumulant la force de la pensée et la force de la volonté. Et quand on a de la volonté, un peut resister, lutter et vaincre, même les souffrances physiques.

---

[32] Draft, n.d. but certainly written on October 1938.

[33] See letters nos. 18.35 and 18.26 in the Régnier-Volterra correspondence.

J'ai vu plusieurs fois M. Régnier et j'ai beaucoup réfléchi sur le problème qui nous intéresse. Permettez-moi de vous exposer en toute franchise les résultats de mes réfléxions. La fin de vos remarques sur la note de M. Régnier a besoin d'être un peu rémaniée. Il ne s'agit pas d'erreur quelconque, mais de simple oubli d'un fait physiologique. Reprenons ma formule sous sa forme simple[34]

$$p_m = \frac{\varepsilon}{h} - \frac{c}{h^2}\log\left(1 + \frac{\varepsilon}{h}\frac{h^2}{c} - p_0\frac{h^2}{c}\right) \tag{1}$$

Vous affirmez 1° que si le paramètre ε/h croît, p_m croît; 2° que ce paramètre croît avec la croissance de la concentration de nourriture; 3° que ε/h croît dans ce cas indéfiniment. Or, tandis que les deux premières affirmations sont absolument exactes, la troisième est démentie par l'expérience: il existe une limite physiologique supérieure pour l'assimilation de la nourriture et pour la vitesse de croissance et de division cellulaires. Donc, en admettant que dans la formule (1) tous les coefficients sont des fonctions de la concentration de nourriture q, on est forcé de leur imposer les limitations suivantes:

$$\lim_{q\to\infty} \varepsilon = \varepsilon_\infty < \infty\,, \quad \lim_{q\to\infty} h = h_\infty < \infty\,, \quad \lim_{q\to\infty} c = c_\infty < \infty \tag{2}$$

Par conséquent, le maximum p_m considéré comme fonction de $\underline{q}$ tend nécessairement vers une limite finie, ce qui concorde très bien avec les observations de M. Régnier, et ces observations ne sont en aucune contradiction, au contraire, avec votre hypothese de l'intoxication accumulée. Vous voyez qu'en changeant très peu la fin de votre note, on peut obtenir un résultat 1° très agréable aux mathématiciens biologistes, car nous ne sommes pas blasés par les confirmations expérimentales, 2° très agréable à M. Regnier et à M. Kostitzin qui vous aiment beaucoup et qui sont heureux de pouvoir vous apporter cette confirmation de votre hypothèse initiale. On pourrait donner à cette fin de votre note la forme suivante: "Nous avons dans la Note citée de M. Kostitzin les maxima donnés par les formules (9). Or, le maximum p_m dépend de trois coefficients $\underline{\varepsilon}$, $\underline{h}$, $\underline{c}$, lesquels à leur tour sont fonctions de la concentration de la nourriture $\underline{q}$. On a des raisons physiologiques pour supposer que le coefficient de multiplication $\underline{\varepsilon}$ est une fonction croissante bornée de $\underline{q}$, et que les coefficients $\underline{h}$ et $\underline{c}$ ne varient que très lentement tout en restant bornés. Dans ces conditions le maximum p_m croît avec $\underline{q}$ et tend vers une limite. Donc, la courbe théorique des maxima obtenue dans l'hypothèse de l'intoxication correspond très bien à la courbe expérimentale. Ainsi cette hypothése trouve dans les expériences de M. Régnier une nouvelle confirmation."

En ce qui concerne ma théorie mathématique de ces phénomenes, elle n'existait pas encore il y a deux semaines et n'a aucun rapport direct à ce problème particulier. Elle est encore in statu nascendi. Permettez moi de vous communiquer quelques résultats préalables. Soit

[34] Here Kostitzin considers the formula of the population maximum referring to the growth equation contained in V. A. Kostitzin 1935b.

$$p' = \varepsilon p - hp^2 - cpP \qquad P = \int_0^t p(s)\,\mathrm{d}s \tag{3}$$

l'équation que vous connaissez. Supposons que les coefficients $\underline{\varepsilon}$, $\underline{h}$, $\underline{c}$, dépendent de la concentration de nourriture q et écrivons une équation complémentaire

$$q' = -\mu(q)P \tag{4}$$

ce qui veut dire que la nourriture se consomme mais ne se réproduit pas. On peut tirer de (4)

$$P = \lambda(q) = \int_q^{q_0} \frac{\mathrm{d}s}{\mu(s)}\ , \tag{5}$$

q_0 étant la valeur initiale de $\underline{q}$. Il est facile d'intégrer le système (3–4). On trouve sans peine que

$$p(t) = F(q) = p_0\, e^{-\int_q^{q_0} \frac{h(z)}{\mu(z)}\mathrm{d}z} + \int_q^{q_0} \left[\varepsilon(s) - c(s)\,\lambda(s)\right] e^{-\int_q^{s} \frac{h(z)}{\mu(z)}\mathrm{d}z}\, \frac{\mathrm{d}s}{\mu(s)} \tag{6}$$

$$t = \int_q^{q_0} \frac{\mathrm{d}s}{\mu(s)F(s)} \tag{7}$$

On peut montrer facilement que le maximum p_{m} donné par la formule

$$p_m = \frac{\varepsilon(q_m) - c(q_m)\,\lambda(q_m)}{h(q_m)} < \frac{\varepsilon(q_m)}{h(q_m)} \tag{8}$$

et que la concentration correspondante q_{m} se determine de l'équation

$$\frac{\varepsilon(q_0)}{h(q_0)} - p_0 = \int_{q_m}^{q_0} \frac{\partial}{\partial s}\left[\frac{\varepsilon(s) - c(s)\,\lambda(s)}{h(s)}\right] e^{-\int_s^{q_0} \frac{h(z)}{\mu(z)}\mathrm{d}z}\, \mathrm{d}s\ . \tag{9}$$

Ainsi, le problème est virtuellement résolu, mais il reste à determiner la forme des fonctions $\varepsilon(q)$, $h(q)$, $c(q)$, $\mu(q)$, et ici les expériences de M. Régnier + quelques considération probabilitaires peuvent être très utiles.

Veuillez agréer, monsieur et cher maître, l'expression de nos sentiments les plus affectueux et de notre vif désir de vous voir bientôt à Paris; veuillez transmettre à Madame Volterra mes hommages les plus respectueux et le souvenir très cordial de la part de ma femme.

V. Kostitzin

12.36 Kostitzin to Volterra, Paris, 2.11.1938 *Cat. no. 220*

Paris, 2 Novembre 1938

Monsieur et cher maître,

j'ai fait quelques calculs préalables qui montrent tout d'abord qu'il ne faut pas être présomptueux: en effet, dans ma lettre précédente j'ai affirmé que les coefficients h et c ne varient que très lentement en fonction de la concentration q, et la table suivante permet de voir que si cela est vrai pour c c'est tout à fait inexact pour h.

concentr.

q	ε	h	c
0,001	0,33	0,19	0,069
0,005	0,37	0,12	0,060
0,020	0,46	0,028	?
0,050	0,63	0,037	?
0,100	0,93	0,026	0,044
0,250	0,99	0,014	?
0,500	1,18	0,0088	0,034
1,000	1,07	0,0050	0,034
2,500	1,18	0,0041	0,029
5,000	1,23	0,0035	0,032
10,000	1,28	0,0035	0,027
20,000	1,24	0,0033	0,029

Ce coefficient décroît très rapidement, mais (et c'est un fait important) se stabilise au niveau 0,003 pour grandes concentrations de peptone. Il est evident a priori que la concurrence faiblit lorsque la nourriture devient abondante, mais la rapidité de cette décroissance m'étonne. Le coefficient de multiplication ε croît du simple au quadruple, et le coefficient d'action toxique c décroît du double au simple, ce qui est très naturel, la nourriture rendant organisme plus résistant. Tous les coefficients ont une allure nettement asymptotique et tendent vers des limites finies. Il est curieux que les coefficients c et ε calculés d'après les premières expériences de M. Régnier (sous le régime de nourriture abondante) étaient égaux à 0,023 et 1,14 respectivement - concordance très intéressante (v. mon livre p. 70–71). En fin de compte, mes calculs confirment entièrement le point de vue exposé dans ma lettre précédente et s'appuyant sur l'existence de la limite supérieure de l'assimilation de la nourriture. Excusez moi d'occuper votre temps par ces calculs et leur discussion, mais je tiens à ce que vous soyez rapidement informé.

Veuillez agréer, mon très cher et vénéré maître, l'expression de mes meilleurs sentiments, avec mes très respectueux hommages à l'intention de M.me Volterra. Ma femme me prie de la rappeler à votre souvenir.

V. Kostitzin

12.37 Volterra to Kostitzin, Ariccia, 5.11.1938 *Cat. no. 373*

Ariccia 5 Novembre 1938

Cher monsieur Kostitzin,

Je suis parfaitement d'accord avec vous sur ce que vous m'écrivez c'est-à-dire que [blank] ne croît pas indefiniment avec [blank] par des raisons qui sont évidentes. C'est pourquoi je trouve qu'il faut modifier de la manière que vous proposez et je serai très honoré de publier la remarque sous nos deux noms ensemble.

Ci-joint vous trouverez un brouillon de la rédation des remarques. Je pense qu'elles correspondent à votre idée. Autrement vous pourez me la renvoyer modifiée.

J'ai mis à la fin vos mots telles que vous me les avez écrit dans votre lettre.

Je n'ajoute pas autre chose parce que aujourd'hui je suis très fatigué et souffrant. Je ne puis pas vous écrire davantage.

Merci de vos paroles si affectueuses et si bienvellantes. J'ai été emu en les lisant. Ma femme et moi désirent être rappelés au bon souvenir de Madame Kostitzin et je vous prie, cher Monsieur Kostitzin d'accepter l'expression de ma plus haute estime et de mon plus sincère attachement.

Veuillez comuniquer cette lettre à Monsieur Régnier avec mes meilleurs salutations. Elle pourra servir comme reponse à sa dernière lettre.

12.38 Kostitzin to Volterra, Paris, 1.12.1938 *Cat. no. 221*

Paris, 1 Décembre 1938

Monsieur et cher maître,

je vous envoie avec un certain retard le texte de la note. Le retard provient de ce qu'il m'a fallu exécuter des calculs sans nombre pour obtenir la petite table très suggestive que vous verrez dans la Note. Cette table donne une preuve frappante en faveur de votre hypothèse de l'action toxique accumulée, et je suis très heureux de pouvoir vous en faire l'hommage[35]. Excusez-moi d'avoir supprimé les éloges à mon adresse – très flatteurs et très agréables pour moi – contenus dans la rédaction précédente de la Note, mais ces éloges sont incompatibles avec ma signature, et il y a trop de gens qui ne manqueraient pas de le constater.

J'espère que votre santé – si chère à moi – va mieux; je voudrais trouver un moyen verbal pour exprimer mon inquiétude à ce propos et pour vous réconforter. Ma femme partage entièrement mes sentiments. Veuillez agréer, mon très cher et très honoré maître, avec mes hommages à l'intention de M.me Volterra, mes meilleurs voeux.

V. Kostitzin

---

[35] V. A.Kostitzin, Volterra 1938.

12.39 Kostitzin to Volterra, Paris, 6.12.1938 *Cat. no. 222*

Paris, 6 Décembre 1938

Monsieur et cher maître,

hier - lundi - j'ai reçu votre lettre et j'ai fait un saut à l'Académie de Sciences; en résultat, en ce moment la note de M. Régnier et M.lle Lambin avec nos remarques se trouve à l'imprimerie où j'irai demain matin pour corriger les epreuves. Comme je ne sais pas quelles sont vos dispositions sur les tirages à part je consulterai à ce propos M. Régnier pour faire au mieux de vos intérêts.

Permettez moi maintenant d'attirer votre attention à une particularité de l'équation générale d'action résiduelle

$$p' = \varepsilon p - hp^2 - cp\int_0^t k(t-\tau)\,p(\tau)\,\mathrm{d}\tau \qquad (1)$$

Il ne résulte nullement de cette équation que la limite finale de la population soit nulle. Supposons par exemple que l'intégrale

$$I = \int_0^\infty k(s)\,\mathrm{d}s \qquad (2)$$

est finie. Dans ce cas la function p a comme limite

$$0 < p_\infty = \frac{\varepsilon}{h + cI} < \frac{\varepsilon}{h} \qquad (3)$$

et la courbe *p(t)* peut avoir trois formes suivantes

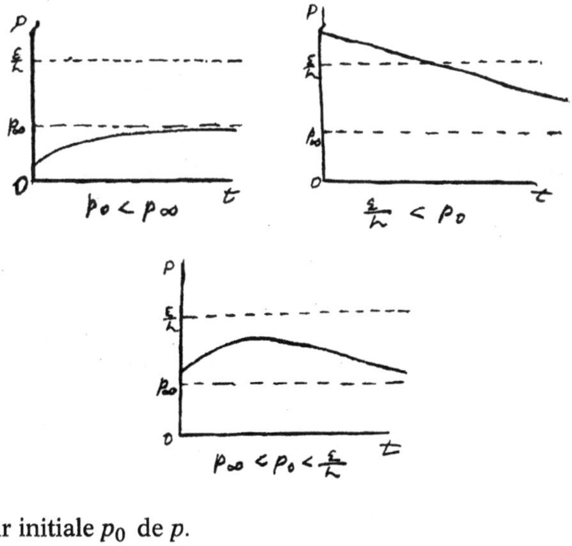

suivant la valeur initiale p_0 de p.

Au contraire, dans le cas de

$$\lim_{t\to\infty} k(t) = \delta > 0 \tag{4}$$

l'équation (1) pour les t grands devient

$$p' = \varepsilon p - hp^2 - c\delta pP\,, \quad P = \int_0^t p \, \mathrm{d}s \tag{5}$$

et on a à la limite

$$\lim_{t\to\infty} P = P_\infty\,, \quad \lim_{t\to\infty} p = 0 \tag{6}$$

comme dans le cas que vous et moi avons étudié.

La même chose peut avoir lieu dans le cas intermédiaire

$$\lim_{t\to\infty} k(t) = 0 \tag{7}$$

$$\lim_{t\to\infty} \int_0^t k(s) \, \mathrm{d}s = \infty \tag{8}$$

qu'il me reste encore à étudier.

J'espère que l'hiver qui vient ne vous sera pas défavorable. Pour nous - je préfère un hiver froid, sec, neigeux, pour vous - un hiver sec à froid moderé, ensoleillé, gai, et la vie privée de soucis et d'émotions - chose irréalisable à notre époque. Nous sommes très inquiets à votre propos et nous vous prions de ne pas nous oublier et de nous donner de vos nouvelles. Veuillez agréer, mon très cher maître, avec mes hommages respectueux à l'adresse de M.me Volterra, l'expression de mon admiration et de mon affection. Ma femme envoie ses meilleurs voeux à M.me Volterra et à vous.

V. Kostitzin

12.40 Kostitzin to Volterra, Paris, 31.12.1938 *Cat. no. 223*

Paris, 31 Décembre 1938

Monsieur et cher maître,

permettez-moi de vous présenter mes voeux les plus sincères à l'occasion de la nouvelle année. Je vous souhaite la santé, le bonheur et la tranquillité si nécessaire pour vous et pour votre famille. Ma femme s'associe à moi.

Il y a quelques jours j'ai vu M. Borel qui a parlé de vous avec beaucoup d'affection et de sollicitude et de désir sincère de vous revoir le plus tôt possible.

Je continue mon travail sur le problème de l'action toxique résiduelle et je vous enverrai bientôt quelques nouveaux résultats.

Veuillez agréer, monsieur et très vénéré maître, avec mes hommages très respectueux à l'intention de M.me Volterra, l'expression de ma très haute considération et de mon affection vive et sincère.

V. Kostitzin

12.41 Kostitzin to Volterra, Paris, 9.3.1939 *Cat. no. 224*

Paris, 9 Mars 1939

Monsieur et cher maître,

je vous envoie simultanément deux notes que j'ai publie entre-temps. Dans ces notes j'etudie la compatibilité des points singuliers stables des équations differentielles et il me semble que mes résultats peuvent être utiles dans les applications[36]. Je prépare actuellement un mémoire ou une suite des notes sur l'équation

$$p' = \varepsilon p - hp^2 - cp\int_0^t k(t-\tau)\,p(\tau)\,\mathrm{d}\tau$$

où j'examine les propriétés de la solution lorsque on admet des hypothèses variées sur le noyau *k(t)* et j'explique ainsi certaines particularités des courbes expérimentales.

Nous pensons souvent à vous avec une inquietude affectueuse et nous souhaitons que le printemps, qui revient, vous apporte une guérison. J'avais un oncle qui attendait toujours avec anxiété le retour du solstice d'hiver, et ce jour-la il disait: “les jours vont s'accroître et avec eux ma santé ira de mieux en mieux”. Et il avait toujours raison. Le pire dans la vie – c'est de se sentir vaincu, mais ce danger ne vous menace pas. Donc, j'espère vous revoir bientôt à Paris avec Madame Volterra.

Veuillez agréer, monsieur et cher maître, avec mes hommages respectueux à l'intention de M.me Volterra, l'expression de mes sentiments très affectueux.

V. Kostitzin

12.42 Kostitzin to Volterra, Paris, 3.6.1939 *Cat. no. 225*

Paris, 3 Juin 1939

Monsieur et cher maître,

je vous envoie avec le même courrier une note que je viens de publier dans les Comptes Rendus[37]. Dans cette note je propose une généralisation, à mon avis nécessaire, de votre équation intégro-différentielle de la théorie de l'action toxique du milieu. J'obtiens un système qui s'intègre facilement dans le cas de

---

[36] V. A. Kostitzin 1939c, d.

[37] V. A. Kostizin 1939b.

l'action toxique constante. On peut de cette façon rendre compte de toutes les particularités (je suis peut-être trop ambitieux) de courbes expérimentales.

Le printemps est revenu, mais vous n'êtes pas venu à Paris, ce qui ne manque pas d'inquiéter tous vos amis et admirateurs. Je ne sais que trop que par les paroles on ne fait pas guérir nos misères physiques mais je crois fermement que notre volonté, notre énergie psychique peuvent agir de façon efficace. Dans ma vie, déja longue, j'en ai vu des exemples. Donc, nous pensons souvent à vous avec inquiétude, avec insistance, avec espoir. Et nous sommes sûrs qu'avec les soins intelligents et attentifs de M.me Volterra votre santé ira mieux.

Veuillez agréer, monsieur et cher Maître, avec nos hommages très respectueux à l'intention de M.me Volterra, tout un ensemble de meilleurs voeux et de meilleurs sentiments de V. et J. Kostitzin.

12.43 Kostitzin to Volterra, Paris, 19.7.1939 *Cat. no. 226*

Paris, 19 Juillet 1939

Monsieur et cher Maître,

en revenant de Genève (Entretiens sur les applications du Calcul des probabilités) c'est avec une grande joie que j'ai trouvé vos tirages à part. Je pars aujourdhui même à Erquy [voici l'adresse complète: Villa Faugères Erquy (Côtes du Nord) France] pour les vacances. Ma femme se trouve à Roscoff[38] et vers la fin du mois elle viendra à Erquy. Je voudrais avoir des nouvelles de votre santé, mais je n'ose pas vous inquiéter. De temps en temps je vois M. Régnier et nous parlons de vous. Borel dans son allocution finale à Genève a parlé de vous avec beaucoup de respect et de tendresse, ce qui m'était infiniment agréable d'entendre.

Veuillez agréer, monsieur et cher Maître, avec mes hommages respectueux à l'intention de M.me Volterra, l'expression de mes sentiments respectueux et affectueux.

Bien à vous

V. Kostitzin

12.44 Kostitzin to Volterra, Erquy, 25.8.1939 *Cat. no. 227*

Erquy (Côtes du Nord), 25. 8. 39

Monsieur et cher Maître,

nous nous trouvons en ce moment au bord de la mer: repos necéssaire mais combien instable... Nous espérons fermement que votre santé, si chère à tous vos amis, va un peu mieux. Veuillez agréer mes meilleurs souhaits et nous rappeler au beau souvenir de M.me Volterra si bonne et si courageuse.

Bien à vous

V. et J. Kostitzin

---

[38] Reference to the Roscoff Biological Station.

12.45 Kostitzin to Volterra, Paris, 4.12.1939 *Cat. no. 228*

Paris, 4 Décembre 1939

Monsieur et cher Maître,

tout ce qui se passe dans le monde est de nature à tuer l'optimisme le plus robuste et à entretenir l'esprit dans la misanthropie noire; il est très difficile de considérer les événements avec la sérénité de l'historien de XXI Siècle; cette histoire est trop visiblement inscrite sur notre peau. "Heureux qui a visité ce monde dans ses minutes fatales"... Heu!...

Nous avons quitté Erquy vers la fin de Septembre et depuis nous sommes à Paris en reprenant peu à peu notre travail scientifique.

Le film "La lutte pour la vie" est terminé, et on peut le voir au Palais de la Découverte. Le public l'avale sans trop réfléchir sur le côté mathématique, en admirant les images dues à l'ingéniosité de Jean Painlevé et en acceptant les graphiques comme une preuve mystico-mathématique de la réalité de l'évolution. En fin de compte, le film n'est pas mauvais, et la seule chose qui manque - c'est votre image[39].

Il y a quelques jours j'ai vu avec plaisir M. Régnier qui est venu pour me consulter à propos de quelques courbes expérimentales. Il s'âgit d'un problème de chimie physique (osmose) peu intéressant du point de vue mathématique mais très important pour les médicins et physiologistes[40].

Ma femme travaille à la Sorbonne. Tout comme moi elle regrette qu'il est impossible de vous voir et de vous parler directement. Plus que jamais nous en sentons la nécéssité. Nous voudrions tellement avoir de vos nouvelles.

Veuillez agréer, monsieur et cher maître, notre souvenir très affectueux et très cordial et de nous rappeler au bon souvenir de M.me Volterra, à laquelle je présente mes plus respectueux hommages

Bien à vous

V. Kostitzin

12.46 Kostitzin to Volterra, Paris, 11.1.1940 *Cat. no. 229*

Paris, 11. 1. 1940

Monsieur et cher maître,

je suis heureux d'apprendre que le séjour à la campagne vous était favorable et je vous remercie de votre lettre qui m'a donné beaucoup de joie. J'ai reçu une lettre de M. Van der Klaauw qui m'apprend que Acta Biotheoretica continuent à paraître malgré la guerre. M. Régnier publiera bientôt une note en collaboration avec moi. Dimanche prochain on va fêter le jubilé scientifique de M. E. Borel.

---

[39] Reference to the film mentioned in note 18.

[40] The membrane permeability issue suggested by Régnier's work was later dealt with in several works by Kostitzin: V. A. Kostitzin 1940b, c. For Régnier's pharmacodynamics studies see Lambin 1969.

Nous travaillons beaucoup et j'espère vous envoyer bientôt qq. resultats intéressants.

Excusez moi de vous écrire sur une carte postale mais les cartes vont plus vite. Veuillez agréer nos meilleurs voeux pour vous et pour M.me Volterra et l'expression de notre affection.

V. Kostitzin

12.47 Kostitzin to Volterra, Paris, 25.4.1940 *Cat. no. 230*

Paris, 25 Avril 1940

Monsieur et cher maître,

je viens d'apprendre par l'intermédiaire de M. Régnier que votre santé, si précieuse pour la Science et pour tous ceux qui vous aiment, s'améliore et que vous reprenez peu à peu votre travail. J'espère que madame Volterra si bonne et si courageuse, peut maintenant être plus tranquille et que sa propre santé, en fonction de la vôtre, ira vers une amélioration. En ce qui nous concerne, rien de nouveau. Nous travaillons beaucoup, et je vous enverrai bientôt de mes nouvelles scientifiques. Bien à vous V. Kostitzin (Ma femme joint aux miens ses meilleurs souvenirs + voeux.)

12.48 Kostitzin to Virginia Volterra, Paris, 16.6.1941 *Cat. no. 231*

Paris, 16 Juin 1941

Chère Madame, il y a quelque temps j'ai reçu une carte de M. D'Ancona, mais il n'avait aucune nouvelle de vous et nous serions heureux d'en avoir. Je prépare actuellement un livre sur l'oeuvre scientifique prodigieux de M. Vito Volterra. Si c'etait possible, je voudrais avoir dans ma disposition une collection plus complète de ses publications et quelques renseignements biographiques. Cependant, comme je n'ai pas reçu son dernier mémoire, que vous m'avez annoncé, peut-être vaut-il mieux attendre un peu avec les envois. Veuillez agréer, chère madame, l'expression de ma profonde et affectueuse considération et meilleurs voeux de la part de ma femme.

V. Kostitzin

12.49 Kostitzin to Virginia Volterra, Paris, 16.4.1942 *Cat. no. 232*

Paris, 16 Avril 1942

Chère madame, le Bureau International des Poids et Mesures m'a transmis tous les matériaux biographiques concernant votre inoubliable époux que je considère comme le plus grand mathématicien de notre époque. J'espère que mon ouvrage sur sa vie et son oeuvre sera bientôt terminé malgré toutes les difficultes et en dépit des nouvelles épreuves qui en ce moment, sont heureusement finies. Veuillez agréer, chère Madame, mes hommages respectueux et mes cordiales salutations.

12.50 Kostitzin to Virginia Volterra, Paris, 17.12.1945 *Cat. no. 233*

Paris, 17 Décembre 1945

Chère madame

Ce n'est pas sans appréhension que je vous écris. Le temps terrible par lequel nous tous sommes passés a accumulé trop de ruines et de malheurs. En ce qui nous concerne, la Gestapo m'a arrêté le jour même de la déclaration de la guerre à ma chère patrie - 22 juin 1941. On m'a mis au camp de prisonniers civils Frontstalag 122 à Compiègne et, heureusement, on m'a liberé neuf mois plus tard. Vous pouvez vous imaginer facilement l'état d'esprit et la situation de ma femme pendant ces neuf mois. Après ma libération nous avons commencé un travail clandestin, ce qui a améné la Gestapo la seconde fois chez nous - 29 Janvier 1944. Cela pourrait être beaucoup plus terrible pour nous, mais nous avons réussi à échapper aux policiers allemands sous leur nez, et cela nous a obligé de nous cacher pendant neuf mois. C'était une très dure epreuve parce que dans la nature humaine il y a de tout - du pire et du meilleur. Nous avons vu du pire chez plusieurs personnages du monde académique qui appréciaient trop leur sécurité pour en risquer au service de leur patrie, et nous avons vu du meilleur chez de gens simples qui, avec une solidarité humaine remarquable ont contribué à la libération. Donc, ces neuf mois de clandestinité étaient très durs, mais aussi très réconfortants. A notre retour chez nous nous avons trouvé notre appartement à moitié pillé: le linge, les couvertures, les draps, tous les objects d'habillement ont disparu, ainsi que plusieurs documents. Enfin, nous pouvons nous estimer heureux parce que il ne manquait que très peu de chose pour que notre retour soit tout à fait impossible.

Il est curieux que tous les documents que j'ai ramassé sur la vie et l'activité scientifique de M. Vito Volterra ont disparu, et il me faut recommencer le travail à partir de zéro. C'est pourquoi, chère madame, je vous prie de bien vouloir m'aider dans cette tache et m'envoyer tous les matériaux que vous jugez utiles pour servir à la biographie de votre cher, grand et inoubliable mari. Nous voudrions aussi savoir comment avez vous veçu pendant l'occupation, la guerre et de nouveau l'occupation de l'Italie.

Veuillez vous agréer, chère madame, mes hommages très respectueux et les souvenirs très affectueux de ma femme.

V. Kostitzin
6, Square de Port-Royal, Paris (13), Francia

12.51 Kostitzin to Virginia Volterra, Paris, 25.1.1946 *Cat. no. 234*

Paris, 25 Janvier 1946

Chère Madame,

c'est avec une très grande joie que j'ai reçu vos lettres et celles de M. D'Ancona. Je suis heureux que vous êtes en bonne santé et que vous continuez votre travail. Nous avions des appréhensions plus que justifiées sur votre sort. La vie humaine ne pèse pas lourd pendant une guerre, et dans cette guerre, avec le banditisme organisé nazi-fasciste, survivre était parfois un problème très grave, et combien nombreux sout ceux qui n'ont pas réussi à le résoudre. Parmi ceux que nous connaissons, très rares sont les gens qui n'ont pas souffert personellement ou qui n'ont perdu personne.

Je vous remercie de votre intention de me faire parvenir vos notices sur votre cher et inoubliable mari. Je les attends avec une très grande ìmpatience. En ce qui concerne M. J. Pérès, je sais très bien qu'il a beaucoup travaillé avec M. Volterra, j'apprecie son talent scientifique, mais, peut-être à tort, je ne sens pas en lui ni de la sympathie, ni de l'indifférence bienveillante pour ma personne, et dans des cas pareils j'évite toujours tout contact personnel, peut-être à tort.

Dans ma lettre précédente, j'ai oublié de vous dire que pendant mon séjour dans le camp allemand Frontstalag 122 à Compiègne (22 juin 1941 - 23 mars 1942) j'ai appris l'existence dans la partie française du camp d'un detenu italien Francesco Volterra. Malgré les barbelés qui nous séparaient des français et des italiens, je réussis à avoir avec lui une conversation. C'était un jeune homme de taille moyenne, roussatre. Il m'a dit qu'il appartenait à la même famille que Vito Volterra. A ma question, s'il avait besoin de quelque chose il repondit négativement. Un peu plus tard les français, ses codétenus, me dirent, qu'on l'a transféré, mais ils ne savaient pas où.

Veuillez agréer, chère madame, mes hommages très respectueux, et de la part de ma femme un souvenir très affectueux.

V. Kostitzin

12.52 Kostitzin to Virginia Volterra, Paris, 14.2.1946 *Cat. no. 235*

Paris, 14 Février 1946

Chère Madame,

je viens de recevoir les matériaux bio-bibliographiques très précieux que vous avez bien voulu m'expédier, et je vous en remercie. Je tiens à vous signaler que la bibliographie, donnée dans les mémoires de Somigliano et surtout de Whittauer a besoin de revisions[41]. Je ne manquerai pas de vous en communiquer les résultats.

A partir d'aujourdhui M. Montel[42] n'est plus doyen de la Faculté des Sciences. Il ne doit pas être très content de ce changement dans sa vie, à tort, à mon avis:

---

[41] This is SOMIGLIANA 1946 and WHITTAKER 1946.

[42] Paul Antoine Aristide Montel (1876–1975), a mathematician, taught in Paris at the École Polytechnique and at the Faculty of Science. He was also director of the École Pratique des

les fonctions administratives accaparent mon seulement tout le temps, mais encore tout le cerveau et un savant ne devrait jamais solliciter cette situation. Dans le cas de M. Montel il faut ajouter, qu'il est resté en fonction tout le temps de l'occupation allemande sous le governement vichissois. Il voulait concilier les choses incompatibles: sauvegarder, autant que cela était possible, les intérêts académiques, et rester bon patriote non collaborationiste. Or, pour la sauvegarde de la Faculté il fallait finasser, louvoyer, faire des concessions, et sur ce chemin Montel est allé un peu plus loin qu'il ne voulait. Résultat: mécontentement parmi ses collégues qui ne lui pardonnent pas les concessions à l'esprit vichissois, et oublient le bien effectif qu'il a réussi à faire. J'espère qu'il n'a pas perdu son talent pendant ces terribles années et qu'on lira bientôt de choses nouvelles et intéressantes.

Vous n'ignorez pas sans doute que pendant les dernières années la biologie mathématique n'était pas en faveur parmi les biologistes et que les méthodes statistiques (R. A. Fisher, Haldane) étaient en vogue. Cette situation, heureusement, ne va pas durer. Les études écologiques ramènent de nouveau les méthodes analytiques au premier plan. Par exemple, les travailleurs de la Station Biologique de Marseille (objet principale la région de la Camargue) étudient attentivement les publications de votre mari, les miennes et s'intéressent beaucoup à l'ouvrage de M. D'Ancona, malheureusement, introuvable en France[43].

Veuillez agréer, chère madame, mes hommages respectueux et les souvenirs très respectueux de ma femme.

V. Kostitzin

12.53 Kostitzin to Virginia Volterra, Paris, 29.4.1947 *Cat. no. 236*

Paris, 29 Avril 1947

Cher Madame

en revenant d'un petit voyage je trouve avec joie votre aimable envoi. Je sais que M. Volterra aimait l'Accademia dei Lincei et je suis très content de voir cet hommage qu'on lui rend dans l'Italie enfin libre. Votre envoi m'apport indirectement des nouvelles rassurantes de vous: ma dernière lettre de 15.12.46 est restée sans réponse. L'année 1946 était très triste: M. Jean Régnier, ami très précieux de M. Volterra (le mien aussi) est mort; il n'était pas vieux, loin de là, mais son organisme était très fragile. Puis deux morts violentes: au début de janvier 46 un camion américan a écrasé M.me Fréchet et au début de janvier 47 un camion français a écrasé M.me Pérès[44].

Hautes Études, president of the Palais de la Découverte and member of the Académie des Sciences. His attitude during the nazi occupation of France was highly controversial (as is shown also by this letter).

[43] D'Ancona 1939 (the Italian version was published in 1942 by Einaudi, Torino). An expanded version of the book is D'Ancona 1954.

[44] He is referring to the wives of the mathematicians Maurice Fréchet and Jean Pérès.

Je m'occupe de mon livre sur votre mari et j'espère le terminer l'année prochain. En passant au Bd. Raspail dans les parages de l'hotel Lutetia je me rappelle toujours les minutes inoubliables passées auprès de vous et de M. Volterra. Vous savez sans doute que pendant l'occupation allemande cet hotel fut occupé par un des bureaux allemands les plus détestables et ses murs doivent être impregnés par les larmes d'innombrables malheureux qui y ont passé.

Chaque fois que j'écoute la Symphonie Pastorale je pense à M. Volterra qui aimait beaucoup cette musique et qui aimait la nature.

Nous continuons comme toujours notre travail scientifique. Ma femme est attachée au laboratoire d'anatomie et d'histologie comparée à la Sorbonne. Je m'occupe beaucoup des problèmes météorologiques et mes recherches biologiques en souffrent, mais ce n'est pas ma faute.

Veuillez agréer, chère madame, mes hommages respectueux et l'expression de l'affection sincère de nous deux.

V. Kostitzin

12.54 Kostitzin to Virginia Volterra, Paris, 4.1.1950 *Cat. no. 237*

Paris, 4 Janvier 1950

Chère Madame et Amie,

votre petit mot m'a apporté beaucoup de joie; je suis très content d'apprendre que chez vous tout va normalement. Tel n'est pas notre cas. Ma chère femme est gravement malade. Son coeur, affaibli par les rhumatismes acquis à Moscou et par les souffrances et émotions ici (mon arrestation et internement par les allemands en 1941–42; vie dans le maquis en 1944), nous donne beaucoup des soucis depuis quinze mois et l'avenir est sombre. Vous m'excuserez donc que cette année-ci je suis en retard pour vous donner de nos nouvelles.

Veuillez agréer, chère Madame et Amie, avec mes hommages respectueux, nos voeux pour la nouvelle année et l'expression de notre vive et sincère affection.

V. Kostitzin

12.55 Kostitzin to Virginia Volterra, Paris, 21.1.1950 *Cat. no. 238*[45]

Paris, 21 Janvier 1950

Chère Madame et Amie,

l'avenir sombre dont je parlais dans ma lettre c'est mon présent actuellement. Ma chère et bien-aimée femme est décédée le 17 Janvier dans des souffrances terribles après un accident bête dans notre salle de bains. C'est l'embolie qui l'a emportée, mais dans l'hémisphère gauche du cerveau, ce qui a transformé pour

---

[45] Mourning postcard.

cinq jours ma femme en un être demi-paralisé et sans parole, mais avec conscience très claire. Je luttais de toutes mes forces pendant ces cinq jours pour la sauver. Elle me regardait avec des yeux infiniment tristes et de temps en temps une petite grimace désabusée et sceptique, que je connaissais si bien, apparaissait sur son cher visage. Vers la fin du cinquième jour de ce calvaire elle m'a fait de sa petite main gauche un geste d'adieu, et quelques minutes plus tard tout était fini.

Excusez moi si je vous fais de la peine en vous donnant ces détails qui n'ont de sens que pour moi, mais en ce moment je souffre et je suis seul. Pendant trente ans nous allions dans la vie ensemble, et dans chaque situation difficile (et il y en avait tant) je sentais dans ma main sa petite main, tendre, franche, honnête, courageuse et sûre. C'était le bonheur.

Veuillez agréer, chère madame, mes hommages respectueux.

V. Kostitzin

12.56 Kostitzin to Virginia Volterra, Paris, 30.12.1950 *Cat. no. 239*

Paris, 30 Décembre 1950

Chère Madame,

je vous envoie mes meilleurs voeux pour la Nouvelle Année. Pour moi l'année qui se termine était la plus terrible de ma vie et cependant je la quitte avec tristesse parce que pendant le 17 premiers jours ma femme était encore avec moi, et un espoir fou était permis.

Veuillez, chère Madame,transmettre mes meilleurs voeux à M. et M.me D'Ancona et agréer mes hommages très respectueux

V. Kostitzin

12.57 Kostitzin to Virginia Volterra, Paris, 28.12.1951 *Cat. no. 240*

Paris, 28 Décembre 1951

Chère Madame, comme chaque année, avec vive émotion je vous envoie mes meilleurs voeux et mes souvenirs affectueux.

Mon ouvrage consacré à M. Volterra, première rédaction, est terminé et je reflechis sur la forme définitive à lui donner.

Dans mon existence il n'y a rien de nouveau. La résignation, l'acceptation ne sont pas venues.

Veuillez agréer, chère Madame, mes hommages respectueux

V. Kostitzin

12.58 Kostitzin to Virginia Volterra, Paris, 26.12.1952 *Cat. no. 241*

Paris, 26. 12. 52

Chère Madame, veuillez trouver ici mes meilleurs voeux pour le Nouvel An, ainsi que l'expression de mes sentiments respectueux et affectueux

V. Kostitzin

12.59 Kostitzin to Virginia Volterra, Paris, 24.12.1953 *Cat. no. 242*

Paris, 24 Décembre 1953

Chère Madame,

Avec mes meilleurs voeux agréez mes remerciements pour l'envoi de la brochure commemorative. Elle est très caractéristique pour notre époque absurde qui ne menage pas les valeurs intellectuelles. Ici aussi on n'a rien fait pour fêter dignement les anniversaires de Victor Hugo, Balzac, Voltaire, Berlioz, etc. etc. Les éditions à bon marché sont publiées en centaines de milliers d'exemplaires, mais c'est à Moscou.

Veuillez agréer, chère Madame, mes hommages respectueux. Bien affectueusement à vous

V. Kostitzin

12.60 Kostitzin to Virginia Volterra, Paris, 28.12.1954 *Cat. no. 243*

Paris, 28 Décembre 1954

Chère Madame, je vous envoie mes

meilleurs voeux pour une bonne et heureuse année [printed]

Je pense sauvant à vous et à votre inoubliable mari avec une affection et reconnaissance

V. Kostitzin

12.61 Kostitzin to Virginia Volterra, Paris, 25.12.1955 *Cat. no. 244*

Paris, 25 Décembre 1955

Chère Madame,

Je vous envoie mes meilleurs voeux pour la Nouvelle Année. J'espère que vous êtes toujours en bonne santé. J'espére aussi que la nouvelle guerre n'est pas pour aujourd'hui.

Veuillez agréer, Madame, mes hommages respectueux et affectueux.

V. Kostitzin

12.62 Kostitzin to Virginia Volterra, Paris, 29.12.1956 *Cat. no. 245*

Paris, 29 Décembre 1956

Chère Madame,

Je vous apporte mes meilleurs voeux pour la Nouvelle Année. L'année qui se termine était assez triste: le monde se transforme, mais ne s'améliore pas, les gens qui constituaient notre univers, disparaissent, amis ou ennemis, mais les uns et les autres nous étaient chers. En ce qui me concerne, j'ai subi une maladie assez grave. Mais à quoi bon se lamenter? Il faut toujours regarder en avant. Je continue mon travail.

Veuillez agréer, chère Madame, mes hommages respectueux et très affectueux.

V. Kostitzin

12.63 Kostitzin to Virginia Volterra, Paris, 17.12.1957 *Cat. no. 246*

Paris, 17 Décembre 1957

Chère Madame,

Avec mes meilleurs voeux pour les fêtes et la nouvelle Année permettez moi de vous apporter le tribut des souvenirs inoubliables et mes hommages très respectueux.

V. Kostitzin

12.64 Kostitzin to Virginia Volterra, Paris, 25.12.1958 *Cat. no. 247*

Paris, 25 Décembre 1958

Chère Madame

Ce tableau chinois moderne représente le fleur de lotus, fleur d'oubli, mais je sais que ce mot n'existe pas pour vous, et moi je garde toujours présent le souvenir de votre inoubliable mari et de vous. Les voeux que je vous envoie à cette époque de l'année n'est jamais pour moi une simple politesse.

Veuillez agréer, chère Madame, mes hommages très respectueux

V. Kostitzin

12.65 Kostitzin to Virginia Volterra, Paris, 16.12.1959 *Cat. no. 248*

Paris, 16 Décembre 1959

Chère Madame,

Je vous envoie mes voeux les plus affectueux pour la Nouvelle Année. Je pense souvent à vous et à l'inoubliable M. Volterra.

Veuillez agréer, chère Madame, mes hommages respectueux

V. Kostitzin

12.66 Kostitzin to Virginia Volterra, Paris, 30.12.1960 ***Cat. no. 249***

Paris, 30.12.60

Chère Madame,

Je vous envoie mes meilleurs voeux avec retard à cause de la mauvaise grippe. Parfois il est assez difficile maintenir l'optimisme vital nécessaire au niveau souhaitable, mais il le faut.

Veuillez croire à mes sentiments respectueux et très affectueux

V. Kostitzin

12.67 Kostitzin to Virginia Volterra, Paris, 24.12.1961 ***Cat. no. 250***

Paris, 24 Décembre 1961

Chère Madame,

Je viens vous apporter mes meilleurs voeux pour la Nouvelle Année: puisse-t-elle n'être pas plus mauvaise que celle qui se terminé!

Je pense souvent à vous et à votre inoubliable mari. Vous savez sans doute que ses travaux biologiques (les miens aussi) étaient très mal côtés dans ma patrie. Mais cela change. A la bonne heure!

Veuillez agréer, chère Madame, mes hommages respectueux et affectueux

V. Kostitzin

12.68 Kostitzin to Virginia Volterra, Paris, 24.12.1962 ***Cat. no. 251***

Paris, 24.12.1962

Chère Madame,

Je vous envoie mes meilleurs voeux pour la Nouvelle Année. Croyez moi que ce n'est pas une formalité saisonnière. Très souvant je reviens en pensée à vous et à votre inoubliable mari.

Veuillez agréer mes meilleurs et très respectueux sentiments.

V. Kostitzin

13 Letters between Joseph Larmor and Vito Volterra

Joseph Larmor was born in Magheragall, County Antrim (Ireland) on 11 July 1857. His family later moved to Belfast and he received his education at the Royal Belfast Academical Institution, later enrolling at Queen's University, Belfast. In 1877 was admitted to St. John's College, Cambridge. He was elected fellow of the College in 1880, after winning the Smith's Prize for his exceptional mathematical skills. He was later awarded the chair of natural philosophy at Queen's College, Galway, where he remained until 1885, the year in which he returned to Cambridge, at St. John's College, first as lecturer and then, from 1903 on, as Lucasian Professor, replacing in this chair the eminent mathematician George Gabriel Stokes (1819–1903).

Larmor was elected fellow of the Royal Society in 1892 and he was the Society's secretary from 1901 to 1912. He was also member of the council of the London Mathematical Society from 1887 to 1912 and became its chairmen for the two-years 1914–15. The numerous prizes and awards received by Larmor include his nomination as Bakerian lecturer, the Royal Medal (1915) and the Copley Medal awarded to him by the Royal Society, as well as the De Morgan Medal (1914) awarded by the London Mathematical Society. A moon crater was also named after him (the Larmor Crater). Larmor was knighted in 1909 and was charged to represent Cambridge University in Parliament from 1911 to 1922 (he mentions this in letter no. 13.4 in the present correspondence). In this position he busied himself enormously with the problem of education and adopted a unionist stance concerning the status of Ireland.

He retired from teaching in 1932 and spent his remaining years in Ireland. He died there, at Holywood, County Down, on 19 May 1942.

Joseph Larmor's scientific activity was focused above all on electromagnetism, optics, analytical mechanics and terrestrial dynamics. He may be considered the last representative of the school of mathematical physics, the leaders of which were William Rowan Hamilton (1805–1865), James Clerk Maxwell (1831–1879) and William Thomson, Baron Kelvin (1824–1907). He kept faith with this classical tradition, maintaining a rather reserved, if not outright sceptical, attitude towards the new developments in physics in the field of relativity theory and quantum theory. The attention he paid to these theories was aimed above all at identifying their critical points and limits, while his main objective was to complete the classical interpretative construction, which he pursued by founding a theory of the electron based on the combination of the concepts of electromagnetism and optics.

His early research in the field of electromagnetism led him to tackle the central issue of the effect of the motion of matter in the ether, which became the main focus of his interests and the topic of a famous book published in 1900, *Aether and Matter* (LARMOR 1900). The book was a collection of the research developed over nearly two decades and set out above all in three publications that appeared in 1894, 1896, 1898 under the title "A Dynamical Theory of the Electric and Luminiferous Media". Among the more important results of this research is the explanation of the socalled FitzGerald-Lorentz contraction - which he gave quite independently of the well-known note by Lorentz - and which led him to the property of the equilibrium of electric charges in the ether in movement. He also arrived independently at a formulation of the Lorentz space-time transformation. He was also the first to calculate the rate of radiation of energy produced by an accelerating electron and to account for the splitting of the spectral lines in a magnetic field. He introduced and studied the socalled "Larmor precession" to which charges orbiting in an electromagnetic field are subjected. As mentioned above, Larmor investigated also terrestrial dynamics, analysing the causes of the irregular motion of the earth's axis and studying the problem of the definition of astronomic time[1].

* * *

13.1 Larmor to Volterra, Cambridge, 25.3.1908 *Cat. no. 253*

St JOHN'S COLLEGE
CAMBRIDGE

March 25. 1908

Dear Prof. Volterra

I must not longer delay my thanks for your great kindness in sending me your treatise on the elasticity of ring shaped bodies[2] and the Address to the Italian Association of Science[3]. The latter I had already read with much interest in the Rivista di Scienza. I can see much in the former that I shall take interest in.

I was delaying my thanks in the hope of meeting you at the Mathematical Congress at Rome. But I have had reluctantly to give up that journey. The affairs entrusted to me by the Royal Society are very exacting when they came in addition to my work here. And at present I am responsible for a biographical notice of Lord Kelvin which must be finished early in May.

---

1 See also CUNNINGHAM 1943, EDDINGTON 1942–3, WOODRUFF 1976.

2 Probably VOLTERRA 1907a.

3 VOLTERRA 1907b.

I have very clear recollection of being introduced to you when you visited Cambridge a few years ago, and I had looked forward to renewed meeting.

Very faithfully yours

J. Larmor

13.2 Larmor to Virginia Volterra, Roma, 11.5.1910 *Cat. no. 254*

GRAND HOTEL DE RUSSIE
ROME

Rome, le Mai 11, 1910

Dear Madame Volterra

I shall have much pleasure in accepting you and Prof. Volterra invitation to dějeuner on Saturday May 14, at 12 1/2

With best thanks

Yours sincerely

J. Larmor

13.3 Larmor to Volterra, London, 1.7.1910 *Cat. no. 255*

THE ROYAL SOCIETY,
BURLINGTON HOUSE, LONDON, W.

July 1.1910

Dear Prof. Volterra

I take the liberty of expressing my personal satisfaction that you will I hope now belong to the Royal Society as a Foreign Member. The election, which has been in progress for some months, has now been completed along with that of Arrhenius, Bornet, Ehrlich and Weismann; and you will be notified in due course by the Foreign Secretary.

With greateful remembrance of all your kindness at Rome and respect to Madame Volterra I remain very truly yours

J. Larmor

13.4 Larmor to Volterra, Cambridge, 9.6.1911 *Cat. no. 256*

(Same heading of letter no. 13.1)

June 9. 1911

Dear Prof. Volterra

Your kind and very welcome letter has just reached me. I must admit with some surprise as regards the main part of it. I have been ashamed for years of being so occupied with administration as not to have had time to keep abreast of mathematics and physics as I would have wished to do. Recently through the action of my friends in the University of Cambridge in sending me to Parliament my chance of doing so is still further reduced. But I hope to get out of the entanglement in time. Meantime the indulgent view of such competent judges as the Lincei on what little I was able to accomplish years ago is a strong additional inducement to try to be worthy of it.

We still recall, Sir Archibald Geikie and I, your kindness to us and that of the entire Academy in Rome last year. We want much desire to renew these pleasant memories in England. Perhaps we may be able to induce you to come again to Cambridge to the Mathematical Congress in August next year and Mme Volterra to whom I desire to send kindest remembrances.

I wish I were able to reciprocate your attention in writing in English: but I am now too old to begin to learn.

With very cordial thanks

Yours very sincerely

J. Larmor

13.5 Larmor to Volterra, Cambridge, 30.1.1913 *Cat. no. 257*

(Same heading of letter no. 13.1)

Jan 30. 1913

Dear Prof. Volterra

I must allow myself a few lines of condolance on the sad nature of the last message from you, and of sorrow at the loss that you and Mme Volterra have sustained.

Believe me yours ever faithfully

J. Larmor

13.6 Larmor to Volterra, Cambridge, 9.11.1913 *Cat. no. 258*

(Same heading of letter no. 13.1)

Nov 9. 1913

Dear Prof. Volterra

I have delayed in thanking you for your valued gift of your Paris lectures until I could read some of it[4]. I have just now read the beginning and the end, much appreciating the historical sketch in which you push the integral calculus beyond Newton and Leibnitz and on to Archimedes. I note your reference to Prof. Vacca[5], who is now I observe your colleague at Rome. I don't know when I shall have the opportunity to continue my reading of the book: so I write to thank you now.

---

[4] VOLTERRA 1913.

[5] Reference to the mathematician Giovanni Vacca (1872–1953), a student of Giuseppe Peano and specialist in the history of mathematics.

I have become entrusted in public affairs as you have been for long: and I find such persistent claims for public administration in which Science is involved that my time for study is almost hopelessly diminished. So it is the greater pleasure to be remembered by the most active workers. With kind remembrance of you and Madame Volterra's hospitality at Rome.

Very truly yours

Joseph Larmor

13.7 Volterra to Larmor, n.p., January 1915 *Cat. no. 378*[6]

Dear Sir

Ho ricevuto il Suo biglietto e La ringrazio dei Suoi auguri che ricambio cordialmente. E vivamente li contraccambio e li estendo a tutto il vostro paese verso il quale le simpatie già così grandi presso di noi sono ancora maggiormente accresciute nelle circostanze attuali.

Ho ricevuto da Sir Archibald Geikie vari opuscoli relativi alla guerra in cui si parla delle ragioni che hanno spinto l'Inghilterra nel conflitto, e della sua condotta verso il Belgio. Fra cui vi è anche la lettera colla quale gli scienziati inglesi hanno risposto ai Tedeschi.

Ho già risposto a Sir Archibald Geikie che per parte mia divido pienamente le idee manifestate dagli scienziati Inglesi nella loro risposta e che è viva in me l'ammirazione per la condotta dell'Inghilterra sia verso il Belgio sia nel voler salvare la Europa dall'aggressione dei due imperi tedeschi. E che la guerra sia derivata da una aggressione voluta e preparata dalla Germania e dall'Austria lo mostrano tutti i documenti comparsi.

Tali idee sono lieto che siano divise dalla gran maggioranza dei miei compatriotti ed io non dubito che esse serviranno di norma alla condotta dell'Italia il cui intervento io mi auguro possa condurre a notevoli conseguenze.

Sono ben lieto di inaugurare l'anno 1915 col voto che i vincoli tra i nostri due Paesi si facciano sempre più stretti.

Yours sincerely

Dear Sir,

I received your work; and I thank you very much for your kind wishes. I wish the same to you, and not only to you, but to all your great country, towards which all our sympathies already so large with us, are greatly increased in the actual state of affairs.

I received from Sir Archibald Geikie several essays about the war, in which it is spoken of the different reasons that have pushed England in the conflict, and of her behaviour towards Belgium. Among them there is the letter written by the Englishmen in answer to the Germans. I already answered to Sir Archibald Geikie that on my part I fully share the ideas expressed by the scientists of England in that letter, and that I have a very strong admiration for the behaviour of England, both towards Belgium and to oppose to the aggression of the two

[6] Draft with English translation.

German Empires. And all documents appear to show that this war is caused by the aggression desidered and prepared by Germany and Austria. I am happy to say that these are also the ideas of most of my countrymen and I am sure that they will be of a guide in the plans of Italy whose interference I hope may lead to considerable consequences.

I am also happy to oper the year 1915 with the wish that ties between our two countries may be still stronger.

Sincerely yours,

13.8 Larmor to Volterra, Cambridge, 13.1.1915 *Cat. no. 259*

(Same heading of letter no. 13.1)

Jan 13. 1915

Dear Professor Volterra

Your kind letter of the New Year was extremely welcome. I will not venture – I regret that it would not be possible for me – to reply in Italian, in contrast with your admirable English.

Five months ago we were all oppressed by the gravity of the decision that had suddenly to be made. But we saw no way out of it, and felt bound in honour and selfrespect to make it and take whatever consequences might ensue.

Since then the nation has become accustomed to the task it has entered upon, and we have on the whole come off hitherto much better than one might have expected. But it remains a great comfort to know that we have not been blamed by the public opinion of the world, and especially may I say of Italy and America.

The public spirit and selfsacrifice amongst us have risen much higher than might have been anticipated in our times of self-degradation which possibly are more obvious to ourselves than to the rest of the world. Last term half the University of Cambridge were away with their regiments; and for nearly all the remainders who were elegible, everything academic gave way to hard military training all day long. There are few left in my department except post-graduate students mainly from other countries.

At present our absent students are replaced in the town by troops, and the chambers and domestic life of the college are occupied by officers to their fullest capacity.

The terrible outlook in a situation otherwise full of hope is the dreadful loss of life and suffering that in a contest long drawn out seems inevitable in this modern warfare.

Meantime, with very cordial thanks for your letter, and hopes for happier times of peace, and of the still more intimate cooperation and understanding with the Italian nation, of which you speak, and with kind remembrance to Madame Volterra, I remain

Very truly yours

Joseph Larmor

13.9 Volterra to Larmor, n.p., May 1915 *Cat. no. 379*[7]

La fotografia rappresenta la seduta del Senato del 20 Maggio mentre l'Onorevole Salandra legge la sua dichiarazione. La vista è quale io avevo dal mio posto

The picture represents the meeting of the Senat on the 20th of May, while Hon. Salandra is reading his declaration. The sight is the same I had from my place.

I am very happy that Italy is fighting with England.

Long life to England

Yours truly

13.10 Larmor to Volterra, Cambridge, 24.5.1915 *Cat. no. 260*

Evviva Italia !

J. Larmor

Cambridge, May 24. 1915

13.11 Volterra to Larmor and Rutheford, n.p., n.d. *Cat. no. 374*[8]

Larmor

Sono lieto di presentarle il Sig. Dott. Pontremoli[9] laureato ultimamente dopo buonissimo esame in Fisica nella nostra Università il quale come ebbi già l'onore di scriverle si reca costà per proseguirei suoi studi di fisica e matematica.

Lo raccomando caldamente a Lei e mentre Le invio i ringraziamenti.

Rutherford

Dear Sir

sono lieto di presentarLe e di raccomandarLe il Sig. Dott. Pontremoli distinto allievo della nostra Università dove si é laureato in Fisica dopo aver sostenuto ottimi esami. Egli si reca costà desiderando di poter frequentare il laboratorio di Fisica e proseguire i suoi studi. Spero che Ella vorrà aiutarlo e mentre La ringrazio anticipatamente.

---

[7] Draft, n. d. but obviously written in May 1915.

[8] Draft.

[9] Reference to the physicist Aldo Pontremoli (1896–1928) who died on the North Pole expedition led by Umberto Nobile.

13.12 Volterra to Larmor, n.p., 1920 *Cat. no. 377*[10]

Dear Sir Joseph

I found at my return to Rome J. Clerk Maxwell's book: Matter and Motion that you have very kindly sent me. Thank you very much. This book with your Note is very interesting.

I have found a copy of Bartoli's paper on the motion provided by light and heat. You will find in the second part (page 22nd) Bartoli's demonstration on the pression of the light. This paper was not published in scientific transactions or journals, but Bartoli published it privately. Therefore few scholars have read the original demonstration. I am happy to send it to you. I hope you will accept as a friendly token of the time spent together in Strasbourg. I remember with great pleasure our conversations from which I have much learned.

My friend [illeg.] found in Newton's principia the following passage:

I think it is very interesting to meditate and to compare it with the principles of the relativity.

Yours very faithfully friend

13.13 Larmor to Volterra, Cambridge, 27.10.1920 *Cat. no. 261*[11]

Cambridge Oct. 27. 1920

Dear Prof. Volterra

I was glad to receive your letter, and the Bartoli paper. I find that Boltzmann in his last papers of 1884 gives an adequate account of B.'s argument. The [illeg.] point is that the presence of natural radiation [illeg.] up to 1/3 E instead of E: and that it is natural radiation that one has to deal with because if a surface [illeg.] arbitrary directed radiation it must be also [illeg.] natural radiation. Thus I think that Boltzmann's solution was [illeg.], though this [illeg.] suggestion was Bartoli's: [illeg.]

[The rest of the letter is completely illegible][12].

---

[10] Draft, n.d. but obviously written in 1920: see following letter.

[11] This letter is practically illegible.

[12] It is very difficult to reconstruct the subject of this letter and of the preceding one both because the hand-writing of the second is practically illegible and because it refers to an unpublished article. Nevertheless, the work by Boltzmann to which Larmor refers is probably BOLTZMANN, "Ableitung des Stefanschen Gesetzes betreffend die Abhängigkeit der Wärmestrahlung von der Temperatur aus der elektromagnetischen Lichttheorie, *Wiedemann Annalen*, **22**, 1884, pp. 291–294.

13.14 Volterra to Larmor, n.p., n.d. *Cat. no. 382*[13]

Dear Sir Joseph,

Thank you very much for your kind letter. Opening with d/dy on my solution the ciclicity happens but I think the solution is not useful for the application of Green's method.

It is very singular that the two cyclic solutions for the equations

$$\frac{\partial^2}{\partial x^2}+\frac{\partial^2}{\partial y^2}+\frac{\partial^2}{\partial t^2}=0$$

and the equations of the double refraction have no applications. What is their physical interpretation?

There are more things in heaven and earth, (specially in the waves's theorie), than are dreamt of in your philosophy.

Can you give me information how was transformed the English's office of inventions after the war? In France and in America were found the Direction des recherches scientifiques et industrielles and the Natural research Council. In Italy we have a project that is waiting for the approbation of the Parliament.

Yours very truly friend

13.15 Volterra to Larmor, n.p., 1925 *Cat. no. 375*[14]

Dear Sir Joseph

I thank you very much for the beautiful Memoirs you sent me and which I read with the greatest pleasure. Many observations and new ideas were brought to me, by that reading, and I hope to be able to speak of them with you.

I have the honour to send you a copy of my last volume on Permutable Functions that I have written with Prof. Perès[15].

I hope you will accept it as an homage and as one of my remembrances.

I have proposed for a fellowship some days ago to the Int. Educ. Board Dr. Persico[16], a clever scholar of our University. So that he may be able to study Mathematics and Physics for next six months at the University of Cambridge, beginning from July. I hope he will be allowed to assist your course at the University and work at the Cavendish laboratory.

---

[13] Draft, n.d. The year 1934 was subsequently annotated on the letter although this dating is not reliable and the most likely position is this one.

[14] Draft, n.d. but probably written in 1925.

[15] VOLTERRA, PÉRÈS 1924.

[16] Reference to the physicist Enrico Persico (1900–1960), a fellow student of Enrico Fermi, who played a fundamental role in the spread of theoretical physics in Italy.

13.16 Larmor to Volterra, Cambridge, 12.6.1925 *Cat. no. 262*

Cambridge, 12.VI.25

Dear Professor Volterra

I have delayed too long an answer to your note pressed by many engagements. I heard from Rutherford that your friend Persico is expected at the Cavendish Laboratory. I hope to make his aquaintance also, but I may be away this summer, having been here continuously since sept. last

I have been thinking again about your curious result that the optical disturbance from a centre in a crystalline medium has a trail unless you plit the medium between the optic axes. Huygens thought otherwise in his theory of a wavefront, which thus requires modification. Also, what I did and fully comprehend when you informed me, that in an isotropic medium of $2n+1$ spatial dimensions there is no trail, but of $2n$ there is.

Some day this kind of physical sciences will come back again into favour.

Very truly yours

Joseph Larmor

13.17 Volterra to Larmor, Roma, 11.6.1926 *Cat. no. 380*[17]

VIA IN LUCINA, 17
ROMA

June 11th 1926

Dear Sir Joseph

I owe you always my heartiest thanks for your kindness towards Dr. Persico. He and we all are very glad of the summer he spent in Cambridge and of the advantage he took from it.

I thank you also for all you have done to let me have the numbers of the Procceding of the London Mathematical Society I missed. I have received them regularly. I am also very grateful to the Editor who sent them to me.

In the last time I have studied a problem of mathematical biology; it leads to results which agree with the observations[18].

Here enclosed I send you a short note on this subject as I wish to let you know it.

I should be very glad if this note (also shortened by the editor if he thinks it necessary) could be published in Nature. As I don't know the editor, I beg you, if you think it is possible, to ask if that journal would pubblish my note.

Perhaps this summer I shall come to England; if so I shall be very glad to see you.

With many thanks and best regards

---

[17] Draft.

[18] Compare the content of this and the following letters with the correspondence with D'Arcy W. Thompson. They show that Larmor was the go-between in the establishment of relations between Volterra and D'Arcy Thompson.

13.18 Larmor to Volterra, Cambridge, 17.6.1925 *Cat. no. 263*

Cambridge, June 17 1926

Dear Prof. Volterra

Your welcome letter has arrived. I have read with much interest the biological mathematics, and ventured to alter the expression of the English in places The main result is however to stimulate one to see the nature of the hypotheses and proofs which establish fluctuation. I observe that these are already on the press.

I doubt whether working biologists would grasp what is not easy to me. But we have here a biologist with a mathematical [illeg.] Prof. D'Arcy Thompson*, of St. Andrews, Scotland, who is on the North Sea Fishery Bound and so in the way of being interested in such work. If you allowed me to send your ms to him, he might quite probably be interested in putting the results into form that the biologists could more readily grasp, - provided you send him the memoir also when it appears. Or you may prefer that it go to <u>Nature</u> at once: I observe that it amounts to 2500 words or more, which is long for that journal.

I think of going to Switzerland soon for a holiday as I am [illeg.]: I fear miss you if you come to England; but [illeg.]

Very truly yours

Joseph Larmor

* His book "Growth and Form" (Cambridge Press) you may have seen

13.19 Volterra to Larmor, Paris, 1926 *Cat. no. 381*[19]

HOTEL LUTETIA
43, BOULEVARD RASPAIL
PARIS 6^E^

Sir Joseph Larmor
St. John's College
Cambridge
England

Dear Sir Joseph

Je vous suis très reconnaissant de votre aimable lettre et de la peine que vous vous êtes donné de lire ma Note et d'en corriger quelques phrases Anglaises.

Je n'aurais aucune difficulté à consentir de l'envoyer au Professeur D'Arcy Thompson. Les ouvrages de ce savant sont très connus et très appréciés chez nous. Mais, comme vous remarquez justement, il faudrait lui envoyer aussi mon mémoire. Or celui-ci est sous presse et aucune feuille n'est encore parue. Dès que le première feuille paraîtra je pourrais l'envoyer pour la presenter au professeur

[19] Draft, n.d. The letter was subsequently annotated as 1926 which appears a reliable dating.

Thompson (aussi qu'à vous même) et ainsi de suite les suivantes. Je serai heureux d'avoir son avis et votre avis sur mes recherches.

Mais en attendant je serais heureux de voir ma Note publiée dans "Nature". Je pense qu'on pourrait la raccourcir en supprimant l'exemple des trois espèces dans une île, et on pourrait, tout en citant Ch. Darwin, renoncer à rapporter ses paroles. Est-ce qu'ainsi réduite la Note ne dépasserait pas les limites ordinaires? Si ces reductions ne seraient pas suffisantes est-ce qu'il serait possible de revoir quel est le nombre des mots qui serait accepté par la rédaction du journal? Je tâcherai alors de me restreindre dans ces limites.

Le but de ma Note est de rappeler l'attention sur les recherches que j'ai faites, si elles peuvent intéresser quelque biologiste ou quelque mathématicien tout en renonçant à pouvoir en donner une idée complète dans un si court aperçu.

Votre lettre m'a rejoint au moment où je quittais Rome, pour me rendre à Paris. D'ici j'irai Dimanche à Bruxelles pour une réunion du Conseil international des recherches et je resterai dans cette ville probablement jusqu'à Jeudi de la semaine prochaine. Je ferai retour après à Rome.

Ce ne serait que plus tard que je me rendrais en Angleterre; si vous y serez à l'époque de mon voyage je ne manquerai pas de vous avertir et je serai heureux de vous voir.

Veuillez accepter mes meilleurs remerciements et veuillez bien me pardonner la liberté que j'ai prise en vous donnant tant d'ennuis.

13.20 Larmor to Volterra, Cambridge, 28.6.1926 *Cat. no. 264*

Cambridge, June 28. 1926

Dear Prof. Volterra

I am sending the MS. to Prof. D'Arcy Thompson asking him to send it on to Nature if he consents, as being forwarded by him and me jointly. It will have more force that way as I cannot pretend to be a biologist.

They are usually overloaded with material nowaday.

Very truly yours,

J. Larmor

13.21 Volterra to Larmor, n.p., n.d. *Cat. no. 376*

Dear Sir Joseph

I regret very much to have not seen you in Edinburgh and to be not able to go to Cambridge.

I am going today to London Grosvenor Hotel and I shall leave London on Monday 19th September at 11 o'clock

I remain

Yours very truly

14 Letters between Alfred J. Lotka and Vito Volterra

Alfred James Lotka was born on 2 March 1880 at Lemberg in Austria (now Lvov in the Ukraine) of American parents. He received his early schooling in France and Germany, followed by university studies in England at the University of Birmingham. These were followed by graduate courses in physics at Leipzig and Cornell (USA). After moving permanently to the United States, he worked in a wide range of different jobs: for the General Chemical Company, the U.S. Patent Office and the National Bureau of Standards. After his residence at the John Hopkins University, during which he wrote the book that was later to make him famous, *Elements of Physical Biology* (published in 1925 and then reprinted in 1956 under the title of *Elements of Mathematical Biology*), he joined the Metropolitan Life Insurance Company of New York. Here he was to work for the rest of his life, in the position of supervisor of the Statistical Bureau. In the period 1938–1939 he was Chairman of the Population Association of America and in 1942 Chairman of the American Statistical Association. Lotka died at Red Bank (New Jersey) on 5 December 1949.

Lotka's work is highly eclectic and touches upon a vast range of topics, including the dynamics of chemical reactions, the theory of evolution – always at the focus of his interest, following a Spencerian approach –, and population dynamics and statistics. The appendix to *Elements of Physical Biology* contains a list of 94 articles that gives some idea of the huge scope of his interests. This eclecticism gave rise to a rather cold response to his work and meant that he never succeeded in obtaining a position in academic circles. His research programme was extremely ambitious: his *Elements of Physical Biology* actually set out to describe the entire biological world viewed as an integrated unity – a vision resembling the one proposed by Russian scientists such as Vernadsky, Kostitzin and Gause. This description was to be pursued using the tools of mathematical analysis and following a physical analogical approach. According to Lotka, "nature must be considered as a whole if we wish to understand it in detail". It is for this reason that he tackles such general issues as the scientific (and if possible, mathematical) definition of the concept of evolution, life, living matter and vital force, the origin of the subjective feeling of the unidirectional nature of time and the relations between consciousness and physical energy. Lotka's interest in these topics emerges also in his correspondence with Volterra, published herein. In his study of oscillating chemical reactions (LOTKA 1910, 1920b, c) Lotka had succeeded in coming up with a system of differential equations that were actually the same as those used by Volterra and himself in the mid 1920s to describe the dynamics of a biological association of prey and predators. This gave birth to the question of the priority to give to the discovery of these equations that were extensively mentioned in the correspondence between him and Volterra and between Volterra and D'Arcy W. Thompson (see also letters nos. 12.9 and 12.10 in the correspondence between Volterra and Kostitzin).

After *Elements of Physical Biology*, Lotka published a book in French in the Hermann collection, in which also the books of Volterra-D'Ancona, Gause and Kostitzin had been published (VOLTERRA, D'ANCONA 1935, GAUSE 1934, V. A. KOSTITZIN 1934, 1935a). The subject of the book was the "analytical theory of biological associations", and it was divided into two parts (LOTKA 1934, 1939). Lotka was preparing an English edition of the book when he died.

Although Lotka's work had little influence during his lifetime it subsequently attracted such passionate readers as Norbert Wiener, Nicholas Rashevsky and Ludwig von Bertalanffy, who deserve credit for transmitting and developing the fundamental ideas expressed in it. Anyone talking of the "thermodynamics of biological systems", "systems theory", of the "holistic" approach, and of "complexity", owes a lot - whether they know it or not - to the ideas introduced by Lotka over seventy years ago[1].

* * *

14.1 Lotka to Volterra, New York, 2.11.1926 *Cat. no. 265*

METROPOLITAN LIFE INSURANCE COMPANY
HALEY FISKE, PRESIDENT
NEW YORK CITY

STATISTICAL BUREAU
[...] Alfred J. Lotka
Supervisor

ce 2 novembre 1926

Monsieur,

C'est avec un vif intérêt que j'ai lu votre article sur les variations dans les nombres d'espèces biologiques, qui a paru dans le numéro du 16 Octobre de Nature[2]. Je n'ai pas encore pu me procurer une copie de votre memoir détaillé.

Je vous ai envoyé l'autre jour un exemplaire de mon livre "Elements of Physical Biology"[3], dans lequel j'ai traité le même problème par des méthodes alliées au votres. Cependant, j'ai admis qu'il y ait des termes de premier degré

[1] See GRIDGEMAN 1976, KINGSLAND 1985, ISRAEL 1982a, 1988, 1993b.

[2] VOLTERRA 1926b.

[3] LOTKA 1925.

dans les membres droits des équations fondamentales. Certains cas qui sont ensuite relevés à titre d'éxemples, indiquent qu'en effet de tels termes doivent se présenter quelquefois, par exemple, quand une espèce subsiste en se nourrissant des produits déchets d'une autre. Ces produits figurent alors parmi les X, car ceux-ci comprennent non seulement les êtres vivants, mais encore certaines autres matières éssentielles pour le maintien de la vie. Pour un X_j qui prend naissance comme produit déchet (excréments, cadavres, feuilles mortes) d'une espece X_i, la dérivée $\mathrm{d}X_j/\mathrm{d}t$ sera proportionnelle a X_i, la masse ou le nombre de l'espèce d'où sort le produit X_j. Il est vrai qu'un interêt tout particulier se rattache aux termes de second degré, vu que la relation la plus connue entre deux espèces est celle de nourriture et consommateur. J'ai traité dans mon livre un ou deux cas de ce genre, avec des résultats semblables aux vôtres. (Voir particulièrement pages 88, 90, 92). Il me semble bien heureux que vous aviez bien voulu vous interesser dans le problème plus général des équations de ce genre en un nombre de variables dépendantes supérieur a deux.

Le but que je m'étais posé dès le commencement de mon enquête est de trouver une éxpression définitive pour la direction du processus de l'évolution. Comme vous, j'ai abouti a certaines formes quadratiques desquelles on peut dire qu'au cours de l'évolution elles tendent à un maximum. Le feu Professeur Mac Mahon, en conversation au sujet de mon article "Evolution and Irreversibility" (Science Progress, janvier 1920, page 406)[4] se servit un jour de l'expression "Evolution Potential" pour désigner la fonction

$$\left(\lambda_1 \xi_1^2 + \lambda_2 \xi_2^2\right)$$

qui figure en forme quadratique caractéristique de l'equilibre de la malaria d'après mon étude sur les équations de Ross[5]. Ce terme est bien expressif, mais il me semble qu'il vaudrait peut-être mieux de le réserver pour un vrai potentiel ayant la dimension d'une énergie, ou plutôt d'une puissance (energie par unité de temps), qu'on devrait chercher à relier aux systèmes évoluants. A cet égard je dois vous renvoyer à page 357 de mon livre.

Quand à cela, il semble que l'intérêt principal ne serait pas dans la cinétique autant que dans la dynamique des systèmes comprenant un grand nombre de transformateurs d'énergie ayant certaines propriétés caractéristiques. Dans la dernière partie de mon livre j'ai tâché de développer cette branche de la science. La discussion nous mène ici a l'étude de l'appareil à l'aide duquel ces transformateurs réussissent à accaparer l'énergie libre nécéssaire pour leur fonctionnement. Il entre dans le discours certaines dérivées partielles

$$\frac{\partial r_i}{\partial X_j}$$

que j'ai reliées à la "valeur économique". Basant sur les principes de Spencer et Jevons, que l'être sentient cherche à faire son plaisir un maximum, en même

[4] LOTKA 1920a.

[5] See ROSS 1911a, 1911b, 1923.

temps qu'il agit dans l'intérêt de l'espèce, une relation se relève entre la valeur économique et les dérivées partielles

$$\frac{\partial r_i}{\partial X_j} .$$

Cette relation serait parfaite chez une espèce parfaitement adaptée. Elle s'éloigne, dans les espèces actuelles, plus ou moins de cet idéal.

Voici les reflexions qui surgissent tout involontairement à la lecture de votre article. Je serais bien content de croire qu'elles ont aussi quelque intérêt pour vous.

Si mon livre vous parait d'interêt suffisant, peut être que vous vous sentiriez prêt a en recommender la revue dans un journal Italien ou international, tel que Scientia. Point nécessaire de dire que je vous en serai bien reconnaissant. Je l'apprécierais aussi si vous vouliez avoir la bonté de me tenir au courant de vos publications au sujet de la biologie physique.

Agréez, Monsieur, l'expression de mes sentiments les plus respectueux.

Alfred J. Lotka

14.2 Volterra to Lotka, n.p., November 1926 *Cat. no. 383*[6]

La ringrazio del cortese invio della Sua importante opera "Physical Biology". Sono dolente che non conoscevo il Suo libro che non ho potuto citare nella mia Memoria, né nel Giornale Nature ma riparo avvertendo oggi stesso l'editore del Giornale Nature.

Ho l'onore di mandarle la mia Memoria pubblicata nella R. Accademia dei Lincei.

Dovendone fare una nuova ristampa nelle Memorie del R. Comitato Talassografico Italiano[7], farò le dovute citazioni dei suoi risultati.

14.3 Volterra to Lotka, n.p., November 1926 *Cat. no. 384*[8]

Dear Sir,

La ringrazio molto dell'invio del Suo interessante libro ... Non mi era noto quanto si trova nella p. 90, da quale ora soltanto ricavo che una delle questioni da Lei trattate ed una di quelle che ho svolto nelle mie memorie coincidono. Le mando una copia delle mie Memorie nella quale troverà lo sviluppo di questa e di altre questioni. Sarei ben lieto se suscitasse il Suo interesse ed Ella potesse farne qualche utile applicazione. Da parte mia studierò con tutta la cura il suo libro del quale ho apprezzato molto la quantità di informazioni in campi svariatissimi. Non mancherò in una prossima pubblicazione di citare l'opera che Ella mi ha mandato.

---

[6] Draft, n.d. but certainly written on November 1926.

[7] VOLTERRA 1927d.

[8] Draft with English translation, n.d. but certainly written on November 1926.

Dear Sir,

I thank you very much for your interesting book. What you say on page 90 was not known to me; I see there only now that one of the problems you have treated coincides with one of those I have developed in my memoir.

I send you a copy of my memoir, where you will find this and other problems developed. I should be very glad if they should interest you, and you clould make some useful application to them.

On my part I shall study your book withe the greatest attention; I have much appreciated the vast amount of information it contains on a great variety of subjects. In one of my next publications I will not fail to mention your work and to acknowledge your priority respecting the problem on page 90 and your results.

Very sincerely yours

14.4 Volterra to Lotka, n.p., 20.11.1926 *Cat. no. 385*[9]

Cher Monsieur,

Je vous suis très reconnaissant de votre aimable lettre et de l'important recueil de vos mémoires que vous avez eu l'obligeance de m'envoyer. J'en ai parcouru la plus grande partie en me repromettant de les approfondir à mon aise, mais en attendant j'ai lu les deux mémoires N° 3, N° 13 et j'ai compris la voie que vous avez suivie pour obtenir les équations dans le cas de deux espèces[10]. C'est l'analogie du cas biologique avec la question chimique qui vous a guidé.

Pour ma part j'ignorais la question chimique et c'est pour repondre à la question que M. D'Ancona me posait que j'ai obtenu les mêmes équations au §3. J'en ai donné une justification par le concept des rencontres des individus dans le §4 et j'ai employé le même concept dans le cas général de n espèces (§5). Je reviens dans le §11 sur la question des rencontres pour remarquer qu'il m'était nécessaire étendre les équations comme est indiqué dans le même §. Les trois lois que j'ai déduites des intégrales au §3 m'ont servi pour la question de la pêche ainsi que ce que j'ai fait dans le §4 et plus spécialement dans le §8 où je tâche d'obtenir les conditions les plus avantageuses pour la pêche[11].

Je crois que la considération des associations conservatives pour lesquels ont peut obtenir une intégrale qui montre la persistance des fluctuations et pour lesquelles on peut étendre les susdites trois lois puisse présenter de l'interêt, mais je pense que ce sont les associations dissipatives qui s'approchent le plus de la réalité.

Il est bien singulier que travaillant d'une manière indépendante partant de points de vue différents nous avons abouti dans le cas de deux espèces, aux mêmes équations. D'un côté j'en suis heureux parce-que cela confirme la justesse des vues sur le sujet, mais d'un autre côté je regrette beaucoup de ne vous avoir pas cité. C'est ce que je ferai bientôt, car j'ai sous presse un résumé de mon

[9] Draft, n.d. but certainly written on November 20, 1926: see letter no. 14.6, Cat. no. 267.

[10] The work referred to clearly consists of two out of the three publications LOTKA 1910, 1920b, c.

[11] Here Volterra is referring to VOLTERRA 1926a.

mémoire pour le journal Scientia[12]. Dès que j'en recevrai les épreuves j'y ferai la citation de votre livre et vos travaux, ainsi que dans un probable remaniement de mon mémoire. Je demanderai aussi au Directeur de Scientia s'il acceptera un compte rendu spécial de votre livre.

Les membres de la commission des études de la pêche dans la mer du Nord se sont intéressés à mon mémoire pour les applications à la pêche; et en particulier M. J. Schmidt m'a écrit en me poussant à la publier en Anglais. Il me donne l'avis qu'il serait possible de le faire publier par la Smithsonian Institution. Puisque j'ai eu la chanche d'entrer en rapport avec vous si l'ensemble de mon travail vous interesse pourriez-vous vous intéresser à le faire publier en Anglais dans les Recueils de la Smithsonian Institution ou ailleurs? Il va sans dire que la citation de vos travaux n'y manquera pas.

Dans votre lettre vous me parlez de la question la plus importante: la question dynamique et énergétique. Elle prime naturellement toutes les autres. J'y avais beaucoup songé et j'y songerai encore. Vous qui avez des idées si vastes et qui avez approfondi un domaine si étendu vous pouvez obtenir là dessus des résultats d'une grande portée. C'est ce que je vous souhaite de tout mon coeur.

14.5 Lotka to Volterra, New York, 1.12.1926 *Cat. no. 266*

(Same heading of letter no. 14.1)

New York, 1 Decem. 1926

Dear Sir,

Please accept my cordial thanks for the copy of your memoir on fluctuations in the number of biological species[13]. I need hardly tell you that I have been reading this with the greatest interest. In addition to what I knew I would find in it from my perusal of your shorter in article in Nature, I have, of course, found in your longer monograph many things of prime importance. I will not now enter in a discussion of the matter; only one point I might speak of in passing, namely that I have been particularly interested in Section 4 in which you broach the question of encounters between individuals of the two species. I believe that here is the starting point for some further developments. The problems of the statistical mechanics of systems of energy transformers (for that is what animals and plants are), hinge just on this question of the collisions or encounters between the individual organisms. All these encounters are determined, not, as in the comparatively simple case of gaseous molecules, merely by velocities and positions, but by an elaborate mechanism comprising sense organs, motor organs and adjustor faculties. If you have been turning over the leaves of my book, you may have found that I have a little to say on the question of collisions on page 359. What I have there said I would like to modify only in one minor particular. The discussion is there conducted as if the motion of the organism extended over the twenty-four hours of the day. In actual fact, of course, only a certain fraction

[12] Volterra 1927f.

[13] Volterra 1926a.

k of the twenty-four hours are so employed, and this fraction should enter in obvious manner into the discussion. The discussion will further have to take into account the kind of thing that is discussed on pages 346 to 358. I am at present giving my thoughts to this phase of the subject and I am hoping that perhaps in a little while I may have something in readiness for publication. There is something unsatisfactory in the rather extreme degree of idealization, or conventionalization, which it is necessary to introduce into the discussion at the present stage of this very new science. I believe that it is necessary for us to deliberately overcome a certain repugnance which one feels towards a such extreme conventionalization and to proceed with the work in the hope that the first crude steps may turn out in time to have been necessary preliminaries for a more perfect treatment of the subject. Owing to immense complexities of the physical systems presented to us in living nature it is necessary, I believe, in order to be able to progress at all, to begin with modest and greatly simplified examples. From your monograph it appears to me that you entertain somewhat similar thoughts. Naturally, I should be greatly interested in a expression of your views.

This morning I received a letter from the editor of Nature telling me that he had forwarded to you a brief note which I sent to him some weeks ago, pointing to the fact that certain points brought out in your paper were also dealt with in my book[14]. While there are many points of difference in the details of your development and mine, it appears to me quite remarkable how closely parallel our efforts run as regards the main issues. Even certain of the details are remarkably alike, I would like to think that this is a sign that we are moving in the right direction.

I trust that my brief note to Nature which the Editor tells me is submitting to you may meet with your approval; but if there is anything about it that does not please you I shall, of course, be only too glad to comply with your desires in the matter. Your letter which reached me a few days ago was, I think, in reply to the receipt of my book, and probably not in reply to the letter which I mailed to you about a week after sending the book. In the meantime I presume that you have received that letter and also some reprints which I sent you.

Very truly yours,

Alfred J. Lotka
Supervisor of Mathematical Research

14.6 Lotka to Volterra, New York, 13.12.1926 *Cat. no. 267*

(Same heading of letter no. 14.1)

New York, 13 Decem. 1926

Dear Professor Volterra

Your letter of November 20th has given me great pleasure. It has very definitively encouraged me to renew the work were I left off and try to push it

[14] For this letter and Volterra's reply see LOTKA, VOLTERRA 1927. On the issue of priority see also the D'Arcy W. Thompson-Volterra and Kostitzin-Volterra correspondence.

further. I am inclined to think I see my way to further advances. Your very kind interest and good wishes are of material assistance to me in renewing my energies on a topic in which there has not always been much encouragement for my work and in which I had almost come to feel that I would not be able to do much more hereafter, but I feel differently since reading your letter.

I have now received both your memoir and the other reprints which you kindly sent me. Among these, particular interest attaches for me to the one on hereditary phenomena and on line functions. The first of these has an obvious relation to the treatment of equations affected with a lag or lead, to which I have had occasion to refer in my book on page 47, 48 and more in detail in my malaria monograph on page 96[15]. It appears that the treatment of differential equations involving a lag or lead is liable to become mathematically quite difficult when certain functions step outside the bonds of linear relations.

Your paper on line functions is of interest in connection with what I have called intra-group evolution. Here the distribution of matter may be defined in terms of frequency functions, so that you have virtually an infinite number of variables. It has appeared to me quite singular how closely parallel in many respects your memoir and my work have run. Of particular interest in this connection is the fact that both you and I have in one way or another placed considerable emphasis on the problem of fisheries. It is true that on page 95 of my book I refer to the matter only incidentally, but in subsequent sections, (pages 171–184), the matter is referred to again in more detail although not along mathematical lines. It looks as if the problems of fisheries might be among the first to lend themselves to analytical treatment; at the same time the principles involved are somewhat general and should surely in time find applications elsewhere. As a matter of fact the problem of fluctuations of land species presents certain points of very particular interest on which I am exerting some of my efforts at the present time.

As regards the publication of your paper in English, perhaps by the Smithsonian Institution, I can only say that one the first thoughts that occurred to me in reading it is that I wish it were available in English. Since then, in the progress of working through the paper, I have acquired some reading facility in Italian so that personally, I no longer feel the need for an English edition as I did at first. Nevertheless, it does seem very desirable that we should have an English edition. Whether I am able to do anything toward realizing such a project it would be hard to say. I have not personally any connection with the Smithsonian Institute. However, I have acquaintances in Washington and I can possibly get away from a day or two, I am going to make an effort in the near future to run up to Washington and I can take the opportunity to mention the matter to one or two men there who may I think, be able to help bring about the desired result. In the meanwhile, if you have means of establishing direct contacts, I do not want to cause you to neglect these, inasmuch as it is uncertain what I may be able to accomplish. I can only say that I shall make every effort in my power.

---

[15] Lotka 1923.

I hope from time to time I may have progress to communicate to you on my problem and in the meantime I expect that I shall see publications from your pen appearing in the scientific press.

You express regret at not having mentioned my work in your memoir. I appreciate your thoughtfulness and I can very easily understand the circumstances which lead to the omission. My work is not very generally known, the field being somewhat aside from the established lines of research. This is a disadvantage which a man has to accept, and it is compensated to some extent by the satisfaction of laboring in a comparatively untilled field. You tell me also in your letter that you will offer a review of my book to Scientia: this would give me a great pleasure. I can think of no one from whose pen I should rather have a review than yours and Scientia would be an excellent organ in which to adress the scientific reading public. With kind regards, I remain

Very truly yours,

Alfred J. Lotka
Supervisor of Mathematical Research

14.7 Lotka to Volterra, New York, 23.1.1927 *Cat. no. 268*

(Same heading of letter no. 14.1)

New York, 23 January 1928

Dear Professor Volterra,

Please accept my very cordial thanks for the two reprints which you kindly sent me and which have just reached my desk. I am very glad indeed to have these as you will readily understand that I wish to keep my files as complete as possible on the subject of your publications relating to biological species. The paper in Scientia I had seen before but the other paper is new to me and I shall peruse it with the greatest interest.

My own work in this direction has been progressing somewhat slowly but has not been altogether at a standstill. I hope in the time – although that may mean after considerable delay – to be in a position to make some further communications.

Very truly yours,

Alfred J. Lotka
Supervisor, Mathematical Research

14.8 Lotka to Volterra, New York, 19.1.1931 *Cat. no. 269*

(Same heading of letter no. 14.1 but with F. H. Ecker as President)

New York, 19 January 1931

Dear Professor Volterra,

Let me acknowledge with cordial thanks the receipt of your recently published book on "The Mathematical Theory of the Struggle for Existence"[16]. I am sure it is quite unnecessary for me to tell you that I am keenly interested. I shall probably have more to say on this subject in the near future of I can find the time to prepare for publication some matters which I have had in hand for some time past.

Very truly yours,

Alfred J. Lotka
Supervisor Mathematical Research

---

[16] Volterra 1931a.

15 Letters between Edouard Monod-Herzen and Vito Volterra

We have very little biographical information about Edouard Monod-Herzen. We know that he was born in Paris on 24 December 1873 and that he obtained his diploma of "licencié ès sciences" and then that of "lauréat" of the Académie des Sciences de Paris. He himself wrote that he had had as teachers Félix Thomas for philosophy, Carlo Bourlet for the exact sciences and Frédéric Houssay for biology. He was also a good friend of the eminent writer Romain Rolland, with whom he had a constant and intense intellectual dialogue. He edited a book containing the correspondence of Romain Rolland (ROLLAND 1948).

Edouard Monod-Herzen's main activity was that of decorator. He was above all a specialist in the field of metal working using the techniques of embossing, hammering, chiselling, sculpting, enamelling, inlaying and studding. He held exhibitions of his work at the Société Nationale des Beaux-Arts, at the Salons d'Automne et des Décorateurs. His works are to be found also at the Musée de Luxembourg, the Musée des Arts Décoratifs and the Musée des Arts et Métiers. For his professional achievements as a decorator, he was awarded the Legion of Honour. He was also librarian at the École Nationale Supérieure des Arts Décoratifs and wrote several books on topics related to his professional activities (MONOD-HERZEN 1914, 1961).

His scientific background, combined with his professional interests, led him to study the relations between science and culture. On this topic, after writing a short treatise (MONOD-HERZEN 1925), he published his one and only scientific book of broad scope and ambition, *Principes de Morphologie Générale* (MONOD-HERZEN 1927), in two volumes. The first is dedicated to an analysis of form – defined forms, families of forms, associations of forms – and to the relationship between form and function, and is concluded with an analysis of the transition from the forms of inert bodies (crystals) to those of living matter. The second volume deals specifically with living matter and human morphology and is concluded by an analysis of the relations between morphology and aesthetics. From his correspondence with Volterra it transpires that Monod-Herzen was planning a further development of the *Principes de Morphologie Générale* of the completion of which there is however no trace. He also translated into French three books describing the reactions of living and non living matter and the nervous mechanism of plants (JAGADISACANDRA VASU 1926, 1931, 1934).

* * *

15.1 Monod-Herzen to Volterra, Paris, 16.11.1927 *Cat. no. 270*[1]

12 rue Antoine Chantin (XIVe) *16-11-27*

Monsieur,

Je reçois à l'instant votre Mémoire sur les Variazioni ... in specie animali conviventi[2].

Je vous en remercie infiniment. Je suis extrêmement heureux de le recevoir de vous même, et de pouvoir l'étudier. Vous avez conçu une recherche entièrement originale et nouvelle et ouvert là une voie des plus remarquables.

J'ai le plaisir de vous announcer que le tome II de mes "Principes des Morphologie"[3] vient de paraître, et qu'un exemplaire va aussitôt partir à votre adresse. Sans comparer en aucune manière, même d'infiniment loin, mes modestes efforts à votre si grand talent, je forme le voeu que mon travail puisse vous sembler un tout petit peu intéressant.

Veuillez agréer, Monsieur, l'expression de ma plus haute et plus vive appréciation.

Edouard Monod-Herzen.

15.2 Monod-Herzen to Volterra, Paris, 10.12.1927 *Cat. no. 271*

12, rue Antoine Chantin
Paris XIV *10.12.27*

Cher Monsieur,

j'ai d'abord à vous transmettre les hommages et les meilleurs souvenirs de mon ami Léon Deshairs qui, pendant la guerre, venu à Rome avec Borel, a eu le grand plaisir de vous connaître et d'être reçu par vous. Je lui envie amicalement ce bonheur!

J'ai été très frappé, parmi bp. d'autres choses, de retrouver, dans votre Mémoire CXXXI, la présence de grandes oscillations, ou longues alternances successives, qui semblent, sous des modalités diverses, se rencontrer constamment dès qu'on on touche à la matière vivante.

[Je ne parle même pas de ce fait admirable: que vous etez arrivé à soumettre ce phénomène au calcul!]

Dans un domaine tout différent, j'ai eu connaissance de travaux récents d'un physiologiste russe, Manoukhine[4], qui complètent d'une manière remarquable ceux de Metchnikoff[5] sur la phagocytose.

---

[1] Postcard

[2] Volterra 1926a.

[3] Monod-Herzen 1927.

[4] Probably a reference to the Russian physiologist Boris Nikolaevich Manukhin (who in 1968 wrote *Fiziologiia adrenoretseptorov*).

Manoukhine a démontré que les globules blancs, ou leucocytes sont alternativement produits en grande quantité (leucocytose), puis détruits, en masse, dans le sang (leucocytolyse).

Ces phénomenes d'apparition et de destruction successives sont étroitement liés aux influences contraires de la rate et du foie. Ils deviennent surtout visibles en cas de maladie - et Manoukhine en a deduit toute une série d'applications thérapeutiques des plus remarquables, en agissant (convenablement) sur la rate par les rayons X. Il est probable que par les moyens d'une analyse mathématique puissante, mise en oeuvre par un savant comme vous, ces phénomènes alternés pourront, eux aussi, être soumis au calculs, et que de très grands espoirs prennent naissance à la suite de votre recherche!

Lorsque les circonstances vous auront permis d'achever de jeter un coup d'oeil dans mon petit essai, je vous serai reconnaissant de m'en dire les critiques. Je sais tout ce qui me manque, et moi-même ai pris soin de signaler toutes mes ignorances. Mais votre avis éclairé, qu'il soit favorable ou defavorable, me sera précieux. Il m'aidera à orienter mes efforts - à insister ici ou là. J'ai beaucoup de peine à obtenir un tel avis en France. Ici, j'ai pour moi (comme on dit) les philosophes, les mathématiciens, et les artistes(1). Mais déjà les physiciens apprehendent d'avoir une opinion, "par principe", pourrait-on dire, parce que je parle sans m'appuyer à chaque pas sur Le Nombre. Et ensuite, j'ai deux groupes d'opposants: ceux des biologistes qui ignorent tout des sciences exactes, c'est à dire presque tous, et, plus loin, les critiques d'art, qui sont furieux de penser qu'il pourrait bien y avoir, un jour, une base sérieuse (qui leur manque!) à la critique d'art...

Tout ceci ne me décourage pas le moindrement. De plus, je sais fort bien tous les reproches auquels on s'expose lorsque l'on formule, ou énonce, un problème, par anticipation en quelque sorte, avant d'être en état de le resoudre. Mais il me semble qu'il faut oser poser certaines questions, parce qu'on ne sait jamais si cela ne pourra pas susciter ou inciter les recherches d'autrui, mieux armés pour les conduire plus loin...?...

Mais d'un autre côte, il est naturel que j'aie des scrupules, et la crainte de faire fausse route. Les indications d'esprits qui ont, à la fois, la connaissance approfondie de la plus exacte des Sciences, et qui ont des idées générales (et pas de préjugés!), peuvent m'être d'un prix infini.

Voilà pourquoi je me suis permis de solliciter vos critiques, - à l'occasion - si les circonstances (je vous sais entièrement occupé) vous en donnent le temps.

Si les questions d'Esthétique scientifique vous intéressent, il me sera facile de vous faire part de ce que je publierai dans l'avenir.

Car mon ouvrage actuel (les 2 tomes) n'est que le premier volet d'un triptyque.

Il y aura aussi une étude des relations entre la Forme et les propriétés de la matière, où les examples concrets seront pris dans l'art du metal, que je connais bien.

---

[5] Elie Metchnikoff (1845–1926), Russian biologist who carried out seminal research on phagocytosis and the problem of immunity and joined in 1886 the Paris Institut Pasteur. For the latter he was awarded the 1908 Nobel Prize. His main works include METCHNIKOFF 1901.

Et ensuite (ou avant, je ne sais) il y aura des "Principes d'Estétique Générale" avec bp. de données nouvelles.

Veuillez, cher Monsieur, excuser cette lettre un peu longue. Vous êtes (je vous prie de croire que je ne dis pas cela pour vous flatter stupidement: ce serait indigne et de vous et de moi-même) un des plus grands savants non seulement de l'Italie, mais du monde, et je n'aurais pas osé m'addresser à vous si longuement si Deshairs ne m'avait dit que vous étiez aussi le plus affable des hommes: excusez moi!

Et recevez, je vous prie, avec tout mon meilleur respect, l'expression de ma plus sincère admiration.

Edouard Monod-Herzen.

P.S. Verriez vous avec plaisir, en principe, votre mémoire "Variazioni .... conviventi" traduit en français? J'ai un grand fils (28 ans) qui s'occupe des sc. exactes, et qui parle l'italien comme le français [Il a traduit "L'Evolution de la Logique", d'Enriques[6]]. Actuellement, il est très pris (il travaille aux labos. de Chimie-Physique de J. Perrin). Mais si un jour il avait le temps, approuveriez vous qu'il envisageât de traduire votre mémoire? Pour ma part, j'estime que sa diffusion présente un immense interêt.

(1) Ceux-ci ont senti l'exactitude de certains de mes énoncés, même lorsque ils ne sont pas en état de saisir le fond des choses.

15.3 Monod-Herzen to Volterra, Paris, 20.1.1929 *Cat. no. 272*[7]

12 rue Antoine Chantin (XIV) Deshairs est prévenu *Mercr.*

Cher Monsieur,

Je n'ai pas vu mon dir., qui vient irrégulièrement à l'Ecole, en ce moment. Mais, ce qui est pire, j'ai un commencement de grippe, et je ne sortirai en ts. les cas pas demain.

Mais si vos conférences sont rédigées je serais heureux de pouvoir en prendre connaissance. Et si mes occupations et celles de mon fils le permettaient, et que vs. n'ayez, jusqu'à present, personne qui se soit chargé de revoir votre texte au pt. de vue du français pur, je voudrais examiner si ns. ne pourrions pas le faire à nous deux.

A bientôt, cher Monsieur, et veuillez, je vous prie, recevoir l'expression de mes sentiments les plus dévoués.

Edouard Monod-Herzen

6 Enriques, Monod-Herzen 1926, French version of Enriques 1922.

7 Postcard.

16 Letters between Raymond Pearl and Vito Volterra

Raymond Pearl was born in Farmington (New Hampshire, USA) on 3 June 1879. He attended secondary school at Farmington and Rochester and followed higher studies at Dartmouth College. In 1902 he received his Ph.D. in zoology at the University of Michigan. He then left for Europe to complete his training, first at the University of Leipzig and then at University College London, in the two years 1905–16. Here he had Karl Pearson as his research director and was so strongly influenced by him that he may well be considered one of Pearson's disciples. Indeed Pearson oriented him definitively towards the application of statistical methods to biological problems and, in particular, to population studies, that is, to a field of research that Pearson himself had defined as "biometrics". Pearl developed the conviction - which he retained for the rest of his life - of the existence of mathematical laws governing population growth that were quite similar to the mathematical laws governing the motion of heavenly bodies.

On his return to the United States, Pearl got a job as zoology lecturer at the University of Pennsylvania (1906–1907) and later headed the Maine Agricultural Experimental Station from 1907 to 1918. He directed the statistics bureau of the U.S. Food Administration from 1917 to 1919. In 1919 he became professor of biometrics and vital statistics at the School of Hygiene and Public Health of the Johns Hopkins University, where he set up the Department of Medical Statistics and Biometrics and continued to hold the chair until 1926. In 1926 he founded the *Quarterly Review of Biology* and in 1929 *Human Biology*. He directed the Institute of Biological Research from 1925 to 1930 and held the chair of biology at the School of Medicine from 1925 to 1940. One of his principal assistants was his wife Maud Mary DeWitt, whom he married in 1903 (see letter no. 16.13 herein). He died at Hershey (Pennsylvania, USA) on 17 November 1940.

Raymond Pearl wrote more than 700 articles and books, including *Introduction to Medical Biometry and Statistics* (PEARL 1923) - a well-known and widely read university coursebook for many years -, *Studies in Human Biology* (PEARL 1924), *The Biology of Population Growth* (PEARL 1925) and *The Natural History of Population* (PEARL 1939). In 1920, Pearl achieved fame when he applied the logistic equation to determining the human population in the year 2100. This equation had already been introduced by the statistician Pierre-François Verhulst (1804–1849), although his research had been ignored. It was Pearl who developed its applications in a completely independent way. He observed the S shape of the logistic curve in many empirical cases and, in particular, he observed it experimentally on the growth of a population of *Drosophila Melanogaster*. It was Volterra who named this behaviour in the growth curve of a population the "Verhulst-Pearl effect". Pearl used the logistic equation also to investigate the growth of the US population and that of several European countries, as well as to predict their steady states. The results he obtained were in good agreement with the real data and aroused great interest. He was assisted in this research by the biostatistician Lowell J. Reed (see PEARL, REED 1920), whom he recruited to the Department of Medical Statistics and Biometrics at John Hopkins University: Reed's was in fact the second chair to be set up at the Department after Pearl's.

Pearl pioneered studies on the correlation between smoking and cancer and alcohol consumption and longevity. He claimed that moderate drinkers of alcohol lived longer than teetotallers and that intellectuals had a good chance of living longer than labourers. He adopted an extremely advanced stance for his time concerning birth control, declaring that he was in favour of contraception and thus drawing bitter criticism. His ideas were disseminated by his great friend, the well-known social critic and columnist of the *Baltimore Sun*, Henry Louis Mencken (1880–1956)[1].

* * *

16.1 Pearl to Volterra, Baltimore, 16.1.1931 *Cat. no. 274*

THE JOHN HOPKINS UNIVERSITY
SCHOOL OF HYGIENE AND PUBLIC HEALTH

Department of biology

1901 East Madison Street
Baltimore, Maryland
U. S. A.

xvi January 1931

Dear Doctor Volterra: Please accept my best thanks for the copy of your new book, "Leçons sur la Théorie Mathematique de la Lutte pour la Vie"[2], which came to hand yesterday. I am looking forward with the greatest interest to reading this book.

Prof. Vito Volterra,
University of Rome,
Rome, Italy

Sincerely yours,
Raymond Pearl

[1] See FREEMAN 1941, MENCKEN 1940, KINGSLAND 1985 and PARKER 1976 (which contain further bibliographic references).

[2] VOLTERRA 1931a.

16.2 Volterra to Pearl, Roma, January 1931 *Cat. no. 386*[3]

Cher Monsieur,

J'ai eu l'honneur de vous envoyer en hommage il y a quelques jours le volume que je viens de publier: Leçons. J'éspère que vous l'aurez reçu.

Je viens d'apprendre que vous avez publié recentement un ouvrage: The Biological[4] dans lequel vous avez cité mes recherches. Il m'intéresserait beaucoup de le connaître car je ne doute pas qu'il pourrait être très utile pour mes études. C'est pourquoi je prends la liberté de vous écrire pour voir comment on pourrait faire pour se procurer cet ouvrage.

Au mois de Juin passé je vous ai écrit une lettre en repondant à certaines questions que vous m'avez faites. J'espère que mes reponses aient pu vous satisfaire.

16.3 Volterra to Pearl, Paris, 28.10.1937 *Cat. no. 387*[5]

Hotel Lutétia
43, Boulevard Raspail
Square du Bon Marché
Paris

Cher Professeur Pearl

J'ai su de la part du Prof. Van der Klaauw de Leiden et du Prof. F. G. Donnan de Londres que vous êtes à Londres où vous donnez un cours de lectures. Je serais heureux de vous voir et de pouvoir fixer un rendez-vous avec vous. Si par hasard vous venez à Paris par exemple pour visiter l'exposition je serais heureux de vous recevoir ici à l'Hôtel où je suis logé, mais si vous ne pouvez pas je tacherai de faire une course à Londres, si vous m'indiquez le jour l'heure et l'endroit où je pourrais vous rencontrer. Je n'aurais que très peu de temps à Londres, et je n'ai pas besoin de vous dire que c'est la première proposition qui me serait plus agréable étant donné mon âge avancé.

J'écris contemporainement au Professor Donnan avec lequel on pourrait se mettre d'accord.

J'espère recevoir une reponse de votre part. Veuillez agréer, cher Prof. Pearl, mes hommages les plus dévoués.

---

[3] Draft, n.d. although certainly written in January 1931.

[4] Owing to the vagueness of the reference it is not clear which publication Volterra is referring to.

[5] Draft, n.d. but certainly written on 28.10.1931, as demonstrated by the following letter.

16.4 Pearl to Volterra, London, 30.10.1937 *Cat. no. 275*

CUMBERLAND HOTEL,
MARBLE ARCH,
LONDON, W.1

30th Oct. 1937

Dear Professor Volterra:

On my return to the hotel last night I found your very gracious letter of the 28th. I am greately honored by your desire to meet me. Of course this desire is reciprocal. Nothing could give me greater pleasure than to become personally acquainted with you.

Unfortunately it is impossible for me to come to Paris at this time. I sail for New York on Nov. 6th. Under the circumstances the best suggestion I can make is that, if you should care to take the very considerable trouble, you to come to this Hotel at a little before 3 P.M. on Thursday, Nov. 4. We could then have a little talk, following which I could take you along to the School of Hygiene, where you could hear, if you wished, my final lecture in the Heath Clark series that I am giving.

It embarasses me to make the suggestion that so old and distinguished a man as yourself should take all this trouble to meet such an unimportant person as myself. But if you really want to do it, I shall be delighted to see you.

With admiration and respect, I am
Sincerely yours
Raymond Pearl

16.5 Volterra to Pearl, Paris, 31.10.1937 *Cat. no. 389*

(Same heading of letter no. 16.3)

31 Octobre 1937

Cher Professeur Pearl

Je vous suis très reconnaissant de votre aimable lettre que j'ai reçu ce matin même. Je pense de partir avec ma femme qui m'accompagne Mercredi et d'arriver le jour même à Londres. Nous descendrons à l'Hôtel Cumberland où nous venons de fixer l'appartement. Votre proposition me convient parfaitement et je serai heureux après avoir causé avec vous de pouvoir assister à votre conférence à l'Ecole d'Hygiène.

16.6 Pearl to Volterra, Baltimore, 10.1.1938 *Cat. no. 276*

HUMAN BIOLOGY
A RECORD OF RESEARCH
EDITORIAL OFFICE
DEPARTMENT OF BIOLOGY
SCHOOL OF HYGIENE AND PUBLIC HEALTH
THE JOHNS HOPKINS UNIVERSITY
1901 EAST MADISON STREET
BALTIMORE, MARYLAND
U. S. A.

EDITOR
RAYMOND PEARL

x January 1938

Dear Professor Volterra: My best thanks for yours of December nineteenth. We have incorporeted your corrections, and herewith I am sending galley proof to you[6]. I hope that you will be satisfied with the way we have put the matter in type.

Please return the proof, with your corrections, at the earliest possible moment, and if possible by fast steamship, because there is very little time to spare if we are to get the paper out in the next number of HUMAN BIOLOGY.

Also please let me know how many reprints you will want.

With best regards, and best wishes for the New Year.

Sincerely Yours

Raymond Pearl

16.7 Volterra to Pearl, Roma, January 1838 *Cat. no. 390*

Via in Lucina, 17
Roma

Caro Professor Pearl

La ringrazio per la vostra del X January che ho ricevuto insieme alle bozze (proofs). Sono molto soddisfatto del modo come è stata stampata la memoria (paper)[7]. La prego di voler fare eseguire le correzioni indicate sulle bozze ed eventualmente correggere qualche errore di stile o di lingua che fosse rimasto.

Rispedisco per espresso le bozze nella speranza che giungano in tempo perché la memoria sia stampata sul prossimo numero di Human Biology. Desidero che vengano stampate 100 reprints della memoria

With best regards

Yours sincerely

---

[6] Reference to the publication VOLTERRA 1938b in *Human Biology*.

[7] See preceding note.

Dear Professor Pearl,

My best thanks for yours of January (I thank you) tenth, which I received jointly with the proofs. I am very satisfied of the way the paper has been printed. Please have the corrections made as pointed out in the proofs and eventually revise the mistakes of stile & language which I may have overlooked. I am returning the proof by fast steamship hoping that you will receive them in time so that the paper can be printed out in the next number of Human biology. I wish to have a hundred reprints of my paper.

With best regards,

Yours sincerely

16.8 Volterra to Pearl, Roma, 17.10.1938 *Cat. no. 388 and 391*[8]

Caro Prof. Pearl,

Ho l'onore di mandarle in un plico raccomandato collo stesso corriere una memoria avente per titolo: Calcolo delle variazioni e equazione di Pearl[9]. In questa memoria mostro come la Sua equazione può farsi dipendere da un problema di minimo, cioè il movimento della popolazione avviene in modo da render minima una certa quantità. Ciò è analogo a quanto avviene per il movimento dei corpi nella dinamica secondo il principio di Hamilton-Maupertuis. Vi è dunque perfetta analogia fra la dinamica dei sistemi demografici e la dinamica dei sistemi materiali. Mi sembra che il resultato che si impernia sulla Sua equazione e quindi sulla Sua curva debba avere un certo interesse. La memoria può in certo modo anche considerarsi come una continuazione dell'altra mia memoria inserita nel Suo giornale[10] (vedi ultimo §).

Le sarò grato se vorrà darmi un cenno di risposta. La autorizzo a migliorare lo scritto in vista della lingua adoperata.

I remain

Yours very faithfully

Roma, 17 October 1938

Dear Professor Pearl

I have the honour to send you in a registered letter a paper entitled "Calculus of Variations. Pearl's Equation". In this paper I demonstrate how your equation can be a result of a problem of minimum. Namely that the motion of population happens [occurs] so as to make a minimum of certain quantity. This is analogical to what happens concerning motions of bodies in dynamics according to the principle of Hamilton-Maupertuis. There is, then a perfect analogy between

---

8 Drafts of a letter, the second of which (Cat. no. 391) is dated 17.10.1938 while the first (Cat. no. 388) is undated and in Italian, but practically identical.

9 The reference here is to Volterra's second article published in *Human Biology*, i.e. VOLTERRA 1939b.

10 VOLTERRA 1938b.

demographic dynamics and the dynamics of material systems. It seems to me important.

I will be very glad if you will publish my paper in "Human Biologie". In a certain way this paper can be considered as a continuation of my other paper which I have insered in your Revue. (See Human Biology, February 1938, Vol. 10. No 1. Volterra: Population Growth ... §8 equation (i)).

I authorize you to improve the style considering the language which has been used.

16.9 Letter of Pearl to Volterra, Baltimore, 1.11.1938 *Cat. no. 277*

(Same heading of letter no. 16.6)

i November 1938

Dear Professor Volterra: My best thanks for your letter of October seventeenth and the accompanying manuscript, which seems at first reading a very interesting and important contribution.

In accordance with your authorization I shall, as soon as possible, work the manuscript over to make the text conform to standard American usages, and then submit the revised manuscript to you for your approval or alterations, and then get it into print as soon as posible.

With best regards to you and Madam Volterra from Mrs. Pearl and myself, I am

Sincerely yours,

Raymond Pearl

16.10 Letter of Volterra to Pearl, Roma, November 1938 *Cat. no. 392*[11]

Dear Professor Pearl

La ringrazio molto della sua lettera del 1 Novembre. Sono lieto che Ella abbia accettato di pubblicare la mia memoria (paper) nel Suo giornale Human Biology. Se Ella vorrà inviarmi il manoscritto over to make the text conform to standard American usage io lo riguarderò e glielo rimanderò subito affinché venga stampato. Le sarò molto obbligato se verrà stampato il più presto possibile. Voglia gradire i migliori riguardi per Lei e per Mrs. Pearl da parte mia e della signora Volterra

I remain

Yours very faithfully

[11] Draft subsequently dated 1938, and certainly written only a few days after the preceding letter.

16.11 Letter of Pearl to Volterra, Baltimore, 17.1.1939 *Cat. no. 279*

(Same heading of letter no. 16.1)

xvii January 1939

Dear Professor Volterra: It was a great pleasure to the Pearls to get the Christmas greeting from you and Madam Volterra. We hope you are both well and enjoying life.

I did receive your letter of november 12th and will return your manuscript to you as soon as I have worked it over from the standpoint of language. It will appear in the May number of HUMAN BIOLOGY. It was impossible to get it into an earlier number after its receipt.

With best regards from the house to house.

Sincerely yours

Raymond Pearl

16.12 Letter of Pearl to Volterra, Baltimore, 7.2.1939 *Cat. no. 280*

(Same heading of letter no. 16.6)

vii February 1939

Dear professor Volterra: In reply to your letter of January twenty-fifth, I may say that the delay about your paper has been occasioned by the fact that there was a large accumulation of manuscripts ahead of yours for publication in HUMAN BIOLOGY.

We are putting your paper forward just as rapidly as possible. There were so few changes necessary in the manuscript that we are having it put into type before sending it to you. I hope that we shall be able to send you off the galley proof within the next week or ten days.

With best regards

Sincerely yours

Raymond Pearl

16.13 Letter of M. De Witt Pearl to Volterra, Baltimore, 24.2.1939 *Cat. no. 273*

(Same heading of letter no. 16.6)

xxiv February 1939

Dear Professor Volterra: Enclosed you will find galley proofs of your paper entitled "Calculus of Variations. Pearl's Equation"[12], which is to appear in Human Biology. We will appreciate the prompt return of this proof to us.

---

[12] The definitive title was "Calculus of variations and the logistic curve" (VOLTERRA 1939b): see following letter no. 16.15.

I am enclosing information relative to reprints. Kindly order the reprints when you return the proof, stating whether or not you wish the reprints with or without covers. Also will you kindly prepare a brief abstract of your paper on the enclosed blank and return it to us with the proof.

Sincerely yours

Maud De Witt Pearl
Managing Editor

16.14 Letter of Pearl to Volterra, Baltimore, 2.3.1939 *Cat. no. 281*

(Same heading of letter no. 16.6)

ii March 1939

Dear Professor Volterra: By the time this reaches you, you should have received the proofs of your article. They were sent during my absence from the city. I hope that they will reach you all right, and that my editing of your manuscript will meet with your approval.

I note that you will want 100 reprints, and I shall take pleasure in seeing that this order is properly filled.

With best regards

Sincerely yours

Raymond Pearl

16.15 Letter of Volterra to Pearl, Roma, 10.4.1939 *Cat. no. 393 e 394*[13]

Dear Professor Pearl

Mi affretto a mandarle qui accluse le proofs of my Paper which is to appear in Human Biology. Prego di mandarmi 100 reprints with covers. Accludo anche a brief abstract of my paper. Coi migliori ringraziamenti e migliori saluti anche per la Signora Pearl che gentilmente mi ha mandato le bozze.

Il titolo che avevo dato alla Memoria era: Calculus of Variations Pearl's Equation ma il titolo stampato è: Calculus of Variations and the Logistic Curve. Per parte mia vi autorizzo a mettere il titolo che volete purché sia lo stesso nella Memoria e nel abstract.

I hasten to send you (here) enclosed, in this letter, the proofs of my Paper which is to appear in Human Biology. (Will you) kindly send me a hundred reprints with covers? I include also a brief abstract of my paper.

My best thanks & kindest regards also to Mrs Pearl, who was so kind as to send me the proofs.

The title that I had given to the Memoriam was "Calculus of Variations Pearl's Equation", whereas the title that has been printed is: "Calculus of

[13] Italian and English drafts of the same letter, n.d. but written on 10 April 1939, as revealed by the content of the next letter.

Variations & the Logistic Curve". On my side I authorize you to put any title you wish, provided that the title should be the same both in the & in the abstract.

16.16 Letter of Pearl to Volterra, Baltimore, spring 1939 *Cat. no. 278*[14]

(Same heading of Letter n. 1)

Dear Professor Volterra: My best thanks for your letter of April tenth.

I hope you were pleased with the appearance of your paper as it finally came out in HUMAN BIOLOGY.

With best regards from house to house.

Sincerely yours

Raymond Pearl

---

[14] N.d. (although it bears the indication 1939, added subsequently) but obviously written in April 1939 or shortly after.

17 Letters between Karl Pearson and Vito Volterra

Karl Pearson was born in London on 27 March 1857. He spent seven years at University College School, London, from which he withdrew for one year in 1873 owing to illness, continuing his studies with a private tutor. In 1875 he was admitted to King's College, Cambridge, where he studied mathematics with teachers of the calibre of Edward John Routh (1831–1907), George Gabriel Stokes (1819–1903), Arthur Cayley (1821–1895) and others, obtaining his Bachelor of Arts in 1879. He then moved to Germany where he studied physics and philosophy at Heidelberg and followed courses on darwinism given by Emil Du Bois Reymond (1818–1896) at Berlin. From 1880 to 1883 he studied law at London but never practised. In 1884 he joined University College, London (where he spent most of his university career) as Goldsmid professor of applied mathematics and later also as Gresham professor of geometry (from 1891 to 1894). In 1907 he was appointed director of the applied mathematics department. It was in this period that his interest became focused on the problem of the application of statistics to biology and, in particular, to inheritance and evolution, under the influence of the writings of Francis Galton (1822–1911) and his friendship with Walter F. R. Weldon (1860–1906). In 1892 he published what is probably his best known work, *Grammar of Science*, in which he dealt with the central problems of science philosophy, and from 1893 to 1912 he wrote a series of articles on "Mathematical Contributions to the Theory of Evolution" in which he set out his basic ideas of statistics and its applications to biological problems. He was elected Fellow of the Royal Society in 1896 and awarded the Darwin Medal in 1898.

In 1911 he left the Goldsmid chair and became the first professor of the Galton chair of eugenics, set up expressly for him according to the wishes of Galton himself: he held this chair until 1933. In 1901, together with Galton and Weldon, he founded the famous review *Biometrika*, directing it until 1936. From 1925 to 1936 he was also editor of *Annals of Eugenics*. Pearson withdrew from active university life in 1933, but continued with his research work at University College until just before his death, at Coldharbour (Surrey) on 27 April 1936.

It is not easy to give an outline of Pearson's interests as they ranged from mathematics to biology, and from philosophy to literature and even political theory. In the latter field he was attracted by the theories of Karl Marx (in whose honour he changed the spelling of his own name from Carl to Karl), becoming a convinced socialist and combining socialism and evolution into a theory of "social darwinism".

As stated above, the *Grammar of Science* is one of Pearson's most important works, which is well known also because it contains an anticipation of the theory of relativity, with the statement of the principle of the relativity of the motion of all bodies. Pearson put forward a positivist view of science, as a classification of facts, aimed at seeking "how" and never "why" with regard to the phenomena studied and the unity of which is defined by the method and not by the object. Furthermore, in the *Grammar of Science*, as well as in other works, he believes in the predominance of the Galtonian concept of "correlation" over that of "causation", insofar as the latter is a broader and less restrictive notion than the former.

Pearson may be considered one of the founders of modern statistics. He made fundamental contributions to regression analysis, and he introduced the socalled statistical significance test and the concept of standard deviation to replace the concept of mean square error. His applications of statistics to population theory, inheritance (in particular to Mendel's theory) and to evolution were of fundamental importance. He also used regression techniques to reconstruct the measurements of extinguished races, also for the purpose of testing the reliability of the predictions of evolutionary theory. This research was part of his studies on eugenics, a branch which he helped to found, and which remains perhaps one of the more questionable aspects of his work.

Pearson's statistical approach to biology may account for his evident coolness to Volterra's views as the latter, on the contrary, subordinated the statistical approach to the analytical approach[1].

* * *

17.1 Pearson to Volterra, London, 24.4.1902 *Cat. no. 282*

University College
London

April 24, 1902

I most sincerely regret that our arrangement with the publishers of Biometrika do not permit of the exchange you propose. The excellent Atti & Rendiconti of the Lincei are, however, I am glad to say accessible to the editors

Karl Pearson

[1] See EISENHART 1976, which contains an extensive bibliography.

17.2 Volterra to Pearson, 1910 *Cat. no. 395*[2]

Dear Sir,

In the months of March or April next year there will be an International Congress of Philosophy held at Bologna.

I have the honour to invite you in the name of the Scientific Section of the Congress to be so good as to come & give a general lecture of your own choice. We hope you will accept the invitation.

I thank you for what you wrote me on the subject of Prof De Helguero.

We would be very pleased if you would write us only a few words regarding your judgement on the researches of Prof. De Helguero published in your journal Biometrics[3].

I remain

Yours faithfully

17.3 Volterra to Pearson, Rome, 21.9.1926 *Cat. no. 396*[4]

Roma, 21 Settembre 1926

Dear Sir

Mi permetto inviarle una copia della mia Memoria: Variations and fluctuations....[5] che spero Ella vorrà gradire in omaggio. Ho tentato di dare una teoria matematica delle fluttuazioni delle specie animali ricavandone leggi che sembrano confermate dalle statistiche.

Sarei molto lieto se questo studio potesse interessarle e se il giornale Biometrika volesse pubblicare la parte essenziale delle suddette memorie convenientemente ridotta e raccorciata, la quale potrebbe essere scritta in Inglese[6].

Io Le sarei molto grato se a tutto Suo agio, presa visione delle Memorie volesse rispondermi una parola in proposito.

Karl Pearson

(l'indirizzo si trova nell'annuario della Royal Society (Studio)). Si può scrivere nella carta Via in Lucina e basta.

---

[2] Draft. It is undated but the International Congress of Philosophy took place in Bologna in 1911, so we can infer that the letter was written in 1910.

[3] Reference to De Helguero 1904.

[4] Draft.

[5] Volterra 1926a.

[6] Not only was nothing written by Volterra ever published in *Biometrika*, but, as we saw in the introductory essay, the journal published a critical review of D'Ancona's experiments under the name of Pearson's son, Egon, who in 1933 took over the direction of the statistics department of the University College of London (E. S. Pearson 1927).

18 Letters between Jean Régnier and Vito Volterra

Jean Lucien Régnier was born on 4 June 1892 at Blanzy-les-Mines (in the French region of Saône-et-Loire). His father Claude was a pharmacist, and formerly an intern at the Paris hospitals. His father was to have an important role in orienting his son towards his same field of interest. After his high school studies at Lyon and Dijon, in 1912, Jean Régnier enrolled in the École de Pharmacie of Paris. His brilliant initial studies were interrupted by the Great War. Called up in 1914, he served first as a military nurse and was later promoted auxiliary pharmacist and assigned as a bacteriologist to a mobile unit attached to a field hospital. This experience had an important influence on his training and was also the origin of the joint and rheumatic complaints from which he suffered for the rest of his life and that brought him to a premature death.

On his demobilisation in 1919, while working as an intern at the Hôpital Laennec, he completed his studies and, in 1922, presented a doctoral thesis in pharmacy on "L'évolution microbienne dans les premières heures de la plaie de guerre" which was awarded a prize by the Académie de Médecine. In 1925 he terminated a doctoral thesis in science concerning the anaesthesia of the cornea and, in 1926, passed his "agrégé" examination at the Faculty of Pharmacy. In 1929, he completed a doctoral thesis in medicine on methods of measuring local anaesthesia which was awarded a prize by the Académie des Sciences and the Académie de Médecine.

Chief pharmacist from 1924 at the Hôpital Ambroise Paré of Paris, he was nominated "agregé" of the Faculty of Pharmacy in 1930, where he directed research in analytical chemistry and held courses in galenic pharmacy and, from 1937 on, as "maître de conférences" in parasitology and bacteriology. In 1925 he established a research centre at the Hôpital Ambroise Paré where a number of researchers were employed and which he directed for over fifteen years. This was the setting in which he carried out his most important research.

The outbreak of World War II interrupted the activity of his research group and Régnier, although exempt from active service on health grounds, was sent to direct the pharmacy service of the Hôpital Laennec to replace his colleague Picon who had been called up. In 1942 the Hôpital Ambroise Paré was practically destroyed by bombing, which put an end to the activity of his research centre. In the meantime, in 1941, he was appointed director of the chair of microbiology. Shortly after the liberation of France and the end of the war, on 13 April 1946, Jean Régnier died suddenly at the age of 54, at the height of his activity.

Jean Régnier's research work developed in two main directions: first in the field of microbiology and then in that of pharmacodynamics. His microbiological research was, as we have seen, suggested by his war experiences. It was directed towards the study of antibacterial chemiotherapeutic substances, which required a preliminary analysis of the normal rate of bacterial proliferation. This analysis enabled him to study the bacteriostatic or bactericidal activity of antibacterial substances and, in this field, he focused his attention mainly on the action of chemical inhibitors on the morphology and vitality of bacteria and on the culture of bacteria in a continuous medium. This research paved the way to the development in vivo of chemiotherapeutic tests to determine the effectiveness of new compounds such as the sulfonamide drugs.

It was this research on the dynamics of bacterial growth that aroused the interest of numerous scholars, in particular of Vito Volterra and Vladimir A. Kostitzin, who intuitively grasped their potential for providing an experimental support for their own studies on the mathematical laws governing population growth and the struggle for existence. This led to research on the bacterial antagonism among different species which Régnier carried on together with Volterra and Kostitzin and to which one of his main collaborators, Suzanne Lambin, also contributed.

The other scientific research field to which Régnier devoted the most energy was that of the pharmacodynamics of local anaesthesia. He developed quantitative methods to measure the activity of local anaesthetics so that they could be compared and their dynamics investigated. He did particularly important research work in the field of alkaloids (cocaine, morphine). One characteristic feature of Régnier's research method was to constantly broaden his approach so as to take into account all the factors involved in the phenomena he studied, particularly the physiological and physico-chemical phenomena. This explains the important contributions he was able to make to the complete definition of the physico-chemical properties of the salts studied, their toxic action and the general problem of cell permeability.

He wrote nearly 250 publications and was elected to many French and international scientific academies and societies. He was also Vice President of the Anaesthesia and Analgesia Society[1].

* * *

18.1 Régnier to Volterra, Paris, 12.3.1932 *Cat. no. 283*

12 Mars 1932

Monsieur le Professeur

Je vous remercie pour votre beau livre[2] dont j'ai pris connaissance avec grand intérêt.

---

[1] See also LAMBIN 1969.

[2] VOLTERRA 1931a.

Comme je vous l'ai dit l'autre jour, il me parait possible d'essayer de vérifier les lois, que vous avez mises en évidence, dans le domaine de la concurrence bactérienne. Dès qu'il me sera possible, je commencerai donc des recherches quantitatives, en mettant en présence soit "deux espèces se disputant la même nourriture", soit "deux espèces dont l'une se nourrit de l'autre", soit "deux espèces dont l'une est nuisible à l'autre". Nous serons, comme je vous l'ai déja expliqué, gêné par le fait que les milieux bactériologiques habituels sont rapidement inaptes à la multiplication bactérienne sans pour cela que les aliments nutritifs soient forcément épuisés. Si donc les phénomènes que nous cherchons n'apparaissent pas rapidement, c'est à dire dans les dix ou douze premières heures après l'ensemencement, nous serons obligés d'avoir recours à des techniques plus compliquées avec lesquelles il me faudra me familiariser (milieux, renouvelés, à composition constante).

En raison des recherches qu'il me faut terminer je ne pourrai pas commencer ces travaux avant quelques mois. Dès que j'aurai des renseignements intéressants je me ferai un plaisir de vous les communiquer.

Veuillez, Monsieur le Professeur, accepter de nouveau tous mes remerciements pour l'honneur que vous me faites en m'associant à vos beau travaux, et veuillez agréer avec l'expression de mon grand respect l'assurance de tout mon dévouement

J. Régnier 1, rue Chardon Lagache. Paris

18.2 Régnier to Volterra, Paris, 11.6.1933 *Cat. no. 284*

11 Juin 1933

Monsieur le Professeur.

J'ai eu hier l'occasion de voir Monsieur le Docteur Roux, et j'ai cru comprendre, en parlant avec lui, que vous étiez surpris de n'avoir pas encore de résultats pour les expériences, portant sur la concurrence bactérienne, que j'ai commencées sur vos conseils. Je vous prie, Monsieur le Professeur, de bien vouloir m'excuser de ce retard involontaire. Mon laboratoire est, en effet, petit, tout à fait personnel. Je ne dispose que de deux aides, deux jeunes filles fort sérieuses, qui ne peuvent malheureusement pas me consacrer entièrement leur temps. Du reste, moi même, je suis obligé de mener de front mes recherches de laboratoire et mes fonctions de professeur agrégé à la Faculté de Pharmacie. Enfin, et surtout, nous avons dû consacrer les dernièrs mois de 1932, et les premièrs de 1933, à l'achèvement de travaux en cours, dont certains, portant sur les mesures des pouvoirs antiseptiques des substances chimiques, sont commencés depuis plus de cinq ans.

Quoiqu'il en soit, depuis deux mois, les recherches qui vous intéressent directement sont commencées. Il nous a fallu cependant perdre encore quelques semaines pour acquérir suffisamment d'habilité dans la numération de deux espèces microbiennes se développant en même temps. Enfin, pour le moment, se continuent des expériences portant, autant que possible, sur le premier cas envisagé: espèces se disputant la même nourriture. J'ai choisi pour débuter le B.

Coli et le Staphylocoque, morphologiquement dissemblables. Mais ces expériences, dont je vous enverrai dès qu'il me sera possible les résultats, seront certainement insuffisantes. En effet, ici, la question est plus complexe qu'ailleurs, car, en milieu limité, la multiplication microbienne s'arrête d'elle même, et les produits mis en liberté, d'une façon ou d'autre, par l'un des microbes peuvant venir gêner l'evolution de l'autre. Il me faudra donc resoudre sans tarder ces deux problèmes, ce qui me demandera certainement des efforts assez longs.

Veuillez donc, Monsieur le Professeur, encore une fois m'excuser, et agréer l'expression de mon très respectueux et entier dévouement.

J. Régnier 1 Rue Chardon-Lagache Paris (16)

18.3 Volterra to Régnier, June-July 1933 *Cat. no. 398*[3]

Cher Monsieur,

Je vous remercie beaucoup d'abord de votre aimable lettre du 11 Juin et après plus particulierement de votre dernière lettre ainsi que des papiers et des tables que vous y avez joint.

Je trouve les esperiences très intéressantes car elles montrent, déjà d'une manière très nette l'action que les individus en presence les uns des autres exercent entre eux.

Je m'empresse à vous écrire pour vous exprimer ma reconnaissance, mais je vais examiner plus en détails les données et les tables et je vous écrirai à nouveau. En attendant j'éspère que vous voudrez bien continuer vos recherches.

18.4 Régnier to Volterra, Paris, 16.7.1933 *Cat. no. 285*

16 Juillet 1933

Monsieur le Professeur

Je vous envoie, ci joints, les premiers résultats obtenus pour les essais entrepris sur votre conseil. Dans ces deux premiers essais apparait, très nettement, le pouvoir inhibiteur du B. Coli sur la multiplication du Staphylocoque. Ce ne sont là, évidemment, comme je vous l'ai déjà dit, que des essais préliminaires; je souhaite cependant que ces quelques chiffres puissent vous donner quelques indications intéressantes. J'ai classé a priori cet antagonisme, Coli-Staphylocoque, (antagonisme d'ailleurs peu étudié), dans la groupe des antagonismes entre espèces se disputant la même norriture. Il me reste à le prouver. Pour ceci, je vais poursuivre mes recherches dans divers sens, et particulièrement dans des milieux à renouvellement continu.

Veuillez agréer, Monsieur le Professeur, l'expression de mes sentiments très respectueusement dévoués.

J. Régnier 1 Rue Chardon-Lagache Paris (16)

---

[3] Draft n.d. but obviously written in June or July 1993.

18.5 Régnier to Volterra, Paris, 14.3.1934 *Cat. no. 286*

14 Mars 1934

Monsieur le Professeur et cher Monsieur

Je vous envoie de nouveaux résultats concernant la concurrence bacterienne: colibacille-staphylocoque. Dans ces essais, nous avons fait varier le nombre de colibacilles ensemencés, ce germe s'étant montré, au cours de nos premiers essais, l'agent antagoniste. Nous avons donc mis en présence d'une quantité determinée de cocci, (staphylocoques, 500.000 germes par cm^3), des quantités de plus en plus petites de colibacilles, soit: 50.000, 5.000, puis 500 par cm^3. Chacune des expériences a été effectuée 2 fois, successivement.

Je vous envoie également, à nouveau, les courbes des 2 premières expériences, dans lesquelles étaient ensemencées des quantités égales des 2 germes (500.000 par cm^3).

J'espère que ces resultats vous intéresseront. De mon côté, je compte presenter à la société de Biologie, avec la collaboration de mademoiselle Lambin, une ou deux courtes notes, exposant la question de l'antagonisme entre les microbes, d'un point de vue général, en insistant particulièrement sur l'antagonisme: colibacille-staphylocoque, montrant que, dans cette concurrence, l'agent "freinateur" est le colibacille, et l'agent gêné le staphylocoque, et indiquant en outre qu'il peut y avoir, dans la seconde phase de la poussée, un ralentissement dans la diminution du nombre des colibacilles, du fait de la présence du staphylocoque. J'espère présenter assez rapidement ces notes qui auront une tendance purement bactériologique. Aussitôt après, j'en réunirai la substance en une note globale qui sera destinée, si vous le voulez bien, à l'Académie des Sciences[4]. Si vous le permettez, je poserai alors le problème du point de vue biologique tel que vous l'envisagez. Je vous soumettrai cette note, vous laissant le choix, soit de la présenter vous même, soit de m'indiquer un biologiste, membre de l'Institut, qui pourra la présenter.

Je vais continuer les essais sur cette antagonisme, particulierement en milieux renouvelés, et j'espère aborder, dans peu de temps, l'étude d'un nouvel antagonisme entre deux espèces dont l'une se nourrit de l'autre: protozoaires et bactéries.

Veuillez, Monsieur le Professeur, agréer l'expression respectueuse de mon entier dévouement.

J. Régnier

---

[4] Clearly RÉGNIER, LAMBIN 1934 who, as shown by the following correspondence, was presented by Volterra to Académie des Sciences for publication. Suzanne Lambin was born in Nantes on 1st August 1902. She studied pharmacy at Nantes and Paris and in 1925 began to work as an intern at the Hôpital Ambroise Paré, where she met Régnier. Under his guidance she completed her doctoral thesis (LAMBIN 1933), and Régnier had a great influence in her scientific work. In 1951 she was appointed professor of microbiology of the Paris Faculty of Pharmacy (the first woman to have a post in this Faculty). She retired from teaching in 1972 and was named in 1973 "professeur honoraire des universités".

18.6 Volterra to Régnier, Paris, n.d. *Cat. no. 399*[5]

Monsieur Jean Regnier
1 Rue Chardon Lagache, Paris.

Cher Monsieur
Nous allons envoyer, Monsieur D'Ancona et moi même, à l'Editeur Hermann le manuscrit de l'ouvrage qui doit paraître dans la Collection Scientifique sur la Biologie Mathematique[6]. Nous y avons inséré une de vos tables en citant vos travaux. Si vous désirez présenter à l'Académie des Sciences une note pour les Comptes Rendus sur nos recherches avant la publication de notre ouvrage je serai heureux de la faire parvenir au secrétaire perpétuel ou directement ou si vous le preferez par l'entremise de Monsieur Joubin à moins que vous ne preferez vous même la remettre à quelque membre de l'Institut avec lequel vous êtes en rapport.

Veuillez accepter, cher Monsieur, l'expression de mes sentiments les meilleurs.

18.7 Régnier to Volterra, Paris, 18.11.1934 *Cat. no. 287*

18 Novembre 1934

Monsieur le Professeur
Veuillez m'excuser pour le retard que j'apporte à vous répondre. Ce mois-ci se traduit pour nous par un grand accroissement de travail, du fait de la rentrée des étudiants à la Faculté, et, dans ces derniers jours, je me suis trouvé aux prises avec de vifs soucis pour la santé d'un de mes collaborateurs mordu par un rat en expériences, et dont la vie a été fort en danger.

Je vous envoie une note, au nom de Melle S. Lambin et de moi même, pour que vous puissiez la présenter, vous même, si cela vous convient, à l'Académie des Sciences[7]. Nous serons extrêmement heureux d'avoir votre patronage, et nous serions fort contents si vous vouliez bien faire suivre cette note bactériologique, d'une communication à votre nom où vous exposeriez les réflexions mathématiques qu'ont pu vous suggérer nos chiffres. C'est là, Monsieur le Professeur, une simple suggestion que je me permets de faire et qui a pour base la grande fierté que nous ressentons, mes collaborateurs et moi-même, en travaillant en liaison avec vous. Il est bien évident que je suis tout près à modifier le texte de ma note si vous le jugez utile.

Un de mes élèves commence l'étude de l'antagonisme Staphylo-Coli, en milieu de culture à renouvellement continu. Nous espérons ainsi avoir des résultats plus nets, que nous vous communiquerons au fur et à mesure

---

[5] Draft, n.d. but apparently written in reply to the preceding letter

[6] Volterra, D'Ancona 1935.

[7] See note 2.

Veuillez, Monsieur le Professeur, agréer l'expression de mon très respectueux dévouement.

J. Régnier

18.8 Régnier to Volterra, Paris, 9.12.1934 *Cat. no. 288*

Paris, 9 Décembre 1934

Monsieur le Professeur

Je vous remercie d'avoir bien voulu faire suivre notre note de vos remarques mathématiques.

Bien que je ne puisse comprendre entièrement le raisonnement mathématique que vous formulez, il me semble qu'il s'appuie sur l'idée que la loi d'accroissement du nombre des bactéries n'est pas dirigée par l'épuisement de la nourriture, mais bien par la mise en liberté, du fait même de la vie microbienne, de substances nuisibles au développement microbien.

C'est là, je pense, l'idée qui tend à s'imposer. Pourtant cette idée ne me donne pas encore entière satisfaction. En effet, dans des expériences qu'il me faut encore vérifier, dans lesquelles j'ai ensemencé d'emblée le nombre maximum des microbes par cm^3, je n'ai pas constaté de développement sensible. En tout cas le nombre des bactéries est bien loin d'avoir doublé, comme il aurait fallu théoriquement s'y attendre, s'il ne s'agissait que d'épuisement de nourriture, ou de mise en liberté de produits métaboliques nuisibles.

Si ces expériences sont confirmées, il faudra donc renoncer à envisager la multiplication des êtres microscopiques avec la même simplicité que l'on envisage la multiplication des autres êtres vivants étudiés jusqu'ici. Il faudra vraisemblablement tenir compte des forces physiques qui, à leur échelle, jouent un très grand rôle: tension superficielle, et surtout charges électriques. Il est fort possible qu'il faille un certain espace libre autour d'une bactérie, et que, à partir d'une concentration donnée, variable selon la bactérie et selon le milieu, la multiplication s'arrête.

Madame Kaplan a montré, il y a quelques années, dans mon laboratoire, que les bactéries semblent être très sensibles au nombre de leurs voisines, et que, dans le même volume de milieu nutritif, elles gouvernent leur multiplication bien plus lentement si elles sont plus nombreuses, et ceci dès le debut de leur développement, c'est à dire à un moment où l'influence de la nourriture (épuisement ou mise en liberté de déchets) n'a pas encore pu intervenir[8].

Peut-être y a t il encore d'autres influences physiques: radiations?

Peut-être y a t il encore apport par la bactérie de substances diffusibles toxiques pour elles-mêmes (toxines...)?

Quoiqu'il en soit, il semble bien qu'il faillera résoudre déjà ce premier problème avant de comprendre vraiment ce qui se passe dans le cas des antagonismes microbiens. En premier lieu, semble t il, il faudra se demander pourquoi le staphylocoque étant gêné par le colibacille, ce dernier ne l'est pas par

[8] See Régnier, David, Kaplan 1932.

le staphylocoque. Dans ce cas particulier les substances nutritives peuvent jouer un rôle.

Pour le moment un de mes èlèves, Mr Léger, installe, dans notre laboratoire, un appareil permettant l'apport continu des matières nutritives et l'enlèvement continu des substances de déchet, les microbes étant maintenus. Il sera extrêmement intéressant de voir, dans ces nouvelles conditions, ce que deviendra le développement: rapidité, nombre maximum, décroissance du nombre etc...

Je serai, Monsieur le Professeur, très heureux de vous envoyer ces nouveaux résultats dans quelques semaines.

Je vous priè d'agréer, avec tous mes remerciements, l'expression de mon grand respect et de tout mon dévouement.

J. Régnier

18.9 Régnier to Volterra, Paris, 30.1.1935 *Cat. no. 289*

30 Janvier 1935

Monsieur le Professeur et cher Monsieur.

Veuillez recevoir quelques tirés à part de la note que j'ai publiée avec Melle Lambin, et de l'analyse mathématique que vous avez bien voulu ajouter[9].

Nous continuons nos expériences en milieu nutritif constamment renouvelé. Je vous enverrai ces nouveaux résultats dans quelques semaines. Il s'agit pour l'instant d'une culture simple, nous commencerons ensuite l'étude d'un antagonisme microbien.

Avec l'expression respectueuse de tout mon dévouement.

J. Régnier

18.10 Régnier to Volterra, Paris, 5.4.1935 *Cat. no. 290*

5 Avril 1935

Monsieur le Professeur et cher Monsieur

J'allais justement vous écrire pour vous demander quel jour vous pourriez me recevoir. Je voulais aller vous voir Lundi après midi à l'Académie des Sciences, mais je m'aperçois que ce jour là je serai pris entièrement à la Faculté de Pharmacie. Voulez vous donc me fixer un rendez vous à l'endroit qui vous conviendra le mieux.

Je vous apporterai des courbes nouvelles, faites avec une méthode opacimétrique, qui sont très belles et régulières.

J'espère, cher Monsieur, que vous êtes complètement remis de vos malaises.

Veuillez, je vous prie, presenter mes très respectueux hommages à Madame Volterra et agréer l'expression respectueuse de tout mon dévouement.

J. Régnier

---

9 This is VOLTERRA 1934a.

Je serai libre Lundi après midi à partir de $5^h.30$, ou Lundi matin, ou Mardi matin, ou Mardi après midi à partir de $3^h.30$.

18.11 Régnier to Volterra, Paris, 19.5.1935 *Cat. no. 291*

Paris, 19 Mai 1935

Monsieur le Professeur et cher Monsieur

Voulez vous m'excuser de n'avoir pas encore demandé de vos nouvelles. Je serai très heureux de savoir que votre voyage de retour ne vous a pas trop fatigué, et que vous pouvez maintenant reprendre votre pleine activité, dans votre beau pays. Je serai également fort satisfait de savoir que Madame Volterra est en bonne santé.

Ici, à Paris, nous avons un mauvais temps, froid, pluie et vent. Les vacances de Pâques n'ont pas été, de ce fait, très gaies. Je pense que le soleil vous est plus favorable en Italie qu'ici.

Vous avez eu l'amabilité de vous intéresser à mes recherches sur les anesthésiques locaux. Veuillez trouver avec ma lettre un tiré à part de la note que Monsieur Desgrez a presenté à l'Istitut le 8 Avril. J'aurai plaisir à vous tenir au courant de ces recherches que j'ai commencées dès 1920.

Par ailleurs nos essais sur la multiplication bactérienne se continuent. Nous avons été amenés à modifier nos appareils pour permettre une meilleure nutrition des bactéries, et en particulier une meilleure oxygénation.

A ce propos voudriez vous me permettre, cher Monsieur, de vous demander une lettre d'introduction pour Monsieur le Professeur Jean Perrin[10], ou plus exactement pour Mademoiselle Choucroun qui travaille dans son laboratoire. Je travaille, en effet, en ce moment, l'action des ions métalliques et l'action des métaux à distance sur le développement des bactéries, et je voudrais faire sur ces microbes des mesures d'électrophorèse, de transport électrique, pour connaître le rôle des charges électriques des bactéries dans ces phénomènes. Je pense que je ne pourrais trouver de meilleur maître que Mademoiselle Choucroun pour m'initier à ces mesures délicates. Je vous remercie bien vivement d'avance de ce que vous pourrez faire pour me permettre d'entrer en relation avec Mademoiselle Choucroun.

Mademoiselle Lambin aura bientôt terminé la rédaction du grand article que nous preparons sur les antagonismes bactériens. Je vous l'enverrai tout aussitôt.

Veuillez, Monsieur le Professeur et cher Monsieur, recevoir l'expression très respectueuse et, si vous le permettez, affectueuse de tous mon dévouement, et

---

[10] This refers to Jean Baptiste Perrin, a well-known French physicist, born in Lille in 1870 and died in New York in 1942. His research bore mainly on colloids, the determination of absolute molecular weights and Brownian motion: for the latter, which provided experimental confirmation of Albert Einstein's theoretical studies, in 1926 he received the Nobel Prize for physics. He was professor of physical chemistry at the University of Paris, member of the Académie des Sciences as from 1923 and its President from 1938 on. His book *Les Atomes* (1913) is a classic in scientific literature.

présenter à Madame Volterra tous mes respectueux hommages. Veuillez aussi recevoir tous mes voeux de complet rétablissement.

J. Régnier

18.12 Volterra to Régnier, N. Choucroun, J. Perrin, 1935 *Cat. nos. 400, 366, 397*[11]

Cher Monsieur

je vous remercie beaucoup de votre aimable lettre du 19 mai et de votre interessant article sur les anesthésiques locaux dont vous m'aviez déjà exposé les résultats lors de notre dernier entretien à Paris. Vous m'interesserez beaucoup en me tenant au courant de vos recherches sur ce sujet. Je suis heureux d'apprendre que vous continuez vos belles expériences sur la multiplication bactérienne et que vous avez perfectionné vos méthodes. Je lirai avec le plus grand plaisir le grand article redigé par M.lle Lambin sur les antagonismes bactériens. Ci-joint vous trouverez le deux lettres d'introduction pour Monsieur Jean Perrin et Mademoiselle Choucroun. Je ne doute pas de leur bon accueil.

Malhereusement ma santé est fort mauvaise depuis mon retour en Italie. Je ne souffre plus du coeur mais je souffre terriblement à la gorge, une pharingite obstinée. Même ici le temps toujours très mauvais et je n'ai pas pu rentrer à la campagne où peut-être l'air pure me serait favorable.

Madame Volterra se porte bien et elle vous est très reconnaissante de votre souvenir et de vos souhaits. Veuillez accepter cher Monsieur l'assurance de ma plus haute estime et l'expression de mon attachement le plus affectueux et le plus dévoué.

Chère Mademoiselle

Je vous ai parlé plusieurs fois des recherches très intéressantes de Monsieur Jean Régnier de la Faculté de Pharmacie. Ces expériences sur les antagonismes bactériens sont en des rapports avec mes recherches. J'ai ajouté dans le Comptes Rendus quelques calculs mathématiques à une de ses Notes.

Maintenant il s'occupe de la multiplication bactérienne et ses études l'ont amené dans le domaine de la physique et spécialement de l'électricité. Vous avez fait des travaux d'un énorme interêt sur ces sujets et je ne doute pas que vos conseils et votre aide seront d'une grande utilité pour Monsieur Régnier. Il m'a écrit qu'il désire vous être presenté ce que je fais avec le plus grand plaisir bien sûr de votre bon accueil à ce savant distingué.

Veuillez accepter, chère Mademoiselle, l'expression de ma plus haute estime et les hommages le plus dévoués.

Mon cher Ami

J'ai l'honneur et le plaisir de te présenter Monsieur Jean Régnier de la Faculté de Pharmacie. Ce savant très distingué est bien connu par ses travaux dans les domaines de la pharmacologie et de la bactériologie. Il s'est occupé entre autres

[11] Drafts.

choses des antagonismes bactériens. Mes études ont eu des rapports avec le siens car j'ai tâché d'appliquer les mathématiques à ses experiences. Maintenant il s'occupe de la multiplication bactérienne et cela l'a amené à des études physiques specialement d'électricité. Il désire à ce propos tes conseils et ton aide ainsi que ceux de M.lle Choucroun et je ne doute pas que tu voudras le récevoir et l'accueillir et lui être utile.

Je te ne remercie d'avance de tout ce que tu voudras faire pour lui. Je te prie, mon cher ami, de vouloir bien accepter l'expression de ma plus fidèle et affectueuse amitié.

18.13 Régnier to Volterra, Paris, 28.5.1935 *Cat. no. 292*

28 Mai 1935

Monsieur le Professeur et cher Monsieur.

Je vous remercie bien vivement des deux lettres de recommandation que vous avez bien voulu m'envoyer, et je fais des voeux que votre pharyngyte soit rapidement terminée. Je pense que le temps irrégulier de ce printemps douteux est pour beaucoup dans ces maux de gorge qui ont été, pour beaucoup de gens, fort douloureux et fort tenaces. Moi même, j'ai fort souvent de tels maux, qui sont du reste à l'origine de mes rhumatismes, et j'ai cette année particulièrement souffert. J'espère donc, bien vivement, que vous êtes dès maintenant guéri, et que votre coeur, si vous voulez bien ne pas le fatiguer, vous laisse tout à fait en repos.

Nous continuons nos essais, j'aurai plaisir à vous en communiquer les résultats. Je vais dans quelques jours pénétrer dans le livre de Gause[12]. Je voudrais non seulement présenter au public français, médecin et pharmacien, le problème des antagonismes microbiens, mais encore montrer quelles sont vos conceptions appuyées sur vos belles recherches mathématiques. De ce point de vue votre mise au point qui paraîtra chez Hermann me sera de grand secours.

Je voudrais, cher Monsieur, aller plus vite dans toutes ces recherches, et vous apporter plus rapidement des résultats numériques sur ces antagonismes, mais nous menons une vie si bousculée, par force, qu'il faut m'en excuser.

Veuillez encore, Monsieur le Professeur et cher Monsieur, recevoir tous mes remerciements, et tous mes voeux de complet rétablissement et agréer l'expression de mon grand respect et de mon affectueux dévouement. Veuillez présenter aussi mes très respectueux hommages à Madame Volterra.

J. Régnier

18.14 Régnier to Volterra, Paris, 6.8.1935 *Cat. no. 293*

6 Août 1935

[12] G. F. Gause 1935.

Monsieur le Professeur et cher Monsieur.

De retour d'un voyage en Grèce je reçois votre lettre. Je vous remercie bien vivement, ainsi que Monsieur D'Ancona de l'envoi de votre bel ouvrage. Je vous remercie également d'avoir bien voulu tenir compte de nos travaux. Je pense que votre livre nous sera fort précieux pour la rédaction de l'article que nous rédigeons en ce moment.

Je suis profondement attristé des maladies qui ont éprouvé votre famille, et du malaise que vous continuez à ressentir. Peut être y a t il, en effet, des mesures thérapeutiques à prendre, qui pourraient être plus efficaces que la simple expectative. J'ai eu, hier, l'occasion d'en parler à mon maître le Professeur Charles Lenormant, chirurgien, de la Faculté et de l'Académie de Médicine. Ce dernier à qui j'ai eu souvent l'occasion de parler de vous-même, et qui par ailleurs était entièrement dévoué au Docteur Roux, votre illustre ami, pense qu'il serait bon de voir les radiographies de votre gorge. Si vous vouliez donc m'envoyer ces radiographies, je pourrais les lui transmettre et lui-même, ainsi, pourrait vous diriger sur tel ou tel spécialiste. Il existe en effet à Paris de forts bon spécialistes des maladies de la gorge ou de l'œsophage, que vous pourriez, je pense consulter avec fruit.

Malheureusement il ne faut pas trop compter trouver à Paris les médicins qualifiés avant Septembre. Mais je pourrai d'avance me renseigner et connaître la date du retour à Paris de tel ou tel spécialiste choisi. Veuillez donc, si vous n'y voyez pas d'inconvénient, cher Monsieur, me faire parvenir vos radiographies, ainsi pourrons nous gagner un peu de temps et frapper tout de suite à la porte du spécialiste le mieux qualifié.

Mais j'espère bien que votre souffrance va s'attenuer d'elle-même, et que votre prochain séjour à Paris se fera en pleine santé. Ces maux de gorge si tenaces sont pour moi bien connus. Ce sont eux qui m'ont réduit en l'état voisin de l'infirmité où vous m'avez vu. Ils me fatiguent surtout en été, où le froid des courants d'air succèdent brusquement à la chaleur étouffante, et, chose curieuse, l'hiver à ce point de vue m'est beaucoup plus favorable. J'espère fort qu'il en sera de même pour vous, et que la disparition de votre mal coincidera avec la fin des grandes chaleurs. Quoiqu'il en soit, veuillez, cher Monsieur, m'envoyer vos radiographies.

Mon travail s'avance péniblement et lentement mais j'espère pouvoir dans quelques semaines vous donner des résultats nouveaux sur l'antagonisme entre colibacille et staphylocoque, étudié en milieu continu, et vous présenter également le mémoire d'ensemble que nous rédigeons.

Veuillez, je vous prie, cher Monsieur, présenter mes respectueux hommages à Madame Volterra, et agréer l'expression très respectueuse de toute mon affection et de tout mon dévouement.

J. Régnier

18.15 Régnier to Volterra, Paris, 3.11.1935 ***Cat. no. 294***

3 Novembre 1935

Monsieur le Professeur et cher Monsieur

Je vous remercie bien vivement de votre aimable lettre. Je serais bien heureux si toutes les visites médicales que vous avez faites vous permettent de retrouver complètement votre santé; et je crois sincèrement que vous étiez en meilleure santé au moment de votre départ de Paris qu'au moment de votre arrivée. Je serais cependant content d'avoir de vous nouvelles et de savoir que votre voyage ne vous a pas trop fatigué.

J'espère encore que vous avez retrouvé toute votre famille en bonne santé, et je fais des voeux pour que toutes les difficultés que rencontrent votre pays et le nôtre s'aplanissent le plus vite possible, sans autres complications.

Je vous suis particulièrement reconnaissant d'avoir bien voulu me présenter à Monsieur Mesnil, et je n'oublie pas que c'est grâce à vous que j'ai fait connaissance avec Mademoiselle Choucroun avec laquelle je vais travailler demain.

Je vais écrire à Monsieur Kostitzine de votre part.

Veuillez, cher Monsieur, présenter mes respectueux hommages à Madame Volterra et recevoir l'assurance de tout mon respectueux et affectueux dévouement.

J. Régnier

Ma femme et Mademoiselle Lambin me chargent de les rappeler à votre bon souvenir ainsi qu'à celui de Madame Volterra.

18.16 Régnier to Volterra, Paris, 2.1.1936 ***Cat. no. 295***

2 Janvier 1936

Monsieur le Professeur et cher Monsieur

J'aurais été très heureux, en ce début d'année, de vous savoir complètement remis. Mais je vois bien que vous souffrez toujours, et je n'ose espérer que, depuis un mois, vous ayez pu constater quelque amélioration. Je vous présente donc, et ma femme se joint à moi, tous nos meilleurs voeux pour 1936, voeux avant tout de bonne santé. Nous souhaitons aussi cette bonne santé pour Madame Volterra et pour toute votre famille. Souhaitons aussi, cher Monsieur, que la paix revienne sur nous tous.

Je suis en ce moment, dans mon pays où j'ai essayé, pendant ces quelques jours de vacances, de terminer la rédaction de mon livre (traduction et complements) sur la perméabilité cellulaire. Je n'ai pas encore réussi à me libérer de ce travail. Pourtant je n'ai pas abandonné notre article sur l'antagonisme microbien[13], et j'espère le donner dans très peu de temps à Monsieur Mesnil, que je rencontre de temps en temps et qui a bien voulu se souvenir de moi.

---

[13] In Volterra 1937c we read: "Mr Régnier and Miss Lambin are about to print the results of their interesting research on microbial antagonism". However, this work was never published, as

J'ai eu également le grand plaisir de transmettre, de leur part, votre souvenir à Monsieur Perrin et à Mademoiselle Choucroun. J'ai à vous trasmettre tous leurs meilleurs voeux de complet rétablissement.

Mon cher Monsieur, je vous suis fort reconnaissant de l'aimable lettre que vous avez bien voulu m'écrire. Je n'ai qu'un faible mérite à vous avoir guidé, un peu, parmi les medicins de Paris. J'ai été bien heureux d'apprendre par eux que votre cas n'était pas grave, mais j'aurais voulu qu'ils puissent, séance tenante, vous débarasser de toute souffrance. Je crois qu'il vous faudra encore de la patience, qu'il vous sera nécessaire de bien suivre votre traitement sédatif, de bien faire en sorte que le calme règne (si non entièrement, du moins le plus possible) dans votre esprit. Ce n'est évidemment pas très gai, très amusant, mais ce n'est rien en comparaison d'une opération que nous avons cru nécessaire pendant quelque temps.

Mon cher Monsieur Volterra, veuillez présenter à Madame Volterra tous nos respectueux compliments, et accepter l'expression de tout mon respectueux et affectueux dévouement

J. Régnier

P. S. J'irai dans très peu de temps rendre visite à M. Kostitzin.

18.17 Régnier to Volterra, Paris, 13.3.1936 *Cat. no. 296*

13 Mars 1936

Monsieur le Professeur et cher Monsieur

Je serais bien content d'avoir de vos nouvelles, et j'éspère fort que votre santé est bonne et que vous viendrez à Paris, comme d'habitude, avant Pâques.

J'ai été hier rendre visite à Monsieur Kostitzin qui m'a reçu fort aimablement. Il a déja essayé de résoudre l'équation de la concurrence microbienne mais sans succès. Il va encore tenter quelques essais.

J'aurais bien voulu donner mon article dès maintenant à Monsieur Mesnil, mais je vais encore attendre un peu (8 ou 15 jours) dans l'espoir que Monsieur Kostitzin réussira.

Cher Monsieur, veuillez, s'il vous plait, me donner de vos nouvelles, et présenter à Madame Volterra tous nos respectueux hommages.

Je vous prie de bien vouloir agréer l'expression respectueuse et affectueuse de tout mon dévouement.

J. Régnier

---

was confirmed directly to Ana Millán Gasca by Suzanne Lambin during their meeting in Paris in 1993. Miss Lambin handed to A. Millán Gasca a manuscript containing the foreword to the work that, in their intention, was to become a book (see also the following letters and in particular letter no. 18.18). The title is "Évolution des cultures microbiennes. Vie des microbes et phénomènes généraux caractérisant la vie des populations"; the authors are Jean Régnier and Suzanne Lambin, and the work was to be preceded by a foreword written by Vito Volterra. This manuscript will be the subject of a specific publication by Ana Millán Gasca. A series of particular works on the topic were indeed published, namely Régnier, Lambin, Szollosi 1937a, b, c, Régnier, Lambin 1938a, b, c, d, as well as certain works by Kostitzin (see bibliography).

Je serais très content si vous pouviez me faire parvenir quelques tirages à part de la note à l'Institut, avec Melle Lambin, suivie de vos considerations mathématiques. Il ne m'en reste plus. Veuillez m'excuser de la peine que je vous demande de prendre ainsi.

Nous serions bien heureux de savoir que vous n'avez pas été trop gêné par votre gorge, et nous faisons des voeux bien sincères pour que tout s'arrange au gré de nos deux nations.

18.18 Régnier to Volterra, Paris, 13.4.1936 *Cat. no. 297*

13 Avril 1936

Monsieur le Professeur et cher Monsieur

J'ai été fort content d'avoir de vos nouvelles, et j'éspère bien que votre santé sera assez bonne pour que vous puissiez venir à Paris dans quelques jours.

J'ai bien reçu les tirages à part et je vous en remercie vivement. Je n'ai pas de nouvelles de M. Kostitzin, mais j'irai le voir dès mon retour à Paris. Bien que connaissant fort peu de choses aux mathématiques, je me rends compte de la difficulté du problème posé. Quoiqu'il en soit je viens de compléter notre article[14], en exposant plus longuement les travaux de Verhulst... Deschamps... qui ont trait au développement des populations et des organismes. J'ai aussi développé plus longuement vos vues sur l'association de deux espèces se disputant la même nourriture. Enfin j'ai examiné les récents travaux de Gause (niche unique, niche diffèrente...) et montré que le cas d'antagonisme entre microbes devrait rentrer dans la 3e catégorie examinée par Gause, c'est à dire exactement dans les conditions definies par vous-même. Comme le Staphylocoque est loin d'être expulsé complètement par le Colibacille, il faut donc en conclure, ce qui apparaissait du reste a priori, que l'antagonisme entre microbes est sous la dépendance de facteurs plus nombreux que l'antagonisme entre d'autres espèces, animales par example. Vous avez, vous même, au surplus, dans vos calculs parfaitement tenu compte de ces divers facteurs (effet Verhulst, Pearl, auto-intoxication d'une espèce par elle-même, intoxication d'une espèce par l'autre...), et il n'y a qu'à regretter que vos équations soient devenues ainsi, par force, si compliquées qu'elles soient si difficiles à résoudre.

J'ai encore insisté sur la coïncidence de la courbe Volterra-Kostitzin, avec la courbe expérimentale dans le cas d'une espèce pure et montré plus largement les renseignements qui, dès maintenant, découlent de cette coincidence et exposé aussi les services que ce rapprochement peut rendre dans la suite. Il n'est pas douteux, en effet, qu'en rapprochant, quantitativement, le mieux possible, les deux courbes (dont la similitude, qualitative, est acquise) on obtiendra pour les équations des valeurs déterminées de l'action des differents facteurs. Il suffira, ensuite, de chercher expérimentalement (tous autres facteurs éliminés si possible) la valeur d'une action dans telle ou telle hypothèse, pour, après comparaison, éliminer l'une ou l'autre de ces hypothèses. Dès qu'il me sera possible je

[14] See previous note.

reprenderai donc l'étude de la multiplication microbienne dans des conditions spéciales: renouvellement continu du milieu, etc. continuant ainsi des expériences non encore publiées.

Mais avec tous ces compléments et discussions notre article est devenu énorme. Il va s'accroître encore de l'exposé de travaux récents: Ludwig, Richards... que Mademoiselle Lambin est en train de traduire. Si bien que je ne sais pas du tout si Monsieur Mesnil pourra le prendre aux Annales de l'Istitut Pasteur. Si non, je le présenterai à d'autres revues qui n'ont pas peur des longs articles, ou même je l'offrirai à Georges Teissier, de qui j'ai fait connaissance, pour la collection de statistique biologique où vous même vous avez présenté vos travaux, avec M^r^ D'Ancona.

Mais peut être aurai je la chance de vous entretenir de tout ceci de vive voix. C'est ce que je souhaite de tout mon coeur, cher Monsieur, car ceci prouverait que votre santé est relativement bonne. Nous en serions tous si heureux

Veuillez, cher Monsieur, présenter mes respectueux hommages à Madame Volterra et recevoir l'expression affectueuse et très respectueuse de tout mon dévouement

J. Régnier

18.19 Régnier to Vito and Virginia Volterra, Paris, 7.1.1937 *Cat. no. 298*

7 Janvier 1937

Cher Monsieur et chère Madame.

Je suis bien en retard pour vous présenter mes vœux et ceux de ma femme. Mais je suis certain que vous voudrez bien nous excuser. Mon excuse est toujours la même: un peu trop de travail.

Nous faisons des vœux très vifs pour que 1937 soit une très bonne année pour vous même et les vôtres. Nous désirons en particulier que votre santé soit parfaite, que Monsieur Volterra ne se ressente plus du tout des troubles qui nous avaient émus, et qu'il puisse continuer son oeuvre pour le plus grand intérêt de tous.

Nous sommes très fiers d'avoir un ami tel que lui, et nous en sommes, si possible, encore plus heureux maintenant que nous savons tout ce que la France lui doit, en ce qui concerne son existence même.

Cher Monsieur, chère Madame, veuillez recevoir encore tous nos remerciements, agréer tous nos vœux, et accepter l'assurance de tout notre respectueux et affectueux dévouement.

J. Régnier

18.20 Régnier to Volterra, Paris, 19.2.1937 *Cat. no. 299*

19 Février 1937

Cher Monsieur Volterra

J'ai été bien heureux de recevoir votre bel opuscule sur les Principes de Biologie Mathématique[15] et j'ai transmis l'exemplaire dedié à Mademoiselle Lambin qui l'a accueilli avec beaucoup d'intérêt et de reconnaissance.

Me voici arrivé à la fin du 1er semestre de l'année scolaire. Les cours et les Travaux pratiques vont prendre fin. Je vais donc reprendre, en m'aidant des idées si simples et si belles que vous apportez, la rédaction de notre livre[16]. Voilà bien longtemps que nous remettons notre travail en rédaction, espérons que nous arriverons à sa fin dans le mois qui viennent. En tous cas votre dernière oeuvre me fait un peu moins regretter notre retard, puisque ce retard me permettra d'exposer votre dernière conception.

Et puis, cher Monsieur, nous serions tous si heureux d'avoir de vos nouvelles, de vos bonnes nouvelles, ainsi que de celle de Madame Volterra. Nous espérons fort que vos ennuis de l'an dernier sont complètement oubliés, et qu'à Pâques vous reverrons en santé absolutement parfaite.

Cher Monsieur, voulez vous recevoir tous nos remerciements et agréer l'expression de toute notre dévouement, pour Madame Volterra et pour vous-même.

J. Régnier

18.21 Régnier to Volterra, Paris, 10.4.1937 *Cat. no. 300*

10 Avril 1937

Monsieur le Professeur et cher Monsieur

J'espérais avoir de vos nouvelles, et apprendre votre arrivée à Paris. Je souhaite bien vivement que tout aille parfaitement bien pour Madame Volterra, votre famille et vous-même, et que nous ayons dans quelque temps le grand plaisir de vous voir.

Je suis en pleine rédaction du livre sur l'antagonisme entre microbes, je le complète par les chapitres explicatifs nécessaires sur la numération des bactéries et sur la multiplication microbienne, puis j'ajouterai aux chapitres déja écrits et que vous connaissez, les données générales recentes que vous avez apportées.

Voudriez vous, cher Monsieur, me faire le grand honneur d'écrire une préface pour ce livre? Je vous demande là une peine assez lourde. Mais ce livre vous devra tant et nous serons Mademoiselle Lambin et moi-même si fiers d'être présentes par vous même[17].

---

[15] VOLTERRA 1937a.

[16] See note 10.

[17] See note 10 again.

A ce propos nous vous remercions bien vivement du mot aimable que vous avez bien voulu consacrer à nos essais dans l'introduction que vous avez écrite pour le livre de M^r. Kostitzin.

Cher Monsieur, nous espérons fort avoir bientôt de bonne nouvelles de Madame Volterra et de vous même.

Veuillez agréer l'expression de notre fidèle et respectueux souvenir et présenter à Madame Volterra tous nos compliments respectueux.

J. Régnier

18.22 Régnier to Volterra, Paris, 25.4.1937 *Cat. no. 301*

25 Avril 1937

Monsieur le Professeur et cher Monsieur

Je suis extrèmement heureux de vous savoir en parfaite santé, et fort content de savoir que vous avez pu faire sans fatigue votre voyage en Haute Egypte. Puisque j'ai été pendant quelques jours un peu votre conseiller médical je me permets cher Monsieur, de vous recommander de ne trop pas vous fatiguer pendant vos futurs voyages, non seulement pour les efforts physiques mais aussi pour les efforts intellectuels. Mais je pense que si vous voulez bien vous surveiller tout ce qui vous distraira sera parfait.

Madame Volterra saura du reste tout à fait où s'arrêtent les limites de la distraction et du devoir, si je puis dire ainsi, et où commence la fatigue.

Excusez moi, cher Monsieur, de m'immiscer un peu dans votre vie, mais c'est, je pense, la simple conséquence de ma respectueuse affection.

J'ai donc repris la rédaction de mon livre avec Mademoiselle Lambin, et vais faire compléter mes expériences non seulement sur l'antagonisme entre microbes, mais encore sur le croît microbien en culture pure. J'arrive peu a peu à l'idée que pour pouvoir faire jouer vos lois et d'une façon genérale des lois mathématiques simples, il me faudra travailler en milieu peu riche en éléments nutritifs.

Enfin tout ceci se fera peu à peu, et je vous en parlerai à votre passage à Paris.

Ma femme me charge de la rappeller à votre bon souvenir et je vous prie de présenter mes respectueux compliments à Madame Volterra que j'espère en très bonne santé, ainsi que toute votre famille.

Veuillez, Monsieur le Professeur et cher Monsieur, agréer avec l'assurance de tout mon dévouement, l'expression de ma très respectueuse affection

J. Régnier

18.23 Régnier to Vito and Virginia Volterra, Paris, 9.2.1938 *Cat. no. 302*

9 Février 1938

Cher Monsieur et cherè Madame

Dès la reception de la lettre de Madame Volterra j'ai été rendre visite au Professeur Laubry. Aux premiers mots d'explication, sans même avoir ouvert le lettre du médécin romain, il était parfaitement au courant, tellement il se souvient, avec netteté, du cas de Monsieur Volterra. La lecture de la lettre et l'inspection du graphique n'ont fait que le confirmer dans son diagnostic: Monsieur Volterra s'est fatigué, il fait les accidents prévus et avec du repos, du calme et la surveillance de son alimentation tout va rentrer dans l'ordre à bref délai.

Monsieur Laubry va du reste écrire au D^r^. Nazari, peut-être conseiller quelques légères modifications au traitement et insister sur la surveillance du régime de notre malade, dès que celui-ci sera tout à fait remis.

En somme, un diagnostic et un pronostic excellents, mais avec obligation d'une sagesse bien plus grande pour l'avenir.

Sans être très savant dans les malades du coeur, on pouvait, il y a quelques mois, prévoir, cher Monsieur, les ennuis que vous subissez actuellement, et il faudra à l'avenir tout faire pour mener une vie plus calme, exempte le plus possible de soucis, presque exempte de fatigues, et aussi plus modérée pour la nourriture. Toutes choses, bien cher Monsieur, très souvent répetées mais dont il faudra, je vous en prie, vraiment tenir compte. Après m'avoir rassuré, monsieur Laubry qui relevait, lui même, de maladie, m'a longuement interrogé sur le travail de Mr Volterra. Bien entendu je n'ai été capable d'expliquer tout ce que l'on doit à votre génie, cher Monsieur. Mais le peu que j'en ai dit a suffisamment montré à Monsieur Laubry que votre gloire était dès maintenant suffisante et qu'il fallait d'abord maintenant, penser à vous.

Je serais bien content, cher Monsieur et chère Madame, si vous veuillez m'envoyer de temps en temps de vos nouvelles. Veuillez, cher Monsieur et cherè Madame, accepter l'expression de notre profond respect très affectueux et de mon dévouement, avec mes souhaits les plus vifs de pront rétablissement.

J. Régnier

18.24 Régnier to Virginia Volterra, Paris, 9.2.1938 *Cat. no. 303*

9 Février 1938

Cherè Madame

Cette lettre ne fera que confirmer celle que vous recevez en même temps à votre domicile. Le diagnostic est formel: Mr Volterra mène une vie trop nerveuse, trop fatiguante, il s'alimente trop. La maladie dont il souffre, grave pour un homme en pleine activité, est d'un pronostic beaucoup moins sombre pour un homme d'un certain âge, d'activité reduite. Il faut donc que Monsieur Volterra mène une vie non plus d'un homme en pleine force, mais la vie calme, exempte le plus possible de soucis, d'un homme au repos, de ce que nous appellons en France “un retraité”. Bien entendu, il ne s'agit pas d'interdire à Monsieur Volterra toute sortie, toute voyage, tout commerce avec ses amis et ses confrères. Mais il ne doit plus se fatiguer. La chose est simple: son coeur est malade, il ne faut pas le surmener. Moyennant quoi de longues années calmes seront la récompense de Monsieur Volterra.

Monsieur Laubry n'a pas voulu écouter mon offre de payer sa consultation. Il s'est montré touché de votre pensée de vous adresser, tout de suite, à lui, et heureux de rendre un peu service à Monsieur Volterra qu'il admire.

Voila, chère Madame; les détails techniques seront trasmis par Monsieur Laubry lui même au Docteur Nazari. Mais je pense vous avoir trasmis exactement la pensée générale de Monsieur Laubry.

Au revoir, chère Madame, rassurez vous, et faites tout pour qu'à l'avenir Monsieur Volterra se fatigue moins.

Veuillez agréer l'expression respectueuse de tout notre dévouement et l'assurance de toute notre affection.

J. Régnier

18.25 Régnier to Vito and Virginia Volterra, Paris, 17.2.1938 *Cat. no. 304*

17 Février 1938

Cher Monsieur et chère Madame.

Je suis bien heureux d'apprendre que Monsieur Volterra va mieux, et que, notamment, il passe de meilleures nuits. Je pense qu'avec une vie calme, bien réglée, tout reviendra très vite en ordre.

Nous espérons avoir la grande joie de vous voir tout à fait remis dans quelques semaines.

Avec nos respectueux compliments, veuillez, cher Monsieur et chère Madame, agréer l'expression de tout mon respectueux et affectueux dévouement.

J. Régnier

18.26 Régnier to Volterra, Paris, 3.5.1938 *Cat. no. 305*

3 Mai 1938

Cher Monsieur et chère Madame

J'ai eu l'occasion de voir Monsieur Laubry hier au soir, qui m'a fixé pour vous le rendez-vous suivant: il vous attendra Vendredi prochain 6 Mai à trois heures et demie. 39 avenue Victor Hugo. Ses journées sont assez chargées, il a donc cru bien faire en fixant lui-même la date. J'aurai plaisir à connaître les encouragements qu'il vous donnera, cher Monsieur. Je téléphonerai donc après, dans les jours qui suivront la consultation.

Nous avons aussi quelques résultats de multiplication microbienne que nous vous montrerons avant votre départ.

Veuillez, cher Monsieur et chère Madame, agréer l'expression de mon très respectueux et affectueux dévouement.

J. Régnier

18.27 S. Lambin and Régnier to Volterra, Paris, 17.7.1938 *Cat. no. 252*

Paris, 17 Juillet 1938

Monsieur le Professeur

Je vous fais enfin parvenir les résultats de nos essais, relatifs à l'influence qu'exerce la quantité de matières nutritives sur la multiplication des bactéries.

Vous trouverez également le texte de la note que nous avons publiée à la Societé de Biologie.

Monsieur Régnier me prie de vous rappeler que nous sommes à votre entière disposition pour publier, à l'Academie des Sciences, une note suivie des déductions mathématiques que vous aurez bien voulu tirer de nos résultats. Nous partons ce soir en vacances, mais nous serons à nouveau au travail en Septembre.

Veuillez, Monsieur le Professeur, transmettre à Madame Volterra mon plus respectueux souvenir, et agréer l'assurance de mes sentiments respectueusement dévoues.

Lambin

Monsieur le Professeur et cher Monsieur. Chère Madame.

J'espère que votre santé est bonne, et je serai bien content d'avoir de bonnes nouvelles de Rome.

Veuillez agréer l'expression respectueuse et affectueuse de tout mon dévouement.

J. Régnier

18.28 Volterra to Régnier, n.p., 1938 *Cat. no. 401*[18]

Mon cher ami

J'ai attendu à repondre à vous et à M.lle Lambin parce que vous m'avez annoncé que vous seriez en voyage jusqu'à Septembre. Je pense que maintenant vous allez rentrer et je m'empresse à vous écrire pour vous remercier d'abord de la Note que vous avez publiée à la Sociéte de Biologie[19] et aussi des tables et dessins que vous m'avez envoyé.

Les études sur la croissance des colybacilles sous l'influence des diverses quantités des substances nutritives sont très intéressants et [illeg.] réprésente bien le phénomène. Je ne doute pas que vous pourrez trouver un jeune mathématicien qui puisse vous aider dans ces calculs. Veuillez me le dire parce que si vous ne reussissez pas à le trouver je pourrai tâcher de chercher quelqu'un. Ma santé n'est mauvaise mais dans ce derniers jours je souffre le matin d'un peu d'oppression et je me sens toujours plus faible. Nous avons énormement de soucis dans ce

---

[18] Draft, n.p., n.d. (probably written on September 1938).

[19] Almost certainly Régnier, Lambin 1938c.

moment[20]. S'il nous sera possible j'espère de pouvoir venir à Paris en Octobre et de vous voir ainsi que Mme Régnier en bonne santé et satisfaits de vos vacances.

Veuillez presenter mes hommages ainsi que ceux de Mme Volterra a Mme Régnier et nous rappeler un bon souvenir de Mlle Lambin. Par le même courrier j'ai l'honneur de vous envoyer la conférénce que j'ai tenue à Paris l'année dernière[21].

Veuillez agréer, mon cher ami, mes salutations les plus dévouées et les plus affectueuses.

18.29 Régnier to Volterra, Paris, 5.9.1938 ***Cat. no. 306***

5 Septembre 1938

Monsieur le Professeur. Mon cher Maître.

Nous avons été bien heureux de recevoir hier votre lettre, et de savoir que, malgré tout, votre santé n'est pas mauvaise. Nous espérons bien avoir le grand plaisir de vous voir dans quelques semaines, en bonne santé et que les événements sur lesquels nous ne pouvons rien nous permettront de nous retrouver[22].

Je fais aussi des voeux bien vifs pour que toute votre famille se porte bien, et que rien d'inquiétant ne vienne vous troubler.

J'aurais voulu, mon cher Maître, que vous examiniez vous même du point de vue mathématique la loi que gouverne l'évolution des nombres maxima de bactéries en fonction de la quantité de nourriture[23]. Ce doit être une loi bien simple. Elle vous permettrait, peut-être, de préciser le rôle du facteur aliment dans les lois compliquées qui cherchent à expliquer le croît microbien dans un milieu très largement pourvu de de substances alimentaires. Si ces calculs ne vous intéressent pas, ou s'ils vous fatiguent, voulez vous que je m'adresse à Kostitzin, je ne connais pas d'autres mathématiciens? Mais je préférerai bien que ce soit vous-même. Excusez-moi d'insister.

Ici, notre vie est toujours semblable. Paris, fort beau en ce moment, est très calme. Nous attendons avec confiance les événements, et éspérons que les causes de conflit s'atténueront peu à peu. Mais s'il le fallait je suis, maintenant, bien certain que nous ferions notre devoir, tous, et avec la plus grand simplicité.

Mais ne soyont pas pessimistes, mon cher Maître, je m'excuse bien vivement auprès de Madame Volterra de n'avoir pas encore demandé de ses nouvelles, et je vous prie de lui transmettre nos très respectueux compliments.

Bien respectueusement et affectueusement.

J. Régnier

---

[20] A clear reference to the racial legislation the fascist regime was about to introduce in those days (5–7 September 1938).

[21] VOLTERRA 1938a.

[22] See preceding letter and note 20.

[23] This refers to the preparatory work of RÉGNIER, LAMBIN 1938d.

Je reçois à l'instant, les deux opuscules sur les "Fluctuations pour la lutte pour la vie..."[24]. Je vous en remercie bien vivement.

18.30 Volterra to Régnier, n.p., September 1938 *Cat. no. 402*[25]

Mon cher ami,

La courbe de maxima que vous m'avez envoyée a une allure qui correspond à celle de Pearl; ors j'ai discuté de la chose avec M. D'Ancona qui remarque que l'allure de la courbe de Pearl correspond à tous les phénomènes dans lesquels lorsque l'abscisse croit infiniment, l'ordonnée se mantient finie. La courbe alors tend vers un asymptote parallel à l'axe des abscisses. Alors j'ai pensé que pour vérifier d'une manière simple et synthétique votre résultat on pourrait suivre cette voie: examiner la courbe théorique des maxima et voir si l'ordonnée tend vers l'infini lorsque l'abscisse dévient infinie. Si cela arrive on peut-être surs que la courbe théorique n'a pas l'allure de celle de Pearl et par suite elle ne suffit pas à vérifier vos résultats expérimentaux. Nous avons dans la note de Kostitzin des Comptes Rendus du 31 mai 1937 les maxima qui sont donnés par les formules (9). (Il faut remarquer qu'à mon avis il y a une faute d'impression car en lieu de p_m^l il faudrait lire p_m). Ors p_m est constitué de deux termes, le premier terme este ε / h, le second terme est

$$-\frac{c}{h} P_m \,.$$

Si ε que l'on peut prendre comme mésure de la concentration de la nourriture croit indéfiniment, M croit également d'une manière lineaire et en vertu de la seconde formule (9) P_m croit infiniment d'une manière logarithmique, par consequent des deux termes de P_m le premier croit davantage que le second terme négatif. On en tire que ces maxima croissent infiniment en croissant infiniment la nourriture. La courbe des maxima théorique dans laquelle on tient compte de la seule intoxication ne serve à vérifier votre courbe. Il faut donc conclure que cette allure de la courbe des maxima ne s'explique pas par l'intoxication. Il y a donc d'autres causes par lesquelles votre belle courbe se produit et il faudra les réchercher. Est-ce l'[illeg.] comme M. D'Ancona soupçonne? Pardonnez moi je ne peux vous dire que si peu de choses. C'est déjà beaucoup pour l'état moral dans lequel je me trouve.

Mon coeur ne va pas bien. Il marche trop vite, il dépasse d'une manière irregulière les 90 pulsations. Mon médicin d'ici m'aurait ordonnée de la digitale. Pourriez vous demander à M. Laubry qu'est qu'il en pense?

Veuillez accepter mon cher ami mes salutations les plus affectueuses et les plus dévouées.

---

24 VOLTERRA 1938a.

25 Draft, n.d., but clearly written in September 1938.

18.31 Régnier to Volterra, Paris, 19.9.1938 *Cat. no. 307*

19 Septembre 1938

Monsieur le Professeur et cher Monsieur

Je suis bien fâché de vous donner tant de mal avec mes résultats, et je vous suis tout à fait reconnaissant d'avoir bien voulu vous y intéresser. Je vous remercie bien vivement.

Je viens d'essayer de théléphoner à Monsieur Laubry. Malheureusement il n'est pas encore de retour de vacances. Dès son retour, dans quelques jours, je lui demanderai ce qu'il pense, pour vous, d'un traitement digitalique. D'ici là, veuillez, cher Monsieur, bien vous conformer au traitement institué par votre médecin ordinaire.

Et surtout, nous voudrions tant que votre esprit soit en repos. Ici, nos aussi, nos avons été inquiets, très inquiets. Nous nous préparions tous pour le pire, mais ce pire n'est pas venu et ne viendra pas tout au moins ce temps-ci. Voilà donc une grande inquietude partir, les autres partiront de même. Dormez donc bien tranquillement et profitez bien de cette fin d'été, tout en obeissant aux prescriptions médicales. Il ne se passera rien.

J'ai essayé de suivre, avec grand intêret, votre raisonnement et celui de Monsieur D'Ancona, je trouve ces déductions précieuses et elles correspondent bien à ce que je pense moi même en simple biologiste: il y a autre chose que la quantité d'aliment, et que l'intoxication. Je serai bien heureux si vous arrivez à mettre ce fait en évidence.

Avez vous lu dans le journal "Le Temps" du 14 Septembre 1938, une étude sur "La lutte pour la vie" présentées, en causerie scientifique, par L. Houllevigne? Cette mise au point est, à mon avis, bien faite, pour le grand public, et fort élogieuse pour vous-même. Je cite par exemple: "C'est sur ces idées directrices que s'est constituée la biologie mathématique. Elle honore comme grand patron le savant italien Vito Volterra dont les "Leçons sur la théorie mathématique de la lutte pour la vie", (Gauthier-Villars 1931) sont l'Evangile du biologiste mathématicien." Je voudrais bien vous envoyer ce feuilleton, mais il vaut mieux que j'attende votre arrivée à Paris.

Cher Monsieur, je souhaite de tout mon coeur que cette lettre vous trouve en meilleure santé, au milieu des vôtres, également en bonne santé, et plus confiant en l'avenir.

J'écris, en ce moment, (et ce sera terminé dans quelques jours) la note que vous avez bien voulu accepter de présenter à l'Istitut[26], note qui précedera et servira de base à vos études mathématiques. Cette note ne contient évidemment que les chiffres que vous possedez déjà. Je vous l'enverrai très vite.

Merci, cher Monsieur. Veuillez présenter mes respectueux compliments et mes voeux de bonne santé à Madame Volterra, et recevoir l'expression très respectueuse de tout mon affectueux dévouement.

J. Régnier

J'espère que votre fils est actuellement complètement rassuré sur l'état de ses yeux, et que tous les membres de votre famille sont également en bonne santé.

---

[26] Régnier, Lambin 1938d.

18.32 Régnier to Volterra, Paris, 21.9.1938 *Cat. no. 308*

21 Septembre 1938

Monsieur le Professeur et cher Monsieur

Voici le texte de la note que vous pourrez, si vous le voulez bien, transmettre à l'Académie, en la faisant suivre de votre note mathématique[27]. J'espère que notre propre note ne dépassera la longueur admise.

Cher Monsieur, je vous remercie bien vivement d'avoir pris cette peine en un tel moment, où vous étiez vous même fatigué. Je vous en suis très reconnaissant.

Veuillez transmettre mes respectueux compliments à Madame Volterra, et agréer l'expression très respectueuse de toute mon affections et de mon dévouement.

J. Régnier

18.33 Régnier to Volterra, Paris, 29.9.1938 *Cat. no. 309*

Paris, 29 Septembre 1938

Monsieur le Professeur et cher Monsieur

Je viens d'écrire à Mr Kostitzin pour lui demander un rendez vous prochain, et je ferai exactement ce que vous indiquez.

Je serai bien heureux d'avoir en quelques mots le résultat de la consultation que vous venez d'avoir, résultat que je transmettrai aussitôt à Mr Laubry.

Beaucoup de parisiens ne sont pas rentrés ou sont partis pour les raisons que vous imaginez. Mais depuis hier soir, grâce en grand partie à votre beau pays, l'éspoir refleurit, et nous espérons bien que notre vie normale va reprendre[28].

Je pense que les bonnes nouvelles, et j'espère bien qu'elles serons encore meilleures lorsque vous recevrez cette lettre, feront plus pour vous remettre qu'un traitement médicamenteux. Nous serons tous si heureux de vous revoir avec Mme Volterra dans quelques semaines.

Bonne santé, cher Monsieur. Soignez vous bien et croyez bien certain que votre santé va refleurir de même que la tranquillité des esprits. Pour l'un et pour l'autre but, ma femme, Melle Lambin et moi-même, nous formulons les voeux les plus vifs.

Veuillez présenter mes respectueuse hommages à Madame Volterra et merci, cher Monsieur, pour votre Note.

Respectueusement et affectueusement à vous

J. Régnier

---

27 The note referred to has already been mentioned RÉGNIER, LAMBIN 1938d, while Volterra's note was then prepared and published in collaboration with Kostitzin (VOLTERRA, V. A. KOSTITZIN 1938), as is shown also by the following letter.

28 This refers to the risk of war which was very strong in those days.

18.34 Régnier to Volterra, Paris, 11.10.1938 *Cat. no. 310*

Paris, 11 Octobre 1938

Monsieur le Professeur et cher Monsieur

Nous serions tous, ici, bien heureux d'avoir de bonnes nouvelles de la récente consultation des médecins romains, et nous faisons des voeux extrêmement vifs pour que tous les événements survenus ces jours ci vous aient apporté une complète tranquillité d'esprit. J'ai vu deux fois Mr Kostitzin. Après avoir longuement examiné nos chiffres et la note mathématique, Mr Kostitzin pense qu'il serait préférable, peut-être, et pour éviter certaines objections, d'envisager à côté du développement mathématique déjà fait, d'autres hypothèses. Devant ses explications, qu'il va vous fournir lui-même, et son désir évident d'examiner en liaison avec vous ce problème, je lui ai fait part de mon projet de vous proposer, pour cette note, sa collaboration. Mr Kostitzin serait certainement heureux et très fier de voir son nom uni au vôtre, pour cette note, qui suivrait notre exposé bactériologique. Peut-être, cher Monsieur, n'y verrez vous pas un grand inconvenient? auquel cas c'est lui-même, Mr Kostitzin, qui vous soumettrait ses projets.

Peut-être, du reste, aurons nous dans peu de temps le plaisir de vous voir, et Mr Kostitzin vous parlera.

Ici, à Paris, la belle saison se termine. Nous avons cependant encore de belles journées. Le jardin du Luxembourg est très beau avec ses arbres jaunis, et Madame Volterra, et vous même, pouvez encore y passer de bonnes heures.

Veuillez, cher Monsieur présenter mes très respectueux compliments à Madame Volterra, et agréer l'expression respectueuse et affectueuse de mon entier dévouement.

J. Régnier

Je forme des vœux, également, pour que toute votre famille soit en bonne santé.

18.35 Volterra to Régnier, Ariccia, 17.10.1938 *Cat. no. 403*

VILLA VOLTERRA
ARICCIA (ROMA)

Ariccia, 17 Ottobre 1938

Mon cher Ami

Merci de votre lettre et de votre empressement pour moi.

J'ai reçu aussi une lettre de Monsieur Kostitzin dans quelle il me montre que dans ma remarque j'avais commis la faute de prendre (1) pour la puissance 1.

Je lui ai demandé si en corrigeant cette bévue mes calculs déviennent exacts dans le cadre de ma précédente Note.

Dans ce cas vous pourriez la publier après votre Note. Mon seul but a été de montrer que l'hypothèse de l'empoisonnement que j'avais fait dans la remarque à votre Note précédente était complètement insuffisante.

Mais, si mon calcul est inexact même en corrigeant l'échange de 1 par (1), il faut supprimer ma remarque complètement.

D'autre part j'apprends par la lettre de Monsieur Kostitzin qu'il a trouvé le moyen d'expliquer mathématiquement vos resultats. Cela est, à mon avis très important. C'est pourquoi il faut qu'il publie à son nom ce calcul dans une Note séparée. Je ne puis m'en approprier aucune partie; ce serait un vol[29]. J'ai proposé à Monsieur Kostitzin que ce serait un honneur pour moi de présenter sa Note à l'Académie, mais que j'ai tellement de confiance en lui que je l'autorise même à présenter directement sa Note au Sécrétariat s'il veut la publier plus rapidement. J'espère que tout cet arrangement ira bien.

Veuillez présenter mes hommages et les souvenirs de ma femme à Mme Régnier.

Bien à vous affectueusement.

18.36 Régnier to Volterra, Paris, 31.10.1938 *Cat. no. 311*

31 Octobre 1938

Monsieur le Professeur et cher Monsieur

J'espère que votre santé est tout à fait meilleure, et j'en veux voir la preuve dans votre dernière lettre, où la recherche scientifique fait l'objet de vos pensées, comme aux meilleurs jours. J'ai vu récemment Monsieur Kostitzin qui pense de plus en plus que votre nom doit être devant le sien pour la Note concernant les développements bactériens. C'est vous, cher Monsieur, qui êtes à l'origine de tous ses travaux, c'est vous qui m'avez encouragé à poursuivre ces recherches, et c'est vous, enfin, qui pour la première fois avez apporté toute la lumière mathématique pour essayer de les mieux éclairer. Dans cette note même c'est vous qu'avez presenté le fil conducteur, c'est donc un honneur que ressentira Mr Kostitzin en mettant son nom à coté du votre. Je crois qu'il faut, donc, accepter, cher Monsieur. Monsieur Kostitzin vous soumettra très prochainement la nouvelle rédaction de votre Note et vous demandera votre approbation.

Ici, le temps est gris et triste. J'espère que vous avez du beau temps, et que votre esprit consent à ne se porter que vers les choses encourageantes. Malgré tout il y en a.

Soignez vous bien, cher Monsieur, ou plutôt consentez à vous laisser soigner et veuillez présenter nos respectueux compliments à Madame Volterra.

Très respectueusement et bien affectueusement à vous

J. Régnier

---

[29] However, Kostitzin did not want to publish the note in his own name alone (see also the following letter). The preparatory work related to this note is clarified also by the correspondence between Volterra and Kostitzin of the time, which should be read together with these letters.

18.37 Régnier to Vito and Virginia Volterra, Paris, 2.12.1938 *Cat. no. 312*

2 Décembre 1938

Cher Monsieur et chère Madame

Nous vivons ici dans une grande inquiètude pour vous mêmes et votre famille, et malgré tout le désir que je dois avoir d'encourager Monsieur Volterra, nous ne pouvons nous empêcher de vous transmettre tous nos sentiments anxieux[30]. Il est infiniment triste pour une personnalité d'aussi grande envergure que notre Maître qu'il soit, dans un pays qu'il a tant honoré, en but à des possibilités de doute sur son propre avenir et sur celui des siens. C'est là une injustice telle, une incompréhension si massive, qu'elle devrait laisser indifférent Monsieur Volterra. Personne ne pourra faire que ses travaux, que ses conceptions n'existent pas, personne ne peut faire qu'il est, de l'avis de toutes les compétences, le plus grand mathématicien actuel. Cette simple idée doit le soutenir et lui permettre de prendre aussi philosophiquement qu'il sera possible toutes les tristesses actuelles.

Par ailleurs il faut, chère Madame, veiller au moral de Monsieur Volterra, le remonter, lui permettre de dormir, pour que son coeur puisse battre normalement.

Je reçois à l' instant une lettre de Mr Kostitzin m'informant que le texte de la Note mathématique est partie pour Rome. Dans très peu de temps les deux notes paraîtront.

Veuillez, chère Madame, prèsenter tous nos respectueux encouragements à Monsieur Volterra, lui présenter l'assurance de tout notre dévouement affectueux et l'expression de notre admiration pour tout ce qu'il représente.

J. Régnier

18.38 Régnier to Volterra, Paris, 26.12.1938 *Cat. no. 313*

26 Décembre 1938

Monsieur le Professeur. Mon cher Maître.

Votre note vient de paraitre, il est probable que vous avez déja le numéro des Comptes Rendus qui la publie[31]. Je suis très heureux que nos essais bactériologiques vous aient donné l'occasion à vous, et à Monsieur Kostitzin, de publier de si belles interprétations mathématiques. Vos études sont en même temps, pour

---

[30] The whole letter refers to the consequences of the introduction of the racial laws and (in the absence of the letters written by Volterra) attests to the anxiety caused to himself and Virginia Volterra experienced by these events. In this connection see also several references contained in the correspondence with D'Ancona. The allusive tone of and absence of detail in the reference to political issues in Volterra's correspondence may be explained by the fact that his correspondence had long been subjected to discreet police control (see also GOETZ 1993).

[31] VOLTERRA, V. A. KOSTITZIN 1938.

nous, d'un interêt personnel très grand, car elles nous montrent dans quel sens il faut que nous dirigions nos essais, et de ceci je vous remercie bien vivement.

Nous sommes dans le froid, sous la neige, et les nouvelles de l'extérieur ne sont pas encourageantes, puis nous connaissons toute la tristesse de vos existences après d'aussi glorieux efforts que les vôtres, et ceci nous attriste.

Nous sommes totalement avec vous, mon cher Maître, avec Madame Volterra et avec votre famille. Veuillez leur présenter nos meilleurs souhaits pour 1939, et recevoir nos voeux pour vous même; et dans ces mots nous mettons tout ce que notre admiration, notre reconnaissance, et notre respectueuse affection peut y mettre. Malgré tout ayons foi en l'avenir. Il paraît certain que les forces divines sont avec nous. Il suffira peut-être d'attendre.

Je fais les voeux les plus ardents pour que vous en ayez la force.

Respectueusement et affectueusement à vous

J. Régnier

18.39 Régnier to Vito and Virginia Volterra, Paris, 7.1.1939 *Cat. no. 314*

7 Janvier 1939

Cher Monsieur et chère Madame

Je suis fort ennuyé que Monsieur Volterra ait en plus de tant de maux, des rhumatismes, et je regrette que le beau climat romain n'ait pas pu l'en protéger. Mais j'espère fort que cette atteinte ne sera passagère et que rapidement, notre malade pourra reprendre la liberté de ses mouvements.

Ici, le temps est triste et froid, très humide. Nous avons eu une longue periode de très basse temperature, qui est venue bouleverser bien des projets, par example nos projets de partir en province, dans notre Bourgogne natale, pour passer les fêtes du Nouvel An. Maintenant la pluie et la neige refont leur apparition, et les pauvres membres des rhumatisants, dont je suis, se soudent et craquent. Je pense que pour l'instant Monsieur Volterra est mieux à Rome, malgré tout.

De tous cotés on me demande des nouvelles de Monsieur Volterra et aussi de vous même, chère Madame. Personne ne vous oublie, tous vous suivent, anxieusement par la pensée, et font de souhaits pour que rapidement, quand'il fera beau ici, nous ayons la grande joie de vous voir.

Ma lettre n'a que ce seul but vous dire notre respectueuse affection, et nous souhaiter, cher Monsieur, un très prompt rétablissement.

Dès que nous aurons des tirés à part de notre note, nous vous les ferons parvenir.

Très respectueusement à vous, cher Monsieur et chère Madame, et très affectueusement et fidèlement votre

J. Régnier

18.40 Régnier to Vito and Virginia Volterra, Paris, 21.2.1939 *Cat. no. 315*

21 février 1939

Chère Madame et Cher Monsieur

Quelle tristesse d'apprendre que Monsieur Volterra est encore atteint dans son activité, et qu'il doit se maintenir au lit malgré la joie qu'il aurait, et que nous aurions tous avec lui, de le voir à Paris.

Bien entendu, il faut vous soigner sérieusement, très serieusement, mon cher Maître. Je sais que vous êtes en très bonnes mains, et il faut obéir. Pour les ondes courtes, seuls sont qualifiés, pour en décider, vos médicins traitants, mais peut-être serait-ce un bien, surtout si cela ne doit vous occasioner aucune gêne, aucun déplacement.

J'ai eu le grand plaisir de faire connaissance avec votre fils qui m'a donné de vive voix de vos nouvelles. Malgré toutes les tristesses votre fils est calme et souriant, je suis bien certain qu'il reussira là ou il va.

Je lui ai malheureusement appris qu'il m'avait été impossible de trouver à Paris la Pancreatic Hormone Grant. Aucune pharmacie, aucune maison d'importation de produits pharmaceutiques étrangers, ne possèdent cet article. Seules les maisons de Londres pourront le fournir.

Chère Madame, mon cher Maître, je serai bien heureux, dans quelques jours, d'avoir de meilleures nouvelles et je suis bien certain que j'aurai de meilleures nouvelles.

Très respectueusement et, aussi, affectueusement à vous

J. Régnier

18.41 Régnier to Vito and Virginia Volterra, Paris, 11.3.1939 *Cat. no. 316*

11 Mars 1939

Chère Madame. Mon cher Maître.

J'ai eu bien du plaisir à faire la connaissance de votre fils Edouard après avoir fait connaissance, trop rapide, de votre fils Henri. J'ai eu ainsi tous ces jours-ci de bonnes nouvelles de vous-même, et je sais que le traitement par la Pancreatic Hormone, et par les radiations courtes, a produit son plein effet. Il vous faut donc encore beaucoup de patience, cher Monsieur, mais dès maintenant vous êtes tout à fait assuré de la guérison. Bien entendu il ne faut pas faire d'imprudences et obéir scrupuleusement aux désirs de médecins dont la science et l'experience sont bien connus.

Votre fils, chère Madame, est certainement en bonne santé, plein de courage et de confiance. Vos lettres, et aussi celles de Monsieur Volterra, où il suit la guérison, lui font le plus grand bien. J'espère le voir souvent, et je suis à sa disposition, comme je suis à la vôtre.

Bonne amelioration sans cesse, mon cher Maître, bon courage et bonne santé, chère Madame. Il y aura encore beaucoup de bons et beaux jours. Les heures tristes passent et passeront maintenant bien vite.

Très respectueusement et affectueusement à vous. Avec l'assurance de mon fidèle dévouement pour vous et les vôtres

J. Régnier

18.42 Régnier to Vito and Virginia Volterra, Paris, 30.4.1939 *Cat. no. 317*

30 Avril 1939

Cher Monsieur et chère Madame

Dès mon retour de vacances, pour avoir des nouvelles j'ai essayé de télephoner à votre fils Edouard. On m'annoncé son départ. Dès qu'il reviendra j'aurai plaisir à le voir.

Mais je voudrais avoir de vos nouvelles, cher Monsieur et chère Madame. J'espère que Monsieur Volterra est en meilleure santé.

Voulez vous agréer l'expression très respectueuse de tout mon dévouement affectueux.

J. Régnier

18.43 Régnier to Vito and Virginia Volterra, Paris, 8.8.1939 *Cat. no. 318*

8 Août 1939

Monsieur le Professeur et cher Monsieur. Chère Madame.

Nous sommes fort heureux de vous savoir à nouveau installés à Ariccia, et nous espérons fort que l'air de la campagne achévera complètement de vous remettre, cher Monsieur. Bientôt sans doute vous obtiendrez le permission de descendre au jardin et vous pourrez profiter complétement de l'air et des arbres.

Nous avons quitté Paris, il y a quelques jours, sans avoir eu le plaisir de voir votre fils Edouard. mais nous le savons en bonne santé et nous espérons que toutes ses difficultés personnelles vont s'atténuer peu à peu.

Nous voici dans notre Bourgogne natale mais il n'y fait pas beau temps. Il en est du reste, ainsi, pour l'instant, sur la France entière. Mais les beaux jours reviendrant certainement pour tous.

Ma femme, Melle Lambin, et moi, serons bien contents, cher Monsieur et chère Madame, d'avoir de vos nouvelles, et nous vous remercions bien vivement d'avance.

Veuillez agréer l'expression très respectueuse de notre bien affectueux dévouement, ainsi que tous nos voeux de très prompt rétablissement.

J. Régnier

18.44 Régnier to Vito and Virginia Volterra, Paris, 19.8.1939 ***Cat. no. 319***

19 Août 1939

Mon cher Maître. Chère Madame.

Votre lettre vient m'attrister, dans l'espoir que je formais et que je forme encore, d'autant mieux, pour votre guérison complète. Certes je ne doute pas de votre retour à une certaine activité, mais je veux espérer que, rapidement, vous reprendrez, cher Monsieur l'usage complete de vos jambes, et que vous pourrez, comme il y a quelques semaines, non seulement profiter d'Ariccia, mais aussi, peut-être, vous déplacer suffisamment, pour venir nous voir, nous, vos amis, qui vous admirons et vous aimons.

Et puis, puisque le sort veut de votre part, pendant un certain temps, une immobilité presque absolue, pourquoi n'en pas profiter, sans toutefois fatiguer votre esprit, pour faire le bilan, mon cher Maître, de tout ce que vous avez apporté à l'Esprit des hommes. Que de travaux, que de créations de votre part. Moi-même qui malheureusement ne puis juger votre oeuvre mathématique princeps, je reste confondu devant les idée géniales que vous avéz soulevées dans la biologie des nombres. Grace à vous, combien de problémes résolus, combien de questions nouvelles posées, combien de discussions scientifiques soulevées. Ainsi votre vie, derrière vous, témoigne du grand honneur, bien rare, qui vous a été fait, d'être un Homme Utile, d'être un homme parmi ceux, exceptionnels, qui ont apporté encore un peu du feu de l'Esprit, à notre epoque de si grande dureté matérielle.

Que cette certitude, mon cher Maître, vous soit consolatrice, mais qu'elle vous apporte aussi l'espoir de voir votre santé suffisamment forte, à nouveau, pour que vous puissiez rendre service, à nouveau.

Mon cher Maître, chère Madame, nous pensons affectueusement à vous, nous y pensons fidèlement et pieusement, et dans ces temps d'incertitude où nos prières montent pour ceux de tous pays qui vont peut-être risquer leur vie, nous vous associons à ces combattants involontaires, ou volontaires, à ces chevaliers servantes d'un Idéal élevé, quelqu'il soit, pour demander à Dieu qu'il vous protège, vous qui n'avez jamais servi que le Bien et la Verité.

Excusez cette lettre. Je la voudrais infiniment respectueuse, capable de vous transmettre mon admiration et mon affection. Veuillez à travers l'insuffisance des mots y voir tout ce que je veux y mettre.

Respectueusement et affectueusement à vous, cher Monsieur. Respectueusement et affectueusement à vous, chère Madame

J. Régnier

18.45 Volterra to Régnier, Ariccia, 25.8.1939 ***Cat. no. 404***

VILLA VOLTERRA
ARICCIA (ROMA)

Ariccia, 25 Août 1939

Mon cher Ami

Je regrette beaucoup de vous avoir écrit ma dernière lettre si desolée.

Mais j'etais si mal ce jour-là, que j'avais perdu tout espoir. Veuillez m'en excuser.

Ce sont les voeux que vous avez formés pour moi qui ont eu l'efficace de produire quelque améliorement dans mon état? Je n'en doute pas et je vous en remercie de tout mon coeur.

Je reste la plupart du temps couché au lit et l'on me fait reposer presque toute la journée, mes nuits sont en général insupportables. Cependant quelques pas je puis maintenant les faire soutenu de deux côtés et mes jambes me soutiennent un peu.

Espérons dans un progrès. Ah! si je pouvais arriver à vous revoir chez vous ainsi que tous mes amis qui vous sont près. L'un des vœux les plus ardents serait accompli. Vous connaissez mes sentiments qui restent immuables et si vous pensez à moi, je vous assure que je pense également à vous qui avez consacré toujours votre vie au bien, à la science et aux soulagements des misères humaines. Vos idéaux vous soulèvent au-dessus de tout et nous vous rappellons avec la plus fidèle et la plus affectueuse amitié et reconnaissance en formant tous les vœux pour l'accomplissement de ce que vous désirez.

Mon fils Edoardo est parti avant sa femme et sa fillette pour reprendre son séjour interrompu.

Veuillez vous rappeller de moi et des miens comme nous vous rappellons à tout instants en pensant à l'avenir proche et éloigné.

Veuillez presenter mes hommages les plus respectueux à Madame Régnier et mes meilleurs salutations à Mademoiselle Lambin.

Bien à vous affectueusement

18.46 Régnier to Vito and Virginia Volterra, Paris, 29.10.1939 *Cat. no. 320*

29 Octobre 1939

Chère Madame et cher Monsieur

Nous avons été bien heureux d'avoir de vos nouvelles, et de savoir que la santé de Monsieur Volterra ne subissait pas d'attaques nouvelles. Je souhaite de tout mon coeur que le printemps prochain nous apporte à tous une santé raffermie et la tranquillité de l'esprit.

Ces derniers mois, comme vous pouvez l'imaginer, ont été évidemment pénibles, bien que la guerre nous ait encore epargnés. Rentrée penible à Paris, mobilisation générale et départ des parents et des amis, inquiétudes de toutes sortes, changements d'existence, déménagements, orientations nouvelles de nos travaux dans un sens plus conforme à la défense nationale, difficultés multiples... Enfin, nous parisiens n'avons pas eu vraiment à craindre pour nos vies, mais que sera l'avenir?

J'ai, au début du mois dernier, téléphoné à votre fils et je le croyais parti en province. Je suis content de le savoir de retour à Paris, et dans peu de temps je me mettrai en liaison avec lui. J'espère que pour lui et sa famille, ainsi que pour vos autres enfants la guerre n'a pas apporté de trop cruelles inquietudes, ni de trop grandes difficultés.

J'allais précisément vous écrire, chère Madame, pour vous demander de vous nouvelles, et j'ai été véritablement très heureux d'avoir les bons renseignements que vous me communiquez sur la santé de Monsieur Volterra.

Veuillez, chère Madame et cher Monsieur agréer l'assurance de nos respectueux et affectueux sentiments et croire à la grande fidelité de tout mon dévouement.

J. Régnier

18.47 Régnier to Vito and Virginia Volterra, Paris, 1.1.1940 *Cat. no. 321*[32]

Jean Régnier
Pharmacien des Hospitaux
Professeur agregé à la faculté de Pharmacie
1, Rue Chandon Lagache, Paris

Chère Madame, mon cher Maître

Nous ne voulons pas laisser passer ce premier jour de l'année sans vous présenter nos meilleurs voeux pour vous et pour les vôtres, vœux de bonne santé, tout particulièrement.

Nous espérons bien avoir dans un peu de temps la visite de vos enfants.

Très respectueusement et affectueusement à vous cher Monsieur et cherè Madame

J.R.

18.48 Régnier to Vito and Virginia Volterra, Paris, 17.2.1940 *Cat. no. 322*

17 Février 1940

Mon cher Maître. Chère Madame.

Nous avons été très heureux il y a quelques semaines, de recevoir de bonnes nouvelles de Rome et nous espérons fort que, la mauvaise saison se terminant, l'état de santé de Monsieur Volterra deviendra encore meilleur.

Avez vous de bonnes nouvelles de vos enfants? J'ai écrit à leur adresse, ici, mais je n'ai eu aucune nouvelle.

Ici, malgré le froid et l'humidité, nous tenons, en relative bonne santé, et nous espérons en des temps moins troublés.

Affectueusement et bien respectueusement à vous chère Madame et cher Monsieur

J. Régnier

[32] Visiting card, n.d. but certainly written on 1st January 1940.

18.49 Régnier to Vito and Virginia Volterra, Paris, 9.3.1940 ***Cat. no. 323***

9 Mars 1940

Mon cher Maître. Chère Madame.

Nous avons tous eu une bien grande joie à retrouver l'écriture même de Monsieur Volterra. Nous sommes particulièrement heureux également de savoir vos enfants en bonne santé, soit autour de vous soit dans des pays amis. Les mauvais jours sont donc passés, et le printemps ne pourra que vous apporter un surcroît de santé. Peut-être nous apportera t il aussi la satisfaction de nos désirs, c'est à dire le retour heureux à la vie calme où tous nous pouvons travailler sans crainte.

Ici notre vie se continue à peu près semblable sauf encore plus de travail. La plupart de mes collaborateurs sont aux armées où ils font simplement et même parfois gaiement leur devoir. Evidemment nous pensons que la demie tranquillité de cet hiver va cesser, mais nous sommes bien certains de tenir, tant à l'avant qu'à l'arrière.

Allons, mon cher Maître, ma chère Madame, je vous exprime encore une fois toute ma joie. Je vous souhaite une bien bonne santé pour vous et les vôtres et je vous prie d'accepter l'expression respectueuse et affectueuse de tout mon dévouement.

J. Régnier

18.50 M.me Régnier to Virginia Volterra, Paris, 30.3.1947 ***Cat. no. 324***

MADAME JEAN RÉGNIER

Chère Madame je suis très touchée de l'envoi du livret retraçant la belle vie de votre cher mari. Hélas mon mari ne le lira pas: il y aura un an le 3 avril que le bon et brave compagnon de ma vie me quittait à jamais. Je vais bientôt quitter Paris pour aller vivre à la campagne avec mon vieux père.

Mon coeur est lourd de peine. Respectueusement à vous.

19 Letters between John Stanley and Vito Volterra

We have incomplete information about the life of John Stanley. We know that he was born in London on 25 June 1905 but we do not know the date of his death (in any case not before the 1980s). Stanley emigrated to Canada in 1913 and attended the University of British Columbia in Vancouver, where he obtained his Bachelor of Arts in 1927. In the same year he moved to the United States where he undertook post-graduate work at the University of Minnesota. He worked as assistant zoologist at this university in the years 1927 and 1928 and as assistant entomologist from 1928 to 1930. He got his Master of Arts in 1929 and his Ph.D in 1931. From the correspondence between Volterra and Chapman we learn that in those years, he worked under the direction of the latter, developing mathematical applications to the population dynamics of *Tribolium*. While working on his doctorate he followed Chapman to the Experiment Station of the Association of Hawaiian Pineapple Canners of the University of Hawaii (Honolulu) where he spent fifteen months as his assistant and continued to develop his research work on *Tribolium*, the results of which became the subject of two publications (STANLEY 1931, 1932). He then returned to Canada, and worked for seven months with the Division of Research Information of the National Research Council in Ottawa. In 1932, he obtained a post as Assistant Professor of biology at Queen's University (Ontario). In 1937 he was promoted do Associate Professor, again of biology, at the same university.

In 1940 Stanley enlisted with the University's Canadian Officer Trading Corp program and was commissioned a Lieutenant in the Reserve Force in the fall of 1941. In November of the same year he applied for admission to the Royal Canadian Air Force and was given the rank of pilot officer. In 1943 he was promoted to Flight Lieutenant and then was made a Squadron Leader.

In 1945 Stanley returned to Queen's University from active service. In May 1946 he resigned and moved to McGill University, first in the capacity of professor of zoology (until 1965) and then as professor of population dynamics. He held this chair until 1970, the year he retired from teaching. At McGill University he was chairman of the department of zoology (from 1947 to 1963) and chairman of the biological science research group (from 1948 to 1963). He was member of the National Research Council of Canada and corresponding member of the Zoological Society of London. He spent the last years of his life at Vancouver.

Stanley's research always had as its main objective the topics of population dynamics and the application of mathematics to the study of the growth of animal population, particularly insects (STANLEY 1931, 1932a, 1932b, 1937, 1942,

1943, 1946). His mathematical skills led him to concern himself also with biometrics, on the subject of which he wrote a book in the autumn of his scientific career (STANLEY 1963), as well as with computer programming.

* * *

19.1 Stanley to Volterra, Kingston, 29.3.1935 *Cat. no. 325*

Prof. Vito Volterra
Via in Lucina 17
Roma.

Dept. of Biology
Queen's University
Kingston Ont.

Mar 29/35

Dear Prof. Volterra

I enclose two copies of my latest paper on Tribolium[1]. Will you be kind enough to send one to D'Ancona, as, owing to damage to my files, I have lost his address.

Yours truly

John Stanley

19.2 Stanley to Volterra, Kingston, 10.10.1936 *Cat. no. 326*

QUEEN'S UNIVERSITY
KINGSTON, ONTARIO
CANADA
Department of Biology

Oct 10, 1936

Dear Professor Volterra,

Please accept my thanks for the reprints which have just arrived from you. I find them of great interest, and regret that I do not have your gift of mathematical insight.

I am hoping to publish further work in the near future, but I do not have the free time to accumulate data as fast as I would like.

I was at a conference recently, at the Dominion Parasite Laboratory at Belleville, Ont, and your work was referred to with great respect. The opinion was expressed often that a compilation of your biological theories, in one volume, in English, would be welcomed by workers in bio-mathematics on this continent.

---

[1] Almost certainly STANLEY 1932.

If you should be favorable to such an idea I would be honoured to be permitted to collaborate and assist in any way, no matter how minor.

Sincerely yours,

J. Stanley

19.3 Volterra to Stanley, Paris, 10.11.1936 *Cat. no. 405*[2]

HOTEL LUTETIA
43 BOULEVARD RASPAIL
PARIS

Cher Professor Stanley

J'ai reçu ici la lettre du 10 Octobre que vous m'avez addressée à Rome, je vous en suis très reconnaissant.

Les nouveaux travaux que vous esperez publier m'interesseront beaucoup.

J'ai envoyé au nouveau journal Acta Biotheoretica de Leiden un mémoire où j'ai mis à côté quelqu'uns de mes anciens résultats avec les nouveaux[3]. Il paraîtra en Janvier et je serai heureux de vous l'envoyer.

J'ai toujours desiré de pouvoir publier un ouvrage en Anglais sur mes recherches de biologie mathematique. C'est pourquoi j'accepte avec la plus grand enthousiasme l'idée que vous me suggerez dans votre lettre. Je serai heureux de pouvoir publier l'ouvrage avec vous et de pouvoir travailler ensemble. En attendant je tiens à vous dire que les travaux que j'ai publiés sont

un mémoire publié à l'Académie des Lincei reproduit dans le bulletin du Comité Talaxographique avec des additions (en italien);

l'ouvrage Leçons sur la theorie mathematique de la lutte pour la vie publié par Gauthier-Villars en Français en 1931;

l'ouvrage Les associations biologiques au point de vue mathématique publié avec M.U. D'Ancona en 1935 par Hermann en Français.

Sous presse le mémoire Principes de Biologie Mathématique dans Acta Biotheoretica de Leiden[4]

En plus beaucoup d'articles sur les mêmes sujets tirés des ouvrages précédents.

Il faudrait réunir ce qui se trouve dans ces ouvrages en supprimant les parties communes et en y ajoutant les résultats et les recherches expérimentales.

Vous m'obligerez beaucoup si vous voudrez bien me faire connaître vos idées la dessus. Je désire aussi savoir si vous connaissez un éditeur prêt à la publication.

Je pars demain pour Rome (Via in Lucina 17) et je vous prie de m'envoyer votre lettre à cette adresse à Rome. Dès que je l'aurai reçue je vous repondrai en vous donnant tous les détails que vous desirerez et qui pourront vous être utiles.

---

[2] Draft, n.d. but obviously written on 10 November 1936, as is shown by the following letter.

[3] VOLTERRA 1937a.

[4] The publications referred to are VOLTERRA 1927d, 1931a, VOLTERRA, D'ANCONA 1935, VOLTERRA 1937a.

19.4 Stanley to Volterra, Kingston, 19.11.1936 *Cat. no. 327*

(Same heading of letter no. 19.2)

Nov 19, 1936

Dear Professor Volterra,

Many thanks for your letter of November 10 written from Paris. I am very glad indeed that you agree to my suggestion of publishing your works in English.

So far I have not done anything in the matter of finding a publisher, but I shall start at once to find out what can be done. I have a number of possible plans in mind, but time does not permit writing at lenght just now as I am leaving for a brief journey in an hour or so. I will write you fully on my return next week.

The main difficulty lies in the fact that technical material of this sort does not have a wide circulation, and hence its publication is expensive. However, I have a wide acquaintance in this country with various publishers and institutions which may be willing to help financially, and I hope we may be able to arrange for something. As I say I will write you more fully in day or two.

Sincerely

John Stanley

19.5 Stanley to Volterra, Kingston, 21.11.1936 *Cat. no. 328*

(Same heading of letter no. 19.2)

Nov 21, 1936

Dear Professor Volterra,

Having now returned to Kingston, I am now writing to you as I promised in my last letter.

I have written to various publishers who publish book by setting type in the usual way, and also to some firms who use “mimeographing” and “planographing” which are methods for reproducing type-written material at much less expense than is necessary when type must be set. I shall have replies from these people in a week or two, and I shall then talk to Dr. Wallace, Principal of Queen's, who has expressed some interest in the matter, and shall endeavor to obtain the University's financial help.

I have in my library the following:

“Leçons sur la théorie mathématique de la lutte pour la vie”

“Les associations biologiques au point de vue mathématique”

“La théorie mathématique de la lutte pour la vie et l' experience”

“Les équations des fluctuations biologiques et le calcul des variations”

“Sur la moindre action vitale”[5]

---

[5] In the following order: VOLTERRA 1931a, VOLTERRA, D'ANCONA 1935, VOLTERRA 1936f, 1936a, 1936e.

Will you kindly send me any reprints that I do not have? Will you also let me know whether you wish to write the English version yourself, or whether you wish me to do it?

I can read French and Italian well enough to translate and have a good acquaintance with your work, and so would be competent to do this. I have also been the assistant Editor of the "Canadian Journal of Research" and so am familiar with publishing and printing methods in this country.

I would suggest that the mathematical developments be given a little more fully than they are at the present, i.e. that the steps be filled in between equations. If you wish me to write the English version I should of course send it to you for approval or alteration before publication.

Hoping you will write me regarding these matters at you earliest convenience, I am,

Very sincerely

John Stanley

P.S. I would appreciate your writing me in French rather than in Italian, as I do not read Italian very well. JS.

19.6 Volterra to Stanley, Roma, 27.12.1936 *Cat. no. 406*[6]

Via in Lucina, 17
Roma

Rome, 27 Décembre 1936

Cher Professeur Stanley,

Veuillez bien m'excuser si je n'ai pu vous écrire qu'avec un si grand rétard. Je vous remercie beaucoup de votre aimable lettre et je vous suis très reconnaissant de vos démarches pour la publication de l'ouvrage.

J'espère que vous réussirez. Pour ma part s'il est nécessaire je renounce à mes droits d'auteur pour faciliter la publication.

J'ai beaucoup cherché de voir quelques ouvrages imprimés de la maniere que vous m'avez indiqué mais je n'ai pas pu en trouver. C'est même à cause de cela que j'ai perdu beaucoup de temps et que j'ai tardé à vous repondre. Vous m'obligerez beaucoup si voudrez bien m'envoyer des modèles de la composition typographique que vous me proposez. Mais, je pense que l'on pourrait réduire l'ouvrage le plus possible pour diminuer les frais et que la meilleure manière pour le publier serait de recourir, s'il est possible, à l'impression de la manière ordinaire.

Je vous donne ici la liste des ouvrages que j'ai publié sur la biologie mathematique. Le dernier qui est sous presse à Leiden va paraître dans les premiers jours de Janvier[7]. Je serai heureux de vous l'envoyer ainsi que les exemplaires des autres travaux que je pourrai me procurer car quelques uns sont épuisés.

---

[6] Draft.

[7] Volterra 1937a.

Veuillez agréer, cher Professeur Stanley, avec les meilleurs voeux pour la nouvelle année l'assurance de ma plus haute considération.

19.7 Stanley to Volterra, Kingston, 11.1.1937 *Cat. no. 329*

(Same heading of letter no. 19.2)

January 11, 1937

Dear Professor Volterra,

I have your very kind letter of December 28, and hasten to give you further information.

I have written to several printing firms, and find that ordinary printing will cost more than I can get money for, but that Planographic printing is more reasonable in cost. I enclose a sample sheet of this, which will not only show you what it looks like, but also tell you how it is done. I have seen several books done in this way, and they look very good.

Prices are: for the binding I have marked

200 copies,	200 pages	\$ 379.00
500 "	200 "	\$ 479.00
200 "	500 "	\$ 823.00
500 "	500 "	\$ 961.00

There will be extra expenses for advertising, some translating, etc, but I think we could do 200 copies of 200 pages for \$ 500. If the book sells well, we can have further copies mode as required. I have asked Dr. Wallace, Principal of Queen's University about money, and I think we may be able to get a grant of \$ 500, but there has been no definite promiss yet. I am getting quotations on costs from some Canadian printers, and will write you later about them.

With kind personal regards and the season's greetings, I am,

Very sincerely yours

John Stanley

19.8 Volterra to Stanley, Paris, 1936 *Cat. no. 407*[8]

Cher Professeur Stanley

J'ai bien reçu les épreuves des pages et des figures et j'attendais encore d'autre détails. Si vous pouvez m'en donner je vous en serai très reconnaissant. Je ne vous cache pas qu'un ouvrage imprimé serait preferable pour moi à un ouvrage tapé. Ce qui me préoccupe le plus ce sont les formules. Je pense qu'il faudrait les écrire à la main ce qui serait peu esthétique. Mais en vue de la dépense et des difficultés pour se procurer l'argent je donne mon avis favorable à la publication de l'ouvrage de la manière que vous proposez. Vous avez donc mon consentement et vous pouvez proceder à faire ce qu'il est necessaire. Je désire seulement

[8] Draft, n.d. but obviously written in 1936.

conserver la proprieté litteraire et le droit de publier l'ouvrage en Anglais et en d'autres langues dans le territoire de l'Empire Britannique et partout ailleurs.

Je suis en voyage. J'ai fait à l'Université de Genève une conference sur la biologie mathématique et je me rends à Paris où je resterai quelques semaines pour les réunions du Comité des poids et mesures. Mon addresse à Paris est

19.9 Stanley to Volterra, Kingston, 9.7.1937 *Cat. no. 330*

CANADA
DEPARTMENT OF AGRICULTURE
ENTOMOLOGICAL BRANCH
ENTOMOLOGICAL LABORATORY

Kelly's Camp, Grand Cascapedia, Gaspé P.Q., Canada

July 9, 1937

Dear Professor Volterra,

Your letter of June 18 has finally reached me here where I am temporarily located working on a statistical study of cocoon populations of spruce saw-flies.

At present it is impossible for me to send you the information you ask for as it is in Kingston, locked up in my office. However, as soon as I return in September, I shall write you again. Future letters should be addressed to me at Queen's University, as I shall not be here later than august.

Regretting that I cannot do more at present,

I am sincerely yours,

John Stanley

19.10 Stanley to Volterra, Kingston, 28.10.1937 *Cat. no. 331*

(Same heading of letter no. 19.2)

Oct 28, 1937

Dear Professor Volterra,

Many thanks for the separate entitled "Leggi delle fluttuazioni e principii di reciprocità in biologia"[9] which arrived a few days ago.

I shall write you soon regarding the English version of your works, but just at present, having been away for many months, I am far behind with everything.

Very sincerely

John Stanley

---

[9] VOLTERRA 1937c.

20 Letters between Georges Teissier and Vito Volterra

Georges Teissier was born in Paris on 19 February 1900. In 1919 he was admitted to the mathematics section of the École Normale Supérieure of Paris. However, he immediately displayed a penchant for natural science, although mathematics was nevertheless to play an important role as he approached biological studies following an interest in population genetics and biometrics, which are highly mathematical disciplines. His early work regarded the theory of biological similarity, animal growth and topics of marine biology. Teissier presented his doctoral thesis in 1931: it was on the subject of the growth of insects and was published by the Biological Station of Roscoff, a biological research centre located in Brittany which depended on the Sorbonne University. He continued to work here throughout the 1930s.

In 1933 the French publishing house Hermann decided to include in its "Actualités scientifiques et industrielles" series, a sub-series entitled "Exposés de biométrie et de statistique biologique", the direction of which was entrusted to Teissier, who wrote the first volume himself (Teissier 1934). His contribution was followed by monographies by Kostitzin (V. A. Kostitzin 1934), Philippe L'Héritier (L'Héritier 1934), Lotka (Lotka 1934) and lastly Volterra and D'Ancona (Volterra, D'Ancona 1934). Also the sub-series begun in 1934 and dedicated to ecology, under the direction of the biologist Marcel Prenant, contributed to enhancing the interest of research circles in the application of mathematics to biology, especially in France.

After the nazi occupation of France, Teissier joined the resistance (1941) and carried on his resistance activities above all in the Sorbonne environment, where he continued to work and to publish articles. This period corresponds to a series of works on mutant forms of Drosophilae populations. In the final phase of the war he joined the "Francs Tireurs et Partisans" group, replacing Marcel Prenant who had been arrested and deported by the nazis.

From 1945 on, after the liberation of France, Teissier played an important role in the reconstruction of the country's scientific institutions. He was appointed director of the new Centre National de la Recherche Scientifique (CNRS), a position originally occupied by the eminent physicist and Nobel Laureate Frédéric Joliot-Curie (1900–1958), and which he held until 1950. He set up the Laboratoire de Génétique Evolutive (Gyf-sur Yvette), and directed it from 1951 to 1965. He also promoted a genetics course at the Sorbonne in collaboration with Philippe L'Héritier. He was nominated member of the Académie des Sciences.

Georges Teissier died in Roscoff on 7 January 1972.

Teissier chose his research topics mainly in the fields of marine biology, animal growth theory and evolutionary genetics. In the first field he was concerned mainly with the growth of hydrozoans. In the second field, Teissier distinguished himself by introducing biometric methods, studying the quantitative representation of the growth of different animal species, in particular certain species of flour insects and crustaceans. The synthesis he made of research on the topic of growth (TEISSIER 1928) served as a reference for Volterra. He also made a detailed study of the topic of biological rhythms and their relationship with size. Teissier's most significant contribution to evolutionary genetics was to seek experimental verification of the population genetics introduced by J. B. Haldane, R. A. Fischer and S. Wright which, in the 1930s, was still a purely theoretical branch of science. One of the empirical contexts in which Teissier (in collaboration with L'Héritier) did most research was that of the genetics of *Drosophila melanogaster*[1].

* * *

20.1 Teissier to Volterra, Paris, 22.12.1933 *Cat. no. 332*

FACULTE DES SCIENCES DE PARIS
STATION BIOLOGIQUE DE ROSCOFF
LABORATOIRE LACAZE DUTHIERS

Paris, 22 Décembre 1933

Monsieur le Professeur

Monsieur Kostitzin m'a informé que vous vouliez bien accepter d'écrire un fascicule pour la série Biometrie des "Actualités Scientifiques"[2]. Je tiens à vous dire combien je me sens encouragé et honoré pour la précieuse approbation que vous voulez bien donner à notre tentative.

Il n'existait jusqu'à présent aucune publication consacrée aux applications très diverses des Mathématiques à la Biologie. Il m'a semblé, lorsque M. Hermann et Cie m'ont demandé de me charger d'une série de leurs fascicules, qu'il avait là une occasion de montrer aux biologistes, par des exemples précis, l'intérêt d'une direction de recherches qui ne peut que se développer largement dans un proche avenir. Votre offre de réserver à la série Biometrie la primeur d'une de vos oeuvres, constitue pour elle une consécration, que je n'aurais pas osé esperer et dont je vous suis infiniment reconnaissant.

Lorsque, dans un délai que je souhaite bien vivement être proche, vous aurez pu préciser le sujet et le titre de votre fascicule, je vous prierai de bien vouloir m'en informer pour que je puisse faire établir par M. Hermann et Cie le contract

---

1 For further details on Teissier's research see PETIT 1976.

2 The entire correspondence refers to the publication of VOLTERRA, D'ANCONA 1935.

d'usage. Je crois pouvoir vous promettre que l'editeur fera son possible pour vous imprimer rapidement.

En vous renouvelant mes remerciements, je vous prie, Monsieur le Professeur, de bien vouloir accepter l'assurance de mes sentiments très respectueusement dévoués.

G. Teissier 2, rue du Val de Grâce, Paris (V^{e})

20.2 Volterra to Teissier, n.p., n.d. *Cat. no. 408*[3]

Cher Monsieur

Je viens de recevoir de la part de la Maison Hermann le projet de contrat. Je suis prêt à le signer et je pense Monsieur D'Ancona aussi, mais il nous faudrait avoir l'assurance préalable que la redaction Française que nous enverrons sera regardée et corrigée au point de vue de la langue et du style et qu'après qu'elle aura été améliorée et mise à point nous pourrons la voir avant l'impression. Je pense qu'il vaut mieux faire des modifications de forme sur le manuscrit que sur les épreuves. J'espère que vous voudrez bien comme directeur de la collection me donner une reponse favorable.

Veuillez accepter, cher Monsieur, l'expression de toute mon estime et de mon sincère dévouement.

20.3 Teissier to Volterra, Paris, 12.1.1934 *Cat. no. 333*

(Same heading of letter no. 20.1)

Paris, 12 Janvier 1934

Monsieur le Professeur

Je viens de recevoir votre lettre du 9 Janvier et vous remercie à nouveau de la précieuse collaboration que vous voulez bien apporter à la série “Biometrie” des Actualités Scientifiques. Je suis tout à votre disposition, si vous voulez bien me faire confiance, pour relire votre manuscrit et pour vous aider à le préparer pour l'impression. Je ne crois d'ailleurs pas, à en juger par vos lettres et par les conférences où j'ai eu le plaisir de vous entendre, que votre rédaction puisse nécéssiter une correction de quelque importance. J'examinerai cependant votre texte avec attention et vous indiquerai, puisque vous le souhaitez, les modifications de qui me paraîtront désirables.

J'éspère que cette solution vous conviendra et je vous prie, Monsieur le Professeur, de bien vouloir agréer l'assurance de mes sentiments très respectueux.

G. Teissier

[3] Draft n.d. but the letter was sent on 9 January 1934 (see the following letter no. 20.3).

20.4 Teissier to Volterra, Roscoff, 19.6.1935 *Cat. no. 334*

(Same heading of letter no. 20.1)

Roscoff, 19 Juin 1935

Monsieur le Professeur

Je viens d'ecrire à la Maison Hermann au sujet de votre fascicule, mais je crois pouvoir vous rassurer. Quand j'ai quitté Paris, il y a une semaine, les exemplaires etaient au hochage. Leur parution ne peut donc tarder.

Je regrette d'ailleurs beaucoup qu'elle ait été aussi tardive, mais M. Freyman a qui j'en ai souvent parlé, explique ce retard par des difficultès économiques dont je ne suis pas juge.

J 'éspère que tout va s'arranger dans très peu de jours et je vous prie, Monsieur le Professeur, de croire à l'assurance de mes sentiments respectueux et reconnaissants.

G. Teissier

10.5 Volterra to Teissier, n.p. 1935 *Cat. no. 409*[4]

Monsieur et cher Collègue

Je tiens à vous informer que je viens de recevoir de la part de la maison Hermann les exemplaires d'auteur du fascicule, qui vient d'être publié. Je tiens aussi à vous remercier de l'aimable lettre que vous m'avez envoyé le 19 ct. Monsieur D'Ancona et moi désirons vous exprimer toute notre reconnaissance pour les soucis que vous étes donnés pour la publication de notre volume et nous espérons qu'il pourra correspondre aux idées qui vous ont guidé dans la direction de l'importante publication qui a été confiée à vos soins et que vous continuez avec tant de succès.

Veuillez accepter, Monsieur, l'assurance de ma plus haute considération et de tout mon dévouement.

[4] Draft n.d. but certainly written in 1935.

21 Letters between D'Arcy W. Thompson and Vito Volterra

D'Arcy Wentworth Thompson was born in Edinburgh (Scotland) on 2 May 1860. His father (who had the same name) was professor of classical literature and greek at Queen's College, Galway and, from 1870 on, classical master at the Edinburgh Academy. It is a fact that his father had a strong influence on the young D'Arcy, as, in his final year, the latter won the prizes for Classics, Greek Testament, Modern Languages, as well as for Mathematics. He enrolled in medicine at Edinburgh University (from 1877 to 1880), and then continued his studies at Trinity College, Cambridge (from 1880 to 1883). He graduated in zoology in 1883 but obtained his Ph.D. only 29 years later, in 1923.

Meanwhile, in 1884, he was appointed professor of biology at University College, Dundee (Scotland), which was incorporated in 1897 in the University of St. Andrews. Here, in addition to his research and lecturing activities, he set up a zoology museum for didactic purposes. In 1917 he transferred to the chair of natural history at the University of St. Andrews, which he held until his death. He thus held some sort of university chair for a good 64 years, an authentic record.

In 1897, D'Arcy Thompson was the leader of a research programme on the fisheries in the Sea of Bering organized by the British government and was appointed to represent the United Kingdom on the Anglo-American Commission of Inquiry into the Bering Sea Fisheries. He was one of the founding fathers of the Conseil Permanent International pour l'Exploration de la Mer, of which he was a council member from 1902 to 1947. He was also chairman of the Statistical Committee and director of *Bulletin Statistique* from 1902 to 1947.

D'Arcy Thompson was elected fellow of the Royal Society in 1916 and, in 1928, member of the Classical Association of England and Wales. From 1934 to 1939 he chaired the Royal Society of Edinburgh and was knighted in 1937. In 1946 he was awarded the Darwin Medal and his mathematical merits were recognized when he was elected honorary member of the Edinburgh Mathematical Society in 1933.

He died at St. Andrews (Scotland) on 21 June 1948.

D'Arcy Thompson is a figure who is difficult to fathom as he does not fit into the usual classifications: he was an authentic humanist scientist who combined a passion for classic learning and philosophy with one for natural and biological science and, in particular, for mathematics. Although he wrote about 300 scientific articles and books, he became famous for his well-known book *On Growth and Form* [THOMPSON 1917]. In this book he moved away from the contemporary approach of zoology, which tended to investigate organic forms in terms of comparative anatomy and evolutionary theory. He instead set out to develop a theory of evolution which he conceived not as the result of small local changes but of global transformations involving the entire organism. The approach he recommended was based on mathematics, as he considered that every biological problem could be described and understood in mathematical terms. Indeed, in his conclusion to *On Growth and Form*, he wrote: "... something of the use and beauty of mathematics I think I am able to understand. I know that in the study of material things number, order and position are the threefold clue to exact knowledge: and that these three, in the mathematician's hand, furnish the first outlines for a sketch of the Universe". Nevertheless D'Arcy Thompson's approach was not a mere resumption of Descartes' *Mathesis Universalis*, as it was not based on any quantitative and reductionist ideal. Rather it proposed a qualitative view of mathematical representation in biology, focused above all on the problem of morphogenesis. By so doing he anticipated topics and views that were to be developed only later and it is no coincidence that the mathematical theories underlying morphogenesis developed by the French mathematician René Thom in the seventies made explicit reference to the thinking and work of D'Arcy Thompson (THOM 1975). More in general, contemporary biological theory owes much to his pioneering work.

On Growth and Form is not only a scientific book but also an extremely powerful literary and poetic text. It successfully achieves a full synthesis between humanist and scientific culture, which was D'Arcy Thompson's greatest goal, and he always held the problem of linguistic style to be of capital importance. He wrote: "The little gift I have for writing English which I possess, and try to cultivate and use, is, speaking honestly and seriously, the one thing I am a bit proud and vain of – the one and only thing"[1].

* * *

21.1 D. W. Thompson to Volterra, St. Andrews, 2.7.1926 *Cat. no. 338*

St. Andrews, 2-7-26

Dear Professor Volterra,

Our friend, Larmor (who has gone to Switzerland) sent me before leaving England your ms. on "Biological Associations"[2] etc. I ventured to make a few alterations; for while your english <u>is most excellent</u>, I thought I could sometimes make it just a little shorter and simpler. Then I sent it to my good friend, Sir Richard Gregory, for publication in "Nature"

---

[1] See DOBELL 1949, R. D. THOMPSON 1976.

[2] Reference to VOLTERRA 1926b. From this letter and the following ones we learn that Volterra's note for *Nature* was presented to D'Arcy Thompson via Joseph Larmor and that Volterra contacted D'Arcy Thompson through Larmor.

Today I have received the proof, which I send you herewith; will you kindly revise it and send it back to me. It is of course necessary that any changes I have made or suggested should have your consent or approval.

I am only a Biologist, but I have some little interest in Mathematics of an elementary kind. One of my little papers I should like to offer you.

May I send my kind regards to those of your collegues whom I had the pleasure of meeting in Russia last year and in Canada the year before, Levi-Civita, Severini and the rest.

Believe me to be

Very faithfully Yours

D'Arcy W. Thompson

21.2 Volterra to D. W. Thompson, n.p., n.d. *Cat. no. 420*[3]

Dear Prof. Thompson

I have received the proofs of my article for "Nature". I thank you very much for having presented it to the Journal. I approve the corrections and modifications you have done. I thank you for the care you have put in making at simpler [?] the article. Your works on biological mathematics are very well known and existimated by us and I am highly honoured that you have read through the proofs of my paper. As soon as the parts of my memoir that is in the press at the Accademia dei Lincei will be ready I shall send it to you. I thank you very much for your important paper that I shall study with the greatest interest.

Prof. Levi Civita and Severi send you their kind regards

With many thanks and best regards

I am yours very faithfully

Caro Prof. Thompson

Ho ricevuto le prove della nota per Nature e La ringrazio moltissimo dell' averle gentilmente fatte pubblicare nel Giornale. Approvo le correzioni e modificazioni che Ella vi ha fatto come La ringrazio della cura che ha dato per semplificare la redazione dell'articolo. Le sue opere di matematica biologica sono molto note ed apprezzate da noi ed io sono molto onorato che la mia nota sia passata sotto i Suoi occhi. Appena pronti i fogli della memoria che è in corso di stampa all'Accademia dei Lincei glieli manderò. La ringrazio moltissimo della Sua importante memoria che studierò col maggiore interesse. I prof. Levi-Civita e Severi ricambiano i Suoi saluti.

---

[3] Italian and English drafts, n.d. but certainly written in 1926.

21.3 D. W. Thompson to Volterra, St. Andrews, n.d. *Cat. no. 335*[4]

Mr. D' Arcy Wentworth Thompson
44 South Street
St. Andrews

Prof. Vito Volterra
Tout se trouve en ordre.
Votre communication aura lieu dans la section A (sous section mathematique) Lundi a 12 heures, on 12.15.

DWT

21.4 Volterra to D. W. Thompson, Ariccia, 20.9.1926 *Cat. nos. 410, 413*[5]

Villino Volterra
Ariccia (Roma)

20-IX-1926

Cher Monsieur

Permettez moi de vous addresser le premier exemplaire, que je reçois aujourd'hui, de mon mémoire publié par l'Académie des Lincei[6].

Veuillez bien l'accepter en hommage avec mes remerciments pour l'interêt que vous avez pris pour mes recherches, et avec l'expression de mon admiration pour vos travaux de biologie et de mathématiques.

Après avoir quitté la Grande Bretagne, j'ai fait un sejour sur les Alpes d'où je me suis rendu ici. Je n'ai plus eu de nouvelles de l'Article écrit pour Nature dont vous aviez les épreuves à Oxford. A-t-il été déjà publié ou quand sera-t-il publié? Pouvez-vous m'en donner des reinseignements?

J'ai promis à la jeune dame qui m'a ramené en auto le jour de ma communication à Oxford un exemplaire de mon mémoire, mais je ne sais pas à quelle adresse je dois le lui envoyer. Si en m'écrivant un mot vous pouvez me l'indiquer je vous en serais très reconnaissant.

Je garde le meilleur souvenir de mon séjour à Oxford et j'ai été heureux de vous y [illeg.] m'entretenir avec vous.

Veuillez accepter, cher monsieur, l'expression de toute mon estime et de tout mon dévouement.

Vito Volterra

---

[4] Postcard, n.d. but probably written in August 1926 (see following letter no. 21.5).

[5] Drafts of the same letter.

[6] VOLTERRA 1926a.

21.5 Volterra to D. W. Thompson, Roma, 24.9.1926 *Cat. no. 414*[7]

Roma, 24th. September 1926

Dear Sir

I have the honour of sending you a short abstract of my lecture made Wednesday August 9 in the section A. I should be very glad if you could publish it in the B. A.[8] report. You will find in it the reference to the publication in extenso of the subject of my lecture. I remain

Yours faithfully

Vito Volterra

21.6 D. W. Thompson to Volterra, St.Andrews, 12.10.1926 *Cat. no. 339*

St. Andrews, 12–10–26

My dear Prof. Vito Volterra,

I am sorry that "Nature" has been so long of publishing your article; for a good many weeks after the British Association, every year, the columns of "Nature" are apt to be very full.

But now Sir Richard Gregory (the editor) tells me that your article will appear this week. You will receive a number of copies, in the form of "slips", just as the thing appears in Nature itself; if they were to be made up in pamphlet form, it would be necessary to pay for them, and the cost is rather high.

I fear that I have never yet thanked you for the copy of your big paper which you sent me about two weeks ago. I have now read it with very great interest, – all except somes of the more difficult parts; and it has interested me very greatly. I am only sorry that we did not enlarge the "Nature" article a little, by including a few more points from the larger paper.

You ask about the little lady who drove you in her auto. She is a very distinguished little woman, and we all have the highest respect and regard for her. Her name is Dr. Dorothy Wrinch, Somerville College, Oxford; she is the wife of Professor John Nicholson, the Physicist, of Balliol College, Oxford, but she is generally known by her own name. She is a most excellent mathematician, by a long way the best woman-mathematician in this country. And besides her mathematical ability, she is also distinguished as a philosopher.

When I was lately in Copenhagen I had the great pleasure of meeting your distinguished collegue Professor Magrini; to make his acquaintance was a pleasure which I had long desired.

I look forward with some hope to coming to Italy next spring, that I may have the pleasure of meeting many friends.

Believe me,

Very faithfully yours,

D' Arcy W. Thompson

---

[7] Draft.

[8] British Association for the Advancement of Science.

21.7 Volterra to D. W. Thompson, Roma, October 1926 *Cat. no. 411*[9]

Mon cher Professeur D'Arcy Thompson

Je vous suis très reconnaissant de votre aimable lettre et de toute la peine que vous vous êtes donnée pour la pubblication de mon article dans Nature. Je suis heureux qu'il soit paru dans le dernier fascicule où je l'ai vu. Je receverai avec le plus grand plaisir les copies de l'article en forme de "slips".

La jeune et très distinguée mathématicienne Madame Wrinch doit avoir reçu mon mémoire. Elle a été si aimable de m'adresser quelques articles qu'elle a publié et qui sont excellents.

Aujourd'hui même j'ai l'honneur de vous envoyer le mémoire de M. D'Ancona sur les statistiques qu'il a fait sur les poissons de la mer Adriatique[10].Vous connaissez déjà quelqu'un des résultats car je vous en ai parlé et ils sont aussi cités dans mes travaux. J'espère que ce mémoire vous intéressera et que vous voudrez bien l'accepter en hommage de la part de son auteur.

Ceux qui ont pris part à la réunion de Copenhagen m'ont informé qu'on y a parlé de mes recherches. Je pense que je dois à vous même si on les a pris en considération et je vous en suis très réconnaissant.

La nouvelle de votre prochain voyage en Italie dans le printemps me rejouit beaucoup. Veuillez bien m'en faire connaître la date exacte lorsque vous l'aurez fixée. Tous ceux qui ont de l'admiration et de l'amitié pour vous s'en rejouiront avec moi.

21.8 Volterra to D. W. Thompson, n.p., November 1926 *Cat. nos. 415, 419*[11]

Cher Monsieur

Je viens de recevoir un volume du Prof. Lotka ayant pour titre ... [in Cat. no. 415: *Physical Biology*][12] où je vois qu'il avait traité en partie le cas de deux espèces [in Cat. no. 415: en arrivant jusqu'au diagramme 2 qui est pareil au mien] [in Cat. no. 419: mais il n'arrive que jusqu'au diagramme 2 qui est pareil au mien]. Je ne connaissais pas le travaux de M. Lotka, c'est pourquoi je n'ai pas pu les citer, mais je le citerai dans les travaux que j'ai sous presse dans le Journal Scientia et dans les Mémoires du Comité Talaxographique [this phrase is eliminated in Cat. no. 415]. M. Lotka part d'une analogie, qui existe avec un problème chimique déjà traité[13] et se borne à integrer les deux équations qu'on a dans le cas de deux espèces que j'ai designé par III, IV, mais il n'en deduit pas les

[9] Draft, n.d. but probably written in October 1926.

[10] D'Ancona 1926.

[11] These are drafts of the same letter, n.d. but obviously written at the end of November 1926 (probably 21 November). Letter no. 419 was handwritten and contains all three parts, while letter no. 415 was typed and contains only the beginning of the first, together with a few changes that we have noted.

[12] Lotka 1925.

[13] Lotka 1910, 1920b, 1920c.

trois lois que j'ai trouvées ni aucune des consequences que j'en ai tirées. Il ne traite non plus tous les autres cas de deux espèces que j'ai examiné dans le § 4.

En outre il n'envisage pas le cas de *n* espèces c'est pourquoi il ne connait pas les fluctuations générales ni l'existence des associations conservatives et dissipatives. D'autre part son volume se rapporte à des questions très générales de statistique de manière que l'on peut dire qu 'il y a dans nos recherches un petit domain commun mais elles s'en détachent complètement dans tout le reste. Quoique ce domaine soit à mon avis assez restreint, je regrette de n'avoir pas cité dans mon travail des Lincei qu'il m'avait precedé au sujet d'une partie de la matière [in Cat. no. 415: que j'ai traité dans le §3]. [eliminated in Cat. no. 415: J'ai voulu vous en écrire tout de suite pour vous tenir au courant].

[Only in Cat. no. 415: Mais je ne connaissais pas le résultats de M. Lotka. J'ai desiré vous en écrire pour vous tenir au courant de cela. Veuillez accepter ...]

Dear Sir

ricevo in questo momento il libro di Lotka "Physical Biology" che io non conoscevo e siccome un caso particolare della teoria generale delle fluttuazioni era stato trattato da lui così ritengo doveroso mandare la unita lettera all'Editore di Nature. La spedisco a Lei con preghiera se Ella la approva di voler avere la gentilezza di mandarla all' Editore di Nature.

Il Dr. Alfred J. Lotka mi manda oggi il suo libro "Physical Biology", pubblicato nel 1925, che io non avevo avuto agio di consultare. Da esso rilevo che egli conosceva prima di me il diagramma rappresentato nella figura 2 del mio articolo di Nature October 16 ed il periodo delle piccole fluttuazioni, però le leggi generali delle fluttuazioni, i casi generali dei sistemi conservativi e dissipativi e le conseguenze di tutte le modalità di incontro non sono stati da lui trattati come ritengo non lo siano stati nemmeno da altri.

21.9 D. W. Thompson to Volterra, St. Andrews, 26.11.1926 *Cat. no. 340*

St. Andrews, Ecosse, le 26 Novembre, 1926

Cher Monsieur et Colleague,

Je viens de recevoir votre aimable communication du 21me Novembre, et je vous remercie.

Sir Richard Gregory, rédacteur en chef de "Nature", m'avait écrit il y a quelques jours au même effet et vous aussi vous avez reçu probablement une communication de lui-même.

C'est à regretter que ni vous ni moi nous n'avions pas vu l'ouvrage de M. Lotka. Moi j'en avais entendu, et je l'avais commandé pour notre bibliothèque; mais malheureusement notre bibliothécaire manquait d'en trouver un exemplaire. De là vient que cet ouvrage me restait inconnu.

Les questions de priorité sont toujours un peu désagréables; mais après tout, parmi le gens de bon coeur, on n'a ni le besoin ni l'habitude de s'en trop fâcher.

M. Lotka a fait quelque chose, il en a tout le mérite, - et vous ne le lui disputez pas! Quant à lui, il reconnaitra sans doute (parce que c'est indisputable) que, partant par la même route, vous y êtes allé beaucoup plus loin que lui. Il est mathématicien, - il s'ensuit donc qu'il appartient aux gens comme il faut! Et vous êtes assez riche en réputation pour pouvoir lui accorder même plus qu'il n'a gagné.

Veuillez agréer cher Monsieur et très honoré Collègue, l'assurance de ma profonde estime et de mon sincère dévouement.

D' Arcy W. Thompson

21.10 Volterra to D. W. Thompson, Roma, 28.11.1926 *Cat. no. 416*[14]

Via in Lucina, 17 *28-XI-1926*
Roma

Cher Monsieur

Après avoir eu l'honneur de vous écrire, j'ai reçu une lettre de l'editeur de Nature où était incluse une note que lui avait été adressé par Dr. Lotka au sujet de mon article et de son ouvrage. L'Editeur de "Nature" me demandait une note de ma part à ajouter à celle de M. Lotka.

Je l'ai preparé tout de suite et je l'ai envoyée. Je prends la liberté de vous envoyer ce que j'ai écrit à l'éditeur de "Nature". Comme vous allez voir il correspond parfaitement à ce que je vous ai écrit dans ma dernière lettre [crossed out: Je pense que entre M. Lotka et moi il n'y a pas de polemique ni de désaccord et que tous les deux avons contribué et contribuerons à developper une branche nouvelle de la science. Veuillez accepter, cher Monsieur, l'expression de ma plus haute estime et de mon devouement. Je serai heureux de connaître votre avis car si vous avez quelque remarque à faire vu... [Addition: D'après votre interêt pour mes recherches je serai heureux de connaître votre avis auquel je donne le plus haute importance]

[A second sheet contains the same text with the following addition crossed out: Je serais heureux cependant si la part qui lui appartient et que je ne connais pas (équations differentielles dans le cas de deux especès dont l'une se nourrit de l'autre, son intégration, diagramme des intégrales, période des petites fluctuations) et la partie nouvelle qui se trouve dans mon mémoire (Les trois lois générales, tous les cas des rencontres, le cas des n espèces, les sistèmes conservatifs et les lois des fluctuations et toutes leurs consequences) furent mises en évidence par quelqu'un en dehors de nos deux.]

14 Draft.

21.11 Volterra to D. W. Thompson, n.p., November 1926 *Cat. nos. 417, 418*[15]

Dear Sir, In answer to your letter of Nov. 22nd (RAG: ET: 2467), I enclose my comment on the letter of Dr. Lotka. I would be glad if it could be printed in Nature together with his letter[16].

I am glad that the publication of my article has enabled me to know the book and the work of Dr. Lotka, which was unknown to me when the article was written.

With best regards, I am yours truly

I have the honour to send you a copy of my paper of the Accademia dei Lincei.

In the above letter from Dr. Lotka, which is in accordance with our proceding correspondance, following upon the pubblication of my article in Nature, he justly observes that he had obtained the differential equations in the case of two species one of which feeds upon the other, that he had given as well the same diagramm of the integral and also the period in the case of small fluctuations. In this I recognize his priority, and am sorry not to have known his work, and therefore not to have been able to mention it.

I did not even know other pubblications of the same kind by other authors, for instance the work of Sir R. Ross on malaria[17], which precedes the writings of Dr. Lotka, who has however found so many new and important results.

The other observations mentioned in the above letter refer to points which I have not treated; but also the sea-fisheries, while I refer to the laws (which I believe to have been the first to formulate) and principally to the third law, which gives in a easy way of calculating the maximum output of fisheries, he, on his part, considers the case of the additions of a third species, which seems to me a different problem. [What follows is contained only in Cat. no. 417]

I also think that the quotation of Darwin's acute intuition referring to my third law, and the quotation from Spencer, which touches only the principle of the existence of fluctuation, are essentially different.

I think that the study of the general case of the convivence of *n* different species, subject to the hypothesis which allow me to distinguish the case of conservative association from that of dissipative association, and to obtain integrals and laws of fluctuation which form the essential and the greatest part of my memory, is absolutely new.

To conclude, I recognize the existence of some common points of Dr. Lotka's work and my own, in which he has the priority, but my work and his diverge in all the rest.

Working indipendently the one from the other, we have found some common results, and this confirms the exactitude and the interest in the position of the problem. I agree with him in his conclusion that these studies and the methods of research deserve to receive greater attention from scholars, and should give rise to important applications.

---

[15] Two drafts of the same letter, n.d. but certainly written at the end of November 1926.

[16] These letters have been published: see LOTKA, VOLTERRA 1927.

[17] Ross 1923.

21.12 D. W. Thompson to Volterra, St. Andrews, 12.7.1930 *Cat. no. 341*

FISHERY BOARD FOR SCOTLAND
EDIMBURGH

St. Andrews, Scotland, 12 th July, 1930

My dear Vito Volterra

I happen to hear (from M.me Eugenie Kraipinsky) that you are - or that you have been in Paris; and it is just possible, therefore, that you may be coming on to England.

I write to tell you that we are having a little Mathematical "Colloquium" here, from July 19th to the end of the month; and I am instructed to tell you that, if you care to come, we shall be greatly honoured and you will be very heartily welcome.

I must tell you that it is a very informal affair. The University is closed for vacation, and our Meeting is more of the nature of a holiday. Besides the few names you see on the enclosed Prospectus, we expect your colleague Levi-Civita, also Wedderburn of Princeton, Williamson of Baltimore, and others.

You were good enough to write me a letter of congratulation some few days ago, and I ought to have replied to it more quickly. But you have been misinformed! The new Knight is another man; - one who has more popular gifts than I have, and who has much better friend to the present Government!

I passed through Rome about the end of last September and called on you at the Via in Lucina. But you had left for Paris, only on half-an-hour before. I had come all the way from the Cape of Good Hope in an Italian ship, which took me on to Venice.

It was a large voyage, and a slow one, but I never enjoyed a voyage more; and I never was so well fed on my life, even on the big Mail Steamers. The captain was a highly educated man, and some of his officers also were delightful company.

I am sending this to the Institute, on the chance of its finding, before you leave Paris.

With kinds regards,
Your very faithfully,

D' Arcy Thompson

21.13 Volterra to D. W. Thompson, n.p., July 1930 *Cat. no. 412*[18]

Cher Professeur D 'Arcy W. Thompson

Je reçois ici avec quelque retard votre aimable lettre du 12 ct.

Après avoir passé presqu'un mois à Paris où j'ai eu des réunions au Bureau des Poids et Mesures et au Conseil international de recherches je suis venu sur les montagnes pour me reposer et soigner ma santé.

Veuillez accepter tous mes remerciements pour l'aimable invitation que vous m'avez faite de prendre part au Congrès de St. Andrews. Je regrette infiniment de ne pouvoir pas intervenir et de ne pouvoir pas me trouver avec des collegues si

[18] Draft, n.d. but probably written in late July 1930.

estimés et de ne pouvoir pas jouir de lectures si intéressantes. Je le regrette d'autant plus que je perd cette année aussi l'occasion de vous voir et de passer quelque temps avec vous. Ce n'est que d'après votre lettre que j'ai appris que vous avez fait un voyage en Italie l'année passée et que vous êtes venu me chercher chez-moi. Ne viendrez vous pas cette année à Padova au Congrés des Zoologistes? Je n'y serai pas car je dois me rendre en Tchecoslovaquie pour y tenir quelques Conferences, mais mon gendre M. D' Ancona y sera. M.M. Joubin, Gravier et d'autres m'ont dit qu'ils interviendront.

Je regrette que ce soit à cause d'un malentendu que je vous ai fait des félicitations. Vos grands mérites et la grande estime que vous jouissez expliquent le malentendu. Veuillez bien me permettre de transformer les felicitations dans un voeu pour un prochain avenir.

21.14 D.W. Thompson to Volterra, St. Andrews, 3.1.1931 *Cat. no. 342*

St. Andrews, 3-1-31

My dear Colleague.

I have received this morning your beautiful new year present, and have begun already to read it - so far as my little mathematics will permit.

I may be able to send you soon some very fine examples of fluctuation in the catch of cod and other fishes, as shown by our statistics of 1903–1929.

You do not mention the very important papers by M. P.-F. Verhulst; v.g. his Rech. math. sur la loi d'accroissement de la population, N. Mém. d. l'Acad. Roy du Belgique, XVIII, 1845[19]. This deals precisely with the subject of yr. Chap. III, - viz "la fonction (linéaire) decroissant de N. d'un espèce vivant seule". But you probably know this paper!

Buon capo d' Anno!

D' Arcy W. Thompson

21.15 Volterra to D. W. Thompson, Roma, January 1931 *Cat. no. 421*[20]

Mon Cher Collègue

Je viens de recevoir votre aimable lettre et je vous suis très reconnaissant. Merci des souhaits que vous m'envoyez pour la nouvelle année. Veuillez accepter les voeux les plus sincères que je forme pour vous et pour les vôtres.

Je suis heureux d'apprendre que mon ouvrage vous a intéressé. Si vous voudrez bien m'envoyer les statistiques que vous avez faites elles seront précieuses pour moi. Je pourrai très bien les insérer dans une nouvelle édition Anglaise si je trouvais un Editeur Anglais. Blackie & Son de Glasgow viennent de publier un ouvrage que j'ai écrit sur "Theorie of Functionals"[21] mais je n'ose pas leur

---

[19] Verhulst 1845.

[20] Draft, n.d. but probably written in late January 1931.

[21] Volterra 1931b.

demander s'ils seraient disposés à cette nouvelle publication. Elle a suscité cependant un certain interêt. Êtes-vous en rapport avec ces éditeurs?

Dans une nouvelle édition j'ajouterai la citation du mémoire de Verhulst dont vous m'avez donné la précieuse indication. Je ne la connaissais pas. J'ai pu consulter bien à l'Académie des Lincei un mémoire de 1847[22] qui est la continuation de son travail de 1845. Celui-ci ne se trouve pas ici. J' ai su qu'il envisage un cas que j'ai aussi traité dans mon livre. Mais dans celui-ci la question est beaucoup plus large et embrasse des cas plus étendus.

Veuillez.......

21.16 D. W. Thompson to Volterra, St. Andrews, 11.2.1931 *Cat. no. 343*

St. Andrews, 11-2-31

My dear Colleague

I have been slow to write you, but I have not neglected the matter on which you wrote me.

I read great part of your book[23], with much admiration. I have spoken of it to one or two mathematical friends, who think (as I do) that it would be very useful to students and others in this country. And I hope to be able to write you, more definitely, in a few days, about the prospects of publication. I expect to hear from Blackie's very soon.

Yours faithfully

D'Arcy W. Thompson

21.17 D. W. Thompson to Volterra, St. Andrews, 28.2.1931 *Cat. no. 344*

St. Andrews, le 28 Fevrier 1931

Mon cher Prof. Vito Volterra.

Tout va bien, mais un peu lentement. Mais vous avez un proverbe italien "Chi va piano va sano!".

Mon Collègue, le Prof. Tumboll [?] a lu votre ouvrage avec une grande sympathie, en ajoutant son avis au mien qu'il serait très utile aux mathématiciens et étudiants anglais. Aussitôt qu'il m'a rendu le volume je l'ai expédié à Blackie, c'est à dire à M. Bisacre, le gérant en chef de la maison. Il en a accusé réception mais j'attends une communication plus définitive. Il est mathématicien lui-même; et votre collègue Levi-Civita l'a rencontré à notre petite réunion du mois d'Aout.

C' est bien possible que M. Bisacre vous écrive directement.

Votre très dévoué

D'Arcy W. Thompson

---

22 VERHULST 1847.

23 VOLTERRA 1931a.

21.18 Volterra to D. W. Thompson, Roma, 24.3.1931 ***Cat. no. 422***[24]

Via in Lucina, 17 *24–III–1931*
Roma

Cher Professeur D' Arcy Thompson

Merci de votre aimable lettre et de toute la peine que vous vous donnez pour moi. Je serai heureux si M. Bisacre voudra bien faire publier par Blackie mon ouvrage. On pourra simplifier les notes mathématiques de M. Brelot qui sont trop longues et pour ma part je suis disposé à faciliter de toutes manière la publication.

Veuillez agréer, cher Professeur D'Arcy Thompson, l'expression de ma plus haute estime et de mon sincère dévouement.

21.19 D. W. Thompson to Volterra, St. Andrews, 11.8.1931 ***Cat. no. 345***

St. Andrews University, Scotland, 11th August, 1931

My dear Colleague,

You must think I have been indeed a long time of writing you about your book.

I believe I told you that Blackie's people, after long consideration, found themselves unable to undertake it.

I then send it to the Clarendon Press (i.e., as you know, the University Press of Oxford), where I have many friends. They have the book in hand for many weeks, and now at last I hear from them today. You will see by the letter which I send you that, while the matter is not finally settled, yet the Secretary holds out no hope of the Press publishing the book. The reason partly is that the French edition is at hand, and accessible to all who need it. But behind that is, of course, the enormous cost of printing, especially mathematical printing, which makes the press refuse very many books which they would like to publish.

I wonder whether we shall have the great pleasure of seeing you in London, during September / October, when so many things will be going on: – the British Association, the Faraday Celebration, and (not least) the Clerk Maxwell Celebration.

Believe me,
Ever yours faithfully
D' Arcy Thompson

THE CLARENDON PRESS-OXFORD *10 th August, 1931*

Dear Professor,

Chapman is away, and the Delegates will not formally consider Volterra's book, about which you wrote on 24th June, until they come back. But I think I ought to tell you at once that on the reports we have had, they will not feel able to undertake it. It is an excellent book, and the reputation of the author is well

[24] Draft.

known to us. But it is the Delegates' rule not to translate very accessible foreign language unless the book can pay its way, and we do not think this one possibly could. The rule is, of course, intended to conserve funds for unpublished work, which these days are not so easy to find.

I hope he will not be unduly disappointed, and will not suppose that the quality of the work is in any way in question. For a translation, it becomes a plain matter of finance.

Yours very truly

Kenneth Sisan

to Professor D'Arcy Thompson
St. Andrews

21.20 D. W. Thompson to Volterra, Cochem a. d. Mosel, n.d. *Cat. no. 336*[25]

Al Illmo Senatore e Professore Vito Volterra
17 Via in Lucina
Roma, Italy

I am glad to have your letter, but sorry indeed that your health is not good. You work too hard!

Here am I, idling on the Moselle, making my way slowly towards Trier, then to Luxemburg, then home. I work every evening in the ancient Inn pictured on the other side.

D'Arcy W. Thompson

21.21 D. W. Thompson to Volterra, St. Andrews, 21.7.1935 *Cat. no. 346*

St. Andrews. 21. VII. 35

My best thank for your little, but invaluable, work on Les Association Biologiques[26].

I am writing a line also to M. George Teissier to congratulate him on these admirable little volumes, of which each one is more interesting than the one before.

D'Arcy Thompson

[25] Postcard showing photo of Alte Historische Thorschenke Cochem a. d. Mosel (Gasthaus).

[26] Volterra, D'Ancona 1935.

21.22 D. W. Thompson to Volterra, St. Andrews, 6.10.1936 *Cat. no. 347*

M^r D'Arcy Thompson
44 South Street
St. Andrews

6. X. 1936

Many thanks for your papers. I read them from beginning to the end in every word.

D'Arcy W. Thompson

21.23 D. W. Thompson to Volterra, St. Andrews, 24.6.1937 *Cat. no. 348*

St. Andrews, 24. VI. 37

Mon cher Collègue

Comme vous êtes aimable! Je suis vraiment touché par vos félicitations, et je vous en remercie bien sincèrement. Le même courrier m'apporte une lettre très cordiale d'A. J. Lotka. C'est intéréssant, n'est-ce-pas, d'avoir ses nouvelles et les votres au même jour!

Je suis en train d'aller à Copenhague quelques jours d'ici, pour notre Conseil des Pèches auquel j'ai partecipé pendant presque quarante ans! Après, je vais faire (avec ma fille) un petit voyage d'agrément dans la mer Egéenne, allant d'une île à l'autre! Et puis on m'a invité à Pékin, pour y donner des conférences!! – Mais à mon age, ça me semble un peu loin!!

Votre très dévoué

D'Arcy W. Thompson

21.24 D. W. Thompson to Volterra, n.p., n.d. *Cat. no. 337*[27]

How very kind of you to send me that beautiful card! I know the very spot when the picture were taken.

I told you, I think, that I went to the Galvani Celebrations at Bologna: it was a splendid and a most friendly meeting. And I found Bologna (which I did not know before) an extraordinarily interesting place. I hope to go there again, to study (inter alia) the Aldrovandi papers. And now a Happy Christmas to you and (what we all want) a peaceful New Year.

D'Arcy W. Thompson

---

27 Postcard, n.d. but certainly written in December 1937. Thompson mentions Luigi Galvani (1737–1798) and Ulisse Aldrovandi (1522–1604).

21.25 D. W. Thompson to Volterra, St. Andrews, 29.12.1937 *Cat. no. 349*[28]

Royal Society
Edimburgh, 2

Buon Capo d'Anno!

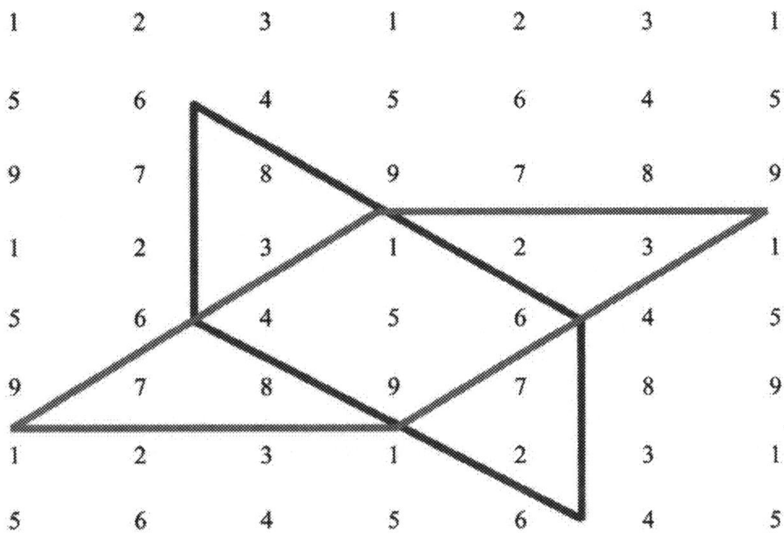

D'Arcy W. Thompson

[28] Postcard containing the following sketch.

22 Letters between William R. Thompson and Vito Volterra

William Robin Thompson was born in London (Canada) on 29 June 1887. He did his secondary schooling at the London Collegiate Institute, subsequently enrolling at the University of Toronto where he followed courses on entomology. In 1908, while still a student, he worked at the Gypsy Moth Parasite Laboratory in Massachusetts. He graduated from Toronto University the following year and was taken on as an entomologist at the U. S. Department of Agriculture. He was sent by the Department to Cornell University to attend graduate courses (here he obtained his Master of Science in 1912), and to Italy, to complete his studies under the guidance of the distinguished entomologist Filippo Silvestri. In 1913 he resigned from the Department in order to carry out research at Cambridge (1914–1915) and Paris where he was employed in the European Parasite Laboratory (1919) and obtained a diploma of docteur-ès-sciences in 1921.

In this period, also influenced by his reading of *On Growth and Form* by D'Arcy Thompson (THOMPSON 1917), he became passionately interested in the possibility of applying mathematics to biology, aiming in particular at the quantitative representation of biological properties such as periodicity and rhythmicity. Thompson experienced considerable difficulty in convincing his American colleagues (with the exception of Lotka) of the usefulness of the mathematical approach, while he found greater receptiveness in the French environment. This explains why most of his mathematically oriented work in this period was published in France (W. R. THOMPSON 1922a, b, c, d, 1923, 1924). It was in the later 1920s that Thompson came across Volterra's work, by which he was strongly attracted.

In 1928 Thompson left France and was appointed Assistant Director of the Imperial Institute of Entomology (a post that he held until 1947), with the special function of superintendent of the Farnham House Laboratory, which had the task of coordinating and directing all the biological control activities in the British Empire. From 1940 to 1947 he was also director of the Imperial Parasite Service. It was during the 1930s that Thompson's attitude towards the usefulness of mathematics in biology underwent a radical change, becoming strongly negative. This change was due partly to the growing gap he observed between mathematical models and actual biological conditions, and partly to the negative impression aroused in him by the work on population genetics by J. B. S. Haldane and R. A. Fisher, which he considered to be unduly abstract and remote from the world of common sense and facts. In 1937 he published a book in which he expounded his critical view of the relationship between mathematics and biology (and in which he also treated general philosophical issues) entitled *Science and Common*

Sense: An Aristotelian Excursion (W. R. THOMPSON 1937). The reference to Aristotelianism (and his interest in the thinking of St. Thomas Aquinas) reveals Thompson's new stance. It is no coincidence that for a time he was also the chairman of the Aquinas Society. However, in 1960, he came back to Volterra's work, mitigating his criticism and proposing a possible integration between mathematical theory and biological facts (W. R. THOMPSON 1960).

In 1949 Thompson returned to Canada where he took up the post of professor at the Institute of the Civil Service of Canada. He died in Ottawa (Ontario, Canada) on 30 January 1972.

Thompson received numerous distinctions, including that of correspondent of the Académie di Bordeaux, honorary member of the Entomological Society of Canada, the Prix Passet of the Société Entomologique de France, the First Harry Scott Smith Award (1967). He chaired the 10th International Congress of Entomology (1956) and was visiting professor at Michigan State University (1959). He directed the *Canadian Entomologist* between 1947 and 1958[1].

* * *

22.1 W. R. Thompson to Volterra, Hyères, 29.7.1927 *Cat. no. 350*

EUROPEAN PARASITE
LABORATORY
BUREAU D'ENTOMOLOGIE
DES ÉTATS UNIS D'AMÉRIQUE

DOMAINE DU MONT FENOUILLET
HYÈRES (VAR) FRANCE

M.le Professeur Vito Volterra *29 juillet 1927*
Rome.

Monsieur: Je me permets de vous envoyer en meme temps que cette lettre quelques petits travaux qui constituent ma tres modeste contribution a la question de l'application des mathematiques (mathematiques elementaires, dans mon cas) a certains problemes de la biologie. Je voudrais aller un peu plus loin dans cette voie et etudier dans un petit livre pour les entomologistes, l'ensemble des problemes dont je me suis occupe dans le travaux ci-inclus.

Je serais donc tres heureux de recevoir si c'est possible une copie de votre magistral travail paru recemment dans le Memoires de l'Academie dei Lincei[2], ainsi que des tirages à part des autres travaux que vous avez publies a ce sujet.

Veuillez agreer, Monsieur, avec mes remerciments anticipes, l'expression de ma tres respectueuse consideration.

W. R. Thompson

---

[1] See also THORPE 1973 and KINGSLAND 1985.

[2] VOLTERRA 1926a.

22.2 Volterra to W. R. Thompson, n.p., n.d. *Cat. no. 423*[3]

Très honoré Monsieur

Je vous suis très reconnaissant des intéressants travaux que vous avez bien voulu m'envoyer et qui constituent une de plus importantes contributions aux recherches mathématiques sur la biologie.

Les tirages à part que je possedais de mon mémoire sont epuisés mais j'écris à l'Académie pour la prier de vous envoyer le fascicule du recueil des Atti où il a paru. D'autre part une seconde édition du même mémoire est sous presse. J'ai déja regardés les derniers [illeg.]. Dès que je pourrai vous en envoyer un exemplaire je m'[illeg.] de le faire.

22.3 W. R. Thompson to Volterra, Hyères, 23.8.1927 *Cat. no. 351*

(Same heading of letter no. 22.1)

M. le Professeur Vito Volterra *23.VIII.1927*
Hotel Ampezzo
Dobbiaco, Italie

Tres honore Monsieur:

Veuillez agreer tous mes remerciements pour votre tres aimable et tres encourageante lettre du 16 aout. Je suis enchante de savoir que les petits travaux que je vous ai envoyes aient pu vous interesser[4]. Je serais tres content de recevoir votre memoire qui me sera de la plus grande utilité pour la redaction du petit livre que je propose de faire cet hiver sur l'application des methodes en question aux problemes de la parasitologie entomologique.

Je vous prie d'accepter, tres honore Monsieur, avec mes remerciements renouvelles, l'expression de ma tres respectueuse consideration.

W. R. Thompson

22.4 W. R. Thompson to Volterra, Hyères, 3.10.1927 *Cat. no. 352*

(Same heading of letter no. 22.1) *3.X.1927*

Honore Monsieur: j'ai bien recu votre grand travail publie par l'Academie dei Lincei. Je n'ai pas eu encore le temps de l'etudier car je suis rentre il y a quelques jours de la montagne pres de Chamonix où j'etais en vacances. Aussitot que j'aurai mis les travaux courants du laboratoire en ordre je vais me mettre a l'etudier car je vois que vos resultats sont de la plus grande importance pour la

[3] Draft contained in the same sheet of letter no. 22.1 (Cat. no. 350), n.d. but clearly written immediately after receiving letter no. 22.1.

[4] Almost certainly some of the works cited in the bibliography of this book.

biologie generale et surtout pour les questions dont nous nous occupons ici. Je me permettrai de vous demander des eclaircissements lorsque je rencontre des sections ou la mathematique est trop difficile pour moi.

Veuillez agreer, Honore Monsieur, avec mes remerciements les plus chaleureux, l'expression de ma consideration la plus distinguee.

W. R. Thompson.

22.5 W. R. Thompson to Volterra, Farnham Royal, 8.4.1930 *Cat. no. 353*

THE IMPERIAL BUREAU OF ENTOMOLOGY
etc.
FARNHAM HOUSE LABORATORY,
FARNHAM ROYAL,
BUCKS.

Superintendent:
W. R. THOMPSON PH.D., D.SC.

8.IV.30

Cher Monsieur:

Vous vous souviendrez peut-être de m'avoir envoyé il y a quelques années votre admirable mémoire sur la mathématique des associations biologiques - sujet que j'avais moi-même abordé mais avec des méthodes bien plus simples. (Je vous envois avec cette lettre quelques autres publications sur ces questions)

Je désire maintenant entreprendre une étude approfondie de vos travaux dans le but d'étendre les résultats aux problèmes de l'entomologie (associations hôtes-parasites). Comme vous le savez, chaque domaine de la biologie présente des problemes particuliers, necessitant des applications speciales d'une théorie générale!

Je vous en serais donc très obligé si vous vouliez bien m'envoyer vos pubblications ultérieures à ce sujet. Je possède déjà le mémoire CXXXI du R. Com. Talassografico[5] et celui de 1927 (Acad. dei Lincei, Ser. 6, Vol. II, fasc. III) mais rien d'autre.

J'ajoute que nous avons ici la station la plus importante dans l'Empire pour l'étude de ces questions de la parasitologie entomologique, aussi bien du p. de vue théorique que du p. de vue pratique.

Veuillez agréer, cher Monsieur, avec mes remerciements anticipés, l'expression de ma consideration très distinguée.

W. R. Thompson

---

[5] VOLTERRA 1927d.

22.6 W. R. Thompson to Volterra, Farnham Royal, 4.5.1930 *Cat. no. 354*

(Same heading of letter no. 22.5)

4.V.1930

Honoré Monsieur:

Je suis en train d'étudier votre travail sur les associations biologiques dans le but d'appliquer vos résultats aux problèmes concrets de la parasitologie entomologique. Mais je viens de rencontrer una difficulté qui me paraît necessiter une modification des formules fondamentales. Voici de quoi il s'agît:

Dans les équations

$$\frac{dN_1}{dt} = (\varepsilon_1 - \gamma_1 N_2)N_1$$

$$\frac{dN_2}{dt} = (\varepsilon_2 - \gamma_2 N_1)N_2$$

on dit que le nombre de rencontres entre individus des deux espèces sera proportionnel au produit $N_1 N_2 dt$ et égal à $aN_1 N_2 dt$ ($a = \text{const.}$); pour une certaine fraction K ($K = \text{const.}$) ces rencontres auront une issue defavorable aux individus de la première espèce et, de ce fait, leur nombre diminuera de $KaN_1 N_2 dt$. Ces mêmes rencontres auront une issue favorables à la deuxième espèce. Vous avez élargi le champ de l'hypothèse fondamentale, mais je ne sais pas si le cas dont je veux vous écrire y entre encore.

Le cas est particulierement net chez les insectes. Il est bien vrai que la probabilité de rencontre augmente avec la multiplication des individus du parasite (prédateur) et de son hôte (proie): mais en même temps augmente ici la probabilité que deux ou plusieurs parasites visitent un même individu de l'hôte. Admettons que chaque rencontre est suivi d'une ponte. Il s'ensuit que plusieurs oeufs du parasites s'accumule chez un même hôte. D'une façon generale on peut dire que si x oeufs sont distribués parmi N hôtes, le nombre d'hôtes parasités est

$$N\left(1 - e^{-\frac{x}{N}}\right).$$

Or, un seulement, ou, en tout cas un maximum de m de ces oeufs peuvent se developper chez un même hôte: les autres meurent. La fraction de rencontres favorables à N_2 n'est donc pas constant: elle diminue au fur et a mésure que les nombres des deux espèces augmentent. D'autre part, le nombre de N_2 exigés pour la destruction d'un individu de N_1 augmente aussi.

Or, le cas est très frequent parmi les insectes et je me demande même si ce n'est pas un cas général. Chez les poissons, par example on peut imaginer l'arrivée simultanée de 2 ou 3 predateurs qui se partagent la proie et dont aucun ne profite donc pas pleinement.

Je vous en serais très obligé si vous vouliez bien me donner votre avis sur ce point qui me paraît d'une importance capitale pour les problèmes qui m'interessent.

Veuillez agréer, Honoré Monsieur, avec mes remerciements anticipés, l'expression de ma respectueuse consideration.

W. R. Thompson

22.7 W. R. Thompson to Volterra, Farnham Royal, 5.2.1931 *Cat. no. 355*

(Same heading of letter no. 22.5)

5, fevrier, 1931

Honoré Monsieur:

Je viens de recevoir votre aimable lettre du 30 Janvier. Je vous en remercie vivement ainsi que pour la copie de votre admirable volume sur la théorie mathématique de la lutte pour la vie[6]. Je compte, des que j'aurai un moment de loisir, de me mettre à l'étudier. J'ai bien reçu également la reponse que vous avez eu l'amabilité de m'envoyer au mois de juin et des explications detaillées concernant divers points. Malhereusement, le mathematicien qui avait bien voulu accepter de collaborer avec nous dans l'étude de ces questions (Monsieur Soper, du Collège de Medicine Tropicale, mathématicien distingué, et aussi aimable que distingué) est mort assez subitement au mois d'août; et depuis ce temps je n'ai trouvé personne pour m'aider. J'espère pourtant faire un arrangement qui nous permettra d'aborder ces questions en détail dans un mois ou deux et je vous demanderais alors, de bien vouloir nous fournir quelques indications supplementaires.

Je vous envois avec plaisir le travail dont vous m'ecrivez ainsi qu'un autre, donnant un résumé général de la question. Un troisième travail, fait en collaboration avec M. Soper, est maintenant sous presse.

Veuillez agréer, Honoré Monsieur, avec mes meilleurs remerciements pour le volume que vous m'avez envoyé ainsi que pour les renseignements que vous avez eu la grande aimabilité de me fournir, l'expression de ma considération très distinguée.

W. R. Thompson

22.8 W. R. Thompson to Volterra, Farnham Royal, 23.3.1931 *Cat. no. 356*

(Same heading of letter no. 22.5)

23rd March, 1931

My dear Professor Volterra,

I have just received your letter of 19th March. I am afraid that through a regrettable oversight the papers promised were not sent to you, but you will, I hope, receive them at the same time as this letter.

It is greatly to be hoped that your paper may be translated into English and published here, for I am sure it will be of great interest to many biologists here.

Yours very sincerely,

W.R.Thompson,
Superintendent

---

[6] VOLTERRA 1931a.

23 Catalogue of Letters

The letters are ordered by correspondent and by date. Undated letters have been placed at the beginning. Reference is also made to the corresponding catalogue number in the Volterra Archive of the Accademia Nazionale dei Lincei (*ANL*). This is followed by the number indicating the order of transcription of each exchange of letters in the book (*H* = *"Here"* = *"in this book"*). Where no explicit reference is made to place or date in the letter it is indicated respectively by *"n.p."* = *no place; "n.d."* = *no date*; nevertheless, in several cases, this information may be reconstructed fairly accurately from the contents of the letters. The transcription of the epistolary exchanges thus contains greater information concerning the letters and often a different ordering of them. Vis-à-vis the original cataloguing system used in the Volterra Archive, in the present catalogue we have merely proceeded to correct material errors referring to the indication of places and dates. It must be borne in mind however that, after the original cataloguing work of the Volterra Archive a further computer-based catalogue was made by L. Dell'Aglio and F. La Teana, under the guidance of G. Paoloni and with the technical consultancy of G. Israel.

1 Baldi Edgardo to D'Ancona Umberto, 1937 Oct 7, Paris. ANL: D'Ancona-Volterra no. 79. H: D'Ancona-Volterra no. 83.

2 Brelot Marcel to D'Ancona Umberto, 1930 Oct 4, Paris. ANL: Brelot-Volterra no. 50. H: Brelot-Volterra no. 54.

3 Brelot Marcel, to Volterra Vito, n.d. (wedesday evening), Roma. ANL: Brelot-Volterra no. 21. H: Brelot-Volterra no. 23.

4 Brelot Marcel to Volterra Vito, n.d. (thursday evening), Roma. ANL: Brelot-Volterra no. 22. H: Brelot-Volterra no. 24.

5 Brelot Marcel to Volterra Vito, n.d. (saturday), Roma. ANL: Brelot-Volterra no. 23. H: Brelot-Volterra no. 22.

6 Brelot Marcel to Volterra Vito, n.d. (friday), Paris. ANL: Brelot-Volterra no. 54. H: Brelot-Volterra no. 21.

7 Brelot Marcel to Volterra Vito, n.d. (saturday morning), Paris. ANL: Brelot-Volterra no. 55. H: Brelot-Volterra no. 69.

8 Brelot Marcel to Volterra Vito, n.d. (monday), Paris. ANL: Brelot-Volterra no. 56. H: Brelot-Volterra no. 70.

9 Brelot Marcel to Volterra Vito, n.d. (friday morning), Paris. ANL: Brelot-Volterra no. 57. H: Brelot-Volterra no. 13.

10 Brelot Marcel to Volterra Vito, n.d. (sunday morning), Paris. ANL: Brelot-Volterra no. 58. H: Brelot-Volterra no. 71.

11 Brelot Marcel to Volterra Vito, n.d. (sunday evening), Roma. ANL: Brelot-Volterra no. 59. H: Brelot-Volterra no. 25.

12 Brelot Marcel to Volterra Vito, n.d. (monday morning), Roma. ANL: Brelot-Volterra no. 60. H: Brelot-Volterra no. 27.

13 Brelot Marcel to Volterra Vito, 1929 May 13, Paris. ANL: Brelot-Volterra no. 1. H: Brelot-Volterra no. 1.

14 Brelot Marcel to Volterra Vito, 1929 Jun 5, Paris. ANL: Brelot-Volterra no. 2. H: Brelot-Volterra no. 2.
15 Brelot Marcel to Volterra Vito, 1929 Jul 22, Paris. ANL: Brelot-Volterra no. 6. H: Brelot-Volterra no. 6.
16 Brelot Marcel to Volterra Vito, 1929 Aug 12, Boisseaux. ANL: Brelot-Volterra no. 7. H: Brelot-Volterra no. 7.
17 Brelot Marcel to Volterra Vito, 1929 Aug 14, Boisseaux. ANL: Brelot-Volterra no. 8. H: Brelot-Volterra no. 8.
18 Brelot Marcel to Volterra Vito, 1929 Sep 9, Boisseaux. ANL: Brelot-Volterra no. 10. H: Brelot-Volterra no. 10.
19 Brelot Marcel to Volterra Vito, 1929 Sep 18, Boisseaux. ANL: Brelot-Volterra no. 11. H: Brelot-Volterra no. 11.
20 Brelot Marcel to Volterra Vito, 1929 Sep 26, Paris. ANL: Brelot-Volterra no. 12. H: Brelot-Volterra no. 12.
21 Brelot Marcel to Volterra Vito, 1929 Oct 19, Boisseaux. ANL: Brelot-Volterra no. 13. H: Brelot-Volterra no. 14.
22 Brelot Marcel to Volterra Vito, 1929 Oct 22, Boisseaux. ANL: Brelot-Volterra no. 14. H: Brelot-Volterra no. 15.
23 Brelot Marcel to Volterra Vito, 1929 Oct 25, Boisseaux. ANL: Brelot-Volterra no. 15. H: Brelot-Volterra no. 16.
24 Brelot Marcel to Volterra Vito, 1929 Oct 31, Boisseaux. ANL: Brelot-Volterra no. 16. H: Brelot-Volterra no. 17.
25 Brelot Marcel to Volterra Vito, 1929 Nov 6, Paris. ANL: Brelot-Volterra no. 17. H: Brelot-Volterra no. 18.
26 Brelot Marcel to Volterra Vito, 1930 Feb 23, Roma. ANL: Brelot-Volterra no. 18. H: Brelot-Volterra no. 19.
27 Brelot Marcel to Volterra Vito, 1930 Mar 16, Boisseaux. ANL: Brelot-Volterra no. 19. H: Brelot-Volterra no. 20.
28 Brelot Marcel to Volterra Vito, 1930 May 7, Roma. ANL: Brelot-Volterra no. 20. H: Brelot-Volterra no. 26.
29 Brelot Marcel to Volterra Vito, 1930 Jun 15, Roma. ANL: Brelot-Volterra no. 24. H: Brelot-Volterra no. 28.
30 Brelot Marcel to Volterra Vito, 1930 Jun 21, Roma. ANL: Brelot-Volterra no. 25. H: Brelot-Volterra no. 29.
31 Brelot Marcel to Volterra Vito, 1930 Jul 5, Roma. ANL: Brelot-Volterra no. 26. H: Brelot-Volterra no. 30.
32 Brelot Marcel to Volterra Vito, 1930 Jul 6, Roma. ANL: Brelot-Volterra no. 27. H: Brelot-Volterra no. 31.
33 Brelot Marcel to Volterra Vito, 1930 Jul 7, Roma. ANL: Brelot-Volterra no. 28. H: Brelot-Volterra no. 32.
34 Brelot Marcel to Volterra Vito, 1930 Jul 29, Capri. ANL: Brelot-Volterra no. 30. H: Brelot-Volterra no. 33.
35 Brelot Marcel to Volterra Vito, 1930 Jul 29, Capri. ANL: Brelot-Volterra no. 29. H: Brelot-Volterra no. 34.
36 Brelot Marcel to Volterra Vito, 1930 Aug 2, Roma. ANL: Brelot-Volterra no. 31. H: Brelot-Volterra no. 35.
37 Brelot Marcel to Volterra Vito, 1930 Aug 11, Boisseaux. ANL: Brelot-Volterra no. 33. H: Brelot-Volterra no. 37.
38 Brelot Marcel to Volterra Vito, 1930 Aug 12, Boisseaux. ANL: Brelot-Volterra no. 34. H: Brelot-Volterra no. 38.
39 Brelot Marcel to Volterra Vito, 1930 Aug 15, Boisseaux. ANL: Brelot-Volterra no. 35. H: Brelot-Volterra no. 39.
40 Brelot Marcel to Volterra Vito, 1930 Aug 20, Boisseaux. ANL: Brelot-Volterra no. 36. H: Brelot-Volterra no. 40.
41 Brelot Marcel to Volterra Vito, 1930 Aug 21, Boisseaux. ANL: Brelot-Volterra no. 37. H: Brelot-Volterra no. 41.
42 Brelot Marcel to Volterra Vito, 1930 Aug 22, Boisseaux. ANL: Brelot-Volterra no. 38. H: Brelot-Volterra no. 42.
43 Brelot Marcel to Volterra Vito, 1930 Aug 24, Paris. ANL: Brelot-Volterra no. 39. H: Brelot-Volterra no. 43.

44 Brelot Marcel to Volterra Vito, 1930 Sep 2, Bruxelles. ANL: Brelot-Volterra no. 41. H: Brelot-Volterra no. 45.
45 Brelot Marcel to Volterra Vito, 1930 Sep 9, Paris. ANL: Brelot-Volterra no. 42. H: Brelot-Volterra no. 46.
46 Brelot Marcel to Volterra Vito, 1930 Sep 13, Boisseaux. ANL: Brelot-Volterra no. 43. H: Brelot-Volterra no. 47.
47 Brelot Marcel to Volterra Vito, 1930 Sep 14, Boisseaux. ANL: Brelot-Volterra no. 44. H: Brelot-Volterra no. 48.
48 Brelot Marcel to Volterra Vito, 1930 Sep 15, Boisseaux. ANL: Brelot-Volterra no. 45. H: Brelot-Volterra no. 49.
49 Brelot Marcel to Volterra Vito, 1930 Sep 17, Paris. ANL: Brelot-Volterra no. 46. H: Brelot-Volterra no. 50.
50 Brelot Marcel to Volterra Vito, 1930 Sep 20, Boisseaux. ANL: Brelot-Volterra no. 47. H: Brelot-Volterra no. 51.
51 Brelot Marcel to Volterra Vito, 1930 Sep 24, Boisseaux. ANL: Brelot-Volterra no. 48. H: Brelot-Volterra no. 52.
52 Brelot Marcel to Volterra Vito, 1930 Sep 27, Boisseaux. ANL: Brelot-Volterra no. 49. H: Brelot-Volterra no. 53.
53 Brelot Marcel to Volterra Vito, 1930 Oct 16, Boisseaux. ANL: Brelot-Volterra no. 51. H: Brelot-Volterra no. 55.
54 Brelot Marcel to Volterra Vito, 1930 Oct 16, Boisseaux. ANL: Brelot-Volterra no. 52. H: Brelot-Volterra no. 56.
55 Brelot Marcel to Volterra Vito, 1930 Nov 7, Berlin. ANL: Brelot-Volterra no. 53. H: Brelot-Volterra no. 57.
56 Brelot Marcel to Volterra Vito, 1931 Aug 10, Boisseaux. ANL: Brelot-Volterra no. 61. H: Brelot-Volterra no. 58.
57 Brelot Marcel to Volterra Vito, 1931 Sep 24, Cepoy. ANL: Brelot-Volterra no. 62. H: Brelot-Volterra no. 59.
58 Brelot Marcel to Volterra Vito, 1931 Oct 24, Berlin. ANL: Brelot-Volterra no. 63. H: Brelot-Volterra no. 60.
59 Brelot Marcel to Volterra Vito, 1931 Dec 26, Berlin. ANL: Brelot-Volterra no. 64. H: Brelot-Volterra no. 61.
60 Brelot Marcel to Volterra Vito, 1932 Jan 10 Berlin. ANL: Brelot-Volterra no. 65. H: Brelot-Volterra no. 62.
61 Brelot Marcel to Volterra Vito, 1932 Feb 26, Berlin. ANL: Brelot-Volterra no. 66. H: Brelot-Volterra no. 63.
62 Brelot Marcel to Volterra Vito, 1932 Jun 3, Berlin. ANL: Brelot-Volterra no. 67. H: Brelot-Volterra no. 64.
63 Brelot Marcel to Volterra Vito, 1932 Jul 19, Berlin. ANL: Brelot-Volterra no. 68. H: Brelot-Volterra no. 65.
64 Brelot Marcel to Volterra Vito, 1933 Jan 3, London. ANL: Brelot-Volterra no. 69. H: Brelot-Volterra no. 66.
65 Brelot Marcel to Volterra Vito, 1938 Nov 26, Bordeaux. ANL: Brelot-Volterra no. 70. H: Brelot-Volterra no. 67.
66 Brelot Marcel to Volterra Virginia, 1947 Mar 21, Grenoble. ANL: Brelot-Volterra no. 71. H: Brelot-Volterra no. 68.
67 Chapman Royal N. to Volterra Vito, 1928 Aug 9, St. Paul, Minnesota. ANL: Chapman-Volterra no. 1. H: Chapman-Volterra no. 1.
68 Chapman Royal N. to Volterra Vito, 1928 Sep 26, St. Paul, Minnesota. ANL: Chapman-Volterra no. 3. H: Chapman-Volterra no. 3.
69 Chapman Royal N. to Volterra Vito, 1929 Jan 28, St. Paul, Minnesota. ANL: Chapman-Volterra no. 4. H: Chapman-Volterra no. 4.
70 Chapman Royal N. to Volterra Vito, 1929 Aug 5, St. Paul, Minnesota. ANL: Chapman-Volterra no. 5. H: Chapman-Volterra no. 5.
71 Chapman Royal N. to Volterra Vito, 1930 Jan 29, St. Paul, Minnesota. ANL: Chapman-Volterra no. 7. H: Chapman-Volterra no. 7.
72 Chapman Royal N. to Volterra Vito, 1931 Apr 4, Honolulu, Hawai. ANL: Chapman-Volterra no. 11. H: Chapman-Volterra no. 9.
73 D'Ancona Umberto to Volterra Vito, n.d., n.p. ANL: D'Ancona-Volterra no. 55. H: D'Ancona-Volterra no. 49.

74 D'Ancona Umberto to Volterra Vito, n.d., n.p. ANL: D'Ancona-Volterra no. 62. H: D'Ancona-Volterra no. 64.
75 D'Ancona Umberto to Volterra Vito, n.d., n.p. ANL: D'Ancona-Volterra no. 78. H: D'Ancona-Volterra no. 81.
76 D'Ancona Umberto to Volterra Vito, n.d., n.p. ANL: D'Ancona-Volterra no. 81. H: D'Ancona-Volterra no. 83.
77 D'Ancona Umberto to Volterra Vito, n.d., n.p. ANL: D'Ancona-Volterra no. 86. H: D'Ancona-Volterra no. 97.
78 D'Ancona Umberto to Volterra Vito, n.d., n.p. ANL: D'Ancona-Volterra no. 89. H: D'Ancona-Volterra no. 91.
79 D'Ancona Umberto to Volterra Vito, n.d., Padova. ANL: D'Ancona-Volterra no. 90. H: D'Ancona-Volterra no. 92.
80 D'Ancona Umberto to Volterra Vito, n.d., n.p. ANL: D'Ancona-Volterra no. 95. H: D'Ancona-Volterra no. 96.
81 D'Ancona Umberto to Volterra Vito, 1926 Jun 21, Roma. ANL: D'Ancona-Volterra no. 1. H: D'Ancona-Volterra no. 1.
82 D'Ancona Umberto to Volterra Vito, 1926 Oct 20, Roma. ANL: D'Ancona-Volterra no. 2. H: D'Ancona-Volterra no. 2.
83 D'Ancona Umberto to Volterra Vito, 1929 Jan 28, Roma. ANL: D'Ancona-Volterra no. 3. H: D'Ancona-Volterra no. 3.
84 D'Ancona Umberto to Volterra Vito, 1929 Feb 10, Roma. ANL: D'Ancona-Volterra no. 4. H: D'Ancona-Volterra no. 4.
85 D'Ancona Umberto to Volterra Vito, 1929 Nov 25, Siena. ANL: D'Ancona-Volterra no. 5. H: D'Ancona-Volterra no. 5.
86 D'Ancona Umberto to Volterra Vito, 1930 Mar 10, Siena. ANL: D'Ancona-Volterra no. 6. H: D'Ancona-Volterra no. 6.
87 D'Ancona Umberto to Volterra Vito, 1930 Mar 15, Siena. ANL: D'Ancona-Volterra no. 7. H: D'Ancona-Volterra no. 7.
88 D'Ancona Umberto to Volterra Vito, 1930 Mar 28, Siena. ANL: D'Ancona-Volterra no. 8. H: D'Ancona-Volterra no. 8.
89 D'Ancona Umberto to Volterra Vito, 1930 Jun 15, Siena. ANL: D'Ancona-Volterra no. 9. H: D'Ancona-Volterra no. 9.
90 D'Ancona Umberto to Volterra Vito, 1930 Jul 26, Roma. ANL: D'Ancona-Volterra no. 10. H: D'Ancona-Volterra no. 10.
91 D'Ancona Umberto to Volterra Vito, 1930 Jul 28, Roma. ANL: D'Ancona-Volterra no. 11. H: D'Ancona-Volterra no. 11.
92 D'Ancona Umberto to Volterra Vito, 1930 Aug 16, Fiume. ANL: D'Ancona-Volterra no. 12. H: D'Ancona-Volterra no. 12.
93 D'Ancona Umberto to Volterra Vito, 1930 Oct 9, Napoli. ANL: D'Ancona-Volterra no. 13. H: D'Ancona-Volterra no. 13.
94 D'Ancona Umberto to Volterra Vito, 1930 Nov 18, Siena. ANL: D'Ancona-Volterra no. 14. H: D'Ancona-Volterra no. 14.
95 D'Ancona Umberto to Volterra Vito, 1930 Dec 7, Siena. ANL: D'Ancona-Volterra no. 15. H: D'Ancona-Volterra no. 15.
96 D'Ancona Umberto to Volterra Vito, 1931 Jan 2, Siena. ANL: D'Ancona-Volterra no. 16. H: D'Ancona-Volterra no. 16.
97 D'Ancona Umberto to Volterra Vito, 1931 Feb 11, Siena. ANL: D'Ancona-Volterra no. 17. H: D'Ancona-Volterra no. 17.
98 D'Ancona Umberto to Volterra Vito, 1931 Feb 18, Siena. ANL: D'Ancona-Volterra no. 18. H: D'Ancona-Volterra no. 18.
99 D'Ancona Umberto to Volterra Vito, 1931 Apr 4, Napoli. ANL: D'Ancona-Volterra no. 19. H: D'Ancona-Volterra no. 19.
100 D'Ancona Umberto to Volterra Vito, 1931 Apr 12, Napoli. ANL: D'Ancona-Volterra no. 20. H: D'Ancona-Volterra no. 20.
101 D'Ancona Umberto to Volterra Vito, 1931 Apr 18, Roma. ANL: D'Ancona-Volterra no. 21. H: D'Ancona-Volterra no. 21.
102 D'Ancona Umberto to Volterra Vito, 1931 Jun 1, Siena. ANL: D'Ancona-Volterra no. 22. H: D'Ancona-Volterra no. 22.
103 D'Ancona Umberto to Volterra Vito, 1931 Jul 24, Siena. ANL: D'Ancona-Volterra no. 23. H: D'Ancona-Volterra no. 23.

104 D'Ancona Umberto to Volterra Vito, 1933 Jul 26, Siena. ANL: D'Ancona-Volterra no. 24. H: D'Ancona-Volterra no. 24.
105 D'Ancona Umberto to Volterra Vito, 1933 Oct 28, Siena. ANL: D'Ancona-Volterra no. 25. H: D'Ancona-Volterra no. 25.
106 D'Ancona Umberto to Volterra Vito, 1933 Nov 3, Siena. ANL: D'Ancona-Volterra no. 26. H: D'Ancona-Volterra no. 26.
107 D'Ancona Umberto to Volterra Vito, 1933 Nov 30, Siena. ANL: D'Ancona-Volterra no. 26bis. H: D'Ancona-Volterra no. 27.
108 D'Ancona Umberto to Volterra Vito, 1934 Feb 7, Siena. ANL: D'Ancona-Volterra no. 27. H: D'Ancona-Volterra no. 28.
109 D'Ancona Umberto to Volterra Vito, 1934 Jun 8, Siena. ANL: D'Ancona-Volterra no. 28. H: D'Ancona-Volterra no. 29.
110 D'Ancona Umberto to Volterra Vito, 1934 Jun 29, Siena. ANL: D'Ancona-Volterra no. 29. H: D'Ancona-Volterra no. 30.
111 D'Ancona Umberto to Volterra Vito, 1934 Jul 3, Siena. ANL: D'Ancona-Volterra no. 30. H: D'Ancona-Volterra no. 31.
112 D'Ancona Umberto to Volterra Vito, 1934 Jul 9, Siena. ANL: D'Ancona-Volterra no. 31. H: D'Ancona-Volterra no. 32.
113 D'Ancona Umberto to Volterra Vito, 1934 Aug 19, Fiume. ANL: D'Ancona-Volterra no. 32. H: D'Ancona-Volterra no. 33.
114 D'Ancona Umberto to Volterra Vito, 1934 Sep 4, Fiume. ANL: D'Ancona-Volterra no. 33. H: D'Ancona-Volterra no. 34.
115 D'Ancona Umberto to Volterra Vito, 1934 Dec 2, Siena. ANL: D'Ancona-Volterra no. 34. H: D'Ancona-Volterra no. 35.
116 D'Ancona Umberto to Volterra Vito, 1935 Jan 20, Siena. ANL: D'Ancona-Volterra no. 35. H: D'Ancona-Volterra no. 36.
117 D'Ancona Umberto to Volterra Vito, 1935 Jan 23, Siena. ANL: D'Ancona-Volterra no. 36. H: D'Ancona-Volterra no. 37.
118 D'Ancona Umberto to Volterra Vito, 1935 Jan 31, Siena. ANL: D'Ancona-Volterra no. 37. H: D'Ancona-Volterra no. 38.
119 D'Ancona Umberto to Volterra Vito, 1935 Feb 6, Siena. ANL: D'Ancona-Volterra no. 83. H: D'Ancona-Volterra no. 39.
120 D'Ancona Umberto to Volterra Vito, 1935 Feb 17, Siena. ANL: D'Ancona-Volterra no. 38. H: D'Ancona-Volterra no. 40.
121 D'Ancona Umberto to Volterra Vito, 1935 Feb 21, Siena. ANL: D'Ancona-Volterra no. 39. H: D'Ancona-Volterra no. 41.
122 D'Ancona Umberto to Volterra Vito, 1935 Feb 23, Siena. ANL: D'Ancona-Volterra no. 40. H: D'Ancona-Volterra no. 42.
123 D'Ancona Umberto to Volterra Vito, 1935 Feb 24; Siena. ANL: D'Ancona-Volterra no. 41. H: D'Ancona-Volterra no. 43.
124 D'Ancona Umberto to Volterra vito, 1935 Feb 25, Siena. ANL: D'Ancona-Volterra no. 42. H: D'Ancona-Volterra no. 44.
125 D'Ancona Umberto to Volterra Vito, 1935 Feb 28, Siena. ANL: D'Ancona-Volterra no. 43. H: D'Ancona-Volterra no. 45.
126 D'Ancona Umberto to Volterra Vito, 1935 Mar 9, Siena. ANL: D'Ancona-Volterra no. 44. H: D'Ancona-Volterra no. 46.
127 D'Ancona Umberto to Volterra Vito, 1935 Mar 13, Siena. ANL: D'Ancona-Volterra no. 45. H: D'Ancona-Volterra no. 47.
128 D'Ancona Umberto to Volterra Vito, 1935 Mar 15, Siena. ANL: D'Ancona-Volterra no. 46. H: D'Ancona-Volterra no. 48.
129 D'Ancona Umberto to Volterra Vito, 1935 Apr 13, Siena. ANL: D'Ancona-Volterra no. 47. H: D'Ancona-Volterra no. 50.
130 D'Ancona Umberto to Volterra Vito, 1935 Apr 17, Siena. ANL: D'Ancona-Volterra no. 48. H: D'Ancona-Volterra no. 51.
131 D'Ancona Umberto to Volterra Vito, 1935 Apr 18, Venezia. ANL: D'Ancona-Volterra no. 49. H: D'Ancona-Volterra no. 52.
132 D'Ancona Umberto to Volterra Vito, 1935 Apr 22, Venezia. ANL: D'Ancona-Volterra no. 50. H: D'Ancona-Volterra no. 53.
133 D'Ancona Umberto to Volterra Vito, 1935 Apr 25, Venezia. ANL: D'Ancona-Volterra no. 51. H: D'Ancona-Volterra no. 54.

134 D'Ancona Umberto to Volterra Vito, 1935 Jun 29, Siena. ANL: D'Ancona-Volterra no. 52. H: D'Ancona-Volterra no. 55.
135 D'Ancona Umberto to Volterra Vito, 1935 Nov 4, Siena. ANL: D'Ancona-Volterra no. 52bis. H: D'Ancona-Volterra no. 56.
136 D'Ancona Umberto to Volterra Vito, 1936, n.p. ANL: D'Ancona-Volterra no. 61. H: D'Ancona-Volterra no. 63.
137 D'Ancona Umberto to Volterra Vito, 1936 Apr 17, Siena. ANL: D'Ancona-Volterra no. 53. H: D'Ancona-Volterra no. 57.
138 D'Ancona Umberto to Volterra Vito, 1936 May 1, n.p. ANL: D'Ancona-Volterra no. 56. H: D'Ancona-Volterra no. 58.
139 D'Ancona Umberto to Volterra Vito, 1936 May 1, n.p. ANL: D'Ancona-Volterra no. 54. H: D'Ancona-Volterra no. 58.
140 D'Ancona Umberto to Volterra Vito, 1936 May 12, Siena. ANL: D'Ancona-Volterra no. 57. H: D'Ancona-Volterra no. 59.
141 D'Ancona Umberto to Volterra Vito, 1936 May 20, Siena. ANL: D'Ancona-Volterra no. 58. H: D'Ancona-Volterra no. 60.
142 D'Ancona Umberto to Volterra Vito, 1936 Jul 14, Siena. ANL: D'Ancona-Volterra no. 59. H: D'Ancona-Volterra no. 61.
143 D'Ancona Umberto to Volterra Vito, 1936 Jul 23, Siena. ANL: D'Ancona-Volterra no. 60. H: D'Ancona-Volterra no. 62.
144 D'Ancona Umberto to Volterra Vito, 1936 Sep 18, Siena. ANL: D'Ancona-Volterra no. 63. H: D'Ancona-Volterra no. 65.
145 D'Ancona Umberto to Volterra Vito, 1936 Sep 24, Siena. ANL: D'Ancona-Volterra no. 64. H: D'Ancona-Volterra no. 66.
146 D'Ancona Umberto to Volterra Vito, 1936 Oct 23, Siena. ANL: D'Ancona-Volterra no. 65. H: D'Ancona-Volterra no. 67.
147 D'Ancona Umberto to Volterra Vito, 1936 Nov 10, Siena. ANL: D'Ancona-Volterra no. 66. H: D'Ancona-Volterra no. 68.
148 D'Ancona Umberto to Volterra Vito, 1937, n.p. ANL: D'Ancona-Volterra no. 67bis. H: D'Ancona-Volterra no. 70.
149 D'Ancona Umberto to Volterra Vito, 1937, n.p. ANL: D'Ancona-Volterra no. 68. H: D'Ancona-Volterra no. 71.
150 D'Ancona Umberto to Volterra Vito, 1937 Mar 10, Pisa. ANL: D'Ancona-Volterra no. 67. H: D'Ancona-Volterra no. 69.
151 D'Ancona Umberto to Volterra Vito, 1937 Mar 14, Pisa. ANL: D'Ancona-Volterra no. 69. H: D'Ancona-Volterra no. 72.
152 D'Ancona Umberto to Volterra Vito, 1937 Mar 18, Pisa. ANL: D'Ancona-Volterra no. 70. H: D'Ancona-Volterra no. 73.
153 D'Ancona Umberto to Volterra Vito, 1937 Apr 21, Pisa. ANL: D'Ancona-Volterra no. 71. H: D'Ancona-Volterra no. 74.
154 D'Ancona Umberto to Volterra Vito, 1937 Apr 25, Pisa. ANL: D'Ancona-Volterra no. 72. H: D'Ancona-Volterra no. 75.
155 D'Ancona Umberto to Volterra Vito, 1937 May 9, Pisa. ANL: D'Ancona-Volterra no. 77. H: D'Ancona-Volterra no. 76.
156 D'Ancona Umberto to Volterra Vito, 1937 May 15, Pisa. ANL: D'Ancona-Volterra no. 73. H: D'Ancona-Volterra no. 77.
157 D'Ancona Umberto to Volterra Vito, 1937 May 20, Pisa. ANL: D'Ancona-Volterra no. 74. H: D'Ancona-Volterra no. 78.
158 D'Ancona Umberto to Volterra Vito, 1937 Jun 2, Pisa. ANL: D'Ancona-Volterra no. 75. H: D'Ancona-Volterra no. 79.
159 D'Ancona Umberto to Volterra Vito, 1937 Sep 22, Siena. ANL: D'Ancona-Volterra no. 76. H: D'Ancona-Volterra no. 80.
160 D'Ancona Umberto to Volterra Vito, 1937 Oct 24, Pisa. ANL: D'Ancona-Volterra no. 82. H: D'Ancona-Volterra no. 84.
161 D'Ancona Umberto to Volterra Vito, 1937 Nov 11, Pisa. ANL: D'Ancona-Volterra no. 80. H: D'Ancona-Volterra no. 86.
162 D'Ancona Umberto to Volterra Vito, 1938 Aug 22, Abbazia. ANL: D'Ancona-Volterra no. 84. H: D'Ancona-Volterra no. 87.
163 D'Ancona Umberto to Volterra Vito, 1938 Oct, n.p. ANL: D'Ancona-Volterra no. 85. H: D'Ancona-Volterra no. 88.

164 D'Ancona Umberto to Volterra Vito, 1939 Jan 4, Padova. ANL: D'Ancona-Volterra no. 87. H: D'Ancona-Volterra no. 89.
165 D'Ancona Umberto to Volterra Vito, 1939 Jan 28, n.p. ANL: D'Ancona-Volterra no. 88. H: D'Ancona-Volterra no. 90.
166 D'Ancona Umberto to Volterra Vito, 1939 May 29, Padova. ANL: D'Ancona-Volterra no. 91. H: D'Ancona-Volterra no. 93.
167 D'Ancona Umberto to Volterra Vito, 1939 Jul 10, Padova. ANL: D'Ancona-Volterra no. 92. H: D'Ancona-Volterra no. 94.
168 Elton Charles S. to Volterra Vito, 1931 Jan 6, Oxford. ANL: Elton-Volterra no. 1. H Elton-Volterra no. 1.
169 Elton Charles S. to Volterra Vito, 1931 Sep 24, Oxford. ANL: Elton-Volterra no. 3. H Elton-Volterra no. 3.
170 Elton Charles S. to Volterra Vito, 1931 Dec 30, Oxford. ANL: Elton-Volterra no. 4. H Elton-Volterra no. 4.
171 Elton Charles S. to Volterra Vito, 1932 Mar 9, Oxford. ANL: Elton-Volterra no. 5. H Elton-Volterra no. 5.
172 Elton Charles S. to Volterra Vito, 1936 Sep 28, Oxford. ANL: Elton-Volterra no. 6. H Elton-Volterra no. 7.
173 Elton Charles S. to Volterra Vito, 1936 Oct 2, Oxford. ANL: Elton-Volterra no. 7. H Elton-Volterra no. 8.
174 Elton Charles S. to Volterra Vito, 1936 Oct 12, Oxford. ANL: Elton-Volterra no. 8. H Elton-Volterra no. 9.
175 Elton Charles S. to Volterra Vito, 1937 Feb 11, Oxford. ANL: Elton-Volterra no. 9. H Elton-Volterra no. 10.
176 Friederichs Karl to Volterra Vito, 1936 Oct 10, Rostock. ANL: Friederichs-Volterra no. 1. H: Friederichs-Volterra no. 1.
177 Gause Georgii F. to Volterra Vito, 1932 Nov 12, Moscow. ANL: Gause-Volterra no. 1. H: Gause-Volterra no. 1.
178 Gause Georgii F. to Volterra Vito, 1933 Oct 12, Moscow. ANL: Gause-Volterra no. 2. H: Gause-Volterra no. 2.
179 Gause Georgii F. to Volterra Vito, 1934 Nov 25, Moscow. ANL: Gause-Volterra no. 3. H: Gause-Volterra no. 3.
180 Gause Georgii F. to Volterra Vito, 1935 Aug 1, Moscow. ANL: Gause-Volterra no. 4. H: Gause-Volterra no. 4.
181 Gause Georgii F. to Volterra Vito, 1935 Dec 23, Moscow. ANL: Gause-Volterra no. 5. H: Gause-Volterra no. 5.
182 Gause Georgii F. to Volterra Vito, 1936 Oct 2, Moscow. ANL: Gause-Volterra no. 6. H: Gause-Volterra no. 6.
183 Graham Samuel A. to Volterra Vito, 1928 Jan 10, Ann Arbor. ANL: Graham-Volterra no. 1. H: Graham-Volterra no. 1.
184 Graham Samuel A. to Volterra Vito, 1928 Nov 1, Ann Arbor. ANL: Graham-Volterra no. 2. H: Graham-Volterra no. 2.
185 Graham Samuel A. to Volterra Vito, 1929 Jan 30, Ann Arbor. ANL: Graham-Volterra no. 3. H: Graham-Volterra no. 3.
186 Kermack William O. to Volterra Vito, 1938 Mar 14, Edinburgh. ANL: Kermack-Volterra no. 1. H: Kermack-Volterra no. 1.
187 Kostitzin Vladimir A. to Volterra Vito, 1933 Mar 16, Paris. ANL: Kostitzin-Volterra no. 1. H: Kostitzin-Volterra no. 1.
188 Kostitzin Vladimir A. to Volterra Vito, 1933 Nov 6, Paris. ANL: Kostitzin-Volterra no. 2. H: Kostitzin-Volterra no. 2.
189 Kostitzin Vladimir A. to Volterra Vito, 1933 Dec 14, Paris. ANL: Kostitzin-Volterra no. 3. H: Kostitzin-Volterra no. 3.
190 Kostitzin Vladimir A. to Volterra Vito, 1935 Jul 11, Paris. ANL: Kostitzin-Volterra no. 4. H: Kostitzin-Volterra no. 4.
191 Kostitzin Vladimir A. to Volterra Vito, 1935 Aug 13, Richelieu. ANL: Kostitzin-Volterra no. 5. H: Kostitzin-Volterra no. 5.
192 Kostitzin Vladimir A. to Volterra Vito, 1935 Aug 24, Richelieu. ANL: Kostitzin-Volterra no. 6. H: Kostitzin-Volterra no. 6.
193 Kostitzin Vladimir A. to Volterra Vito, 1935 Sep 19, Richelieu. ANL: Kostitzin-Volterra no. 7. H: Kostitzin-Volterra no. 7.

194 Kostitzin Vladimir A. to Volterra Vito, 1935 Nov 6, Paris. ANL: Kostitzin-Volterra no. 8. H: Kostitzin-Volterra no. 8.
195 Kostitzin Vladimir A. to Volterra Vito, 1935 Dec 31, Paris. ANL: Kostitzin-Volterra no. 10. H: Kostitzin-Volterra no. 10.
196 Kostitzin Vladimir A. to Volterra Vito, 1936 Jul 8, Paris. ANL: Kostitzin-Volterra no. 11. H: Kostitzin-Volterra no. 11.
197 Kostitzin Vladimir A. to Volterra Vito, 1936 Jul 15, Paris. ANL: Kostitzin-Volterra no. 12. H: Kostitzin-Volterra no. 12.
198 Kostitzin Vladimir A. to Volterra Vito, 1936 Oct 11, Paris. ANL: Kostitzin-Volterra no. 13. H: Kostitzin-Volterra no. 13.
199 Kostitzin Vladimir A. to Volterra Vito, 1936 Oct 14, Paris. ANL: Kostitzin-Volterra no. 14. H: Kostitzin-Volterra no. 14.
200 Kostitzin Vladimir A. to Volterra Vito, 1936 Oct 16, Paris. ANL: Kostitzin-Volterra no. 15. H: Kostitzin-Volterra no. 15.
201 Kostitzin Vladimir A. to Volterra Vito, 1936 Nov 4, Paris. ANL: Kostitzin-Volterra no. 16. H: Kostitzin-Volterra no. 16.
202 Kostitzin Vladimir A. to Volterra Vito, 1936 Dec 30, Paris. ANL: Kostitzin-Volterra no. 17. H: Kostitzin-Volterra no. 17.
203 Kostitzin Vladimir A. to Volterra Vito, 1937 Mar 4, Paris. ANL: Kostitzin-Volterra no. 18. H: Kostitzin-Volterra no. 18.
204 Kostitzin Vladimir A. to Volterra Vito, 1937 Apr 5, Paris. ANL: Kostitzin-Volterra no. 19. H: Kostitzin-Volterra no. 19.
205 Kostitzin Vladimir A. to Volterra Vito, 1937 Jun 12, Paris. ANL: Kostitzin-Volterra no. 20. H: Kostitzin-Volterra no. 20.
206 Kostitzin Vladimir A. to Volterra Vito, 1937 Jun 24, Paris. ANL: Kostitzin-Volterra no. 21. H: Kostitzin-Volterra no. 21.
207 Kostitzin Vladimir A. to Volterra Vito, 1937 Jul 3, Paris. ANL: Kostitzin-Volterra no. 22. H: Kostitzin-Volterra no. 22.
208 Kostitzin Vladimir A. to Volterra Vito, 1937 Jul 6, Paris. ANL: Kostitzin-Volterra no. 23. H: Kostitzin-Volterra no. 23.
209 Kostitzin Vladimir A. to Volterra Vito, 1937 Jul 10, Paris. ANL: Kostitzin-Volterra no. 24. H: Kostitzin-Volterra no. 24.
210 Kostitzin Vladimir A. to Volterra Vito, 1937 Jul 22, Paris. ANL: Kostitzin-Volterra no. 25. H: Kostitzin-Volterra no. 25.
211 Kostitzin Vladimir A. to Volterra Vito, 1937 Nov 23, Paris. ANL: Kostitzin-Volterra no. 26. H: Kostitzin-Volterra no. 26.
212 Kostitzin Vladimir A. to Volterra Vito, 1937 Dec 30, Paris. ANL: Kostitzin-Volterra no. 27. H: Kostitzin-Volterra no. 27.
213 Kostitzin Vladimir A. to Volterra Vito, 1938 Mar 16, Paris. ANL: Kostitzin-Volterra no. 28. H: Kostitzin-Volterra no. 28.
214 Kostitzin Vladimir A. to Volterra Vito, 1938 May 4, Paris. ANL: Kostitzin-Volterra no. 29. H: Kostitzin-Volterra no. 29.
215 Kostitzin Vladimir A. to Volterra Vito, 1938 May 5, Paris. ANL: Kostitzin-Volterra no. 30. H: Kostitzin-Volterra no. 30.
216 Kostitzin Vladimir A. to Volterra Vito, 1938 May 15, Paris. ANL: Kostitzin-Volterra no. 31. H: Kostitzin-Volterra no. 31.
217 Kostitzin Vladimir A. to Volterra Vito, 1938 Aug 13, Etelley par Samoens. ANL: Kostitzin-Volterra no. 32. H: Kostitzin-Volterra no. 32.
218 Kostitzin Vladimir A. to Volterra Vito, 1938 Oct 12, Paris. ANL: Kostitzin-Volterra no. 33. H: Kostitzin-Volterra no. 33.
219 Kostitzin Vladimir A. to Volterra Vito, 1938 Oct 27, Paris. ANL: Kostitzin-Volterra no. 35. H: Kostitzin-Volterra no. 35.
220 Kostitzin Vladimir A. to Volterra Vito, 1938 Nov 2, Paris. ANL: Kostitzin-Volterra no. 36. H: Kostitzin-Volterra no. 36.
221 Kostitzin Vladimir A. to Volterra Vito, 1938 Dec 1, Paris. ANL: Kostitzin-Volterra no. 38. H: Kostitzin-Volterra no. 38.
222 Kostitzin Vladimir A. to Volterra Vito, 1938 Dec 6, Paris. ANL: Kostitzin-Volterra no. 39. H: Kostitzin-Volterra no. 39.
223 Kostitzin Vladimir A. to Volterra Vito, 1938 Dec 31, Paris. ANL: Kostitzin-Volterra no. 40. H: Kostitzin-Volterra no. 40.

224 Kostitzin Vladimir A. to Volterra Vito, 1939 Mar 9, Paris. ANL: Kostitzin-Volterra no. 41. H: Kostitzin-Volterra no. 41.
225 Kostitzin Vladimir A. to Volterra Vito, 1939 Jun 3, Paris. ANL: Kostitzin-Volterra no. 42. H: Kostitzin-Volterra no. 42.
226 Kostitzin Vladimir A. to Volterra Vito, 1939 Jul 19, Paris. ANL: Kostitzin-Volterra no. 43. H: Kostitzin-Volterra no. 43.
227 Kostitzin Vladimir A. to Volterra Vito, 1939 Aug 25, Erquy. ANL: Kostitzin-Volterra no. 44. H: Kostitzin-Volterra no. 44.
228 Kostitzin Vladimir A. to Volterra Vito, 1939 Dec 4, Paris. ANL: Kostitzin-Volterra no. 45. H: Kostitzin-Volterra no. 45.
229 Kostitzin Vladimir A. to Volterra Vito, 1940 Jan 11, Paris. ANL: Kostitzin-Volterra no. 46. H: Kostitzin-Volterra no. 46.
230 Kostitzin Vladimir A. to Volterra Vito, 1940 Apr 25, Paris. ANL: Kostitzin-Volterra no. 47. H: Kostitzin-Volterra no. 47.
231 Kostitzin Vladimir A. to Volterra Virginia, 1941 Jun 16, Paris. ANL: Kostitzin-Volterra no. 48. H: Kostitzin-Volterra no. 48.
232 Kostitzin Vladimir A. to Volterra Virginia, 1942 Apr 16, Paris. ANL: Kostitzin-Volterra no. 49. H: Kostitzin-Volterra no. 49.
233 Kostitzin Vladimir A. to Volterra Virginia, 1945 Dec 17, Paris. ANL: Kostitzin-Volterra no. 50. H: Kostitzin-Volterra no. 50.
234 Kostitzin Vladimir A. to Volterra Virginia, 1946 Jan 25, Paris. ANL: Kostitzin-Volterra no. 51. H: Kostitzin-Volterra no. 51.
235 Kostitzin Vladimir A. to Volterra Virginia, 1946 Feb 14, Paris. ANL: Kostitzin-Volterra no. 52. H: Kostitzin-Volterra no. 53.
236 Kostitzin Vladimir A. to Volterra Virginia, 1947 Apr 29, Paris. ANL: Kostitzin-Volterra no. 53. H: Kostitzin-Volterra no. 53.
237 Kostitzin Vladimir A. to Volterra Virginia, 1950 Jan 4, Paris. ANL: Kostitzin-Volterra no. 54. H: Kostitzin-Volterra no. 54.
238 Kostitzin Vladimir A. to Volterra Virginia, 1950 Jan 21, Paris. ANL: Kostitzin-Volterra no. 55. H: Kostitzin-Volterra no. 55.
239 Kostitzin Vladimir A. to Volterra Virginia, 1950 Dec 30, Paris. ANL: Kostitzin-Volterra no. 56. H: Kostitzin-Volterra no. 56.
240 Kostitzin Vladimir A. to Volterra Virginia, 1951 Dec 28, Paris. ANL: Kostitzin-Volterra no. 57. H: Kostitzin-Volterra no. 57.
241 Kostitzin Vladimir A. to Volterra Virginia, 1952 Dec 26, Paris. ANL: Kostitzin-Volterra no. 58. H: Kostitzin-Volterra no. 58.
242 Kostitzin Vladimir A. to Volterra Virginia, 1953 Dec 24, Paris. ANL: Kostitzin-Volterra no. 59. H: Kostitzin-Volterra no. 59.
243 Kostitzin Vladimir A. to Volterra Virginia, 1954 Dec 28, Paris. ANL: Kostitzin-Volterra no. 60. H: Kostitzin-Volterra no. 60.
244 Kostitzin Vladimir A. to Volterra Virginia, 1955 Dec 25, Paris. ANL: Kostitzin-Volterra no. 61. H: Kostitzin-Volterra no. 61.
245 Kostitzin Vladimir A. to Volterra Virginia, 1956 Dec 29, Paris. ANL: Kostitzin-Volterra no. 62. H: Kostitzin-Volterra no. 62.
246 Kostitzin Vladimir A. to Volterra Virginia, 1957 Dec 17, Paris. ANL: Kostitzin-Volterra no. 63. H: Kostitzin-Volterra no. 63.
247 Kostitzin Vladimir A. to Volterra Virginia, 1958 Dec 25, Paris. ANL: Kostitzin-Volterra no. 64. H: Kostitzin-Volterra no. 64.
248 Kostitzin Vladimir A. to Volterra Virginia, 1959 Dec 16, Paris. ANL: Kostitzin-Volterra no. 65. H: Kostitzin-Volterra no. 65.
249 Kostitzin Vladimir A. to Volterra Virginia, 1960 Dec 30, Paris. ANL: Kostitzin-Volterra no. 66. H: Kostitzin-Volterra no. 66.
250 Kostitzin Vladimir A. to Volterra Virginia, 1961 Dec 24, Paris. ANL: Kostitzin-Volterra no. 67. H: Kostitzin-Volterra no. 67.
251 Kostitzin Vladimir A. to Volterra Virginia, 1962 Dec 24, Paris. ANL: Kostitzin-Volterra no. 68. H: Kostitzin-Volterra no. 68.
252 Lambin Suzanne and Régnier Jean to Volterra Vito, 1938 Jul 17, Paris. ANL: Régnier-Volterra no. 29. H: Régnier-Volterra no. 27.
253 Larmor Joseph to Volterra Vito, 1908 Mar 25, Cambridge. ANL: Larmor-Volterra no. 1. H: Larmor-Volterra no. 1.

254 Larmor Joseph to Volterra Virginia and Vito, 1910 May 11, Roma. ANL: Larmor-Volterra no. 2. H: Larmor-Volterra no. 2.
255 Larmor Joseph to Volterra Vito, 1910 Jul 1, London. ANL: Larmor-Volterra no. 3. H: Larmor-Volterra no. 3.
256 Larmor Joseph to Volterra Vito, 1911 Jun 9, Cambridge. ANL: Larmor-Volterra no. 4. H: Larmor-Volterra no. 4.
257 Larmor Joseph to Volterra Vito, 1913 Jan 30, Cambridge. ANL: Larmor-Volterra no. 5. H: Larmor-Volterra no. 5.
258 Larmor Joseph to Volterra Vito, 1913 Nov 9, Cambridge. ANL: Larmor-Volterra no. 6. H: Larmor-Volterra no. 6.
259 Larmor Joseph to Volterra Vito, 1915 Jan 13, Cambridge. ANL: Larmor-Volterra no. 9. H: Larmor-Volterra no. 8.
260 Larmor Joseph to Volterra Vito, 1915 May 24, Cambridge. ANL: Larmor-Volterra no. 11. H: Larmor-Volterra no. 10.
261 Larmor Joseph to Volterra Vito, 1920 Oct 27, Cambridge. ANL: Larmor-Volterra no. 13. H: Larmor-Volterra no. 13.
262 Larmor Joseph to Volterra Vito, 1925 Jun 12, Cambridge. ANL: Larmor-Volterra no. 15. H: Larmor-Volterra no. 16.
263 Larmor Joseph to Volterra Vito, 1926 Jun 17, Cambridge. ANL: Larmor-Volterra no. 17. H: Larmor-Volterra no. 18.
264 Larmor Joseph to Volterra Vito, 1926 Jun 28, Cambridge. ANL: Larmor-Volterra no. 19. H: Larmor-Volterra no. 20.
265 Lotka Alfred J. to Volterra Vito, 1926 Nov 2, New York. ANL: Lotka-Volterra no. 2. H: Lotka-Volterra no. 1.
266 Lotka Alfred J. to Volterra Vito, 1926 Dec 1, New York. ANL: Lotka-Volterra no. 4. H: Lotka-Volterra no. 5.
267 Lotka Alfred J. to Volterra Vito, 1926 Dec 13, New York. ANL: Lotka-Volterra no. 6. H: Lotka-Volterra no. 6.
268 Lotka Alfred J. to Volterra Vito, 1928 Jan 23, New York. ANL: Lotka-Volterra no. 7. H: Lotka-Volterra no. 7.
269 Lotka Alfred J. to Volterra Vito, 1931 Jan 19, New York. ANL: Lotka-Volterra no. 8. H: Lotka-Volterra no. 8.
270 Monod-Herzen Edouard to Volterra Vito, 1927 Nov 16, Paris. ANL: Monod-Herzen-Volterra no. 1. H: Monod-Herzen-Volterra no. 1.
271 Monod-Herzen Edouard to Volterra Vito, 1927 Dec 10, Paris. ANL: Monod-Herzen-Volterra no. 2. H: Monod-Herzen-Volterra no. 2.
272 Monod-Herzen Edouard to Volterra Vito, 1929 Jan 20, Paris. ANL: Monod-Herzen-Volterra no. 3. Monod-Herzen-Volterra no. 3.
273 Pearl Maud de Witt to Volterra Vito, 1939 Feb 24, Baltimore. ANL: Pearl-Volterra no. 15. H: Pearl-Volterra no. 13.
274 Pearl Raymond to Volterra Vito, 1931 Jan 16, Baltimore. ANL: Pearl-Volterra no. 1. H: Pearl-Volterra no. 1.
275 Pearl Raymond to Volterra Vito, 1937 Oct 30, London. ANL: Pearl-Volterra no. 4. H: Pearl-Volterra no. 4.
276 Pearl Raymond to Volterra Vito, 1938 Jan 10, Baltimore. ANL: Pearl-Volterra no. 7. H: Pearl-Volterra no. 6.
277 Pearl Raymond to Volterra Vito, 1938 Nov 1, Baltimore. ANL: Pearl-Volterra no. 10. H: Pearl-Volterra no. 9.
278 Pearl Raymond to Volterra Vito, 1939, Baltimore. ANL: Pearl-Volterra no. 20. H: Pearl-Volterra no. 16.
279 Pearl Raymond to Volterra Vito, 1939 Jan 17, Baltimore. ANL: Pearl-Volterra no. 12. H: Pearl-Volterra no. 11.
280 Pearl Raymond to Volterra Vito, 1939 Feb 7, Baltimore. ANL: Pearl-Volterra no. 13. H: Pearl-Volterra no. 12.
281 Pearl Raymond to Volterra Vito, 1939 Mar 2, Baltimore. ANL: Pearl-Volterra no. 16. H: Pearl-Volterra no. 14.
282 Pearson Karl to Volterra Vito, 1902 Apr 24, London. ANL: Pearson-Volterra no. 1. H: Pearson-Volterra no. 1.
283 Régnier Jean to Volterra Vito, 1932 Mar 12, Paris. ANL: Régnier-Volterra no. 1. H: Régnier-Volterra no. 1.

284 Régnier Jean to Volterra Vito, 1933 Jun 11, Paris. ANL: Régnier-Volterra no. 2. H: Régnier-Volterra no. 2.
285 Régnier Jean to Volterra Vito, 1933 Jul 16, Paris. ANL: Régnier-Volterra no. 4. H: Régnier-Volterra no. 4.
286 Régnier Jean to Volterra Vito, 1934 Mar 14, Paris. ANL: Régnier-Volterra no. 5. H: Régnier-Volterra no. 5.
287 Régnier Jean to Volterra Vito, 1934 Nov 18, Paris. Lincei. ANL: Régnier-Volterra no. 7. H: Régnier-Volterra no. 7.
288 Régnier Jean to Volterra Vito, 1934 Dec 9, Paris. ANL: Régnier-Volterra no. 8. H: Régnier-Volterra no. 8.
289 Régnier Jean to Volterra Vito, 1935 Jan 30, Paris. ANL: Régnier-Volterra no. 9. H: Régnier-Volterra no. 9.
290 Régnier Jean to Volterra Vito, 1935 Apr 5, Paris. ANL: Régnier-Volterra no. 10. H: Régnier-Volterra no. 10.
291 Régnier Jean to Volterra Vito, 1935 May 10, Paris. ANL: Régnier-Volterra no. 11. H: Régnier-Volterra no. 11.
292 Régnier Jean to Volterra Vito, 1935 May 28, Paris. ANL: Régnier-Volterra no. 15. H: Régnier-Volterra no. 13.
293 Régnier Jean to Volterra Vito, 1935 Aug 6, Paris. ANL: Régnier-Volterra no. 16. H: Régnier-Volterra no. 14.
294 Régnier Jean to Volterra Vito, 1935 Nov 3, Paris. ANL: Régnier-Volterra no. 17. H: Régnier-Volterra no. 15.
295 Régnier Jean to Volterra Vito, 1936 Jan 2, Paris. Lincei. ANL: Régnier-Volterra no. 18. H: Régnier-Volterra no. 16.
296 Régnier Jean to Volterra Vito, 1936 Mar 13, Paris. ANL: Régnier-Volterra no. 19. H: Régnier-Volterra no. 17.
297 Régnier Jean to Volterra Vito, 1936 Apr 13, Paris. ANL: Régnier-Volterra no. 20. H: Régnier-Volterra no. 18.
298 Régnier Jean to Volterra Vito and Virginia, 1937 Jan 7, Paris. ANL: Régnier-Volterra no. 21. H: Régnier-Volterra no. 19.
299 Régnier Jean to Volterra Vito, 1937 Feb 19, Paris. ANL: Régnier-Volterra no. 22. H: Régnier-Volterra no. 20.
300 Régnier Jean to Volterra Vito, 1937 Apr 10, Paris. ANL: Régnier-Volterra no. 23. H: Régnier-Volterra no. 21.
301 Régnier Jean to Volterra Vito, 1937 Apr 25, Paris. ANL: Régnier-Volterra no. 24. H: Régnier-Volterra no. 22.
302 Régnier Jean to Volterra Vito and Virginia, 1938 Feb 9, Paris. ANL: Régnier-Volterra no. 25. H: Régnier-Volterra no. 23.
303 Régnier Jean to Volterra Virginia, 1938 Feb 9, Paris. ANL: Régnier-Volterra no. 26. H: Régnier-Volterra no. 24.
304 Régnier Jean to Volterra Vito and Virginia, 1938 Feb 17, Paris. ANL: Régnier-Volterra no. 27. H: Régnier-Volterra no. 25.
305 Régnier Jean to Volterra Vito and Virginia, 1938 May 3, Paris. ANL: Régnier-Volterra no. 28. H: Régnier-Volterra no. 26.
306 Régnier Jean to Volterra Vito, 1938 Sep 5, Paris. ANL: Régnier-Volterra no. 32. H: Régnier-Volterra no. 29.
307 Régnier Jean to Volterra Vito, 1938 Sep 19, Paris. ANL: Régnier-Volterra no. 33. H: Régnier-Volterra no. 31.
308 Régnier Jean to Volterra Vito, 1938 Sep 21, Paris. ANL: Régnier-Volterra no. 34. H: Régnier-Volterra no. 32.
309 Régnier Jean to Volterra Vito, 1938 Sep 29, Paris. ANL: Régnier-Volterra no. 35. H: Régnier-Volterra no. 33.
310 Régnier Jean to Volterra Vito, 1938 Oct 11, Paris. ANL: Régnier-Volterra no. 36. H: Régnier-Volterra no. 34.
311 Régnier Jean to Volterra Vito, 1938 Oct 31, Paris. ANL: Régnier-Volterra no. 38. H: Régnier-Volterra no. 36.
312 Régnier Jean to Volterra Vito and Virginia, 1938 Dec 2, Paris. ANL: Régnier-Volterra no. 39. H: Régnier-Volterra no. 37.
313 Régnier Jean to Volterra Vito, 1938 Dec 26, Paris. ANL: Régnier-Volterra no. 40. H: Régnier-Volterra no. 38.

314 Régnier Jean to Volterra Vito and Virginia, 1939 Jan 7, Paris. ANL: Régnier-Volterra no. 41. H: Régnier-Volterra no. 39.
315 Régnier Jean to Volterra Vito and Virginia, 1939 Feb 21, Paris. ANL: Régnier-Volterra no. 42. H: Régnier-Volterra no. 40.
316 Régnier Jean to Volterra Vito and Virginia, 1939 Mar 11, Paris. ANL: Régnier-Volterra no. 43. H: Régnier-Volterra no. 41.
317 Régnier Jean to Volterra Vito and Virginia, 1939 Apr 30, Paris. ANL: Régnier-Volterra no. 44. H: Régnier-Volterra no. 42.
318 Régnier Jean to Volterra Vito and Virginia, 1939 Aug 8, Paris. ANL: Régnier-Volterra no. 45. H: Régnier-Volterra no. 43.
319 Régnier Jean to Volterra Vito and Virginia, 1939 Aug 19, Paris. ANL: Régnier-Volterra no. 46? H: Régnier-Volterra no. 44.
320 Régnier Jean to Volterra Vito and Virginia, 1939 Oct 29, Paris. ANL: Régnier-Volterra no. 48. H: Régnier-Volterra no. 46.
321 Régnier Jean to Volterra Vito and Virginia, 1940 Jan 1, Paris. ANL: Régnier-Volterra no. 49. H: Régnier-Volterra no. 47.
322 Régnier Jean to Volterra Vito and Virginia, 1940 Feb 17, Paris. ANL: Régnier-Volterra no. 51. H: Régnier-Volterra no. 48.
323 Régnier Jean to Volterra Vito and Virginia, 1940 Mar 9, Paris. ANL: Régnier-Volterra no. 50. H: Régnier-Volterra no. 49.
324 Régnier M.me. to Volterra Virginia, 1947 Mar 30, Paris. ANL: Régnier-Volterra no. 52. H: Régnier-Volterra no. 50.
325 Stanley John to Volterra Vito, 1935 Mar 29, Kingston. ANL: Stanley-Volterra no. 1. H: Stanley-Volterra no. 1.
326 Stanley John to Volterra Vito, 1936 Oct 10, Kingston. ANL: Stanley-Volterra no. 2. H: Stanley-Volterra no. 2.
327 Stanley John to Volterra Vito, 1936 Nov 19, Kingston. ANL: Stanley-Volterra no. 4. H: Stanley-Volterra no. 4.
328 Stanley John to Volterra Vito, 1936 Nov 21, Kingston. ANL: Stanley-Volterra no. 5. H: Stanley-Volterra no. 5.
329 Stanley John to Volterra Vito, 1937 Jan 11, Kingston. ANL: Stanley-Volterra no. 7. H: Stanley-Volterra no. 7.
330 Stanley John to Volterra Vito, 1937 Jul 9, Gaspé P.Q. ANL: Stanley-Volterra no. 9. H: Stanley-Volterra no. 9.
331 Stanley John to Volterra Vito, 1937 Oct 29, Kingston. ANL: Stanley-Volterra no. 10. H: Stanley-Volterra no. 10.
332 Teissier Georges to Volterra Vito, 1933 Dec 22, Paris. ANL: Teissier-Volterra no.1. H: Teissier-Volterra no. 1.
333 Teissier Georges to Volterra Vito, 1934 Jan 12, Paris. ANL: Teissier-Volterra no. 2. H: Teissier-Volterra no. 3.
334 Teissier Georges to Volterra Vito, 1935 Jun 19, Roscoff. ANL: Teissier-Volterra no. 3. H: Teissier-Volterra no. 4.
335 Thompson D'Arcy W. to Volterra Vito, n.d., St.Andrews. ANL: Thompson D.W. -Volterra no. 4. H: Thompson D.W. -Volterra no. 3.
336 Thompson D'Arcy W. to Volterra Vito, n.d., Cochem a.d. Mosel. ANL: Thompson D.W.-Volterra no. 23. H: Thompson D.W. -Volterra no. 20.
337 Thompson D'Arcy W. to Volterra Vito, n.d., n.p. ANL: Thompson D.W.-Volterra no. 27. H: Thompson D.W. -Volterra no. 24.
338 Thompson D'Arcy W. to Volterra Vito, 1926 Jul 2, St. Andrews. ANL: Thompson D.W. - Volterra no. 2. H: Thompson D.W. -Volterra no. 1.
339 Thompson D'Arcy W. to Volterra Vito, 1926 Oct 12, St. Andrews. ANL: Thompson D.W.-Volterra no. 6. H: Thompson D.W. -Volterra no. 6.
340 Thompson D'Arcy W. to Volterra Vito, 1926 Nov 26], St. Andrews. ANL: Thompson D.W.-Volterra no. 10. H: Thompson D.W. -Volterra no. 9.
341 Thompson D'Arcy W. to Volterra Vito, 1930 Jul 12, St. Andrews. ANL: Thompson D.W.-Volterra no. 16. H: Thompson D.W. -Volterra no. 12.
342 Thompson D'Arcy W. to Volterra Vito, 1931 Jan 3, St. Andrews. ANL: Thompson D.W.-Volterra no. 17. H: Thompson D.W. -Volterra no. 14.
343 Thompson D'Arcy W. to Volterra Vito, 1931 Feb 11, St. Andrews. ANL: Thompson D.W.-Volterra no. 19. H: Thompson D.W. -Volterra no. 16.

344 Thompson D'Arcy W. to Volterra Vito, 1931 Feb 28, St. Andrews. ANL: Thompson D.W.-Volterra no. 20. H: Thompson D.W. -Volterra no. 17.
345 Thompson D'Arcy W. to Volterra Vito, 1931 Aug 11, St. Andrews. ANL: Thompson D.W.-Volterra no. 22. H: Thompson D.W. -Volterra no. 19.
346 Thompson D'Arcy W. to Volterra Vito, 1935 Jul 21, St. Andrews. ANL: Thompson D.W.-Volterra no. 24. H: Thompson D.W. -Volterra no. 21.
347 Thompson D'Arcy W. to Volterra Vito, 1936 Oct 6, St. Andrews. ANL: Thompson D.W.-Volterra no. 25. H: Thompson D.W. -Volterra no. 22.
348 Thompson D'Arcy W. to Volterra Vito, 1937 Jun 24, St. Andrews. ANL: Thompson D.W.-Volterra no. 26. H: Thompson D.W. -Volterra no. 23;
349 Thompson D'Arcy W. to Volterra Vito, 1937 Dec 29, Edinburgh. ANL: Thompson D.W.-Volterra no. 28. H: Thompson D.W. -Volterra no. 25.
350 Thompson William R. to Volterra Vito, 1927 Jul 29, Hyères. ANL: Thompson W.R.-Volterra no. 1. H: Thompson W.R. -Volterra no. 1.
351 Thompson William R. to Volterra Vito, 1927 Aug 23, Hyères. ANL: Thompson W.R.-Volterra no. 3. H: Thompson W.R. -Volterra no. 3.
352 Thompson William R. to Volterra Vito, 1927 Oct 3, Hyères. ANL: Thompson W.R.-Volterra no. 4. H: Thompson W.R. -Volterra no. 4.
353 Thompson William R. to Volterra Vito, 1930 Apr 8, Farnham Royal. ANL: Thompson W.R.-Volterra no. 6. H: Thompson W.R. -Volterra no. 5.
354 Thompson William R. to Volterra Vito, 1930 May 4, Farnham Royal. ANL: Thompson W.R.-Volterra no. 5. H: Thompson W.R. -Volterra no. 6.
355 Thompson William R. to Volterra Vito, 1931 Feb 5, Farnham Royal. ANL: Thompson W.R.-Volterra no. 7. H: Thompson W.R. -Volterra no. 7.
356 Thompson William R. to Volterra Vito, 1931 Mar 23, Farnham Royal. ANL: Thompson W.R.-Volterra no. 8. H: Thompson W.R. -Volterra no. 8.
357 Volterra Vito to Brelot Marcel, 1929 Jul 17, Ariccia. ANL: Brelot-Volterra no. 4. H: Brelot-Volterra no. 4.
358 Volterra Vito to Brelot Marcel, 1929 Jul 19, Ariccia. ANL: Brelot-Volterra no. 5. H: Brelot-Volterra no. 5.
359 Volterra Vito to Brelot Marcel, 1929 Aug 20, Ariccia. ANL: Brelot-Volterra no. 9. H: Brelot-Volterra no. 9.
360 Volterra Vito to Brelot Marcel, 1930 Aug 5, Saint Gervais les Bains. ANL: Brelot-Volterra no. 32. H: Brelot-Volterra no. 36.
361 Volterra Vito to Brelot Marcel, 1930 Aug 26, Saint Gervais les Bains. ANL: Brelot-Volterra no. 40. H: Brelot-Volterra no. 44.
362 Volterra Vito to Chapman Royal N., n.d., Napoli. ANL: Chapman-Volterra no. 2. H: Chapman-Volterra no. 2.
363 Volterra Vito to Chapman Royal N., n.d., n.p. ANL: Chapman-Volterra no. 6. H: Chapman-Volterra no. 6.
364 Volterra Vito to Chapman Royal N., n.d., Roma. ANL: Chapman-Volterra no. 8. H: Chapman-Volterra no. 8.
365 Volterra Vito to Chapman Royal N., n.d., n.p. ANL: Chapman-Volterra no. 9. H: Chapman-Volterra no. 10.
366 Volterra Vito to Choucroun N., 1935, n.p. ANL: Régnier-Volterra no. 13. H: Régnier-Volterra no. 12.
367 Volterra Vito to D' Ancona Umberto, 1937 Nov 5, London. ANL: D'Ancona-Volterra no. 93. H: D'Ancona-Volterra no. 85.
368 Volterra Vito to Elton Charles S., n.d., n.p. ANL: Elton-Volterra no. 10. H: Elton-Volterra no. 6.
369 Volterra Vito to Elton Charles S., 1931 Jan 20, n.p. ANL: Elton-Volterra no. 2. H: Elton-Volterra no. 2.
370 Volterra Vito to Kermack William O., 1938 Sep 11, n.p. ANL: Kermack-Volterra no. 2. H: Kermack-Volterra no. 2.
371 Volterra Vito to Kostitzin Vladimir A., n.d., Ariccia. ANL: Kostitzin-Volterra no. 34. H: Kostitzin-Volterra no. 34.
372 Volterra Vito to Kostitzin Vladimir A., 1935 Nov 13, Roma. ANL: Kostitzin-Volterra no. 9. H: Kostitzin-Volterra no. 9.
373 Volterra Vito to Kostitzin Vladimir A., 1938 Nov 5, Ariccia. ANL: Kostitzin-Volterra no. 37. H: Kostitzin-Volterra no. 37.

374 Volterra Vito to Larmor Joseph, n.d., n.p. ANL: Larmor-Volterra no. 12. H: Larmor-Volterra no. 11.
375 Volterra Vito to Larmor Joseph, n.d., n.p. ANL: Larmor-Volterra no. 14. H: Larmor-Volterra no. 15.
376 Volterra Vito to Larmor Joseph, n.d., n.p. ANL: Larmor-Volterra no. 21. H: Larmor-Volterra no. 21.
377 Volterra Vito to Larmor Joseph, n.d., n.p. ANL: Larmor-Volterra no. 22. H: Larmor-Volterra no. 12.
378 Volterra Vito to Larmor Joseph, 1915 Jan, n.p. ANL: Larmor-Volterra no. 8. H: Larmor-Volterra no. 7.
379 Volterra Vito to Larmor Joseph, 1915, n.p. ANL: Larmor-Volterra no. 10. H: Larmor-Volterra no. 9.
380 Volterra Vito to Larmor Joseph, 1926 Jun 11, Roma. ANL: Larmor-Volterra no. 16. H: Larmor-Volterra no. 17.
381 Volterra Vito to Larmor Joseph, 1926, Paris. ANL: Larmor-Volterra no. 18. H: Larmor-Volterra no. 19.
382 Volterra Vito to Larmor Joseph, 1934, n.p. ANL: Larmor-Volterra no. 20. H: Larmor-Volterra no. 14.
383 Volterra Vito to Lotka Alfred J., n.d., n.p. ANL: Lotka-Volterra no. 1. H: Lotka-Volterra no. 2.
384 Volterra Vito to Lotka Alfred J., n.d., n.p. ANL: Lotka-Volterra no. 3. H: Lotka-Volterra no. 3.
385 Volterra Vito to Lotka Alfred J., n.d., n.p. ANL: Lotka-Volterra no. 5. H: Lotka-Volterra no. 4.
386 Volterra Vito to Pearl Raymond, n.d., n.p. ANL: Pearl-Volterra no. 2. H: Pearl-Volterra no. 2.
387 Volterra Vito to Pearl Raymond, n.d., Paris. ANL: Pearl-Volterra no. 3. H: Pearl-Volterra no. 3.
388 Volterra Vito to Pearl Raymond, n.d., Roma. ANL: Pearl-Volterra no. 14. H: Pearl-Volterra no. 8.
389 Volterra Vito to Pearl Raymond, 1937 Oct 31, Paris. ANL: Pearl-Volterra no. 5. H: Pearl-Volterra no. 5.
390 Volterra Vito to Pearl Raymond, 1938 Jan, n.p. ANL: Pearl-Volterra no. 8. H: Pearl-Volterra no. 7.
391 Volterra Vito to Pearl Raymond, 1938 Oct 17, Roma. ANL: Pearl-Volterra no. 9. H: Pearl-Volterra no. 8.
392 Volterra Vito to Pearl Raymond, 1938 Nov, Roma. ANL: Pearl-Volterra no. 11. H: Pearl-Volterra no. 10.
393 Volterra Vito to Pearl Raymond, 1939 Apr 10, Roma. ANL: Pearl-Volterra no. 17. H: Pearl-Volterra no. 15.
394 Volterra Vito to Pearl Raymond, 1939 Apr 10, Roma. ANL: Pearl-Volterra no. 19. H: Pearl-Volterra no. 15.
395 Volterra Vito to Pearson Karl, n.d., n.p. ANL: Pearson-Volterra n. 2. H: Pearson-Volterra no. 2.
396 Volterra Vito to Pearson Karl, 1926 Sep 21, n.p. ANL: Pearson-Volterra no. 3. H: Pearson-Volterra no. 3.
397 Volterra Vito to Perrin Jean, 1935, n.p. ANL: Régnier-Volterra no. 14. H: Régnier-Volterra: no. 12.
398 Volterra Vito to Régnier Jean, n.d., n.p. ANL: Régnier-Volterra no. 3. H: Régnier-Volterra: no. 3.
399 Volterra Vito to Régnier Jean, n.d., n.p. ANL: Régnier-Volterra no. 6. H: Régnier-Volterra: no. 6.
400 Volterra Vito to Régnier Jean, 1935]. n.p. ANL: Régnier-Volterra no. 12. H: Régnier-Volterra: no. 12.
401 Volterra Vito to Régnier Jean, 1938, Roma. ANL: Régnier-Volterra no. 30. H: Régnier-Volterra: no. 28.
402 Volterra Vito to Régnier Jean, 1938, Roma. ANL: Régnier-Volterra no. 31. H: Régnier-Volterra: no. 30.
403 Volterra Vito to Régnier Jean, 1938 Oct 17, Ariccia. ANL: Régnier-Volterra no. 37. H: Régnier-Volterra: no. 35.

404 Volterra Vito to Régnier Jean, 1939 Aug 25, Ariccia. ANL: Régnier-Volterra no. 47. H: Régnier-Volterra: no. 45.

405 Volterra Vito to Stanley John, n.d., Paris. ANL: Stanley-Volterra no. 3. H: Stanley-Volterra no. 3.

406 Volterra Vito to Stanley John, 1936 Dec 27, Roma. ANL: Stanley-Volterra no. 6. H: Stanley-Volterra no. 6.

407 Volterra Vito to Stanley John, n.d., Paris. ANL: Stanley-Volterra no. 8. H: Stanley-Volterra no. 8.

408 Volterra Vito to Teissier Georges, n.d., n.p. ANL: Teissier-Volterra no. 4. H: Teissier-Volterra no. 2.

409 Volterra Vito to Teissier Georges, n.d., n.p. ANL: Teissier-Volterra no. 5. H: Teissier-Volterra no. 5.

410 Volterra Vito to Thompson D'Arcy W., 1926 Sep 20, Ariccia. ANL: Thompson D.W.-Volterra no. 1. H: Thompson D.W.-Volterra no. 4.

411 Volterra Vito to Thompson D'Arcy W., n.d., n.p. ANL: Thompson D.W.-Volterra no. 3. H: Thompson D.W.-Volterra no. 7.

412 Volterra Vito to Thompson D'Arcy W., n.d., n.p. ANL: Thompson D.W.-Volterra no. 7. H: Thompson D.W.-Volterra no. 13.

413 Volterra Vito to Thompson D'Arcy W., 1926 Sep 20, Ariccia. ANL: Thompson D.W.-Volterra no. 5. H: Thompson D.W.-Volterra no. 4.

414 Volterra Vito to Thompson D'Arcy W., 1926 Sep 24, n.p. ANL: Thompson D.W.-Volterra no. 8. H: Thompson D.W.-Volterra no. 5.

415 Volterra Vito to Thompson D'Arcy W., 1926 Nov 21, Roma. ANL: Thompson D.W.-Volterra no. 9. H: Thompson D.W.-Volterra no. 8.

416 Volterra Vito to Thompson D'Arcy W., 1926 Nov 28, Roma. ANL: Thompson D.W.-Volterra no. 11. H: Thompson D.W.-Volterra no. 10.

417 Volterra Vito to Thompson D'Arcy W., n.d., n.p. ANL: Thompson D.W.-Volterra no. 12. H: Thompson D.W.-Volterra no. 11.

418 Volterra Vito to Thompson D'Arcy W., n.d., n.p. ANL: Thompson D.W.-Volterra no. 13. H: Thompson D.W.-Volterra no. 11.

419 Volterra Vito to Thompson D'Arcy W., n.d., n.p. ANL: Thompson D.W.-Volterra no. 14. H: Thompson D.W.-Volterra no. 8.

420 Volterra Vito to Thompson D'Arcy W., n.d., n.p. ANL: Thompson D.W.-Volterra no. 15. H: Thompson D.W.-Volterra no. 2.

421 Volterra Vito to Thompson D'Arcy W., n.d., n.p. ANL: Thompson D.W.-Volterra no. 18. H: Thompson D.W.-Volterra no. 15.

422 Volterra Vito to Thompson D'Arcy W., 1931 Mar 24, Roma. ANL: Thompson D.W.-Volterra no. 21. H: Thompson D.W.-Volterra no. 18.

423 Volterra Vito to Thompson William R., n.d., n.p. ANL: Thompson W.R.-Volterra no. 2. H: Thompson W.R.-Volterra no.2.

424 Volterra Vito to Tisdale, W. E., n.d., n.p. ANL: Brelot-Volterra no. 3. H: Brelot-Volterra no. 3.

425 Volterra Vito to unknown, n.d., n.p. ANL: D'Ancona-Volterra no. 94. H: D'Ancona-Volterra no. 95.

24 References

ADAMS, M. B., 1968, "The Founding of Populations Genetics: Contributions of the Chetverikov School, 1924–1934", *Journal of the History of Biology*, **1**: 23–39.

ADAMS, M. B., 1980a, "Svertsov and Schmalhausen: Russian Morphology and the Evolutionary Synthesis," in: MAYR, E., PROVINE, W. B. (eds.) 1980: 193–225.

ADAMS, M. B., 1980b, "Sergei Chetverikov, the Kol'tsov Institute, and the Evolutionary Synthesis," in: MAYR, E., PROVINE, W. B. (eds.) 1980: 242–278.

BAILES, K., 1990, Science and Russian Culture in the Age of Revolutions: V. I. Vernadsky and His Scientific School, 1863–1945, Bloomington, Indiana University Press.

BAILEY, V. A., 1931, "The Interaction between Host and Parasites", *Quarterly Journal of Mathematics*, **2**: 68–77.

BALADINE, R., 1982, La vie et l'œuvre de Vernadski, Moscou, Éditions MIR.

BODENHEIMER, F. S., 1928, "Welche Faktorem regulieren die Individuenzahl einer Insektenart in der Natur?", *Biologisches Zentralblatt*, **48**: 714.

BODENHEIMER, F. S., 1938, Problems of Animal Ecology, Oxford, Oxford University Press.

BODENHEIMER, F. S., 1932a, "Der Massenwechsel in der Tierwelt. Grundriss einer allgemeinen tierischen Bevölkerrungslehre", *Atti del XI Congresso Internazionale di Zoologia, Padova, Archivio Zoologico Italiano*, **16**: 98.

BODENHEIMER, F. S., 1932b, "Ueber den Massenwechsel der Selachierbevölkerung im oberadriatischen Benthos. Eine stunde ueber tierische Fluktuationen", *Arch. Hydrobiol.*, **24**: 667.

BOESIGER, E., 1980, "Evolutionary biology in France at the time of the evolutionary synthesis," in: MAYR, E., PROVINE, W. B. (eds.) 1980: 309–320

BRELOT, M., 1931, "Sur le problème biologique héréditaire de deux espèces dévorante et dévorée," *Annali di Matematica*, 2 serie, **9**: 57.

BRELOT, M., 1932, "Ueberblick über die biologisch-matematischen Untersuchungen von Volterra," *Jahresbericht der Deutsche Mathematiker-Vereinigung*, **42**: 1–14.

BRILLOUIN, L. N., 1935, Notions élémentaires de mathématiques pour les sciences expérimentales, à l'usage des candidats au certificat d'études physiques, chimiques et biologiques et à la licence ès sciences, Paris, Masson (reprint 1939).

BUICAN, D., 1984, Histoire de la génétique et de l'évolutionnisme en France, Paris, Presses Universitaires de France.

BUNGE, M., 1979, Causality and modern science, New York, Dover, 3rd ed.

BUNYARD, P., GOLDSMITH, E. (eds.), 1988, Gaia, the Thesis, the Mechanisms, and the Implications, Camelford, Wadebridge Ecological Centre.

BURGESS, R. L., 1981, "Sources of Biographical Information on American Ecologists", *Bulletin of Ecological Society of America*, **62**: 236–255.

CARR-SAUNDERS, A., 1922, The Population Problem: A Study in Human Evolution, Oxford, Clarendon Press.

CAULLERY, M., 1931, Le problème de l'évolution, Paris, Payot.

CAULLERY, M., 1935, Les conceptions modernes de l'hérédité, Paris, Flammarion.

CHAPMAN, R. N., 1925, Animal ecology, with especial reference to insects, Minneapolis, Minn., Burgess Brooke; enlarged version: CHAPMAN R. N. 1931, Animal ecology, with special reference to insects, New York & London, Mc-Graw Hill (reprint, New York, Arno Press, 1977).

CHAPMAN, R. N., 1928, "The Quantitative Analysis of Environmental Factors," *Ecology*, **9**: 111–122.

CHAPMAN, R. N., 1933, "The Causes of Fluctuations in Populations of Insects," *Proceedings of the Hawaiian Entomological Society*, **8**: 279–292.

CHAPMAN, R. N., 1939, "Insect Population Problems in Relation to Insect Outbreak", *Ecological Monographs*, **9**: 261–269.

CHOQUET, G., 1987, "Vie et œuvre de Marcel Brelot (1903–1987) - Conférence de Gustave Choquet au Séminaire de Théorie du Potentiel du jeudi 8 octobre 1987", reprint in BRELOT, M., 1999, Théorie classique du potentiel, Association Laplace-Gauss.

CUNNINGHAM, E., 1943, "Sir Joseph Larmor", *Journal of London Mathematical Society*, **18**: 57–64.

D'ANCONA, U., 1926, "Dell'influenza della stasi peschereccia nel periodo 1914–18 sul patrimonio ittico dell'Alto Adriatico", *Memorie del Regio Comitato Talassografico Italiano*, Mem. 126.

D'ANCONA, U., 1927, "Intorno alle associazioni biologiche e a un saggio di teoria matematica delle stesse con particolare riguardo all'idrobiologia", *Int. Rev. Ges. Hydrobiol. u. Hydrograph.*, **17**: 189.

D'ANCONA, U., 1939, Der Kampf ums Dasein. Eine biologisch-mathematische Darstellung der Lebensgemeinschaften und biologische Gleichgewichte, Berlin, Borntraeger.

D'ANCONA, U., 1954, The Struggle for Existence. "Biblioteca Biotheoretica", **6**, Brill, Leiden.

D'ANCONA, U., 1964, "Dagli equilibri biologici alla teoria della pesca", *La Ricerca Scientifica*, **34**: 251–268.

DARWIN, CH., 1882, The Origin of Species by Means of Natural Selection, or the Preservation of Favoured in the Struggle for Life, London.

DAVIDSON, J. N., YATES, F., MCCREA, W. H., 1971, "Kermack, Willian Ogilvy", *Biographical Memoirs of Fellows of the Royal Society*, **17**.

DE HELGUERO, F., 1904, "Sui massimi delle curve dimorfiche," *Biometrika*, **3**: 84–98.

DELEAGE, J. P., 1991, Histoire de l'écologie. Unie science de la nature et de l'homme, Paris, Éditions de la Découverte.

DOBELL, C., 1949, "D'Arcy Wentworth Thompson", *Obituary Notices of Fellows of the Royal Society*, **6**: 599–617.

EDDINGTON, A. S., 1942–3, "Joseph Larmor", Obituary Notices of Fellows of the Royal Society of London, **4**: 197–207.

EISENHART, C., 1976, "Pearson, Karl", *Dictionary of Scientific Biography* (Ch. C. Gillispie, ed.), New York, Scribners, sub voce.

ELTON, C. S., 1924-5, "Periodic Fluctuations in the Number of Animals: Their Causes and Effects", *Journal of Experimental Biology* (British), **2**: 119–163.

ELTON, C. S., 1925, "Plague and the Regulation of Numbers in the Wild Mammals", *Journal of Hygiene*, **24**: 138.

ELTON, C. S., 1927, Animal ecology, London, Sidgwick and Jackson; 2nd ed. with additional notes, New York, Macmillan, 1935.

ELTON, C. S., 1929a, The Relation of Animal Numbers to Climate, Conference of Empire Meteorologists, Agricultural Sections.

ELTON, C. S., 1929b, "Review of JOHNSTONE J. 1928", *Journal of Ecology*, **17**: 402.

ELTON, C. S., 1930, Animal ecology and evolution, Oxford, Clarendon Press.

ELTON, C. S., 1931, "Epidemies among Sledge Dogs in the Canadian Artic and Their Relation to Disease in the Artic Fox", *Canadian Journal of Research*, **5**: 637.

ELTON, C. S., 1942, Voles, mices and lemmings: problems in population dynamics, Oxford, Clarendon Press.

ELTON, C. S., 1946, "Competition and Structure of Animal Ecologies", *Journal of Animal Ecology*, **15**: 54–68.

ELTON, C. S., 1954, The Control of Rats and Mice, London, Methuen.

ELTON, C. S., 1958, The Ecology of Invasions of Animals and Plants, London, Methuen.

ELTON, C. S., 1966, The Pattern of Animal Communities, London, Methuen.

ENRIQUES, F., 1922, Per la storia della logica, Bologna, Zanichelli (reprint 1987).

ENRIQUES, F., MONOD-HERZEN, G. E., 1926, L'évolution de la logique, Paris, Chiron.

FALCONE, M., ISRAEL, G., 1985, "Qualitative and Numerical Analysis of a Class of Prey-Predator Models", *Acta Applicandae Mathematicae*, **4**: 225–258.

FAURÉ-FRÉMIET, E., 1925, La cinétique du développement, multiplication cellulaire et croissance, Paris, Presses Universitaires de France.

FREEMAN, A. W., 1941, "Raymond Pearl, 1879–1940", *American Journal of Public Health*, **31**: 81–82.

FREGUGLIA, P., 1998, "Considerazioni sul modello di Giovanni V. Schiaparelli per una interpretazione geometrica delle concezioni darwiniane", in: P. Freguglia (ed.) Modelli matematici nelle scienze biologiche, Urbino, Quattro Venti: 85–112.

FRIEDERICHS, K., 1906, Zur Biologie der Embiiden: neue Untersuchungen und Übersicht des Bekannten; mit Beiträgen über die Systematik und postembryonale Entwicklung mediterraner Arten, Berlin, Friedländer.

FRIEDERICHS, K., 1918, Beobachtungen über einige solitäre Wespen in Madagaskar, Berlin, Friedländer.

FRIEDERICHS, K., 1923, Ökologische Beobachtungen über Embiiden, 's-Gravenhage, Nijhoff.

FRIEDERICHS, K., 1929, "Welche Faktoren egeln die Individuenzahl einer Insektenart in der Natur? Ein kritisches Referat", *Anz. f. Schädlingskunde*, **5**: 119.

FRIEDERICHS, K., 1930a, "Noch einiges über die Ursachen stärkeren oder schwächeren Anftretens von Insektenarten", *Anz. f. Schädlingskunde*, **6**: 25.

FRIEDERICHS, K., 1930b, Die Grundfragen und Gesetzmässigkeiten der land-und forstwirschaftlichen Zoologie insbesondere der Entomologie, Berlin, Parey.

FRIEDERICHS, K., 1935, Check list of the Embiidae (Embioptera) of Oceania, Honolulu, Hawaii, The Museum.

FRIEDERICHS, K., 1937, Ökologie als Wissenschaft von der Natur, oder Biologische Raumforschung, Leipzig, J. A. Barth.

FRIEDERICHS, K., 1955, Die Selbstgestaltung des Lebendigen; synoptische Theorie des Leben als ein Beitrag zu den philosophischen Grundlagen der Naturwissenschaft, München, E. Reinhardt.

FROMM, H., 1931a, "Grenzen des elastischen Verhaltens beanspruchter Stoffe" in: Handbuch der physikalischen und technischen Mechanik, vol. 4, part I, Leipzig, Johann Ambrosius Barth Verlag: 359–435.

FROMM, H., 1931b, "Nachwirkung und Hysteresis", in: Handbuch der physikalischen und technischen Mechanik, vol. 4, part I, Leipzig, Johann Ambrosius Barth Verlag: 436–550.

GAUSE, G. F., 1930, "Studies on the Ecology of the Orthoptera", *Ecology*, **11**: 307–325.

GAUSE, G. F., 1931, "The Influence of Ecological Factors on the Size of Population", *American Naturalist*, **65**: 70–76.

GAUSE, G. F., 1932a, "Experimental Studies on the Struggle for Existence. I. Mixed Populations of Two Species of Yeast", *Journal of Experimental Biology*, **9**: 389.

GAUSE, G. F., 1932b, "Ecology of Populations", *Quarterly Review of Biology*, **7**: 27–48.

GAUSE, G. F., 1934a, The Struggle for Existence, Baltimore, Williams &Wilkins (reprint New York, Dover 1971).

GAUSE, G. F., 1934b, "Ueber die Konkurrenz zwischen zwei Arten", *Zoologischer Anzeiger*, **105**.

GAUSE, G. F., 1934c, "Experimental Analysis on Vito Volterra's Mathematical Theory of the Struggle for Existence", *Science*, **79**: 16.

GAUSE, G. F., 1935, Vérifications expérimentales de la théorie mathématique de la lutte pour la vie, Paris, Hermann.

GAUSE, G. F., 1936, "The Principles of Biocoenology", *Quarterly Review of Biology*, **11**: 320–336.

GAUSE, G. F., 1941, "The Effect of Natural Selection in the Acclimatization of Euplotes to Different Salinities of the Medium", *Journal of Experimental Zoology*, **87**: 85–100.

GAUSE, G. F., 1947, "Problems of Evolution", *Transactions of the Connecticut Academy of Arts and Sciences*, **37**: 17–68.

GAUSE, G. F., 1960, The search for new antibiotics, problems and perspectives, New Haven, Yale University Press.

GAUSE, G. F., 1983, Opredelitel' aktinomitsetov: rody Streptomyces, Streptoverticillium, Chainia, Moskva, Izd-vo "Nauka".

GAUSE, G. F., ALPATOV, V. V., 1931, "Die logistische Kurve von Verhulst-Pearl und ihre Anwendung im Gebiet der quantitativen Biologie", *Biologisches Zentralblatt*, **51**: 1–14.

GAUSE, G. F., ALPATOV, V. V., 1945, "On the Inverse Relation between Inherent and Acquired Properties of Organisms", *American Naturalist*, **79**: 478–480.

GAUSE, G. F., SMARAGDOVA, N. P., ALPATOV, V. V. 1941, "Geographic Variation in Paramecium and the Role of Stabilizing Selection in the Origin of Geographic Differences", *American Naturalist*, **76**: 63–74.

GAUSE, G. F., WITT, A. A., 1935, "Behavior of Mixed Populations and the Problem of Natural Selection", *American Naturalist*, **69**: 596–609.

GLASS, L., MACKEY, M. C., 1988, From Clocks to Chaos. The Rhythms of Life, Princeton, Princeton University Press.

GOETZ, H., 1993, Der freie Geist und seine Widersacher, Frankfurt am Main, Haag & Herchen Verlag.

GRAHAM, L. R., 1972, Science and Philosophy in the Soviet Union, New York, Knopf.

GRAHAM, S. A., 1929, Principles of forest entomology, New York & London, McGraw-Hill (2nd ed. 1939; 3rd ed. 1952 (publ. under the title Forest entomology), 4th ed. 1965 with KNIGHT, F. B., 5th ed. 1980 with KNIGHT, F. B., HEIKKENEN, H. J.).

GRAHAM, S. A., 1935, The spruce budworm on Michigan pine, Ann Arbor, University of Michigan Press.

GRAHAM, S. A., 1937, The walking stick as a forest defoliator, Ann Arbor, University of Michigan Press.

GRAHAM, S. A., 1939, "Forest Insect Populations", *Ecological Monographs*, **9**: 301–310.

GRAHAM, S. A., 1941, "Royal Norton Chapman (1889–1939)", *Annals of the Entomological Society of America*, **34**: 521–524.

GRAHAM, S. A., 1943, Causes of hemlock mortality in northern Michigan, Ann Arbor, University of Michigan Press.

GRAHAM, S. A., HARRISON, R. P. Jr., WESTELL, C. E. Jr., 1963, Aspens, phoenix trees of the Great Lakes region, Ann Arbor, University of Michigan Press.

GRAHAM, S. A., O'ROURKE, E. C., 1943, On your own; how to take care of yourself in wild country, a manual for field and service men, Minneapolis, The University of Minnesota Press.

GRAHAM, S. A., 1956a, "The larch sawfly in the Lake States", *Forest Science*, **2**: 132–160.

GRAHAM, S. A., 1956b, "Forest insects and the law of natural compensation", *Canadian Entomologist*, **48**: 45–55.

GRANT, P. T., 1961–1970, "Kermack, William Ogilvy", *Dictionary of National Biography*, sub voce.

GRIDGEMAN, N. T., 1976, "Lotka, Alfred James", *Dictionary of Scientific Biography* (Ch. C. Gillispie, ed.), New York, Scribners, sub voce.

GRINEVALD, J., 1988, "Sketch for a history of the idea of the Biosphere," in: BUNYARD P., GOLDSMITH E. (eds.) 1988: 1–32.

HELGUERO, F. DE, 1904, "Sui massimi delle curve dimorfiche", *Biometrika*, **3**: 84–98.

HUXLEY, J. S., 1942, Evolution: The Modern Synthesis, London, Allen & Unwin (American edition, New York, Harper, 1943).

IANNIELLO, M. G., ISRAEL, G., 1993, "Boltzmann's concept of 'Nachwirkung' and the 'mechanics of heredity'", Proceedings of the International Symposium on Ludwig Boltzmann (Rome, February 9–11, 1989), ed. by G. Battimelli, M.G. Ianniello, O. Kresten, Wien, Verlag der Österreichischen Akademie der Wissenschaften, 1993: 113–133.

INGRAO, B., ISRAEL, G., 1990, The Invisible Hand. Economic Equilibrium in the History of Science, Cambridge, Mass., The MIT Press, 1990 (paperback: 2000; italian ed. La mano invisibile. L'equilibrio economico nella storia della scienza, Bari, Editori Laterza, 1987).

ISRAEL, G., 1982a, "Volterra Archive at the Accademia Nazionale dei Lincei", *Historia Mathematica,* **9**: 229–238

ISRAEL, G., 1982b, "Le equazioni di Volterra e Lotka: una questione di priorità", in: O. Montaldo, L. Grugnetti (eds.), Atti del Convegno su "La Storia delle Matematiche in Italia", Cagliari 29–30 Settembre-1 Ottobre 1982, Cagliari, Università di Cagliari, Istituti di Matematica della Facoltà di Scienze e Ingegneria: 495–502.

ISRAEL, G., 1984a, "L'interminabile crisi del meccanicismo," *Rivista di Storia della Scienza,* **1**: 73–99.

ISRAEL, G., 1984b, "Vito Volterra: un fisico matematico di fronte ai problemi della fisica del Novecento", *Rivista di Storia della Scienza,* **1**: 39–72.

ISRAEL, G., 1988, "On the contribution of Volterra and Lotka to the development of modern biomathematics", *History and Philosophy of Life Sciences,* **10**: 37–49.

ISRAEL, G., 1990, "Volterra e la dinamica delle popolazioni biologiche", in: Il pensiero scientifico di Vito Volterra, ed. by A. Roccheggiani, Ancona, La Lucerna Editrice: 87–113.

ISRAEL, G., 1991a, "Il determinismo e la teoria delle equazioni differenziali ordinarie. Un'analisi retrospettiva a partire dalla meccanica ereditaria", *Physis, Rivista Internazionale di Storia della Scienza,* **28** (Nuova Serie): 305-358.

ISRAEL, G., 1991b, "Volterra's Analytical Mechanics of Biological Associations" (two parts), *Archives Internationales d'Histoire des Sciences,* **41**: 57–104; 307–352.

ISRAEL, G., 1991c, "Approccio descrittivo e approccio ermeneutico nella modellistica matematica contemporanea," in: S. Petruccioli (ed.), Immagini, linguaggi, concetti, Roma/Napoli, Theoria, 1991: 139–166.

ISRAEL, G., 1993a, "The Two Paths of the Mathematization of the Social and Economic Sciences", *Physis, Rivista Internazionale di Storia della Scienza,* **30** (Nuova Serie): 27-78.

ISRAEL, G., 1993b, "The Emergence of Biomathematics and the Case of Population Dynamics: A Revival of Mechanical Reductionism and Darwinism", *Science in Context,* **6**: 469–509.

ISRAEL, G., 1996, La mathématisation du réel. Essai sur la modélisation mathématique, Paris, Editions du Seuil (italian transl.: La visione matematica della realtà, Introduzione ai temi e alla storia della modellistica matematica, Roma-Bari, Laterza, 1996, 2nd ed. 1997).

ISRAEL, G., 1998, "Balthazar Van der Pol e il primo modello matematico del battito cardiaco", in: P. Freguglia (ed.) Modelli matematici nelle scienze biologiche, Urbino, Quattro Venti: 133–162.

ISRAEL, G., MILLÁN GASCA, A., 1993, "La correspondencia entre Vladimir A. Kostitzin y Vito Volterra (1933–1962) y los inicios de la biomatematica", *LLULL, Revista de la Sociedad Española de Historia de las Ciencias,* **16**: 159–224.

ISRAEL, G., NASTASI, P., 1998, Scienza e razza nell'Italia fascista, Bologna, Il Mulino.

JAGADISACANDRA VASU (Sir JAGADIS CHUNDER BOSE) 1926, Réactions de la matière vivante et non vivante, (transl. by E. Monod-Herzen), Paris, Gauthier-Villars.

JAGADISACANDRA VASU (Sir JAGADIS CHUNDER BOSE) 1931, Le Mécanisme nerveux des plantes, (transl. by E. Monod-Herzen), Paris, Gauthier-Villars.

JAGADISACANDRA VASU (Sir JAGADIS CHUNDER BOSE) 1934, Les Autographes des plantes et leurs révélations, (transl. by E. and G. Monod-Herzen), Paris, Gauthier-Villars.

JENNINGS, H. S., 1935, Genetics, New York, W.W. Norton.

JENSEN, A. J. C., 1932, "The effect of the plaice fishery in the stock of undersized plaice and its influence on the yeld of the plaice fishery in the North Sea", *Cons. Perm. Int. Expl. Mer. Rapp. Proc. Verb,* **LXXX**: 4.

JENSEN, A. J. C., 1933, "Periodic fluctuations in the size of various stocks of fish and their causes", Fiskeri, **9**, n. 5, København, Meddelelser fra Kommissionen for Havundersøgelser.

JOHNSTONE, J., 1928, "On periodicities in the Abundance of young fishes in the Mersey Estuary Region", *Proceedings and Transactions of Liverpool Biological Soc*iety, **42**: 42–68.

KINGSLAND, S. E., 1985, Modeling Nature: Episodes in the History of Population Ecology, Chicago, University of Chicago Press.

KNIGHT, F. B., HEIKKENEN, H. J., 1980, Principles of forest entomology, New York & London, McGraw-Hill (5th rev. ed. of GRAHAM S. A. 1929).

KOLMOGOROV, A. N., 1936, "Sulla teoria di Volterra della lotta per l'esistenza," *Giornale dell'Istituto Italiano degli Attuari,* **7**: 74–80.

KOSTITZIN, J., KOSTITZIN, V. A., 1931, "Sur la statistique d'infestation des Pagures par les Chlorogaster," *Comptes Rendus de l'Académie des Sciences,* **193**: 86–88.

KOSTITZIN, V. A., 1932, "Une application géologique des équations différentielles," *Comptes Rendus de l'Académie des Sciences,* **195**: 1219–1222.

KOSTITZIN, V. A., 1933, "Sur les solutions asymptotiques des équations différentielles de la théorie de croissance des organismes", *Comptes rendus de l'Académie des Sciences de Paris,* **196***:* 841–843.

KOSTITZIN, V. A., 1934, Symbiose, parasitisme et evolution (étude mathématique), Paris, Hermann.

KOSTITZIN, V. A., 1935a, Evolution de l'atmosphère, circulation organique, époques glaciaires, Paris, Hermann.

KOSTITZIN, V. A., 1935b, "Sur l'intoxication d'un milieu par les produits cataboliques d'une population," *Comptes Rendus de l'Académie des Sciences,* **201**: 516–518.

KOSTITZIN, V. A., 1935c, "Sur la relation entre le sexe et le nombre de parasites dans le même hôte", *Comptes rendus de l'Académie des Sciences,* **211**: 624–626.

KOSTITZIN, V. A., 1936, "Sur les équations différentielles du problème de la sélection mendélienne", *Comptes Rendus de l'Académie des Sciences,* **203**: 156–157.

KOSTITZIN, V. A., 1937a "Sur une généralisation des équations biologiques dans le cas d'une population intoxiquée par les produits de son activité chimique," *Comptes Rendus de l'Académie des Sciences,* **204**: 1683–1685.

KOSTITZIN, V. A., 1937b, Biologie mathématique, Paris, A. Colin.

KOSTITZIN, V. A., 1938a, "Equations différentielles générales du problème de la sélection naturelle," *Comptes Rendus de l'Académie des Sciences,* **206**: 570–572.

KOSTITZIN, V. A., 1938b, "Sur les coefficients mendéliens d'hérédité", *Comptes Rendus de l'Académie des Sciences,* **206**: 883–885.

KOSTITZIN, V. A., 1938c, "Sur les points singuliers des équations différentielles du problème de la selection naturelle," *Comptes Rendus de l'Académie des Sciences*, **206**: 976-978.

KOSTITZIN, V. A., 1938d, "Sur les équations différentielles du problème de la sélection naturelle dans le cas de mutation d'un chromosome sexuel," *Comptes Rendus de l'Académie des Sciences*, **206**: 1273-1275.

KOSTITZIN, V. A., 1938e, "Sélection naturelle et transformation des espèces du point de vue analytique, statistique et biologique," *Comptes Rendus de l'Académie des Sciences*, **206**: 1442-1444.

KOSTITZIN, V. A., 1939a, *Mathematical Biology*, London, G. G. Harrap & Co.

KOSTITZIN, V. A., 1939b, "Sur les équations integro-différentielles de la théorie de l'action toxique du milieu," *Comptes Rendus de l'Académie des Sciences*, **208**: 1545-1547.

KOSTITZIN, V. A., 1939c, "Sur la compatibilité des points singuliers stables des équations différentielles algébriques", *Comptes Rendus de l'Académie des Sciences*, **208**: 411-414.

KOSTITZIN, V. A., 1939d, "Sur la compatibilité des points singuliers stables des équations différentielles", *Comptes Rendus de l'Académie des Sciences*, **208**: 554-555.

KOSTITZIN, V. A., 1940a, "Sur la ségrégation physiologique et la variation des espèces". *Acta Biotheoretica*, **5**: 160-168.

KOSTITZIN, V. A., 1940b, "Sur la perméabilité des membranes", *Comptes Rendus de l'Académie des Sciences*, **211**: 62-65

KOSTITZIN, V. A., 1940c, "Sur le passage du chlorhydrate de novocaine à travers une membrane inerte (cellophane)", *Comptes Rendus de l'Académie des Sciences*, **211**: 353-355.

KOSTITZIN, V. A., 1956, "Sur le développement des populations bactériennes," *Comptes Rendus de l'Académie des Sciences*, **242**: 611-612.

KOSTITZIN, V. A., 1981, *Seibutsu suriron*, Tokyo, Koseisho Jinko Mondai Kenkyujo.

KOSTITZIN, V. A., VOLTERRA, V., 1938, "Remarques sur l'action toxique du milieu à propos de la note de M. Régnier et Mlle. Lambin", *Comptes Rendus de l'Académie des Sciences*, **207**: 1146-1148.

LAMBIN, S., 1933, Méthodes de mesure de l'activité antimicrobienne des substances chimiques, Paris, Masson.

LAMBIN, S., 1969, Jean Régnier (1892-1946), Académie de Pharmacie, Séance annuelle du 8 janvier 1969, Paris.

LARMOR, J., 1900, Aether and Matter, Cambridge, Cambridge University Press.

LEDANOIS, É. (ed.), 1926, Commission internationale pour l'exploration scientifique de la mer Méditerranée, Rapports et procés-verbaux des réunions, Vol. I (nouvelle série), contenant les rapports de MM. F. de Buen, J. Galbis, J. Giral, H. Heldt, R. Issel, P. Marti et Rollet de L'Isle (préface de M. Vito Volterra), Paris, Blondel La Rougery.

LEIGH, E. G., 1968, "The Ecological Role of Volterra's Equations" in: Some Mathematical Problems in Biology, Providence, American Mathematical Society: 1-61.

LEVIN, S., (ed.) 1994, Frontiers in Mathematical Biology, Berlin, Springer.

L'HÉRITIER, P., 1934, Génétique et évolutions. Analyse de quelques études mathématiques sur la sélection naturelle, Paris, Hermann.

L'HÉRITIER, P., TEISSIER, G., 1933, "Étude d'une population de Drosophiles en équilibre", *Comptes Rendus de l'Académie des Sciences de Paris*, **197**: 1765–1767.

L'HÉRITIER, P., TEISSIER, G., 1935, "Recherches sur la concurrence vitale: Études de populations mixtes de *Drosophila Melanogaster* et de *Drosophila funebris*", *Comptes Rendus de la Société Biologique*, **118**: 1396–1398.

LIMOGES, C., 1980, "A second glance at evolutionary biology in France," in: MAYR, E., PROVINE, W. B. (eds.), 1980: 322-328.

LOTKA, A. J., 1910, "Contribution to the Theory of Periodic Reactions", *Journal of Physical Chemistry*, **14**: 271.

LOTKA, A. J., 1920a, "Evolution and Irreversibility", *Science Progress*, **14**: 406.

LOTKA, A. J., 1920b, "Undamped Oscillations Derived from the Law of Mass Action", *Journal of the American Chemical Society*, **42**: 1595.

LOTKA, A. J., 1920c, "Analytical Note on Certain Rythmic Relations in Organic Systems", *Proceedings of the American Academy of Arts and Sciences*, **6**: 410.

LOTKA, A. J., 1923, "Contribution to the Analysis of Malaria Epidemiology", *American Journal of Hygiene* (Suppl.), **3**: 1–121 (in 5 parts, part 4 with F. R. SHARPE).

LOTKA, A. J., 1925, Elements of Physical Biology, Williams & Wilkins, Baltimore (reprint Elements of Mathematical Biology, New York, Dover, 1956).

LOTKA, A. J., 1934, Théorie analytique des associations biologiques. Première Partie. Principes, Paris, Hermann.

LOTKA, A. J., 1939, Théorie analytique des associations biologiques. Analyse démographique avec application particulière à l'espèce humaine, Paris, Hermann.

LOTKA, A. J., VOLTERRA, V., 1927, "Fluctuations in the Abundance of a Species considered Mathematically" (two letters), *Nature*, **119**: 12–3.

LUDWIG, W., 1928, "Ueber den funktionellen Zusammenhang zwischen Populationsdichte, Nahrugnsdichte und Teilungsrate bei Protisten und über die Zunahme der Bevölkerungsdichte überhaupt", *Biologia Generalis*, **4**: 351.

LUR-SALUCES, MARQUIS DE, 1933, Lomonossof, le prodigieux moujik, Paris, Emile-Paul frères.

MACLULICH, D. A., 1937, Fluctuations in the number of the varying hare Lefus americanus, Toronto, University of Toronto (Ph. D. Thesis).

MARCHI, C., 1929a, "Osservazioni sulla statistica della pesca di mare e di stagno a Cagliari dal 1912 al 1927", in: *Scritti Biologici*, **4**: 37.

MARCHI, C., 1929b, "Verifica pratica delle leggi teoriche di Vito Volterra sulle fluttuazioni del numero di individui in specie animali conviventi", Regio Comitato Talassografico Italiano, Mem. CLIV.

MAY, R. M., 1972, "Will a large complex system be stable?", *Nature*, **238**: 413.

MAY, R. M., 1974, Stability and Complexity in Model Ecosystems, Princeton, Princeton University Press.

MAYR, E., 1980, "The Arrival of Neo-Darwinism in France," in: MAYR, E., PROVINE, W. B. (eds.), 1980: 321–322.

MAYR, E., 1982, The Growth of Biological Thought. Diversity, Evolution and Inheritance, Cambridge, Cambridge University Press.

MAYR, E., PROVINE, W. B. (eds.), 1980, The Evolutionary Synthesis: Perspectives on the Unification of Biology, Cambridge, Mass., Harvard University Press.

MAZZOCUT-MIS, M., 1995, "D'Arcy Thompson, la forme et le vivant", *Alliage*, **22**: 24- 31

MENCKEN, H. L., 1940, Obituary, *Sun* (Baltimore), november 24.

METCHNIKOFF, E., 1901, L'immunité dans les maladies inféctieuses, Paris, Masson.

MILLÁN GASCA, A., 1993, "Immagini matematiche della vita", *Prometeo*, **10**: 88–96.

MILLÁN GASCA, A., 1996, "Mathematical Theories versus biological facts: A debate on mathematical population dynamics in the 1930s", *Historical Studies in the Physical and Biological Sciences*, **26**: 347–403.

MILLÁN GASCA, A., 1996, "El ideal de la matematización. La introducción de la matemática en las ciencias biológicas, humanas y sociales", *Arbor, ciencia, pensamiento y cultura*, **606**: 79–102.

MONOD-HERZEN, E., 1914, La gravure au marteau, Paris, Extrait de la Gazette des Beaux-Arts.

MONOD-HERZEN, E., 1925, Science et esthétique, Paris, Gauthier-Villars.

MONOD-HERZEN, E., 1927, Science et esthétique. Principes de morphologie générale, 2 voll., Paris.

MONOD-HERZEN, E., 1961, Le problème de la "Sala delle asse" (lambris) (à Milan) de Leonardo da Vinci, communication faite à l'Académie des Beaux-Arts (25 janvier 1961), Paris, Gauthier-Villars.

NICHOLSON, A. J., 1933, "The Balance of Animal Populations", *Journal of Animal Ecology*, **2**: 132–78.

NICHOLSON, A. J., BAILEY, V. A., 1935, "The Balance of Animal Populations," *Proceedings of the Zoological Society of London*, **3**: 551–598.

OSTER, G., GUCKENHEIMER, J., 1976, "Bifurcation phenomena in population models," in: J. E. Marsden, M. McCracken (eds.) The Hopf Bifurcation and its Applications, New York, Springer.

PAOLONI, G. (ed.), 1990, Vito Volterra e il suo tempo (1860–1940), Mostra storico-ducumentaria, Catalogo a cura di G. Paoloni, Roma, Accademia Nazionale dei Lincei, Consiglio Nazionale delle Ricerche, Archivio Centrale dello Stato.

PARKER, F., 1976, "Pearl, Raymond", *Dictionary of Scientif Biography* (Ch. C. Gillispie, ed.), New York, Scribners, sub voce.

PAVÉ, A., 1994, Modélisation en biologie et en écologie, Paris, Aléas.

PEARL, R., 1923, Introduction to Medical Biometry and Statistics, Philadelphia-Boston.

PEARL, R., 1924, Studies in Human Biology, Baltimore, Williams & Wilkins.

PEARL, R., 1925, The Biology of Population Growth, New York, A. Knopf.

PEARL, R., 1927, "The Growth of Populations", *Quarterly Review of Biology*, **2**: 532–48.

PEARL, R., 1932, "The Influence of Density of Population upon Egg Production in Drosophila melanogaster," *Journal of Experimental Zoology*, **63**: 57–84.

PEARL, R., 1932, "The Influence of Density of Population upon Egg Production in the *Drosophila Melanogaster*", *Journal of Experimental Biology*, **63**: 57–84.

PEARL, R., 1939, The Natural History of Population, London-New York, Oxford University Press.

PEARL, R., PARKER, S. L., 1922, "On the Influence of Density of Population upon the Rate of Reproduction in *Drosophila*", *Proceedings of the National Academy of Sciences of the U.S.A.*, **8**: 212–218.

PEARL, R., REED, L. J., 1920, "On the Rate of Growth of the Population of the United States since 1790 and its Mathematical Representation", *Proceedings of the National Academy of Sciences of the U.S.A.*, **6**: 275–288.

PEARL, R., REED, L. J., 1927, "On the Summation of Logistic Curves", *Journal of the Royal Statistical Society of London*, **90**: 729–746.

PEARSON, E. S., 1927, "The Application of the Theory of Differential Equations to the Solution of Problems Connected with the Interdependence of Species," *Biometrika*, **19**: 216–222, 442.

PEARSON, K., 1900, *The Grammar of Science, 2nd ed.*, London, A. & Ch. Black.

PEARSON, K., 1910, "Darwinism, Biometry and Some Recent Biology", *Biometrika*, **7**: 368–385.

PÉRÈS, J., 1927, "Une application nouvelle des mathématiques à la biologie. La théorie des associations biologiques", *Revue génerale des sciences pures et appliquées*, **37**: 295.

PETIT, C., 1976, "Teissier, Georges", *Dictionary of Scientific Biography* (Ch. C. Gillispie, ed.), New York, Scribners, sub voce.

POLUNIN, N., GRINEVALD, J., 1988, "Vernadsky and Biospheral Ecology", *Environmental Conservation*, **15**: 117–122.

PRENANT, M., 1934, Adaptation, écologie et biocoenotique, Paris, Hermann.

PROVINE, W. B., 1971, The Origins of Theoretical Population Genetics, Chicago, Univeristy of Chicago Press, 1971.

RÉGNIER, J., DAVID, R., KAPLAN, A., 1932, "Contribution à l'étude numérique de la multiplication microbienne. De l'existence de la phase de latence", *Comptes Rendus de l'Académie des Sciences*, **194**: 323–325.

RÉGNIER, J., LAMBIN, S., 1934, "Etude d'un cas d'antagonisme microbien (Bacillus Coli-Staphylococcus aureus)," *Comptes Rendus de l'Académie des Sciences*, **199**, 1682–1683.

RÉGNIER, J., LAMBIN, S., 1938a, "Des variations morphologiques des bactéries en fonction de l'âge de la culture. Essais sur le Colibacille cultivé en bouillon nutritif", *Comptes Rendus de la Société de Biologie*, **128**: 742–3.

RÉGNIER, J., LAMBIN, S., 1938b, "Des variations de résistance des bactéries en fonction de l'âge de la culture. Essais sur le Colibacille cultivé en bouillon nutritif", *Comptes Rendus de la Société de Biologie*, **128**: 744.

RÉGNIER, J., LAMBIN, S., 1938c, "Influence de la quantité de substance nutritive des milieux de culture sur la multiplication des bactéries. Culture du Colibacille en milieu liquide contenant des quantités variables de peptone", *Comptes Rendus de la Société de Biologie*, **128**: 883.

RÉGNIER, J., LAMBIN, S., 1938d, "Etude sur le croît microbien en foction de la quantité de substance nutritive des milieux de culture," *Comptes Rendus de l'Académie des Sciences*, **207**:1263.

RÉGNIER, J., LAMBIN, S. SZOLLOSI, 1937a, "Données générales sur la mésure, le calcul et l'expression du pouvoir toxique des substances médicamenteuses", *Journal de Physiologie et de Pathologie générale*, **35**: 329.

Régnier, J., Lambin, S. Szollosi, 1937b, "De la mésure et de l'interprétation de la toxicité des substances dont on ne peut pas déterminer directement la dose minima mortelle individuelle", *Journal de Physiologie et de Pathologie générale*, **35**: 709.

Régnier, J., Lambin, S. Szollosi, 1937c, "De quelques données mathématiques théoriques qui sont à la base des exposés précédents", *Journal de Physiologie et de Pathologie générale*, **35**: 950.

Richards, O. W., 1928, "The Growth of the Yeast Saccharomyces cerevisiae. I. The Growth Curve, its Mathematical Analysis, and the Effect of Temperature on the Yeast Growth," *Annals of Botany*, **42**: 271–283.

Richards, O. W., 1936, Analysis of the constant differential growth ratio, Washingon, Carnegie Institute.

Rolland, R., 1948, Choix de lettres à Malwida von Meysenburg, (introd. by É. Monod-Herzen), Paris, Coll. des Bibliolâtres de France.

Ross, R., 1911a, "Some Quantitative Studies in Epidemiology", *Nature*, **87**: 466–7.

Ross, R., 1911b, The prevention of malaria (2nd ed.), London, J. Murray.

Ross, R., 1923, Memoirs, With a Full Account of the Great Malaria Problem and Its Solution, London, J. Murray.

Scudo, F. M., 1971, "Vito Volterra and Theoretical Ecology", *Theoretical Population Biology*, **2**: 1–23.

Scudo, F. M., 1984, "The 'Golden Age' of Theoretical Ecology: A Conceptual Appraisal", *Revue Européenne des Sciences Sociales*, **22**: 11–64.

Scudo, F. M., 1988, "Umberto D'Ancona e Vito Volterra: le ragioni di un'amicizia", in: La vita, le forme, i numeri, Biologica (M. Galzigna, ed.), Bologna, Transeuropa: 47–77.

Scudo, F. M., Ziegler, J. R., 1976, "Vladimir Alexandrovich Kostitzin and Theoretical Ecology," *Theoretical Population Biology*, **10**: 395–412.

Scudo, F. M., Ziegler, J. R., 1978, The "Golden Age" of Theoretical Ecology, 1923–1940, Berlin, Springer 1978.

Sermonti, A., 1986, "D'Ancona, Umberto", Dizionario Biografico degli Italiani, Roma, Istituto della Enciclopedia Italiana, vol. 32, sub voce.

Shelford, V. E., 1913, Animal Communities in Temperate America, Chicago, University of Chicago Press.

Smith, J. M., 1974, Models in Ecology, Cambridge.

Somigliana, C., 1946, "Vito Volterra", *Rendiconti del Seminario matematico e fisico di Milano,* **17**: 3–61.

Stanley, J., 1931, "A Mathematical Theory of the Growth of Populations of the Flour Beetle, *Tribolium confusum* Duval," *Canadian Journal of Research*, **6**: 632.

Stanley, J., 1932a, "Is Evolution only a Detail of the Dynamics of Populations?", *Science*, **75**.

Stanley, J., 1932b, "A Mathematical Theory of the Growth of Populations of the Flour Beetle, *Tribolium confusum* Duval II. The Distribution by Ages in the Early Stages of Population Growth," *Canadian Journal of Research*, **7**: 126.

Stanley, J., 1937, "A note on the Reduction in the Fecundity of *Tribolium Confusum* due to Eating its Own Eggs", *Canadian Journal of Research*, **15**.

Stanley, J., 1939, "A Device for Administering Impact Stimuli to Experimental Animals", *Journal of Comparative Psychology*, **28**.

Stanley, J., 1942, "A Mathematical Theory of the Growth of Populations of the Flour Beetle, *Tribolium confusum* Duval VI. The Relation between the Limiting Value of Egg Populations in the Absence of Hatching and the Sex-Ration of the Group of Adult Beetles used in a Culture," *Ecology*, **23**.

Stanley, J., 1943, "A Mathematical Theory of the Growth of Populations of the Flour Beetle, *Tribolium confusum* Duval V. Egg Populations in which the Initial Number of Eggs is above the Limitng Value," *Ecology*, **24**.

Stanley, J., 1946, "A Method of Growing Dense Cultures of Paramecium", *Science*, **103**: 115–6.

Stanley, J., 1963, The essence of biometry, Montreal, McGill University Press.

Sutherland, A., 1898, The Origin and Growth of the Moral Instinct, 2 vols., London, Longmans.

Teissier, G., 1928, "Croissance des populations et croissance des organismes. Examen historique et critique de quelques théories", *Annals of Physiology and Physiochemistry*, **4**: 342.

Teissier, G., 1934, Dysharmonies et discontinuités dans la croissance, Paris, Hermann.

Thom, R., 1975, Structural Stability and Morphogenesis, Reading, Mass., W.A. Benjamin.

Thompson, D. W., 1917, *On Growth and Form*, Cambridge, Cambridge University Press.

Thompson, D. W., 1915, "Morphology and Mathematics", *Transactions of the Royal Society of Edinburgh*, **50**: 857–895.

Thompson, R. D., 1976, "Thompson, D'Arcy Wentworth", *Dictionary of Scientific Biography* (Ch. C. Gillispie, ed.), New York, Scribners, sub voce.

Thompson, W. R., 1922a, "Théorie de l'action des parasites entomophages. Accroissement de la proportion d'hôtes parasités dans le parasitisme cyclique", *Comptes Rendus de l'Académie des Sciences*, **174**: 65.

Thompson, W. R., 1922b, "Théorie de l'action des parasites entomophages. Les formules mathématiques du parasitisme cyclique" *Comptes Rendus de l'Académie des Sciences*, **174**: 1201.

Thompson, W. R., 1922c, "Étude mathématique de l'action des parasites entomophages. Durée du cycle parasitaire et accroissement de la proportion d'hôtes parasités", *Comptes Rendus de l'Académie des Sciences*, **174**: 1433.

Thompson, W. R., 1922d, "Étude de quelques cas simples de parasitisme cyclique chez les insectes entomophages", *Comptes Rendus de l'Académie des Sciences*, **174**: 1647.

Thompson, W. R., 1923, "La théorie mathématique de l'action des parasites entomophages", *Revue générale des Sciences*, **34**: 202.

Thompson, W. R., 1924, "La théorie mathématique de l'action des parasites entomophages et le facteur du hasard", *Annales de la Faculté de Sciences de Marseille*, ser. 2, **2**: 69.

Thompson, W. R., 1927, "A method for the approximate calculations of the progress of introduced parasites of insect pests", *Bulletin of Entomological Research*, **17**: 273.

Thompson, W. R., 1928, "A contribution to the study of biological control and parasitic introduction in continental area", *Parasitology*, **20**: 90.

THOMPSON, W. R., 1929a, "On the relative value of parasites and predators in the biological control of insect pests", *Bulletin of Entomological Research*, **19**: 343.

THOMPSON, W. R., 1929b, "On the effect of random oviposition on the action od entomophagous parasites as agents of natural control", *Parasitology*, **21**: 180.

THOMPSON, W. R., 1929c, "On natural control", *Parasitology*, **21**: 269.

THOMPSON, W. R., 1929d, "On the part played by parasites in the control of insects living in protected situations", *Bulletin of Entomological Research*, **20**: 457.

THOMPSON, W. R., 1930a, "The Principles of Biological Control", *Annals of Applied Botany*, **17**: 306–338.

THOMPSON, W. R., 1930b, "The Utility of Mathematical Methods in Relation to Work on Biological Control", *Annals of Applied Biology*, **17**: 641.

THOMPSON, W. R., 1930c, The Biological Control of Insect and Plant Pests: A Report on the Organisation and Progress of the Work of Farnham House Laboratory, London, King's Printer.

THOMPSON, W. R., 1937, Science and Common Sense: An Aristotelian Excursion, London, Longmans & Green.

THOMPSON, W. R., 1939, "Biological Control and the Theories of the Interactions of Populations", *Parasitology*, **31**: 299–388.

THOMPSON, W. R., 1948, "Can Economic Entomology be and Exact Science?", *Canadian Entomologist*, **80**: 49–55.

THOMPSON, W. R., 1956, "The Fundamental Theory of Natural and Biological Control", *Annual Review of Entomology*, **1**: 379–402.

THOMPSON, W. R., 1960, "Notes on the Volterra Equations", *Canadian Entomologist*, **92**: 582–594.

THOMPSON, W. R., PARKER, H.L., 1927, "The problem of host relations with special reference to entomophagous parasites", *Parasitology*, **19**: 1.

THORPE, W. H., 1973, "William Robin Thompson, 1887–1972", *Biographical Memoirs of the Royal Society of London*, **19**: 655–678.

VAN DER POL, B. L., VAN DER MARK, J., 1928, "The Heartbeat Considered as a Relaxation Oscillation, and an Electrical Model of the Heart," *London, Edinburgh and Dublin Philosophical Magazine and Journal of Science*, **6**: 763–775

VERHULST, P.-F., 1845, "Recherches mathématiques sur la loi d'accroissement de la population", *Nouveaux Mémoires de l'Académie Royale des Sciences et Belles-Lettres de Bruxelles*, **18**: 3–38.

VERHULST, P.-F., 1847, "Deuxième mémoire sur la loi d'accroissement de la population", *Nouveaux Mémoires de l'Académie Royale des Sciences et Belles-Lettres de Bruxelles*, ser. 2, **2**: 3–32.

VERNADSKY, V. I., 1924, La géochemie, Paris, Hermann.

VERNADSKY, V. I., 1929, La biosphère, Paris, Felix Alcan.

VIGNOLI, T., SCHIAPARELLI, G. V., 1898, Peregrinazioni antropologiche e fisiche. Studio comparativo fra le forme organiche naturali e le forme geometriche pure, Milano, 1898.

VOLTERRA, E., 1976, "Volterra, Vito", Dictionary of Scientific Biography (Ch. C. Gillispie, ed.), New York, Scribners, sub voce.

VOLTERRA, V., 1901, "Sui tentativi di applicazione delle Matematiche alle scienze biologiche e sociali", Discorso inaugurale, *Annuario della R. Università di Roma*, pp. 3–28; also in: VOLTERRA, V., Opere Matematiche, Vol. III: 14–29 (republ. in: *Giornale degli Economisti*, s.II, **23**, 1901: 436–458; in french, La Revue du Mois, 1906: 1–20; in: *Archivio di Fisiologia*, **3**, 1906; 175–191; and in: VOLTERRA, V., 1920).

VOLTERRA, V., 1906, "Les mathématiques dans les sciences biologiques et sociales", (traduction par Ludovic Zoretti), *La Revue du Mois*, **1**: 1–20.

VOLTERRA, V., 1907a, "Sur l'équilibre des corps élastiques multiplement connexes", *Annales de l'École Normale Supérieure*, s. III, **24**: 401–518.

VOLTERRA, V., 1907b, "Il momento scientifico presente e la nuova Società Italiana per il Progresso delle Scienze", *Rivista di Scienza*, **II**: 225–237.

VOLTERRA, V., 1913, Leçons sur les fonctions de ligne (professées à la Sorbonne en 1912, recueillies et rédigées par Joseph Pérès), Paris, Gauthier-Villars.

VOLTERRA, V., 1920, Saggi scientifici, Bologna, Zanichelli (reprint 1990).

VOLTERRA, V., 1926a, "Variazioni e fluttuazioni del numero d'individui in specie animali conviventi", *Memorie della R. Accademia dei Lincei*, S.VI, **II**: 31–113.

VOLTERRA, V., 1926b, "Fluctuations in the abundance of a species considered mathematically", *Nature*, **118**: 558–560.

VOLTERRA, V., 1927a, "Sulle fluttuazioni biologiche," *Rendiconti della R. Accademia dei Lincei*, 6 serie, **5**: 3–10.

VOLTERRA, V., 1927b "Leggi sulle fluttuazioni biologiche," *Rendiconti della R. Accademia dei Lincei*, 6 serie, **5**: 61–67.

VOLTERRA, V., 1927c, "Sulla periodicità delle fluttuazioni biologiche", *Rendiconti della R. Accademia dei Lincei*, 6 serie, **5**: 465–470.

VOLTERRA, V., 1927d, Variazioni e fluttuazioni del numero d'individui in specie animali conviventi, Memorie del R. Comitato Talassografico, Mem. 131; also in: VOLTERRA V., *Opere Matematiche*, Vol. V: 1–111.

VOLTERRA, V., 1927e, "Essai mathématique sur les fluctuations biologiques", *Bulletin de la Société d'océanographie de France*.

VOLTERRA, V., 1927f, "Una teoria matematica sulla lotta per l'esistenza", *Scientia*, **41**: 85–102; also in: VOLTERRA V., *Opere Matematiche*, Vol. V: 112–124.

VOLTERRA, V., 1927g, "Une théorie mathématique de la lutte pour la vie", *Scientia*, **41**, Supplemento: 33–48 (French translation of VOLTERRA V. 1927f by M. Thiers).

VOLTERRA, V., 1928a, "Variations and Fluctuations of the Number of Individuals in Animal Species Living Together," *Journal du Conseil international pour l'exploration de la mer*, **3**: 3–51.

VOLTERRA, V., 1928b, Mathematical Theory of Struggle for Existence (in Russian), Moscow.

VOLTERRA, V., 1931a, Leçons sur la théorie mathématique de la lutte pour la vie, (redigées par Marcel Brélot), Paris, Gauthier-Villars (reprint Paris, Gabay, 1990).

VOLTERRA, V., 1931b, Theory of functionals and of integral and integro-differential equations (edited by Professor Luigi Fantappié, authorised translation by Miss M. Long), London & Glasgow, Blackie.

VOLTERRA, V., 1932, Le calcul des variations, son évolution et ses progrès, son rôle dans la physique-mathématique, Praha-Brno, Facultés des sciences de L'Université Charles et de l'Université Masaryk; also in: VOLTERRA V., *Opere Matematiche*, Vol. V: 217–267.

VOLTERRA, V., 1934a, "Remarques sur la note de M. Régnier et Mlle. Lambin," *Comptes Rendus de l'Académie des Sciences*, **199**: 1684–1686.

VOLTERRA, V., 1934b, "Equations aux dérivées partielles et théorie des fonctions", *Annales de l'Institut H. Poincaré*, **4**: 273–352.

VOLTERRA, V., 1936a, "Les équations des fluctuations biologiques et le calcul des variations", *Comptes rendus de l'Académie des Sciences*, **202**: 1953–1957; also in: VOLTERRA, V., *Opere Matematiche*, Vol. V: 392–395.

VOLTERRA, V., 1936b, "Les équations canoniques des fluctuations biologiques", *Comptes rendus de l'Académie des Sciences*, **202**: 2023–2026; also in: VOLTERRA, V., *Opere Matematiche*, Vol. V: 396–398.

VOLTERRA, V., 1936c, "Sur l'intégration des équations des fluctuations biologiques", *Comptes rendus de l'Académie des Sciences*, **202**: 2113–2116; also in: VOLTERRA, V., *Opere Matematiche*, Vol. V: 399–401.

VOLTERRA, V., 1936d, "Le principe de la moindre action en biologie", *Comptes rendus de l'Académie des Sciences*, **203**: 417–421; also in: VOLTERRA V. *Opere Matematiche*, Vol. V: 402–405.

VOLTERRA, V., 1936e, "Sur la moindre action vitale", *Comptes rendus de l'Académie des Sciences*, **203**: 480–481; also in: VOLTERRA, V., *Opere Matematiche*, Vol. V: 406–407.

VOLTERRA, V., 1936f, "La théorie mathématique de la lutte pour la vie et l'expérience (À propos de deux ouvrages de C.F. Gause)", *Scientia*, **60**: 169–174; also in: Volterra, V., *Opere Matematiche*, Vol. V: 408–413.

VOLTERRA, V., 1937a, "Principes de biologie mathématique", *Acta Biotheoretica*, **3**, (I): 6–39; also in: VOLTERRA, V., *Opere Matematiche*, Vol. V: 414–447.

VOLTERRA, V., 1937b, "Applications des mathématiques à la biologie", (text of a lecture at the University of Geneva in June 1937), *L'Enseignement mathématique*, **36**: 297–330; also in: VOLTERRA, V., *Opere Matematiche*, Vol. V: 448–471.

VOLTERRA, V., 1937c, "Leggi delle fluttuazioni e principi di reciprocità in biologia", *Rivista di Biologia*, **22**: 365–380.

VOLTERRA, V., 1938a, "Fluctuations dans la lutte pour la vie: leurs lois fondamentales et de réciprocité", Société mathématique de France (Conférence de la Réunion internationale des mathématiciens tenue à Paris en juillet 1937); also in: VOLTERRA, V., *Opere Matematiche*, Vol. V: 482–485.

VOLTERRA, V., 1938b "Population Growth, Equilibria and Extinction under Specified Breeding Conditions: A Development and Extension of the Theory of the Logistic Curve," *Human Biology*, **10**: 1–11

VOLTERRA, V., 1939a, "The General Equations of Biological Strife in the Case of Historical Actions," *Proceedings of the Edinburgh Mathematical Society*, II series, **6**: 4–10.

VOLTERRA, V., 1939b, "Calculus of Variations and the Logistic Curve", *Human Biology*, **2**, (2): 173–178; also in: VOLTERRA, V., *Opere Matematiche*, Vol. V: 503–507.

VOLTERRA, V., 1954–62, Opere Matematiche, Memorie e Note, Roma, Accademia Nazionale dei Lincei, 5 voll.

VOLTERRA, V., D'ANCONA, U., 1935, Les associations biologiques au point de vue mathématique, "Actualités scientifiques et industrielles", n. 243, Hermann, Paris.

VOLTERRA, V., D'ANCONA, U., 1995, Le associazioni biologiche studiate dal punto di vista matematico (italian translation of VOLTERRA V., D'ANCONA U. 1935), (G. Israel, ed.), Roma, Teknos.

VOLTERRA, V., KOSTITZIN, V. A., 1938, "Remarques sur l'action toxique du milieu à propos de la note de M. Régnier et M.lle Lambin, *Comptes Rendus de l'Académie des Sciences de Paris*, **207**: 1146–1148.

VOLTERRA, V., PÉRÈS, J., 1924, Leçons sur la composition des fonctions permutables, Paris, Gauthier-Villars.

WEINER, D., 1988, Ecology, Conservation and Cultural Revolution in Soviet Russia, Bloomington.

WHITTAKER, E., 1946, "Vito Volterra", *Obituary Notices of Fellows of the Royal Society of London*, **3**: 691–729.

WOODRUFF, A. E., 1976, "Larmor, Joseph", *Dictionary of Scientific Biography* (Ch. C. Gillispie, ed.), New York, Scribners, sub voce.

MIX
Papier aus verantwortungsvollen Quellen
Paper from responsible sources
FSC® C105338

If you have any concerns about our products,
you can contact us on
ProductSafety@springernature.com

In case Publisher is established outside the EU,
the EU authorized representative is:
Springer Nature Customer Service Center GmbH
Europaplatz 3, 69115 Heidelberg, Germany

Printed by Libri Plureos GmbH
in Hamburg, Germany